Open Channel Hydraulics

Open Channel Hydraulics

by

Richard H. French

Water Resources Publications, LLC  **Highlands Ranch**

For Information and Correspondence:

Water Resources Publications, LLC
P. O. Box 630026, Highlands Ranch, Colorado 80163-0026, USA

Open Channel Hydraulics

by

Richard H. French

ISBN-13: 978-1-887201-44-5
ISBN-10: 1-887201-44-0

U.S. Library of Congress Control Number: 2006932023

PREFACE

From the viewpoint of both quantity and quality, water resource projects are of paramount importance to the maintenance and progress of civilization as it is known today and has been historically known throughout the world. The knowledge of open channel hydraulics is fundamental to water resources development, the preservation of water quality, and the enhancement of the environment. Water resources engineering and hydrology has, along with other areas of engineering knowledge, exploded in the past four decades. To some degree, this explosion is due to the advent of the high-speed digital computer and the movement of computational power to the desks of every practicing professional and student. Problems that would not have been computationally tractable thirty years can be addressed and solved from almost anywhere. In addition, the concerns of society with both the preservation and restoration of the aquatic environment has led hydrologic and hydraulic engineers to consider problems that were not seriously considered before. However, the dependence of professionals and students on computer hardware and software has not been and is not always positive; that is, just because the computer produced a numerical answer that does not mean it is correct. There is still the need for a sound understanding of fundamental principles and the exercise of what has traditionally been called engineering perhaps, now more than ever, since accurate computational answers (both right and wrong) are so easily available.

This book is primarily a text for graduate or undergraduate students in civil and agricultural engineering and hydrologic science. As is the case with all books, including the first edition of this book published by McGraw-Hill, it is a statement of what the author believes important.

As is the case with any work of this extent, there are many people and organizations that both need to be acknowledged and deserve thanks. First, there is my wife who has twice supported this effort. Second, there is my editor and publisher, Ms. Branka McLaughlin. Third, there is Debora Noack, the Division of Hydrologic Sciences, Desert Research Institute, Nevada System of Higher Education, who provided graphics support. Fourth, there is the Board of Regents of the University of the Nevada System of Higher Education who authorized a sabbatical leave for me for the Academic Year of 2001-2002. Fifth, there are my new employer, the Department of Civil and Environmental Engineering at the University of Texas at San Antonio, who have provided every encouragement to complete this effort. Finally, sixth, I would like to thank all the people I have contacted during the course of writing this book that have responded to questions regarding their work. I apologize in advance for anyone that I may have forgotten to thank.

Chapter 1 explains the types of flow encountered in open channels are classified with respect to time, space, viscosity, density, and gravity. In addition, the types of channels commonly encountered and their geometric properties are defined. In Section 1.2, the equations which govern flow in open channels, i.e., conservation of mass, momentum, and energy, are developed along with the equations for the energy and momentum correction coefficients. In Section 1.4, theoretical concepts such as the scaling of partial differential equations, boundary layers, and velocity distributions in both homogeneous and stratified flows are briefly discussed. In Section 1.5, the basic concepts of geometric and dynamic similarity are developed in relation to their application to the design and use of physical hydraulic models.

In Chapter 2, the application of the law of conservation of energy to open-channel flows is discussed. Section 2.1 defines specific energy. The effect of streamline curvature on the pressure distribution within the flow field is also considered. Section 2.2 defines subcritical, critical, and supercritical flow. In this section, a number of techniques - trial and error, graphical, and explicit - for estimating critical flow in channels of various cross-sectional shapes are presented. Section

2.3 discusses upstream and downstream controls and the accessibility of various points on the specific energy curve. Section 2.4 introduces the reader to various applications of the energy principle in practice including the use of dimensionless curves, computations associated with channel transitions, and problems encountered in channels of compound section.

Chapter 3 considers the application of the law of conservation of momentum to open-channel flow. Section 3.1 defines specific momentum. In Section 3.2, the occurrence and characteristics of hydraulic jumps in rectangular and nonrectangular channels are discussed. Among the characteristics considered are the energy losses incurred in a hydraulic jump and the length of the hydraulic jump. The types of jumps treated include free jumps, submerged jumps, and jumps which occur in channels with a significant slope. In Section 3.3 the occurrence and characteristics of internal hydraulic jumps are discussed.

Chapter 4 defines uniform flow and develops the Chezy and Manning equations for uniform flow. Both theoretical and applied methods of estimating the resistance coefficients used in these equations are then discussed.

Chapter 5 primarily emphases techniques of computing the normal depth of flow in open channels. The most tedious and difficult normal-flow calculation occurs when the Manning resistance equation cannot be solved explicitly for the normal depth of flow. Solution techniques including trial and error, and graphical for the solution of this implicit problem are discussed. In the third section of this chapter, the problem of estimating composite or average boundary roughness in designed, natural, and laboratory channels is treated. Several techniques for estimating composite flow resistance coefficients in designed and natural channels are considered. In the case of laboratory flumes with artificially roughened bottoms and hydraulically smooth sidewalls, a technique of estimating the bottom boundary resistance coefficient is presented. In the fourth section of this chapter, a number of applications of uniform flow concepts to practice are considered, e.g., slope-area peak flood flow computations, ice-covered channels, and normal discharge in channels of compound section.

Chapter 6 considers the theory and analysis of gradually and spatially varied flow. When the depth of flow in an open-channel flow varies with longitudinal distance, the flow is termed gradually varied. Such situations are found both upstream and downstream of control sections. In this chapter, tabular, digital, and graphical methods of estimating the depth of flow as a function of longitudinal distance in prismatic, nonprismatic, and natural channels are considered.

Steady, spatially varied flow is by definition a flow in which discharge varies with longitudinal distance. Such situations occur in side channel spillways, gutters collecting and conveying storm water runoff, channels with permeable boundaries, and drop structures in the bottom of channels. Tabular solutions of the differential equations governing spatially varied flow for both increasing and decreasing flow are considered. In conclusion, a number of practical considerations are discussed including methods of computing the steady flow around islands in rivers.

In Chapter 7, the design of lined, unlined, and grass-lined channels is considered, and design procedures for each type of channel are discussed and demonstrated. In the case of lined channels, a technique of minimizing lining material costs for rectangular, trapezoidal, and triangular-shaped channels is presented. Such a technique is particularly useful for long channel sections were construction procedures can be oriented to minimizing material costs or in situations in which labor costs are low relative to the lining material costs. In the case of unlined channel design, techniques based on the principles of maximum permissible velocity and the threshold of movement are discussed, but the recommended procedure is based on the principle of tractive force. Using the principle of tractive force, the equations defining the stable hydraulic section are developed and their use demonstrated with an example. Finally, both empirical and theoretical methods of estimating the leakage from an unlined channel are considered. In the final section, the Soil Conservation Service method of designing grass-lined channels is discussed. In addition, a procedure for designing grass-lined channels with the center lined with gravel is presented.

The transport processes known as turbulent diffusion and dispersion are discussed in Chapter 8. First, the governing equations are developed for both one and two dimensions, and then techniques for estimating the vertical and transverse turbulent diffusion coefficients and the longitudinal dispersion coefficient are discussed. Numerical methods of solving the governing equations are presented and the problem of numerical dispersion considered. The chapter concludes with a brief discussion of techniques for estimating the vertical turbulent diffusion coefficient in a stratified environment.

In Chapter 9, the subject of gradually varied unsteady flow is treated. In the initial section, both hydraulic and hydrologic methodologies are discussed. Subsequent sections emphasize numerical solutions of the St. Venant equations. Appropriate boundary and initial conditions are presented, and the subjects of calibration and verification of gradually varied, unsteady flow models are discussed.

Chapter 10 discusses the design, construction, and use of physical models to examine open-channel flow phenomena are discussed. Among the types of models considered are geometrically distorted and undistorted, fixed and movable bed, and ice. In the case of fixed-bed models, special attention is given to the problem of simulating dispersion in a geometrically distorted model. It is demonstrated that in the general case, dispersion cannot be accurately modeled with a distorted physical model. In the case of movable-bed models both empirical and theoretical techniques of model design are discussed. Consideration is also given to the materials and methods which are used to construct physical models. The dual problems of calibration and verification are also discussed.

In the final section of the chapter, physical models of ice problems in open channels are discussed. Two types of models are considered. In the first type both hydrodynamics and hydroelasticity are modeled. Such models must be used when, for example, the forces developed by an ice sheet in contact with a structure are studied. In the second type of model only the hydrodynamics must be simulated correctly, for example, models to study drifting-ice problems.

R. H. French

TABLE OF CONTENTS

CHAPTER 1 - CONCEPTS OF FLUID FLOW 1
    1.1.   INTRODUCTION 1
    1.2.   DEFINITIONS 2
    1.3.   GOVERNING EQUATIONS 15
    1.4.   THEORETICAL CONCEPTS 28
    1.5.   SIMILARITY AND PHYSICAL MODELS 39
    1.6.   QUANTIFYING UNCERTAINTY 41
    1.7.   BIBLIOGRAPHY 45
    1.8.   PROBLEMS... 46

CHAPTER 2 - ENERGY PRINCIPLE...53
    2.1.   DEFINITION OF SPECIFIC ENERGY 53
    2.2.   SUBCRITICAL, CRITICAL AND SUPERCRITICAL FLOW ... 57
    2.3.   ACCESSIBILITY AND CONTROLS 67
    2.4.   APPLICATION OF THE ENERGY PRINCIPLE TO PRACTICE 75
    2.5.   BIBLIOGRAPHY 87
    2.6.   PROBLEMS... 88

CHAPTER 3 - THE MOMENTUM PRINCIPLE 93
    3.1.   DEFINITION OF SPECIFIC MOMENTUM............................... 93
    3.2.   THE HYDRAULIC JUMP.. 96
    3.3   HYDRAULIC JUMPS AT DENSITY INTERFACES............................. 127
    3.4.   APPLICATION OF THE MOMENTUM PRINCIPLE TO
             PRACTICE... 131
    3.5.   BIBLIOGRAPHY 136
    3.6.   PROBLEMS... 138

CHAPTER 4 - DEVELOPMENT OF UNIFORM FLOW CONCEPTS 143
    4.1.   ESTABLISHMENT OF UNIFORM FLOW 143
    4.2.   THE CHEZY AND MANNING EQUATIONS............................ 144
    4.3.   RESISTANCE COEFFICIENT ESTIMATION............................ 147
    4.4.   BIBLIOGRAPHY 218

CHAPTER 5 - COMPUTATION OF UNIFORM FLOW 221
    5.1.   CALCULATION OF NORMAL DEPTH AND VELOCITY 221
    5.2.   NORMAL AND CRITICAL SLOPES 226
    5.3.   CHANNELS OF COMPOSITE ROUGHNESS............................ 231

5.4.   APPLICATION OF UNIFORM FLOW CONCEPTS TO
PRACTICE 239

5.5.   BIBLIOGRAPHY 253

5.6.   PROBLEMS 255

**CHAPTER 6 - THEORY AND ANALYSIS OF GRADUALLY AND
SPATIALLY VARIED FLOW 261**

6.1.   BASIC ASSUMPTIONS AND THE EQUATION OF
GRADUALLY VARIED FLOW 261

6.2.   CHARACTERISTICS AND CLASSIFICATION OF
GRADUALLY VARIED FLOW PROFILES 262

6.3.   COMPUTATION OF GRADUALLY VARIED FLOW 267

6.4.   SPATIALLY VARIED FLOW 304

6.5.   APPLICATION TO PRACTICE 318

6.6   BIBLIOGRAPHY 334

6.7   PROBLEMS 335

CHAPTER 7 - DESIGN OF CHANNELS 339

7.1.   INTRODUCTION 339

7.2.   DESIGN OF LINED CHANNELS 345

7.3.   DESIGN OF STABLE, UNLINED, EARTHEN CHANNELS: A
GENERAL TRACTIVE FORCE DESIGN METHODOLOGY ... 357

7.4.   DESIGN OF CHANNELS LINED WITH GRASS 410

7.5.   BIBLIOGRAPHY 425

7.6.   PROBLEMS 428

**CHAPTER 8 - TURBULENT DIFFUSION AND DISPERSION IN OPEN
CHANNEL FLOW 431**

8.1.   INTRODUCTION 431

8.2.   GOVERNING EQUATIONS 432

8.3.   VERTICAL AND TRANSVERSE TURBULENT DIFFUSION
AND LONGITUDINAL DISPERSION 443

8.4.   NUMERICAL DISPERSION 477

8.5.   VERTICAL, TURBULENT DIFFUSION IN A
CONTINUOUSLY STRATIFIED ENVIRONMENT 480

8.6.   BIBLIOGRAPHY 485

8.7.   PROBLEMS 488

**CHAPTER 9 - UNSTEADY FLOW: HYDROLOGIC AND HYDRAULIC
APPROACHES 493**

9.1.   INTRODUCTION 493

9.2.   HYDROLOGIC APPROACHES 499

9.3.   HYDRAULIC APPROACHES 513

9.4. BOUNDARY AND INITIAL CONDITIONS 537
9.5. CALIBRATION AND VERIFICATION 538
9.6. BIBLIOGRAPHY 541
9.7. PROBLEMS..542

CHAPTER 10 - HYDRAULIC MODELS..**549**
10.1. INTRODUCTION 549
10.2. FIXED-BED RIVER OR CHANNEL MODELS 555
10.3. MOVABLE-BED MODELS 563
10.4. MODEL MATERIALS AND CONSTRUCTION......................................579
10.5. PHYSICAL MODEL CALIBRATION AND VERIFICATION 584
10.6. SPECIAL-PURPOSE MODELS 586
10.7. BIBLIOGRAPHY 590
10.8. PROBLEMS..592

APPENDIX 1 ....**595**

APPENDIX 2 ....**613**

SUBJECT INDEX **625**

AUTHOR INDEX...**635**

CHAPTER 1

CONCEPTS OF FLUID FLOW

1.1.   INTRODUCTION

By definition, an open channel is a conduit for flow which has a free surface; i.e. a boundary exposed to the atmosphere. The free surface is an interface between two fluids of a different density. In the case of the atmosphere, the density of air is much lower than the density for a liquid such as water; and in addition, the pressure is relatively constant. In the case of the flowing fluid, the motion is usually caused by gravitational effects, and the pressure distribution within the fluid is generally hydrostatic. Open channel flows are almost always turbulent and unaffected by surface tension; however, in some cases of practical importance, such flows are density stratified. The interest in the mechanics of open channel flows stems from their importance to what we have come to term civilization. As defined above, open channels include flows occurring in channels ranging from rivulets flow across a field to gutters along urban streets and continental highways to partially filled sewers carrying wastewater to irrigation water halfway across a continent to vital rivers such as the Mississippi, Nile, Rhine, Yellow, Ganges, Amazon, and Mekong. Without exception, one of the primary requirements for the development, maintenance, and advance of civilization is access to a plentiful and economic supply of water.

In the material that follows, it is assumed that the reader is familiar with the basic principles of modern fluid mechanics and hydraulics, calculus, numerical analysis, and computer science. It is the purpose of this chapter to review, briefly, a number of basic definitions, principles, and laws with the focus being their application to the study of the mechanics of open channel flow.

1.2.   DEFINITIONS

1.2.1.   Types of Flow

Change of Depth with Respect to Time and Space As will be demonstrated in this section, it is possible to classify the type of flow occurring in an open channel on the basis of many different criteria. The primary criteria of classification is the variation of the depth of flow y in time t and space x. If time is the criterion, then a flow can be classified as either **steady**, which implies that the depth of flow does not change with time ($\partial y/\partial t = 0$), or **unsteady**, which implies that the depth does change with time ($\partial y/\partial t \neq 0$). The differentiation between steady and unsteady flows depends on the viewpoint of the observer and is relative rather than an absolute classification. For example, consider a surge that is, a singular wave with a sharp front moving either up or down a channel. To a stationary observer on the bank of the channel, the flow is unsteady since this observer will note a change in the depth of flow with time. However, to an observer who moves with the wave front, the flow is steady since no variation of depth with time can be noted. If water is added or subtracted along the channel reach under consideration, which is the case with gutters, drop inlets, and side channel spillways, then the flow may be steady; but it is non-uniform. Specifically, this type of flow is termed **spatially varied** or **discontinuous flow**.

If space is used as the classification criterion, then a flow can be classified as **uniform** if the depth of flow does not vary with distance ($\partial y/\partial x = 0$) or as **nonuniform** if the depth varies with distance ($\partial y/\partial x \neq 0$). Although conceptually an unsteady uniform flow is possible, that is, the depth of flow varies with time but remains constant with distance, from a practical viewpoint such a flow is nearly impossible. Therefore, the terminology **uniform** and **nonuniform** usually implies that the flow is also steady. Nonuniform flow, also termed **varied flow**, is further classified as being either **rapidly varied** - the depth of flow changes rapidly over a relatively short distance such as is the case with a hydraulic jump - or **gradually varied** - the depth changes rather slowly with distance such as is the case of a reservoir behind a dam.

It should be noted that from a rigorous theoretical viewpoint the classification of steady and uniform are very restrictive. Therefore, in practice, the above definitions are extended or relaxed to a point where they are useful. For example, time and space flow classifications are commonly done on the basis of gross flow characteristics. If the spatially averaged velocity of flow

$$\bar{u} = \iint_A u \, dA$$

does not vary significantly with time, then the flow is classified as
steady. Similarly, if the average depth of flow is constant in space, then
the flow is considered uniform. Figure 1.1 schematically provides
examples of the above definitions applied to field situations.

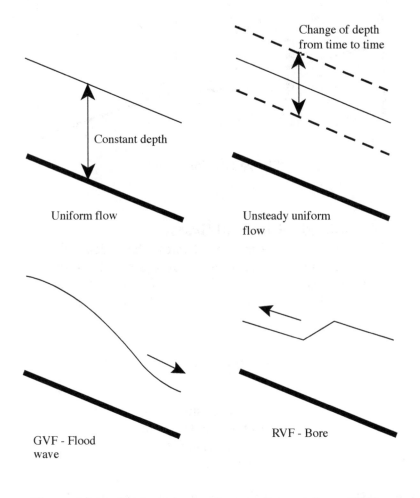

Figure 1.1.   **Various types of open channel flow; GVF =
gradually varied flow, RVF = rapidly varied
flow.**

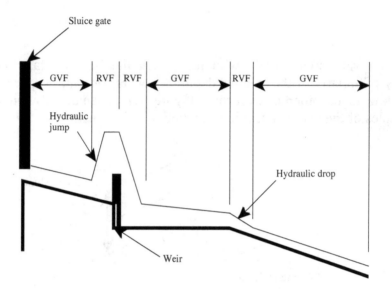

Figure 1.1. Continued.

1.2.2. Viscosity, Density, and Gravity

Recall from elementary fluid mechanics that, depending on the magnitude of the ratio of the inertial forces to the viscous forces, a flow may be classified as laminar, transitional, or turbulent. The basis for this classification is a dimensionless parameter known as the Reynolds number, or

$$Re = \frac{uL}{v} \tag{1.2.1}$$

where $u =$ characteristic velocity of flow, often taken as the average velocity of flow
$L =$ characteristic length
$v =$ kinematic viscosity.

A **laminar flow** is one in which the viscous forces are so large relative to the inertial forces that the flow is dominated by the viscous forces. In such flows, the fluid particles move along definite smooth paths in a coherent fashion. In a **turbulent flow**, the inertial forces are large relative to the viscous forces; hence the inertial forces dominate the situation. In this type of flow, the fluid particles move in an incoherent or apparently random fashion. A **transitional flow** is one which can be classified as neither laminar nor turbulent. In open channel flow, the characteristic length commonly used is the hydraulic radius which is the ratio of the flow area, A, to the wetted perimeter, P. Then

$$Re \leq 500 \qquad\qquad \text{laminar flow}$$

$$500 \leq Re \leq 12,500 \qquad\qquad \text{transitional flow}$$

$$12,500 \leq Re \qquad\qquad \text{turbulent flow}$$

The state of flow base on the ratio of inertial to viscous forces is a critical consideration when resistance to flow is considered.

Flows are classified as homogeneous or stratified on the basis of the variation of density within the flow. If in all spatial dimensions the density of flow is constant, then the flow is said to be **homogeneous**. If the density of the flow varies in any direction, then the flow is termed **stratified**. The absence of a density gradient in most natural open channel flows demonstrates that either the velocity of flow is sufficient to completely mix the flow with respect to density, or that the phenomena which induce density gradients are unimportant. The importance of density stratification is that when stable density stratification exists; that is, density increases with depth or lighter fluid overlies heavier fluid, reducing the effectiveness of turbulence as a mixing mechanism. In two dimensional flow of the type normally encountered in open channels, a commonly accepted measure of the strength of the density stratification is the gradient Richardson number or

$$\text{Ri} = \frac{g \dfrac{\partial \rho}{\partial y}}{\rho \left(\dfrac{\partial u}{\partial y} \right)^2} \qquad (1.2.2)$$

where $\quad g =$ acceleration of gravity
$\qquad\quad \rho =$ fluid density
$\qquad\quad y =$ vertical coordinate
$\qquad\quad \partial u / \partial y \quad =$ gradient of velocity in vertical direction
$\qquad\quad \partial \rho / \partial y \quad =$ gradient of density in the vertical direction.

When $\partial u/\partial y$ is small relative to $\partial \rho /\partial y$, Ri is large and the stratification is stable. When $\partial u/\partial y$ is large relative to $\partial \rho /\partial y$, Ri is small, and as $Ri \rightarrow 0$, the flow system approaches a homogeneous or neutral condition. There are a number of other parameters which are used to measure the stability of a flow; these will be discussed in subsequent chapters.

Depending on the magnitude of the ratio of inertial to gravity forces, a flow is classified as subcritical, critical, or supercritical. The parameter on which this classification is based is known as the **Froude number** or

$$F = \frac{u}{\sqrt{gL}} \qquad (1.2.3)$$

where u = characteristic velocity of flow and L = characteristic length. In an open channel, the characteristic length is taken as the **hydraulic depth**, which by definition is the flow area, A, divided by the width of the free surface, T, or

$$D = \frac{A}{T} \qquad (1.2.4)$$

If $F = 1$, the flow is in a critical state with the inertial and gravitational forces in equilibrium. If $F < 1$, the flow is in a subcritical state, and the gravitational forces are dominant. If $F > 1$, the flow is in a supercritical state, and the inertial forces are dominant.

The denominator of the Froude number is the celerity of an elementary wave in shallow water. In Fig. 1.2, an elementary wave of height Δy has been created by the movement of an impermeable plate from the left to right at a velocity of Δv. The wave has a celerity of c, so the velocity of flow cannot be analyzed by elementary techniques. However, as indicated in an earlier section of this chapter, some unsteady flow situations can be transformed to steady flow problems (Fig. 1.2). In this case, the transformation is accomplished by adapting a system of coordinates that moves at a velocity c. This is equivalent to changing the viewpoint of the observer; i.e., in Fig. 1.2, upper depiction, the observer is stationary while in Fig. 1.2, lower depiction, the observer is moving at the velocity of the wave.

Application of the steady, one dimensional equation of continuity to the situation described in Fig. 1.2, lower depiction yields

$$cy = (y + \Delta y)(c - \Delta u)$$

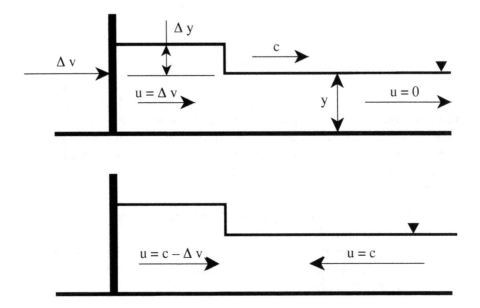

Figure 1.2. Propagation of an elementary wave.

and simplifying

$$c = y \frac{\Delta u}{\Delta y} \tag{1.2.5}$$

Application of the steady, one dimensional momentum equation yields

$$\frac{1}{2}\gamma y^2 - \frac{1}{2}\gamma(y + \Delta y)^2 = \rho c y \left[(c - \Delta u) - c \right]$$

or

$$\frac{\Delta u}{\Delta y} = \frac{g}{c} \tag{1.2.6}$$

Substitution of Eq. (1.2.6) in Eq. (1.2.5) yields

$$c = \frac{gy}{c}$$

or

$$c = \sqrt{gy} \qquad (1.2.7)$$

If it can be assumed that $y \approx d$ (where d is the depth of flow), which is a valid assumption if the channel is wide and not at a steep slope, then it has been proved that the celerity of an elementary gravity wave is equal to the denominator of the Froude number. With this observation, the following interpretation can be applied to the subcritical and supercritical depths of flow:

1. When the flow is subcritical, $F < 1$, the velocity of flow is less than the celerity of an elementary gravity wave. Therefore, such a wave can propagate upstream against the flow, and the upstream areas are in hydraulic communication with the downstream areas.

2. When the flow is supercritical, $F > 1$, the velocity of flow is greater than the celerity of an elementary gravity wave. Therefore, such a wave cannot propagate upstream against the flow, and the upstream areas of the channel are not in hydraulic communication with the downstream areas.

Thus, the possibility of an elementary gravity wave propagating upstream against the flow can be used as a criterion for differentiating between subcritical and supercritical flows.

In the case of density stratified open channel flows, it is often convenient to define an overall but inverted form of the gradient Richardson number. The internal or densimetric Froude number, F_D, is defined by

$$F_D = \frac{u}{\sqrt{gL\dfrac{\Delta\rho}{\rho}}} = \frac{u}{\sqrt{g'L}} \qquad (1.2.8)$$

where with reference to Fig. 1.3, $\Delta\rho = \rho_1 - \rho_2$, $g' = g(\rho_1 - \rho_2)/\rho$, and $L = $ a characteristic length which is usually taken as the depth of flow in the lower layer, y_1. The interpretation of F_D is analogous to that of F; e.g., in an internally supercritical flow a wave at the density interface cannot propagate upstream against the flow.

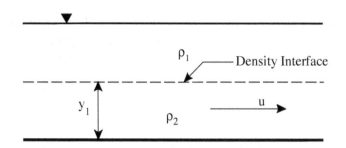

Figure 1.3. Notation for densimetric Froude number.

1.2.3.   Channel Types

Open channels can be classified as either natural or artificial. The terminology **natural channel** refers to all channels which have been developed by natural processes and have not been significantly improved by humans. Within this category are creeks, large and small rivers, and tidal estuaries. The category of **artificial channels** includes all channels which have been developed by human efforts. Within this category are navigation channels, power and irrigation canals, gutters, and drainage ditches. Although the basis principles in this book are applicable to natural channels, a comprehensive understanding of flow in natural channels is an interdisciplinary effort requiring knowledge of several technical fields; i.e., open channel hydraulics, hydrology, geomorphology, and sediment transport. Therefore, although flow in natural channels will be discussed, an extensive treatment of the subject is outside the scope of this book. Since many of the properties of artificial channels are controlled by design, this type of channel is much more amenable to analysis.

Within the broad category of artificial channels are the following subdivisions:

1. **Prismatic:** A prismatic channel has both a constant cross-sectional shape and bottom slope. Channels which do not meet this criterion are termed **nonprismatic**.

2. **Canal:** The term canal refers to a rather long constructed channel of mild slope. These channels may be either unlined or lined with concrete, cement, grass, wood, bituminous materials, or an artificial membrane.

3. **Flume:** In practice, the term **flume** refers to a channel built above the ground surface to convey flows across a depression.

Flumes are usually constructed of wood, metal, masonry, or concrete material. The term flume is also applied to laboratory channels constructed for basic and applied research.

4. **Chute and Drop:** A chute is a channel having a steep slope. A drop channel also has a steep slope but is much shorter than a chute.

5. **Culvert:** A culvert flowing only partially full is an open channel primarily used to convey flow under highways, railroad embankments, or runways.

Finally, it is noted that in this book the terminology **channel section** refers to the cross section of channel normal to the direction of flow.

1.2.4.   Section Elements

The properties of a channel section are wholly determined by the geometric shape of the channel and the depth of flow. Channel section properties are defined as follows:

1. **Depth of flow, y:** The depth of flow is the vertical distance from the lowest point of channel section to the water surface. In most cases, this terminology can be used interchangeably with the terminology **depth of flow of section d**, which is the depth of flow measured perpendicular to the channel bottom. The relationship between y and d is

$$y = \frac{d}{\cos\theta}$$

where θ is the slope angle of the channel bottom with a horizontal line. If θ is small, then

$$y \approx d$$

Only in the case of steep channels is there a significant difference between y and d. In some places in this book y is also used to designate the vertical coordinate of a Cartesian coordinate system. Although this dual definition of one variable can be confusing, it is unavoidable if a traditional notation system is to be used.

2. **Stage:** The stage of a flow is the elevation of the water surface relative to a datum. If the lowest point of a channel section is taken as the datum, then the stage and depth of flow are equal.

3. **Top Width T:** The top width of channel is the width of the channel section at the water surface.

4. **Flow Area A:** The flow area is the cross sectional area of the flow taken normal to the direction of flow.

5. **Wetted Perimeter P:** The wetted perimeter is the length of the line which is the interface between the fluid and the channel boundary.

6. **Hydraulic Radius R:** The hydraulic radius is the ratio of the flow area to the wetted perimeter or

$$R = \frac{A}{P} \qquad (1.2.9)$$

7. **Hydraulic Depth D:** The hydraulic depth is the ratio of the flow area to the top width or

$$D = \frac{A}{T} \qquad (1.2.10)$$

Table 1.1 summarizes the equations for the basic channel elements for the channel shapes normally encountered in practice.

Irregular channels are often encountered in practice, and in such cases, values of the top width, flow area, and the location of the centroid of the flow area must be interpolated from values of these variables which are tabulated as a function of the depth of flow. Franz (1982) developed a rational and consistent method for performing these calculations which define

$$A_y = \int_0^y T_\phi d\phi \qquad (1.2.11)$$

and

$$\bar{y}A_y = \int_0^y (y - \phi) T_\phi d\phi = \int_0^y A_\phi d\phi \qquad (1.2.12)$$

where ϕ =   dummy variable of integration

T_ϕ =   top width of the channel N above the origin (Fig. 1.4)

A_y =   flow area corresponding to a depth of flow y

$\overline{y}$ =   distance from the water surface to the centroid of the flow area.

Franz (1982) defined consistency of interpolation to mean: (1) both the tabular and interpolated values are consistent with Eqs. (1.2.11) and (1.2.12), and (2) the variables vary monotonically between the tabulated values. Then, given tables of T and A as functions of y such that $y_i \leq y \leq$ Franz (1982) defined consistency of interpolation to mean: (1) both the tabular and interpolated values are consistent with Eqs. (1.2.11) and (1.2.12), and (2) the variables vary monotonically between the tabulated values. Then, given tables of T and A as functions of y such that $y_i \leq y \leq$ y_{i+1} for $i = 0, 1, ..., n-1$, and $n + 1$ = number of depth values tabulated,

$$T_y = T_i + \frac{y - y_i}{y_{i+1} - y_i}\left(T_{i+1} - T_i\right) \tag{1.2.13}$$

$$A_y = A_i + 0.5(y - y_i)(T_y + T_i) \tag{1.2.14}$$

and

$$\overline{y}A_y = \overline{y}_i A_i + 0.5\left(y - y_i\right)\left(A_y + A_i\right) - \frac{1}{12}\left(y - y_i\right)^2\left(T_y - T_i\right) \tag{1.2.15}$$

It must be noted that the accuracy of the values interpolated by Eqs. (1.2.13) to (1.2.15) hinges on the accuracy of the top width approximation because all other interpolated values are exact if the top width approximation is exact; conversely, errors in the top width propagate to all other variables.

Table 1.1   Channel section geometric elements.

Channel Type	Area A	Wetted Perimeter P	Hydraulic Radius R	Top Width T	Hydraulic Depth D	Section Factor Z
Rectangle Fig. A	by	$b+2y$	$\dfrac{by}{b+2y}$	b	y	$by^{1.5}$
Trapezoid with equal side slopes Fig. B	$(b+zy)y$	$b+2y\sqrt{1+z^2}$	$\dfrac{(b+zy)y}{b+2y\sqrt{1+z^2}}$	$b+2zy$	$\dfrac{(b+zy)y}{b+2zy}$	$\dfrac{\left[(b+zy)y\right]^{1.5}}{\sqrt{b+2zy}}$
Trapezoid with unequal side slopes Figure C	$by+0.5y^2(z_L+z_R)$	$b+2y\left(\sqrt{1+z_L^2}+\sqrt{1+z_R^2}\right)$	$\dfrac{by+0.5y^2(z_L+z_R)}{b+y\left(\sqrt{1+z_L^2}+\sqrt{1+z_R^2}\right)}$	$b+y(z_L+z_R)$	$\dfrac{b+0.5y^2(z_L+z_R)}{b+y(z_L+z_R)}$	
Triangle with equal side slopes Figure D	zy^2	$2y\sqrt{1+z^2}$	$\dfrac{zy}{2\sqrt{1+z^2}}$	$2zy$	$0.5y$	$0.707zy^{2.5}$
Triangle with unequal side slopes Figure E	$0.5y^2(z_L+z_R)$	$y\left(\sqrt{1+z_L^2}+\sqrt{1+z_R^2}\right)$	$\dfrac{0.5y^2(z_1+z_2)}{y\left(\sqrt{1+z_L^2}+\sqrt{1+z_R^2}\right)}$	$y(z_L+z_R)$	$0.5y$	
Circular Figure F	$0.125(\theta-\sin\theta)d_o^2$	$0.5\theta d_o$	$0.25\left(1-\dfrac{\sin\theta}{\theta}\right)d_o$	$2\sqrt{y(d_o-y)}$	$0.125\left(\dfrac{\theta-\sin\theta}{\sin(0.5\theta)}\right)$	

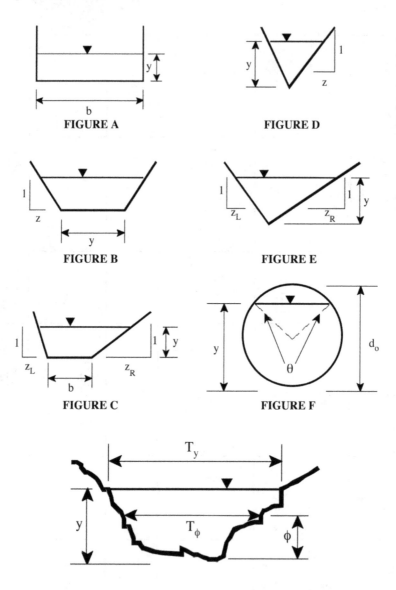

Figure 1.4.   Definition of cross-sectional terms for interpolation.

1.2.5.   Abstract Definitions

A **streamline** is a line constructed such that at any instant it has the direction of the velocity vector at every point; i.e., there can be no flow across a streamline. A **stream tube** is a collection of streamlines. It is noted and emphasized that neither streamlines nor stream tubes have any physical meaning; rather, they are convenient abstractions.

1.3.   GOVERNING EQUATIONS

1.3.1.   Introduction

In most open channel flows of practical importance the Reynolds number exceeds 12,500 and the flow regime is turbulent. Therefore, for the most part, laminar flow regimes are not treated in this book. The apparently random nature of turbulence has led many investigators to assume that this phenomena can best be described in terms of statistics. On the basis of this assumption, it is convenient to define the instantaneous velocity in terms of a time-averaged velocity and a fluctuating velocity component. For a Cartesian co-ordinate system, the instantaneous velocities in the x, y, and z directions are, respectively

$$u = \bar{u} + u'$$

$$v = \bar{v} + v' \tag{1.3.1}$$

and

$$w = \bar{w} + w'$$

Note: The average velocities used above may be determined by being averaged either over time at a point in space or over a horizontal area at a point in time. In this book, the symbolism $\overline{u_T}$ indicates an average in time while $\bar{u}$ indicates an average in space. From this point forward, the pertinent statistics will be defined in the x direction only with the tacit understanding that these definitions also apply in the other two Cartesian coordinate directions. The time averaged velocity is defined as

$$\overline{u_T} = \frac{1}{T} \int_0^T u\, dt \tag{1.3.2a}$$

where T indicates a time scale which is much longer than the time scale of the turbulence. The spatially average velocity is given by

$$\bar{u} = \frac{1}{A} \iint_A u\, dA \tag{1.3.2b}$$

Then, since the turbulent velocity fluctuations are random in terms of a time average

$$\overline{u'} = \frac{1}{T}\int_0^T u'dt \tag{1.3.3}$$

Then the statistical parameters of interest are:

1. Root mean square (rms) value of the velocity fluctuations

$$rms(u') = \left[\frac{1}{T}\int_9^T (u')^2 dt\right]^{1/2} \tag{1.3.4}$$

2. Average kinetic energy (KE) of turbulence per unit mass

$$\frac{\text{Average KE turbulence}}{\text{Mass}} = 0.5\left[\left(\overline{u'}\right)^2 + \left(\overline{v'}\right)^2 + \left(\overline{w'}\right)^2\right] \tag{1.3.5}$$

3. Variable correlations measure the degree to which two variables are interdependent. In the case of v the velocity fluctuations in the *xy* plane, the parameter

$$\overline{u'v'} = \frac{1}{T}\int_0^T u'v'dt \tag{1.3.6}$$

measures the correlation which exists between *u'* and *v'*. In a turbulent shear flow, $\overline{u'v'}$ is finite; therefore it is concluded that *u'* and *v'* are correlated.

1.3.2.   Conservation of Mass

Regardless of whether a flow is laminar or turbulent, every fluid flow must satisfy the equation of conservation of mass or, as it is commonly termed, the equation of continuity. Substituting the expressions for the instantaneous velocities defined in Eq. (1.3.1) into the standard equation of continuity (density is assumed constant) (see for example, Streeter and Wylie, 1975) yields

$$\frac{\partial\left(\overline{u} + u'\right)}{\partial x} + \frac{\partial\left(\overline{v} + v'\right)}{\partial y} + \frac{\partial\left(\overline{w} + w'\right)}{\partial z} = 0$$

or

$$\frac{\partial \overline{u}}{\partial x} + \frac{\partial \overline{v}}{\partial y} + \frac{\partial \overline{w}}{\partial z} + \frac{\partial u'}{\partial x} + \frac{\partial v'}{\partial y} + \frac{\partial w'}{\partial z} = 0 \qquad (1.3.7)$$

Taking the time or space average of each term in Eq. (1.3.7)

$$\frac{\partial \overline{\overline{u}}}{\partial x} + \frac{\partial \overline{\overline{v}}}{\partial y} + \frac{\partial \overline{\overline{w}}}{\partial z} = 0$$

and

$$\frac{\partial \overline{u'}}{\partial x} + \frac{\partial \overline{v'}}{\partial y} + \frac{\partial \overline{w'}}{\partial z} = 0$$

Therefore, Eq. (1.3.7) yields two equations of continuity, one for the mean or time-averaged motion and one for the turbulent fluctuations

$$\frac{\partial \overline{u}}{\partial x} + \frac{\partial \overline{v}}{\partial y} + \frac{\partial \overline{w}}{\partial z} = 0 \qquad (1.3.8)$$

and

$$\frac{\partial u'}{\partial x} + \frac{\partial v'}{\partial y} + \frac{\partial w'}{\partial z} = 0 \qquad (1.3.9)$$

In most applications of practical importance in open channel hydraulics, Eq. (1.3.8) is the only equation of continuity that is used. At this point and for the remainder of this section, the overbar indicating a time average will be dropped.

Other forms of the equation of continuity can be derived by specifying appropriate boundary conditions. For example, assume that the bottom boundary of the flow is given by the function

$$F = y - y_0(x, z) \qquad (1.3.10)$$

where $y = y_0$ is the elevation of the bottom above a datum. Then, if it is assumed (1) the velocity of the bottom is either zero or tangential to the surface and (2) a particle of fluid in contact with the bottom surface remains in contact with the surface, the condition which must be satisfied at the bottom boundary is

$$\frac{\partial F}{\partial t} + u\frac{\partial F}{\partial x} + v\frac{\partial F}{\partial y} + w\frac{\partial F}{\partial z} = 0 \qquad (1.3.11)$$

Substitution of Eq. (1.3.10) into Eq. (1.3.11) yields

$$u\frac{\partial y_0}{\partial x} + w\frac{\partial y_0}{\partial z} - v(x, y_0, z) = 0 \qquad (1.3.12)$$

In an analogous fashion, the free surface is defined by the function

$$F = y - y^*\left(x, y^*, z\right)$$

and

$$\frac{\partial y^*}{\partial t} + u\frac{\partial y^*}{\partial x} + w\frac{\partial y^*}{\partial z} - w\left(x, y^*, z\right) = 0 \qquad (1.3.13)$$

where y^* is the elevation of the free surface above a datum. Integrating Eq. (1.3.8) over the interval y_0 to y^*

$$\int_{y_0}^{y^*}\frac{\partial u}{\partial x}dy + \int_{y_0}^{y^*}\frac{\partial v}{\partial y}dy + \int_{y_0}^{y^*}\frac{\partial w}{\partial z}dy = 0$$

yields

$$\int_{y_0}^{y^*}\frac{\partial u}{\partial x}dy + \int_{y_0}^{y^*}\frac{\partial w}{\partial z}dy + v\left(x, y^*, z\right) - v\left(x, y_0, z\right) = 0 \qquad (1.3.14)$$

Then

$$\frac{\partial}{\partial x}\int_{y_0}^{y^*}u\,dy = u\left(x, y^*, z\right)\frac{\partial y^*}{\partial x} - u\left(x, y_0, z\right)\frac{\partial y_0}{\partial x} + y^*\frac{\partial u\left(x, y^*, z\right)}{\partial x} - y_0\frac{\partial u\left(x, y_0, z\right)}{\partial x}$$

and it is noted that

$$\int_{y_0}^{y^*}\frac{\partial u}{\partial x}dy = y^*\frac{\partial u\left(x, y^*, z\right)}{\partial x} - y_0\frac{\partial u\left(x, y_0, z\right)}{\partial x}$$

Combining the above equations yields

$$\frac{\partial}{\partial x}\int_{y_0}^{y^*} u\,dy = u\left(x,y^*,z\right)\frac{\partial y^*}{\partial x} - u\left(x,y_0,z\right)\frac{\partial y_0}{\partial x} + \int_{y_0}^{y^*}\frac{\partial u}{\partial x}\,dy$$

or

$$\int_{y_0}^{y^*}\frac{\partial u}{\partial x}\,dy = \frac{\partial}{\partial x}\int_{y_0}^{y^*} u\,dy + u\left(x,y_0,z\right)\frac{\partial y_0}{\partial x} - u\left(x,y^*,z\right)\frac{\partial y^*}{\partial x} \qquad (1.3.15)$$

In an analogous fashion

$$\int_{y_0}^{y^*}\frac{\partial w}{\partial x}\,dy = \frac{\partial}{\partial z}\int_{y_0}^{y^*} w\,dy + w\left(x,y_0,z\right)\frac{\partial y_0}{\partial x} - w\left(x,y^*,z\right)\frac{\partial y^*}{\partial z} \qquad (1.3.16)$$

Combining Eqs. (1.3.12) to (1.3.16) yields

$$\frac{\partial}{\partial x}\int_{y_0}^{y^*} u\,dy + \frac{\partial}{\partial z}\int_{y_0}^{y^*} w\,dy + \frac{\partial y^*}{\partial t} = 0 \qquad (1.3.17)$$

Equation (1.3.17) is a form of the equation of continuity from which a number of forms used in the field of open channel hydraulics can be derived. Some examples follow.

1. Define

$$q_x = \int_{y_0}^{y^*} u\,dy$$

and

$$q_z = \int_{y_0}^{y^*} w\,dy$$

where q_x and q_z are the components of the volume flux at the point (x,y) per unit time and unit widths of the vertical cross sections in the x and z directions, respectively, measured between the surface and bottom. With these definitions, Eq. (1.3.17) becomes

$$\frac{\partial q_x}{\partial x} + \frac{\partial q_z}{\partial z} + \frac{\partial y^*}{\partial t} = 0 \qquad (1.3.18)$$

2. In many cases, it may be convenient to introduce value of u and w which are vertically average values; i.e., spatially averaged values, defining

$$\bar{u} = \frac{1}{y^* - y_0} \int_{y_0}^{y^*} u\,dy$$

and

$$\bar{w} = \frac{1}{y^* - y_0} \int w\,dy$$

Note: In this case u and w are average velocities in space rather than time. Rewriting Eq. 1.3.17)

$$\frac{\partial}{\partial x}\left(\frac{y^* - y_0}{y^* - y_0} \int_{y_0}^{y^*} u\,dy \right) + \left(\frac{y^* - y_0}{y^* - y_0} \int_{y_0}^{y^*} w\,dy \right) + \frac{\partial y^*}{\partial t} = 0$$

And substituting the definitions of and yields

$$\frac{\partial}{\partial x}\left[\left(y^* - y_0 \right)\bar{u} \right] + \frac{\partial}{\partial z}\left[\left(y^* - y_0 \right)\bar{w} \right] + \frac{\partial y^*}{\partial t} = 0 \qquad (1.3.19)$$

3. In the case of a prismatic rectangular channel

$$Q = \bar{u}A(t) = \left(y^* - y_0 \right)b\bar{u}$$

where b = channel width. Since the channel is prismatic, the flow can be considered one-dimensional; and therefore,

$$\frac{\partial}{\partial x}\left[\left(y^* - y_0 \right)\bar{u} \right] + \frac{\partial y^*}{\partial t} = 0 \qquad (1.3.20)$$

or

$$\frac{\partial Q}{\partial x} + b\frac{\partial y^*}{\partial t} = 0 \qquad (1.3.21)$$

In this rather lengthy discussion of the equation of continuity, an attempt has been made to derive the forms of this equation which will be utilized throughout this book and which are found in the modern literature.

1.3.3. Conservation of Momentum

A very useful approach to the derivation of an appropriate conservation of momentum equation for open channel flow can be obtained from a consideration of the one-dimensional control volume in Fig. 1.5. The principle of conservation of momentum states

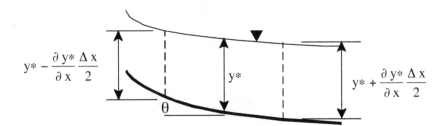

Figure 1.5. One-dimensional control volume.

$$\left\{ \begin{array}{c} Rate\ of\ accumulation \\ of\ momentum\ within \\ the\ control\ volume \end{array} \right\} = \left\{ \begin{array}{c} net\ rate\ of \\ momentum \\ entering \\ control\ volume \end{array} \right\} + \left\{ \begin{array}{c} sum\ of\ forces \\ acting\ on \\ control\ volume \end{array} \right\} \quad (1.3.22)$$

The rate of momentum entering the control volume is the product of the mass flow rate and the velocity or

$$Momentum\ entering = \rho u y^{*}\left(u\right) - \frac{\Delta x}{2}\frac{\partial}{\partial x}\left(\rho u^{2} y^{*}\right)$$

$$Momentum\ leaving = \rho u y^{*}\left(u\right) + \frac{\Delta x}{2}\frac{\partial}{\partial x}\left(\rho u^{2} y^{*}\right)$$

Then, the net rate at which momentum enters the control volume is

$$\rho u y^{*}\left(u\right) - \frac{\Delta x}{2}\frac{\partial}{\partial x}\left(\rho u^{2} y^{*}\right) - \left[\begin{array}{c} \rho u y^{*}\left(u\right) + \\ \frac{\Delta x}{2}\frac{\partial}{\partial x}\left(\rho u^{2} y^{*}\right) \end{array} \right] = \Delta x\frac{\partial}{\partial x}\left(\rho u^{2} y^{*}\right) \quad (1.3.23)$$

The forces acting on the control volume shown in Fig. 1.5 are: (1) gravity, (2) hydrostatic pressure, and (3) friction. The body force due to gravity is the weight of the fluid within the control volume acting in the direction of the x axis or

$$F_x = \rho g y^* \Delta x \sin\theta = \rho g y^* \Delta x S_x \qquad (1.3.24)$$

where θ = the angle the x axis makes with the bottom of the channel. It is assumed that

$$S_x = \sin\theta$$

The pressure force on any vertical section of unit width and water depth y^* is

$$F_P = \int_0^{y^*} p\,dy = \int_0^{y^*} \rho g\left(y^* - y\right)dy = \frac{1}{2}\rho g\left(y^*\right)^2 \qquad (1.3.25)$$

The frictional force which is assumed to act on the bottom and sides of the channel is given by

$$F_f = \rho g y^* \Delta x S_f \qquad (1.3.26)$$

where S_f = slope of the energy grade line or the friction slope
    Combining Eqs. (1.3.23) to (1.3.26) with the law of conservation of momentum, Eq. (1.3.22) yields

$$\Delta x \frac{\partial}{\partial t}\left(\rho u y^*\right) = -\Delta x \frac{\partial}{\partial x}\left(\rho u^2 y^*\right) + g y^* \rho\left(S_x - S_f\right) - \frac{g}{2}\Delta x \frac{\partial\left(y^*\right)^2}{\partial x}$$

Assuming ρ is constant and dividing both sides of the above equation by $\rho\Delta x$,

$$\frac{\partial\left(u y^*\right)}{\partial t} + \frac{\partial\left(u^2 y^*\right)}{\partial x} + \frac{g}{2}\frac{\partial\left(y^*\right)^2}{\partial x} = g y^*\left(S_x - S_f\right) \qquad (1.3.27)$$

which is known as the conservation form of the momentum equation. If this equation is expanded,

$$y^* \frac{\partial u}{\partial t} + u \frac{\partial y^*}{\partial t} + u \frac{\partial\left(u^*\right)}{\partial x} + u y^* \frac{\partial u}{\partial x} + g y^* \frac{\partial y^*}{\partial x} = g y^*\left(S_x - S_f\right) \qquad (1.3.28)$$

In an analogous fashion, the law of conservation of momentum can, when required, be written in a two-dimensional form or

$$\frac{\partial\left(uy^*\right)}{\partial y} + \frac{\partial\left(u^2y^*\right)}{\partial x} + \frac{\partial\left(uvy^*\right)}{\partial y} + \frac{g}{2}\frac{\partial\left(y^*\right)^2}{\partial x} = gy^*\left(S_x - S_{fx}\right) \qquad (1.3.29)$$

and

$$\frac{\partial\left(vy^*\right)}{\partial t} + \frac{\partial\left(v^2y^*\right)}{\partial y} + \frac{\partial\left(uvy^*\right)}{\partial x} + \frac{g}{2}\frac{\partial\left(y^*\right)^2}{\partial y} = gy^*\left(S_y - S_{fy}\right) \qquad (1.3.30)$$

where S_x and S_y are the channel bottom slopes in the x and y directions, respectively; and S_{fx} and S_{fy} are the friction slopes in the x and y directions, respectively. These equations can also be expanded to yield forms similar to Eq. (1.3.28).

In a prismatic, rectangular channel where the equation of continuity is given by Eq. (1.3.20), the corresponding conservation of momentum equation is

$$\frac{\partial u}{\partial t} + u\frac{\partial u}{\partial x} + g\frac{\partial y^*}{\partial x} = g\left(S_x - S_f\right) \qquad (1.3.31)$$

At this point, it is pertinent to observe that the conservation of momentum in turbulent flow is governed by a set of equations known as the Reynolds equations which can be derived from the Navier-Stokes equations for laminar flow by the substitution of Eq. (1.3.1); and the result in a general Cartesian coordinate system is

$$\rho\left(\frac{\partial\overline{u}}{\partial t} + \overline{u}\frac{\partial\overline{u}}{\partial x} + \overline{v}\frac{\partial\overline{u}}{\partial y} + \overline{w}\frac{\partial\overline{u}}{\partial z}\right) = -\frac{\partial\left(\overline{p} + \gamma h\right)}{\partial x}$$

$$+ \mu\left(\frac{\partial^2\overline{u}}{\partial x^2} + \frac{\partial^2\overline{u}}{\partial y^2} + \frac{\partial^2\overline{u}}{\partial z^2}\right) - \rho\left(\frac{\overline{\partial u'^2}}{\partial x} + \frac{\overline{\partial u'v'}}{\partial y} + \frac{\overline{\partial u'w'}}{\partial x}\right)$$

$$\rho\left(\frac{\partial\overline{v}}{\partial t} + \overline{u}\frac{\partial\overline{v}}{\partial x} + \overline{v}\frac{\partial\overline{v}}{\partial y} + \overline{w}\frac{\partial\overline{v}}{\partial z}\right) = -\frac{\partial\left(\overline{p} + \gamma h\right)}{\partial y} \qquad (1.3.32)$$

$$+\mu\left(\frac{\partial^2\overline{v}}{\partial x^2}+\frac{\partial^2\overline{v}}{\partial y^2}+\frac{\partial^2\overline{v}}{\partial z^2}\right)+\rho\left(\frac{\partial\overline{u'v'}}{\partial x}+\frac{\partial\overline{v'^2}}{\partial y}+\frac{\partial\overline{v'w'}}{\partial z}\right)$$

$$\rho\left(\frac{\partial\overline{w}}{\partial t}+\overline{u}\frac{\partial\overline{w}}{\partial x}+\overline{v}\frac{\partial\overline{w}}{\partial y}+\overline{w}\frac{\partial\overline{w}}{\partial z}\right)=-\frac{\partial\left(\overline{p}+\gamma h\right)}{\partial z}$$

$$+\mu\left(\frac{\partial^2\overline{w}}{\partial x^2}+\frac{\partial^2\overline{w}}{\partial y^2}+\frac{\partial^2\overline{w}}{\partial z^2}\right)+\rho\left(\frac{\partial\overline{w'v'}}{\partial x}+\frac{\partial\overline{w'v'}}{\partial y}+\frac{\partial\overline{w'^2}}{\partial z}\right)$$

where   h   =   vertical distance
         ρ =   density
         p   =   pressure
         μ =   absolute or dynamic viscosity

The above equations, along with the law of conservation of mass, govern all turbulent flows. A solution of this set of equations is not possible without an assumption to quantify the velocity fluctuations.

A popular method of quantifying the turbulent fluctuations terms in Eq. (1.3.32) is the Boussinesq assumption which defines the eddy viscosity such that the above equations become

$$\rho\left(\frac{\partial\overline{u}}{\partial t}+\overline{u}\frac{\partial\overline{u}}{\partial x}+\overline{v}\frac{\partial\overline{u}}{\partial y}+\overline{w}\frac{\partial\overline{u}}{\partial z}\right)=\frac{\partial\left(\overline{p}+\gamma h\right)}{\partial x}$$

$$+\left(\mu+\eta\right)\left(\frac{\partial^2\overline{u}}{\partial x^2}+\frac{\partial^2\overline{u}}{\partial y^2}+\frac{\partial^2\overline{u}}{\partial z^2}\right) \qquad (1.3.33)$$

$$\rho\left(\frac{\partial\overline{v}}{\partial t}+\overline{u}\frac{\partial\overline{v}}{\partial x}+\overline{v}\frac{\partial\overline{v}}{\partial y}+\overline{w}\frac{\partial\overline{v}}{\partial z}\right)=\frac{\partial\left(\overline{p}+\gamma h\right)}{\partial y}$$

$$+\left(\mu+\eta\right)\left(\frac{\partial^2\overline{v}}{\partial x^2}+\frac{\partial^2\overline{v}}{\partial y^2}+\frac{\partial^2\overline{v}}{\partial z^2}\right)$$

$$\rho\left(\frac{\partial\overline{w}}{\partial t}+\overline{u}\frac{\partial\overline{w}}{\partial x}+\overline{v}\frac{\partial\overline{w}}{\partial y}+\overline{w}\frac{\partial\overline{w}}{\partial z}\right)=-\frac{\partial\left(\overline{p}+\gamma h\right)}{\partial z}$$

$$+ \left(\mu + \eta \right) \left(\frac{\partial^2 \overline{w}}{\partial x^2} + \frac{\partial^2 \overline{w}}{\partial y^2} + \frac{\partial^2 \overline{w}}{\partial z^2} \right)$$

where η = eddy viscosity. It must be noted that the Boussinesq assumption treats η as if it were a fluid property similar to μ ; however, η is not a fluid property but is a parameter which is a function of the flow and density. In practice, in turbulent flows, μ is neglected since it is much smaller than η.

Prandtl (see, for example, Schlicting, 1968), in an effort to relate the transport of momentum to the mean flow characteristics in a turbulent flow, introduced a characteristic length which is termed the *mixing length*. Prandtl claimed that

$$u' \approx v' \approx 1 \frac{d\overline{u}}{dy} \tag{1.3.34}$$

where l = mixing length. Then

$$\tau = \rho l^2 \left(\frac{d\overline{u}}{dy} \right)^2 \tag{1.3.35}$$

where τ = shear stress. A comparison of the Boussinesq and Prandtl theories of turbulence yields a relationship between the eddy viscosity and the mixing length or

$$\eta = \rho l^2 \frac{d\overline{u}}{dy}$$

and

$$\frac{\eta}{\rho} = 1^2 \frac{d\overline{u}}{dy} = \varepsilon \tag{1.3.36}$$

where η / ρ is a kinematic turbulence factor similar to the kinematic viscosity, and ε is a direct measure of the transport or mixing capacity of a turbulent flow. In a homogeneous flow, ε refers to all transport processes; e.g., momentum, heat, salinity, and sediment. In a density stratified flow, the following inequality is believed to be valid

$$\varepsilon > \varepsilon_M^s > \varepsilon_{Ma}^s \tag{1.3.37}$$

where ε_M^S = stratified momentum transport coefficient and ε_{Ma}^S = stratified mass transport coefficient.

1.3.4.   Energy Equation

From elementary hydraulics and fluid mechanics, recall that the total energy of a parcel of fluid traveling at a constant speed on a streamline is equal to the sum of the elevation of the parcel above a datum, the pressure head, and the velocity head. The one-dimensional equation which quantifies this statement is the Bernoulli energy equation or

$$H = z_A + d_A \cos\theta + \frac{u_A^2}{2g} \tag{1.3.38}$$

where subscript A = point on streamline in open channel flow
z_A = elevation of point A above an arbitrary datum
d_A = depth of flow section
θ = slope angle of channel
u_A = velocity at point A

For small values of θ, Eq. (1.3.38) reduces to

$$H = z + y + \frac{\overline{u}^2}{2g} \tag{1.3.39}$$

where $\overline{u}$ = spatially averaged velocity of flow and y = depth of flow.

1.3.5.   Energy and Momentum Coefficients

In many open channel problems, it is both convenient and appropriate to use the continuity, momentum equations in a one-dimensional form. In such cases, the laws of conservation are:

1.   *Conservation of Mass (Continuity)*

$$Q = \overline{u}A$$

2.   *Conservation of Momentum*

$$\sum F = \rho Q \left(\overline{u}_2 - \overline{u}_1 \right)$$

3.   *Conservation of Energy*

$$H = z + y + \frac{\overline{u^2}}{2g}$$

Of course, no real open channel flow is one dimensional and therefore, the true transfer of momentum through a cross section

$$\iint_A \rho u^2 dA$$

is not necessarily equal to the spatially averaged transfer

$$\rho Q \overline{u}$$

Thus, in situations where the velocity profile varies significantly in the vertical and/or transverse directions, it may be necessary to define a momentum correction coefficient, or

$$\beta \rho Q \overline{u} = \iint_A \rho u^2 dA$$

and solving

$$\beta = \frac{\iint_A \rho u^2 dA}{\rho Q \overline{u}} = \frac{\iint_A \rho u^2 dA}{\rho \overline{u}^2 A} \qquad (1.3.40)$$

where β = momentum correction coefficient. In an analogous fashion, the kinetic energy correction coefficient is

$$\alpha \gamma \frac{\overline{u^3}}{2g} A = \iint_A \gamma \frac{u^3}{2g} dA$$

and solving

$$\alpha = \frac{\iint_A \gamma u^3 dA}{\gamma \overline{u}^3 A} \qquad (1.3.41)$$

where α = kinetic energy coefficient. The following properties of β and α are noted:

1.   The are both equal to unity when the flow is uniform. In all other cases, β and α must be greater than unity.

2.   A comparison of Eqs. (1.3.40) and (1.3.41) demonstrates that for a given channel section and velocity distribution, α is much more sensitive to the variation in velocity than β.

3.   In open channel hydraulics, β and α are generally only used when the channel consists of a main channel with sub-channels and/or berms and floodplains (Fig. 1.6). In such cases, the large variation in velocity of section to section effectively masks all gradual variations in velocity, and it is appropriate to consider the velocity in each of the sub-sections as constant. In channels of compound section, the value of α may exceed 2.

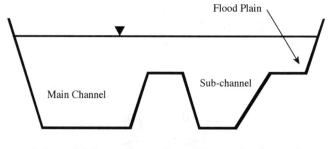

Figure 1.6.   Main   channel   with   sub-channels   and floodplains.

1.4.   THEORETICAL CONCEPTS

1.4.1.   Scaling Equations

In general, the terminology of scaling of equations refers to a process by which a complex differential equation is examined in a rational manner to determine if it can be reduced to a simpler form given a specific application. The first step in the process is to nondimensionalize the equation. As an example, consider an incompressible, laminar, constant density and viscosity flow occurring in a gravity field. The Navier-Stokes equations along with the equation of continuity govern the behavior of this flow. For simplicity, only the x component of the

Navier-Stokes equations will be treated or

$$\frac{\partial u}{\partial t} + u\frac{\partial u}{\partial x} + v\frac{\partial u}{\partial y} + w\frac{\partial u}{\partial z}$$

$$= -g\frac{\partial h}{\partial x} - \frac{1}{\rho}\frac{\partial p}{\partial x} + \frac{\mu}{\rho}\left(\frac{\partial^2 u}{\partial x^2} + \frac{\partial^2 u}{\partial y^2} + \frac{\partial^2 u}{\partial z^2}\right) \tag{1.4.1}$$

Define the following dimensionless quantities

$$\hat{x} = \frac{x}{L} \qquad \hat{v} = \frac{v}{U}$$

$$\hat{y} = \frac{y}{L} \qquad \hat{w} = \frac{w}{U}$$

$$\hat{z} = \frac{z}{L} \qquad \hat{t} = \frac{tU}{L}$$

$$\hat{h} = \frac{h}{L} \qquad \hat{p} = \frac{p}{\rho U^2}$$

$$\hat{u} = \frac{u}{U} \tag{1.4.2}$$

where L and U are a characteristic length and velocity, respectively. Substitution of these dimensionless variables in Eq. (1.4.1) yields

$$\frac{U^2}{L}\frac{\partial\hat{u}}{\partial t} + \frac{U^2}{L}\hat{u}\frac{\partial\hat{u}}{\partial\hat{x}} + \frac{U^2}{L}\hat{v}\frac{\partial\hat{u}}{\partial\hat{y}} + \frac{U^2}{L}\hat{w}\frac{\partial\hat{w}}{\partial\hat{z}}$$

$$= -g\frac{\partial\hat{h}}{\partial\hat{x}} - \frac{U^2}{L}\frac{\partial\hat{p}}{\partial\hat{x}} + \frac{\mu}{\rho}\frac{U^2}{L}\left(\frac{\partial^2\hat{u}}{\partial x^2} + \frac{\partial^2\hat{u}}{\partial y^2} + \frac{\partial^2\hat{u}}{\partial z^2}\right) \tag{1.4.3}$$

The substitution has altered neither the form nor units of Eq. (1.4.1). Equation (1.4.3) can be made dimensionless by dividing both sides by U^2/L with the result being

$$\frac{\partial\hat{u}}{\partial t} + \hat{u}\frac{\partial\hat{u}}{\partial\hat{x}} + \hat{v}\frac{\partial\hat{u}}{\partial\hat{y}} + \hat{w}\frac{\partial\hat{u}}{\partial\hat{z}}$$

$$= -\left(\frac{gL}{U^2}\right)\frac{\partial \hat{h}}{\partial \hat{x}} - \frac{\partial \hat{p}}{\partial \hat{x}} + \frac{\mu}{\rho UL}\left(\frac{\partial^2 \hat{u}}{\partial x^2} + \frac{\partial^2 \hat{u}}{\partial y^2} + \frac{\partial^2 \hat{u}}{\partial z^2}\right) \qquad (1.4.4)$$

Equation (1.4.4) is dimensionless, and the two groups of dimensionless variables, which appear, have been previously defined as the Froude number; i.e.,

$$F^2 = \frac{U^2}{gL}$$

and the Reynolds number; i.e.,

$$Re = \frac{\rho UL}{\mu}$$

The dimensionless equation may be rewritten as

$$\frac{\partial \hat{u}}{\partial \hat{t}} + \hat{u}\frac{\partial \hat{u}}{\partial \hat{x}} + \hat{v}\frac{\partial \hat{u}}{\partial \hat{y}} + \hat{w}\frac{\partial \hat{u}}{\partial \hat{z}}$$

$$= -\left(\frac{1}{F^2}\right)\frac{\partial \hat{h}}{\partial \hat{x}} - \frac{\partial \hat{p}}{\partial \hat{x}} + \frac{1}{Re}\left(\frac{\partial^2 \hat{u}}{\partial x^2} + \frac{\partial^2 \hat{u}}{\partial y^2} + \frac{\partial^2 \hat{u}}{\partial z^2}\right) \qquad (1.4.5)$$

After an equation or set of equations has been scaled; e.g., Eq. (1.4.5), it is examined to determine if significant mathematical simplification can be achieved on the basis of the specifics of a particular problem. For example in Eq. (1.4.5), if *Re* is large, then the last group of terms on the right-hand side of the equation can be neglected.

1.4.2.   Boundary Layers

The principle of scaling can be used to reduce the Navier-Stokes equations to a mathematically tractable set of equations known as the **boundary layer equations** (Schlichting, 1968). A well-established principle of fluid mechanics is that a particle of fluid in contact with a stationary solid boundary has no velocity. Thus, all flows of stationary boundaries exhibit velocity profiles through which the drag force caused by the boundary is transmitted outward. For example, consider a flat surface isolated in a flow. If the viscosity of the fluid is small, then the effect of the surface is confined to a thin layer of fluid in the immediate vicinity of the surface. Outside this "boundary layer," the fluid behaves as if it had no viscosity. In addition, since the boundary layer is thin compared to the typical longitudinal length scale, the pressure difference across the layer is negligible.

Consider a flat surface with U = velocity of flow outside the boundary layer. Experimental and theoretical evidence demonstrates that δ, the boundary layer thickness, depends on U, ρ, μ, and x. For a laminar boundary layer, the Blasius solution (Schlichting, 1968) yields

$$\delta = \frac{5x}{\sqrt{Re_x}} \quad \text{at} \quad \frac{u}{U} = 0.99 \qquad (1.4.6)$$

where x = distance from the leading edge of the surface, and Re_x = $\rho U x / \mu$ = a Reynolds number based on longitudinal distance. In Eq. (1.4.6), it is noted that δ is proportional to $x^{1/2}$, and thus as x increases, δ also increases. As the thickness of the laminar boundary layers increases, the boundary layer becomes unstable and transforms into a turbulent boundary layer (Fig. 1.7a). This transition occurs in the range

$$500,000 < Re_x < 1,000,000$$

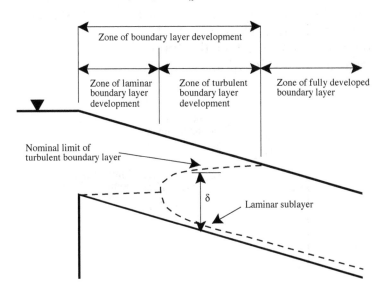

Figure 1.7a. **Stages of boundary layer development in open channels.**

The thickness of a turbulent boundary layer is given by

$$\delta = \frac{0.37x}{Re_x^{0.2}} \quad \text{at} \quad \frac{u}{U} = 0.99 \qquad (1.4.7)$$

Even in a turbulent boundary layer there is a very thin layer near the boundary which remains, for the most part, laminar, and is known as the *laminar sublayer*. With regard to boundary layers, especially in relation to open channel hydraulics:

1. As mentioned, the thickness of a boundary layer, whether laminar or turbulent, is a function of U, ρ, μ, and x. In general, the relationship that exists is

$$x \uparrow, \mu \uparrow, \rho \downarrow, U \downarrow \Rightarrow \delta \uparrow$$

$$x \downarrow, \mu \downarrow, \rho \uparrow, U \uparrow \Rightarrow \delta \downarrow$$

    where $\uparrow$ indicates an increase in value and $\downarrow$ indicates a decrease in value.

2. Boundary layers quite often grow within other boundary layers. In most open channel flows, the boundary layer intersects the free surface; thus, the total depth of flow is the thickness of the boundary layer. A change in the shape of the channel or channel roughness may result in the formation and growth of a new boundary layer (Fig. 1.7b).

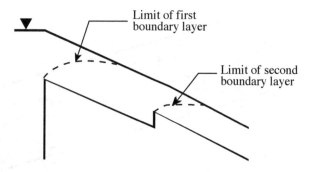

Figure 1.7b. **Growth of boundary layers within boundary layers.**

3. Boundary surfaces are classified as hydraulically smooth or rough on the basis of a comparison of the thickness of the laminar sublayer and the roughness height. If the boundary roughness is such that the roughness elements are covered by the

laminar sublayer, then the boundary is by definition hydraulically smooth. In this case, the roughness has no effect on the flow outside the sublayer. However, if the boundary roughness elements project through the sublayer, then by definition, the boundary is hydraulically rough; and the flow outside the sublayer is affected by the roughness. Schlichting (1968), in connection with flat surfaces and pipes, experimentally determined the following criteria for classifying boundary surfaces:

Hydraulically Smooth Boundary

$$0 \leq \frac{k_s u_*}{\nu} \leq 5 \qquad\qquad (1.4.8a)$$

Transition Boundary

$$5 \leq \frac{k_s u_*}{\nu} \leq 70 \qquad\qquad (1.4.9a)$$

Hydraulically Rough Boundary

$$70 \leq \frac{k_s u_*}{\nu} \qquad\qquad (1.4.10a)$$

where $u^* = (gRS)^{1/2}$ and k_s = roughness height. If the Chezy resistance equation (Chap. 4) or

$$\bar{u} = C\sqrt{RS}$$

where C = Chezy resistance coefficient, is used to calculate u, then Eqs. (1.4.8a) to (1.4.10a) become

$$0 \leq \frac{k_s \bar{u}\sqrt{g}}{C\nu} \leq 5 \qquad\qquad (1.4.8b)$$

$$5 \leq \frac{k_s \bar{u}\sqrt{g}}{C\nu} \leq 70 \qquad\qquad (1.4.9b)$$

$$70 \le \frac{k_s \bar{u} \sqrt{g}}{Cv} \qquad (1.4.10b)$$

In addition, it must be emphasized that defining k_s as a roughness height conveys an improper impression. Where k_s is a parameter which characterizes not only the vertical size of the roughness elements, but also their orientation, geometric arrangement, and spacing. For this reason, the roughness scale rarely corresponds to the height of the roughness elements. Table 1.2 summarizes some approximate values of k_s for various materials.

Table 1.2. Approximate values of k_s (Chow, 1959).

Material	k_s range, ft	k_s range, m
Brass, copper, lead, glass	0.0001 - 0.0030	0.00003 - 0.0009
Wrought iron, steel	0.0002 - 0.0080	0.00006 - 0.002
Asphalted cast iron	0.0004 - 0.0070	0.0001 - 0.002
Galvanized iron	0.0005 - 0.0150	0.0002 - 0.0046
Cast iron	0.0008 - 0.0180	0.0002 - 0.0055
Wood stave	0.0006 - 0.0030	0.0002 - 0.0009
Cement	0.0013 - 0.0040	0.0004 - 0.001
Concrete	0.0015 - 0.0100	0.0005 - 0.003
Drain tile	0.0020 - 0.0100	0.0006 - 0.003
Riveted steel	0.0030 - 0.0300	0.0009 - 0.009
Natural river bed	0.1000 - 3.000	0.3 - 0.9

1.   Flow around a bluff body; e.g., a sphere, usually results in what is termed separation of the boundary layer. In the case of sphere, flow in the vicinity of the leading edge is similar to that over a flat surface. However, downstream where the surface curves sharply away from the flow, the boundary layer separates from the surface (Schlicting, 1968). Boundary layer separation

results in what is known as *form drag* which may be the primary component of drag on a bluff body; such as a bridge pier. Recall from basic fluid mechanics that the total drag; i.e., surface place form drag, is given by

$$F = C_D A \rho \frac{\overline{u}^2}{2}$$

(1.4.11)

where F = total drag force on the body

C_D = drag coefficient

A = area of body projected onto a plane normal to the flow

As noted above, the most common case in open channel flow is that of a completely or fully developed boundary layer; i.e., the boundary layer fills the complete channel section. There are many cases in the chapters which follow where a basic understanding of boundary layer concepts is important.

Velocity Distributions
Within a turbulent boundary layer, Prandtl (see Schlichting, 1968) demonstrated that the vertical velocity profile is approximately logarithmetic. Equation (1.3.35) states that the shear stress at any point in the flow is

$$\tau = \rho \, l \left(\frac{d\overline{u}}{du} \right)^2$$

or

$$du = \sqrt{\frac{\tau}{\rho l^2}} dy$$

(1.4.12)

It is usually assumed that

$$l = ky$$

(1.4.13)

where k was originally called von Karman's turbulence constant. However, it is perhaps more accurate to term k a coefficient, since there is evidence that k may vary over a range of values or Reynolds numbers (see Slotta, 1963, Hinze, 1964, and Vanoni, 1946). If it is assumed that k can be approximated as 0.4, then substitution of Eq. (1.4.13) in Eq.

(1.4.12) yields, after integration

$$u = 2.5\sqrt{\frac{\tau_0}{\rho}} \ln\frac{y}{y_0}$$

or

$$u = 2.5u_* \ln\frac{y}{y_0} \tag{1.4.14}$$

where it is assumed that $\tau = \tau_0$ = shear stress on the bottom boundary, u = turbulent average velocity at a distance y above the bottom [Eq. (1.3.1)], and y_0 = a constant of integration. Equation (1.4.14) is known as the Prandtl-von Karman universal velocity distribution law. The constant of integration, y_0, is of the same order of magnitude as the viscous sublayer thickness and is a function of whether the boundary is hydraulically smooth or rough. If the boundary is hydraulically smooth, then y_0 depends solely on the kinematic viscosity and shear velocity or

$$y_0 = \frac{m\nu}{u_*} \tag{1.4.15}$$

where m is a coefficient equal to approximately 1/9 for smooth surfaces (Chow, 1959). Substitution of Eq. (1.4.15) in Eq. (1.4.14) yields

$$u = 2.5u_* \ln\frac{9yu_*}{\nu} \tag{1.4.16}$$

When the boundary surface is hydraulically rough, y_0 depends only on the roughness height or

$$y_0 = mk_s \tag{1.4.17}$$

where in this case m is a coefficient approximately equal to 1/30 for sand grain roughness. Substitution of Eq. (1.4.17) in Eq. (1.4.14) yields

$$u = 2.5u_* \ln\frac{30y}{k_s} \tag{1.4.18}$$

The vertical velocity profile equations derived above apply to flows occurring in very wide channels. Using the Prandtl-von Karman velocity law [Eq. (1.4.14)], Keulegan (1938) derived an equation for the average velocity of flow in a channel of arbitrary shape. Following Chow's (1959) modification of the Keulegan derivation, and with reference to Fig. 1.8, the total discharge through a channel section is given by

$$Q = \bar{u}A = \int_0^d \bar{u}\,dA = \int_0^A \bar{u}\beta\,dy \qquad (1.4.19)$$

where d = depth of flow, and β = length of isovel (the curve of equal velocity). It is assumed that β is proportional to the distance from the boundary or

$$\beta = P - \Gamma y$$

where Γ is a function which depends on the shape of the channel. Then

$$A = \int_0^d \beta\,dy = Pd - \frac{\Gamma d^2}{2} \qquad (1.4.20)$$

Substituting Eqs. (1.4.14) and (1.4.20) into Eq. (1.4.19), integrating and simplifying yields

$$u = 2.5u_* \ln\left[\frac{R}{y_0}\frac{d}{R}\exp\left(-1 - \frac{\Gamma d^2}{4A}\right)\right] \qquad (1.4.21)$$

where y_0 is defined by Eqs. (1.4.15) and (1.4.17). In Eq. (1.4.21) the parameter

$$\frac{d}{R}\exp\left(-1 - \frac{\Gamma d^2}{4A}\right)$$

is a function of the shape of the channel cross section. Both Keulegan and Chow claimed that the variation in this parameter with sections of various shapes is small; and thus, this parameter can, in general, be represented by an overall constant ϕ. With this assumption, Eq. (1.4.21) becomes

$$\bar{u} = u_*\left(\phi + 2.5\ln\frac{mR}{y_0}\right)$$

which is a theoretical equation for the average velocity of flow in an open channel.   Keulegan (1938) arrived at the following specific conclusions:

For Hydraulically Smooth Channels

$$\overline{u} = u_* \left(3.25 + 2.5 \ln \frac{Ru_*}{v} \right) \qquad (1.4.22)$$

For Hydraulically Rough Channels

$$\overline{u} = u_* \left(6.25 + 2.5 \ln \frac{R}{k_s} \right) \qquad (1.4.23)$$

The foregoing equations for the velocity distribution in open channel flow are one-dimensional equations; i.e., they consider only the variation of the velocity in the vertical dimension.   Most channels encountered in practice exhibit velocity distributions which are usually strongly two dimensional and, in many cases, three dimensional.   For example, the Prandtl-von Karman velocity distribution law predicts that the maximum velocity occurs at the free surface.   However, both laboratory and field measurements demonstrate that the maximum velocity usually occurs below the free surface.   Although in wide, rapid, and shallow flows or flows in very smooth channels, the maximum velocity may be at the free surface.   These observations demonstrate that a one-dimensional velocity distribution law cannot completely describe flows which are two and three dimensional.   In general, the velocity distribution in a channel of arbitrary shape is believed to depend on the shape of the cross section, the boundary roughness, the presence of bends, and changes in cross-sectional shape. The secondary currents caused by bends and changes in channel shape are termed *strong secondary currents*.   The shape of the cross section and the distribution of boundary roughness cause a nonuniform distribution of Reynolds stresses in the flow, and result in weak secondary currents. The terminology, *secondary current* or *flow*, refers to a circulatory motion of the fluid about an axis which is parallel to the primary current or flow, which is a translation of the fluid in the longitudinal direction.
   Einstein and Li (1958) attempted to identify the causes of weak secondary currents and concluded that: (1) secondary currents do not occur in either laminar or isotropic turbulent flow; (2) secondary currents in turbulent flow occur only when the isovels (the lines or

contours of equal velocity) of the primary flow are not parallel to each other or to the boundaries of the channel; and (3) the Reynolds stresses contribute to the secondary flow only when their distribution is not linear. Tracy (1965) directed an investigation along similar lines and came to approximately the same conclusions. The greatest progress in this area of research has been made by Liggett *et al.* (1965) and Chiu *et al.* (1967, 1971, 1976, 1978), who began investigating weak secondary currents which occurred in corners (Liggett *et al.*, 1965). These initial methods were later extended to flows in triangular channels (Chiu and Lee, 1971) and finally to channels of an arbitrary shape (Chiu *et al.*, 1976, 1978).

1.5.   SIMILARITY AND PHYSICAL MODELS

The number of important modern problems in open channel hydraulics, which can be solved satisfactorily by purely analytic techniques, is limited. Most problems must be solved by a combination of numerical and analytic techniques, field measurements, and physical modeling. In a physical model, two criteria must be satisfied: (1) The model and prototype must be geometrically similar; and (2) the model and prototype must be dynamically similar. The requirement for geometric similarity can be met by establishing a length scale ratio between the prototype and the model. The requirement of dynamic similarity requires that two systems with geometrically similar boundaries have geometrically similar flow patterns at corresponding instants in time. In turn, this requires that all the individual forces acting on corresponding fluid elements have same ratios in the two systems. The primary problem in developing physical models is not meeting the requirement of geometric similarity, but ensuring dynamic similarity.

One approach to developing appropriate parameters to ensure dynamic similarity is the scaling of the governing equations. For example, Eq. (1.4.5) has the same solution for two geometrically similar flow systems, as long as F and Re are numerically the same for the two systems. In some cases, one of these parameters may be unimportant; however, for exact dynamic similarity, F and Re for both the model and prototype must be equal. Equality of Froude numbers requires

$$\frac{U_M}{\sqrt{g_M L_M}} = \frac{U_P}{\sqrt{g_P L_P}} \qquad (1.5.1)$$

or

$$U_R = \sqrt{g_R L_R} \qquad (1.5.2)$$

where the subscript M designates the model, P the prototype, and R the ratio of the model-to-prototype variables. The requirement of equality of Reynolds numbers yields

$$U_R = \frac{\mu_R}{\rho_R L_R} \qquad (1.5.3)$$

Combining Eqs. (1.5.2) and (1.5.3) and recognizing $g_R = 1$ results in

$$L_R = \left(\frac{\mu_R}{\rho_R}\right)^{2/3} = \nu^{2/3} \qquad (1.5.4)$$

where ν = kinematic viscosity. Thus, if an exact physical model of an open channel flow is to be constructed, the modeler has only one degree of freedom - the choice of model fluid. Since the range of kinematic viscosities among available fluids is rather limited , the requirement imposed by Eq. (1.5.4) usually results in a model almost the same size as the prototype. The conclusion is that exact physical models of open channel flows are virtually impossible.

In many open channel flows, the effects of viscosity are negligible relative to the effects of gravity. Thus, in most cases, physical models are constructed only on the basis of Froude number equality. If only Froude number similarity is required, then

$$U_R = \sqrt{g_R L_R} \qquad (1.5.5)$$

and

$$T_R = \sqrt{\frac{L_R}{g_R}} \qquad (1.5.6)$$

where T_R = time scale ratio. In requiring only Froude number similarity, the modeler must be aware of two problems

1.   Care must be exercised to ensure that the model does not let the viscous effect become dominant. For example, if flow in the prototype is turbulent, then flow in the model cannot be laminar.

2.   In some cases, fluid friction is important but molecular viscosity is rather unimportant; e.g., high Reynolds number flows. In such cases, the similitude of frictional effects is one of similitude of boundary roughness rather than equality of Reynolds numbers.

In the case that density stratification is an important effect, then the model should be based on the equality of the densimetric Froude number [Eq. (1.2.8)].

In a specific situation, additional dimensionless parameters must be considered. For example, in a model in which small waves are generated, surface tension must be considered. Surface tension must also be considered in situations in which small waves must be prevented. The parameter which measures the relative magnitude of the inertial and capillary forces is the Weber number or

$$W = \frac{\rho U^2 L}{\sigma} \qquad (1.5.7)$$

where σ = surface tension forces per unit length.

1.6.   QUANTIFYING UNCERTAINTY

1.6.1.   Concept of Uncertainty

In the analysis of open channel flow, there are many quantities of interest that are functionally related to variables whose values are uncertain. For example, a commonly used equation is the Manning equation for uniform flow (Chapter 4) or

$$Q = \frac{\phi}{n} A R^{2/3} \sqrt{S} \qquad (1.6.1)$$

where ϕ = 1.49 for English units ϕ = 1.0 for SI units, n = Manning resistance coefficient, A = flow area, R = hydraulic radius, S = longitudinal slope, and Q = flow rate. Because the values of n, A, R, and S, the calculated value of Q is also uncertain. One approach to quantifying uncertainty is the first order second moment (FOSM) approach (Benjamin and Cornell, 1970 or Mays and Tung, 1992).

1.6.2.   FOSM - Single Variate

By definition, if x is a random variable, then

$$E\left(X - \bar{x}\right) = 0 \tag{1.6.2}$$

and

$$E\left(X - \bar{x}\right)^2 = \sigma_x^2 \tag{1.6.3}$$

where $E()$ indicates the expected value, $\sigma_x^2 =$ variance of x, and $\bar{x} =$ mean value of x. For the purposes of discussion, consider that the following functional relationship exists between x and Y

$$Y = g(x)$$

where $g(x)$ is a well-behaved function and the variance of x is not large. As observed by Benjamin and Cornell (1970), the concepts of functional behavior and smallness of variance are relative judgments. How large the variance of x may be depends on both the behavior of $f(x)$ in the vicinity of the value of x of interest and the degree of approximation in the estimated uncertainty that is acceptable. If σ_x^2 is small, x is likely lie close to its expected or average value ($\bar{x}$); and a Taylor series expansion about the expected value can be written

$$g\left(X\right) = g\left(\bar{x}\right) + \left(X - \bar{x}\right)\frac{d\,g\left(x\right)}{dx}\bigg|_{\bar{x}} + \frac{\left(X - \bar{x}\right)^2 d^2 g\left(x\right)}{2\,dx^2}\bigg|_{\bar{x}} + \dots \tag{1.6.4}$$

Retaining the first three terms of Eq. (1.6.4), taking the expectation of both sides, and applying Eq. (1.6.2) yields an expression for the expected value of $g(X)$ or

$$E\left[g\left(X\right)\right] \approx g\left(\bar{x}\right) + \frac{1}{2}\frac{d^2 g\left(x\right)}{dx^2}\bigg|_{\bar{x}}\ \sigma_x^2 \tag{1.6.5}$$

Equation (1.6.5) is a second-order approximation of the expected value of Y, when the uncertainty in the independent variable is σ_x^2. If only the first two terms are retained

$$E\left[g(X)\right] \approx g\left(\overline{x}\right) \tag{1.6.6}$$

Similarly, keeping only the first two terms and taking the variance of both sides and noting

$$\text{Var}\left[g\left(\overline{x}\right)\right] = 0$$

and

$$\text{Var}\left[\frac{dg(x)}{dx}\bigg|_{\overline{x}}\left(X - \overline{x}\right)\right] = \left[\frac{dg(x)}{dx}\bigg|_{\overline{x}}\right]^2 \sigma_x^2$$

yields

$$\text{Var}\left[g(X)\right] = \left[\frac{dg(x)}{dx}\bigg|_{\overline{x}}\right]^2 \sigma_x^2 \tag{1.6.7}$$

1.6.3. FOSM - Multivariate

The foregoing analysis can be extended to multivariate functions by using a multidimensional Taylor series expansion. The second-order approximation of the expected value of Y where

$$Y = g(X_1, X_2, ..., X_N)$$

is

$$E(Y) \approx g\left(\overline{x}_1, \overline{x}_2, ..., \overline{x}_N\right) + \frac{1}{2}\sum_{i=1}^{N}\sum_{j=1}^{N}\frac{\partial^2 g}{\partial x_i \partial x_j}\bigg|_{\overline{x}_i \overline{x}_j} \text{Cov}\left(X_i, X_j\right) \tag{1.6.8}$$

where Cov is the covariance of the variables X_i, and X_j. If the independent variables are not correlated, Eq. (1.6.8) becomes

$$E(Y) \approx g\left(\overline{x}_1, \overline{x}_2 ..., \overline{x}_N\right) + \frac{1}{2}\sum_{i=1}^{N}\frac{\partial^2 g}{\partial x_i^2}\bigg|_{\overline{x}_i} \sigma_{x_i}^2 \tag{1.6.9}$$

The first-order approximation of the variance of Y is

$$\text{Var}(Y) \approx \sum_{i=1}^{N}\sum_{j=1}^{N}\frac{\partial g}{\partial x_i}\bigg|_{\overline{x}_i}\frac{\partial g}{\partial x_j}\bigg|_{\overline{x}_j} \text{Cov}\left(X_i, X_j\right) \tag{1.6.10}$$

and if the X_i are uncorrelated

$$\operatorname{Var}(Y) \approx \sum_{i=1}^{N} \left(\frac{\partial g}{\partial x_i}\Big|_{\overline{x_i}} \right)^2 \sigma_{x_i}^2 \tag{1.6.11}$$

Given the form of Eq. (1.6.11), it is observed that in each of the N random variables, X_i contributes to the variance (uncertainty) of Y, proportional to its own variance $\sigma_{x_i}^2$ and proportional to a factor $(\partial g/\partial x_i)^2$, evaluated at the expected value of x_i, which is related to the sensitivity of Y to changes in X_i. This observation can be used to reduce the analytic effort required to estimate the uncertainty in the dependent variable. That is, in many cases, little accuracy is lost by treating the independent variables as deterministic rather than stochastic, if either the variance or sensitivity of Y is small enough that their product is negligible, compared to the contributions of other variables in the function.

EXAMPLE 2.1

Develop an expression for the uncertainty in the estimated discharge rate (Q) due to uncertainty in the Manning resistance coefficient (n).

Solution
Manning equation

$$Q = \frac{\phi}{n} A R^{2/3} \sqrt{S}$$

Then

$$\frac{dQ}{dn} = - \frac{\phi}{n^2} A R^{2/3} \sqrt{S}$$

Therefore, from Eq. (1.6.7)

$$\sigma_Q^2 = \left(\frac{dQ}{dn}\Big|_{\overline{n}} \right)^2 \sigma_n^2 = \left(-\frac{\phi}{n^2} A R^{2/3} \sqrt{S} \right)^2 \sigma_n^2$$

1.7.   BIBLIOGRAPHY

Benjamin, J.R. and Cornell, C.A., Probability, Statistics, and Decision for Civil Engineers, McGraw-Hill Book Company, New York, 1970.

Chiu, C. L., "Factors Determining the Strength of Secondary Flow," Proceedings of the American Society of Civil Engineers, Journal of the Engineering Mechanics Division, vol. 93, no. EM4, August 1967, pp. 69-77.

Chiu, C.L., and Lee, T.S., "Methods of Calculating Secondary Flow," *Water Resources Research.* Vol. 7, no. 4, August 1971, pp. 834-844.

Chiu, C.L., Lin, H.C., and Mizumura, K., "Simulation of Hydraulic Processes in Open Channels," *Proceedings of the American Society of Civil Engineers, Journal of the Hydraulics Division*, vol. 102, no. HY2, February 1976, pp. 185-205.

Chiu, C.L., Hsiung, D.E., and Lin, H., "Three-Dimensional Open Channel Flow," *Proceedings of the American Society of Civil Engineers, Journal of the Hydraulics Division*, vol. 104, no. HY8, August 1978, pp. 1119-1136.

Chow, V.T., *Open Channel Hydraulics*, McGraw-Hill Book Company, New York, 1959.

Einstein, H.A., and Li, H., "Secondary Currents in Straight Channels," *Transactions of American Geophysical Union*, vol. 39, 1958, p. 1085.

Franz, D.D., "Tabular Representation of Cross-Sectional Elements," *Proceedings of the American Society of Civil Engineers, Journal of the Hydraulics Division*, vol. 108, no. HY10, October 1982, pp. 1070-1081.

Hinze, J.O., "Turbulent Pipe Flow," in *The Mechanics of Turbulence*, A. Favre, ed., Gordon Breach, New York, 1964, pp. 129-165.

Keulegan, G.H., "Laws of Turbulent Flow in Open Channels," Research Paper RP 1151, *Journal of Research*, U.S. Bureau of Standards, vol. 21, December 1938, pp. 707-741.

Liggett, J.A., Chiu, C.L., and Miao, L.S., "Secondary Currents in a Corner," *Proceedings of the American Society of Civil Engineers, Journal of the Hydraulics Division*, vol. 91, no. HY6, November 1965, pp. 99-117.

Mays, L.W. and Tung, Y., *Hydrosystems: Engineering & Management*, Water Resources Publications, LLC, 1992.

Schlichting, H., *Boundary -Layer Theory*, 6th ed., McGraw-Hill Book Company, New York, 1968.

Slotta, L.S., "A Critical Investigation of the Universality of Karman's Constant in Turbulent Flow," in *Studies of the Effects of Variations on Boundary Conditions on the Atmospheric Boundary Layer*, H.H. Lettau, ed., University of Wisconsin Annual Report, Madison, WI, 1963, pp. 1-36.

Streeter, V.L., and Wylie, E.B., *Fluid Mechanics*, 6th ed., McGraw-Hill Book Company, New York, 1975.

Tracy, H..J., "Turbulent Flow in a Three Dimensional Channel, " *Proceedings of the American Society of Civil Engineers, Journal of the Hydraulics Division*, vol. 91, no. HY6, November 1965, pp. 9-35.

Vanoni, V.A., "Transportation of Suspended Sediment by Water," *Transactions of the American Society of Civil Engineers*, 1946, vol. III, pp. 67-133.

1.8.   PROBLEMS

1. Steady flow occurs when
    a.   Conditions at all point of interest steadily change with time
    b.   Conditions are the same at adjacent points at any instant
    c.   Conditions do not change with time at any point

2. Turbulent flow occurs in situations involving
    a.   Very viscous fluids
    b.   Very small velocities of flow
    c.   Capillary tubes
    d.   None of the above

3. Eddy viscosity, defined in Eq. (1.3.36), is
    a.   A physical property of the fluid
    b.   The viscosity divided by the density of the fluid
    c.   Dependent on the flow and density
    d.   Independent of the nature of the flow

4. Viscous forces are weak relative to inertial forces in
    a.   Laminar flows
    b.   Turbulent flows

5. The Reynolds number may be defined as the ratio of
    a.   Viscous forces to inertial forces
    b.   Viscous forces to gravity forces
    c.   Gravity forces to inertial forces
    d.   Pressure forces to elastic forces
    e.   Inertial forces to gravity forces
    f.   None of the above

6. The Froude number may be defined as the ratio of
    a.   Viscous forces to inertial forces
    b.   Viscous forces to gravity forces
    c.   Gravity forces to inertial forces
    d.   Pressure forces to elastic forces
    e.   Inertial forces to gravity forces

f.   None of the above

7. A canoe is floating in a lake of constant volume. In the canoe is a 1-ft^3 block (0.028 m^3) with a specific gravity of 2. If the concrete block is taken out of the canoe and dropped in the lake, does the average depth of the lake
    a.   Increase?
    b.   Decrease?
    c.   Stay the same?
    d.   Justify you answer with calculations.

8. Classify the following flow situations as either steady or unsteady from the viewpoint of the specified observer:

Case	Observer
I. A boat crossing a reservoir at a constant velocity	a. Standing on the boat b. Standing on the shore
II. Flow of a river around bridge piers	a. Standing on the bridge b. In a boat drifting with the current
III. A ship moving upstream at a constant velocity in a river	a. Standing on the bank b. In a boat drifting with the current c. Standing on the ship
IV. Movement of a flood down a river valley after the collapse of a dam	a. Standing on the river banks at the dam site b. Running downstream on a the river bank at the velocity of the dambreak surge

9. Water has an average velocity of 10 ft/s (3.0 m/s) through a 24-in (0.61 m) pipe flowing full. Estimate the discharge through the pipe in cubic feet per second.

10. Given the schematic figure below of a constant width rectangular channel, answer the following questions
    a.   Estimate the net force on the sluice gate if all losses are neglected.
    b.   Sketch the pressure distribution on the surface AB (note that the pressures at points A and B are zero gage pressure).

   c.   Is the pressure distribution sketched in Part (b) a hydrostatic pressure distribution?

   d.   How is the pressure distribution sketched in Part (b) related to the force estimated in Part (a)?

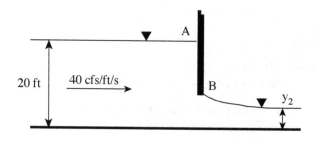

11. A bridge has its piers at a center distance of 7 m (23 ft). A short distance upsteam of the bridge, the water depth is 3 m (9.8 ft) and the velocity is 4 m/s (13 ft/s); when the flow has gone far enough downstream to even out again after the disturbance caused by the piers, the depth of flow is 2.7 m (8.9 ft). Neglecting the bed slope and boundary resistance, and assuming that the pressure variation is given by the hydrostatic law, estimate the thrust (force) on *each* pier.

12. A cylinder of infinite length is place in a uniform parallel flow steam of water as shown in the figure below. Using the information provided in the figure, calculate the force per unit length on the cylinder.

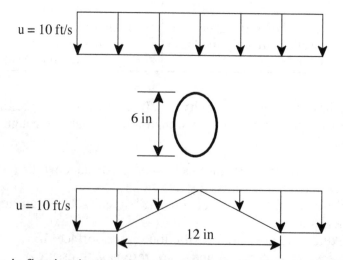

13. Water is flowing in a rectangular channel of unit as width shown in

the figure below. Neglecting all losses, determine the possible depths of flow at Section B.

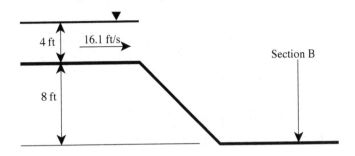

14. Given the data regarding a natural channel below, estimate
    a. The top width (English units)
    c. The flow area (English units)
    d. The centroid of the flow area (English units) when the river stage is 11.6 m.

River stage above an arbitrary datum (ft)	Distance to first perimeter intersection from South bank (ft)	Distance to second perimeter intersection from South bank (ft)
15	330	330
20	240	460
25	200	530
30	170	600
35	150	710
40	130	850
45	110	1200
50	80	1220
55	20	1230

15. Given the data provided in the table below, estimate the top width, the flow area, and the centroid of the flow area when the depth of flow is 45 ft (13.7 m).

Depth (ft)	Top Width (ft)	Area (ft²)
0	93	0
10	110	1015
20	208	2065
30	319	5240
40	410	8885
50	494	13405
60	575	18750

16. Given the data regarding a natural channel below, estimate
   a.   The kinetic energy correction factor
   b.   The momentum correction factor

Element	Average Velocity (m/s)	Area (m²)	Element Flow (m³/s)	Depth, left side of element (m)
1	0.24	12.5	3.0	0
2	0.30	34.4	10	8.2
3	0.37	49.2	18	14
4	0.40	56.7	23	18
5	0.49	57.1	28	19
6	0.43	54.8	24	18
7	0.43	48.8	21	18
8	0.40	40.9	16	14
9	0.37	35.3	13	12
10	0.34	30.2	10	11
11	0.37	26.5	9.8	9.1
12	0.30	23.7	7.1	8.2
13	0.27	21.4	5.8	7.3
14	0.27	20.0	5.4	6.7
15	0.24	19.0	4.6	6.4
16	0.09	17.2	1.5	6.1
17	0.03	7.90	0.23	5.2

17. Given the data in the figure below, and knowing

   $u_1 = 0.5$ ft/s (0.15 m/s)
   $u_2 = 10$ ft/s (3.0 m/s)
   $u_3 = 1$ ft/s (0.30 m/s)

   estimate the momentum correction factor and the kinetic energy correction using Eqs. (1.3.40) and (1.3.41).

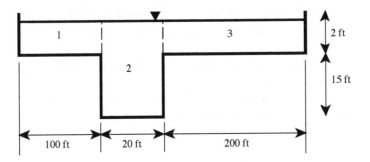

18. Using Eq. 1.4.14, prove that the average velocity of flow $\bar{u}$ can be estimated as follows

   a.   Measurements at 0.2 and 0.8 of the depth of flow with $\bar{u}$, the average velocity is given by

$$\bar{u} = \frac{u_{0.2D} + u_{0.8D}}{2}$$

   where $u_{0.2D}$ is the measured velocity at 0.2 of the depth (D) and $u_{0.8D}$ is the measured velocity at 0.8 of the depth.

   b.   A single measurement is made at 0.6 of the depth of flow. Is the distance measured up from the bottom of the channel or down from the water surface?

19 A laboratory channel 0.5 m (1.6 ft) wide conveys a uniform flow of 0.01 m³/s (0.35 ft³/s) at a depth of 0.05 m (0.16 ft). If the longitudinal slope is 0.0009 m/m and the roughness height k_s of the channel bed is 7.6 x 10^{-4} m (0.0025 ft), estimate the length of channel required for the full development of the boundary layer if the boundary layer is turbulent at the channel entrance. Also, plot the profile of the boundary layer as a function of distance along the channel bottom.

20. A 0.5 in (1.27 cm) high snail is cleaning the bottom of a hydraulically smooth chute that has a roughened entrance section and conveys sewage from a flow equalization pool to the sewage treatment plant.
    a. If the free-stream velocity entering this chute is 17 ft/s (5.2 m/s), how close can the snail crawl to the entrance if the snail is not to encounter any of the free-stream flow? Assume the kinematic viscosity of the sewage is 1.2 x 10^{-5} ft²/s (1.1 x 10^{-6} m²/s).
    b. How will this situation be affected if (1) the free-stream velocity is increased, or (2) the sewage temperature increases?

21. Give two *good* reasons why manhole covers are round.

22. Given the results stated in Eq. (1.5.4) and assuming that the fluid in the prototype is water, what is a common fluid that could be used in the model?

23. Uniform flow occurs in an open channel. The flow area is 90 ft² (8.4 m²), with a wetted perimeter of 35 ft (10.7 m) and a longitudinal slope of 0.0016 ft/ft. If the expected value of the Manning resistance coefficient ($\bar{n}$) is 0.017 and the variance associated with this value is $1.16x10^{-5}$, what is the expected flow rate (1st and 2nd order estimates) and the uncertainty in the expected value of the flow rate?

24. Uniform flow occurs in an open channel. The flow area is 10 m² (108 ft²) with a wetted perimeter of 10 m (32.8 ft). If the expected value of the Manning resistance coefficient ($\bar{n}$) is 0.0016 with $\sigma_S^2 =$ $1.16x10^{-5}$; and the expected value of the longitudinal slope is 0.0016 m/m with $\sigma_n^2 = 2.30x10^{-7}$, what is the expected flow rate (1st and 2nd order estimates) and the uncertainty in the expected value of the flow rate?

CHAPTER 2

ENERGY PRINCIPLE

2.1. DEFINITION OF SPECIFIC ENERGY

A central principle in any treatment of the hydraulics of open channel flow is the law of conservation of energy. From fluid mechanics, recall that the total energy of a parcel of water traveling on a streamline is given by the Bernoulli equation or

$$H = z + \frac{p}{\gamma} + \frac{u^2}{2g} \tag{2.1.1}$$

where
H	=	total energy,
z	=	elevation of the streamline above a datum,
p	=	pressure,
γ	=	fluid specific weight,
p/γ	=	pressure head,
u	=	streamline velocity,
$u^2/2g$	=	velocity head, and
g	=	local acceleration of gravity.

The sum $\left[z + \left(p / \gamma\right)\right]$ defines the elevation of the hydraulic grade line above the datum; and, in general, the value of this sum varies from point to point along the streamline. To examine the variation of this sum under various circumstances, consider a particle of cross-sectional area δA, length δs, density ρ, and mass $\rho \, \delta A$, moving along an arbitrary streamline in the $+S$ direction (Figure 2.1). If it is assumed that the fluid is frictionless, then there are no shear forces and only the gravitational body force and surface forces on the ends of the particle must be considered. The gravitational force is $\rho g \, \delta A \, \delta s$, the pressure force on the upstream face is $p \, \delta A$, and the pressure force on the downstream face is $[p+(\partial p/\partial s)\delta s]\delta A$. Applying Newton's second law of motion in the

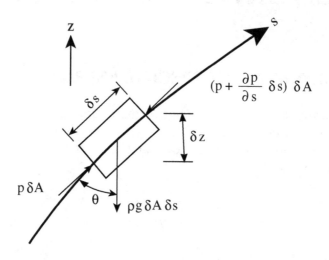

Figure 2.1.   Component forces action on a particle moving along a streamline.

direction of flow yields

$$F_s = a_s \delta m$$

or

$$\rho a_s \delta A \delta s = p \delta A - \left(p + \frac{\partial p}{\partial s} \delta s \right) \delta A - \rho g \delta A \delta s \cos\theta$$

where a_s = acceleration of the fluid particle along the streamline. Simplifying the equation

$$\frac{\partial p}{\partial s} + \rho g \frac{\partial z}{\partial s} + \rho a_s = 0$$

where

$$\frac{\partial z}{\partial s} = \cos\theta$$

Therefore

$$\frac{\partial}{\partial s}\left(p+\gamma z\right)+\rho A_s = 0 \tag{2.1.2}$$

Equation (2.1.2) is known as the Euler equation for motion along a streamline. If $a_s = 0$, Eq. (2.1.2) can be integrated to yield the hydrostatic law; that is, pressure varies linearly with depth.

The implications of Eq. (2.1.2) in open-channel flow are significant. First, if minor fluctuations due to turbulence are ignored and the streamlines have no acceleration components in the plane of the cross section, i.e., the streamlines have neither substantial curvature nor divergence, then the flow is termed parallel and a hydrostatic pressure distribution prevails. In practice, most uniform flows (Chapters 4 and 5) and gradually varied flows (Chapter 6) may be regarded as parallel flows with hydrostatic pressure distributions, since the divergence and curvature of the streamlines in these cases are negligible. In a parallel flow, the sum $z + p/\gamma$ is constant and equal to the depth of flow y, if the datum is taken as the channel bottom. Then by definition, the specific energy of an open-channel flow relative to the bottom of the channel is

$$E = y + \alpha \frac{u^2}{2g} \tag{2.1.3}$$

where α = kinetic energy correction factor, which is used to correct for the non-uniformity of the velocity profile and u = average velocity of flow ($u = Q/A$ where A is the flow area and Q is the flow rate). The assumption inherent in Eq. (2.1.3) is that the slope of the channel is small, or $\cos\theta \cong 1$, and $y \cong d\cos\theta$ (Fig. 2.2a). In general, if $\theta < 10°$ or $S < 0.18$, where S is the slope of the channel, Eq. (2.1.3) is valid.

If θ is not small, then the pressure distribution is not hydrostatic since the vertical depth of flow is significantly different from the depth measured perpendicular to the channel bed. In addition, in channels of large slope, e.g., spillways, the flow may entrain air which will change both the density of the fluid and the depth of flow. In the subsequent material, unless it is specifically stated otherwise, it is tacitly assumed that the channel slope is such that a hydrostatic pressure distribution exists.

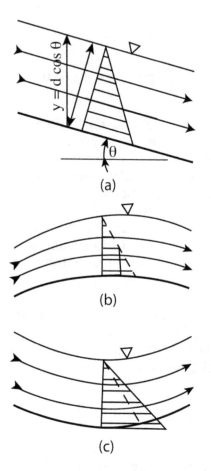

Figure 2.2. (a) **Pressure distribution in parallel flow with** $\cos\theta \cong 1$. (b) **Pressure distribution in concave flow.** (c) **Pressure distribution in convex flow.**

Furthermore, if $a_s \neq 0$, then the streamlines of the flow either will have a significant amount of curvature or will diverge, and the flow is termed *curvilinear*. Such situations may occur when the bottom of the channel is curved at sluice gates and at free overfalls. In such cases, the pressure distribution is not hydrostatic, and a pressure correction factor must be estimated. In concave flow situations (Fig. 2.2b), the forces resulting from streamline curvature reinforce the gravitational forces. In the case of convex flow (Fig. 2.2c), the forces resulting from the curvature of the streamlines act against the gravitational forces. If a channel has a curved longitudinal profile, then the deviation of the pressure distribution from the hydrostatic condition can be estimated from an application of Newton's second law to the *situation*

$$c = \frac{y}{g} \frac{u^2}{r} \qquad (2.1.4)$$

where r = radius of curvature of the channel bottom. The true pressure distribution at a section is then

$$p = y \pm \frac{y}{g} \frac{u^2}{r} \qquad (2.1.5)$$

where the plus and minus signs are used with concave and convex flows, respectively. In many cases, it is convenient to define a pressure coefficient such that the pressure head in a curvilinear flow can be defined as $\alpha' y$ where α' = the pressure coefficient. It can be demonstrated that

$$\alpha' = 1 + \frac{1}{Qy} \iint_A cu\, dA \qquad (2.1.6)$$

where Q = total discharge and dA = an incremental area. Then $\alpha' > 1$ for concave flow, $\alpha' = 1$ for parallel flow, and $\alpha' < 1$ for convex flow. For complex curvilinear flows, the pressure distribution can be estimated from either flow nets or model tests (see, for example, Streeter and Wylie, 1975).

2.2. SUBCRITICAL, CRITICAL AND SUPERCRITICAL FLOW

An examination of Eq. (2.1.3) demonstrates that if $cos\ \theta \approx 1, \alpha = 1$, and the channel section and discharge are specified, then the specific energy is only a function of the depth of flow. If y is plotted against E (Fig. 2.3), a curve with two branches results. The limb AC approaches the E axis asymptotically, and the branch AB asymptotically approaches the line $y = E$. For all the points on the E axis greater than point A, there are two possible depths of flow, known as the alternate depths of flow. Since A represents the minimum specific energy, the coordinates of this point can be found by taking the first derivative of Eq. (2.1.3) with respect to y and setting the result equal to zero or

$$E = y + \alpha \frac{u^2}{2g} = y + \alpha \frac{Q^2}{2gA^2}$$

$$\frac{dE}{dy} = 1 - \alpha \frac{Q^2}{gA^3} \frac{dA}{dy} = 0 \qquad (2.2.1)$$

where A = flow area. In Figure 2.3, the differential water area dA near the free surface is equal to $T\,dy$, where T = top width. By definition, the hydraulic depth D is

$$D = \frac{A}{T} \qquad (2.2.2)$$

Assuming $\alpha = 1$ and substituting Eq. (2.2.2) in Eq. (2.2.1), yields

$$1 - \frac{Q^2}{gA^3} \frac{dA}{dy} = 1 - \frac{Q^2}{gA^3} \frac{T}{A} = 1 - \frac{u^2}{gD} = 0$$

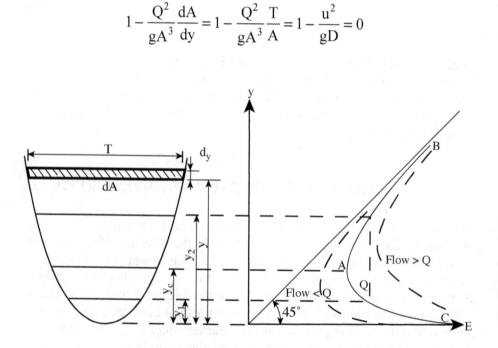

Figure 2.3. Specific energy curve.

or

$$\frac{u^2}{2g} = \frac{D}{2} \tag{2.2.3}$$

$$\frac{u}{\sqrt{gD}} = F = 1 \tag{2.2.4}$$

which is the definition of critical flow presented previously. Thus, minimum specific energy occurs at the critical hydraulic depth. With this knowledge, the branch AC can be interpreted as representing supercritical flows while the branch AB must represent subcritical flows. With regard to Fig. 2.3 and Eq. (2.2.4), the following should be considered. First, for channels of large slope angle θ and $\alpha \neq 1$, it can be easily demonstrated that the criterion for minimum specific energy is

$$F = \frac{u}{\sqrt{\dfrac{gD\cos\theta}{\alpha}}} \tag{2.2.5}$$

Second, in Fig. 2.3, E-y curves for flow rates greater than Q lie to the right of the curve for Q, and E-y curves for flow rates less than Q lie to the left of the curve for Q. Third, in the case of a rectangular channel of width b, Eq. (2.2.4) can be reduced to forms more suitable for computation. For example, define the flow per unit width q as

$$q = \frac{Q}{b}$$

The average velocity is then

$$u = \frac{q}{y}$$

and for a rectangular channel, $y = D$. With these definitions, Eq. (2.2.4) can be rearranged to yield

$$y_c = \left(\frac{q^2}{g}\right)^{1/3} \tag{2.2.6}$$

where y_c = critical depth. Substitution of the above definitions in Eq. (2.2.3) yields

$$\frac{u_c^2}{2g} = \frac{1}{2}y_c \tag{2.2.7}$$

And using the definition of specific energy, Eq. (2.1.3) and Eq. (2.2.7)

$$y_c = \frac{2}{3}E_c \tag{2.2.8}$$

where E_c = specific energy at critical depth and velocity.

At this point, it is appropriate to discuss methods of calculating the critical depth of flow in non-rectangular channels when the shape of the channel and the flow rate are specified. For simple channels, the easiest and most direct approach to this computation is an algebraic solution involving trial-and-error solutions of either Eq. (2.2.3), (2.2.4), or (2.2.5).

EXAMPLE 2.1

For a trapezoidal channel with a base width $b = 6.0$ m (20 ft) and side slope $z = 2$, calculate the critical depth of flow if $Q = 17$ m^3/s (600 ft^3/s).

Solution

From Table 1.1

$$A = \left(b + zy\right)y = \left(6.0 + 2y\right)y$$

$$T = b + 2zy = 6.0 + 4y$$

$$D = \frac{A}{T} = \frac{\left(3 + y\right)y}{3 + 2y}$$

and

$$u = \frac{Q}{A} = \frac{17}{2y(3+y)}$$

Substitution of the above in Eq. (2.2.3) yields

$$\frac{\left[\dfrac{17}{(6+2y)y}\right]^2}{g} = \frac{(3+y)y}{3+2y}$$

Simplifying

$$7.37(3+2y) = \left[y(3+y)\right]^3$$

Then, by trial and error or using an equation solver

$$y_c = 0.84\,\text{m}\,(2.8\,\text{ft})$$

and the corresponding critical velocity is

$$u_c = \frac{17}{\left[6+2(0.84)\right]0.84} = 2.6\,\text{m/s}\,(8.5\,\text{ft/s})$$

A second method of approaching the critical depth computation problem is through semiempirical equations. A set of equations for this purpose was developed by Straub (1982) for a number of common channel shapes, these equations are summarized in Table 2.1.

EXAMPLE 2.2

For a trapezoidal channel with $b = 6.0$ m (20 ft) and $z = 2$, find the critical depth of flow if $Q = 17$ m³/s (600 ft³/s).

Solution

From Table 2.1

$$y_c = 0.81 \left(\frac{\psi}{z^{0.75} b^{1.25}} \right)^{0.27} - \frac{b}{30z} \qquad \text{for } 0.1 \le \frac{Q}{b^{2.5}} \le 0.4$$

where

$$\psi = \alpha \frac{Q^2}{g}$$

As a check,

$$\frac{Q}{b^{2.5}} = \frac{17}{6^{2.5}} = 0.19$$

confirming that the equation from Table 2.1 applies.   Substituting appropriate values,

$$\psi = \frac{1(17)^2}{9.8} = 29.5$$

$$y_c = 0.81 \left(\frac{29.5}{2^{0.75} 6^{1.25}} \right)^{0.27} - \frac{6}{30((2))} = 0.86 \, \text{m} \, (2.8 \, \text{ft})$$

Comparing this result with that obtained in Example 2.1 for the same problem, one notes that the answers are very comparable and further that the method of Example 2.1 requires very little computation.

A third method of determining the critical depth when the shape of the channel section and the flow rate are specified involves the use of a design chart. In developing a chart for this purpose, it is convenient to define the section factor for critical flow computation. Substituting $u = Q/A$ in Eq. (2.2.3), after simplification, yields

$$\frac{Q}{\sqrt{\dfrac{g}{\alpha}}} = Z = \sqrt{\frac{A^3}{T}} \qquad (2.2.9)$$

The left hand side of Eq. (2.2.9) is, by definition, the section factor for critical flow Z, and the right hand side of the equation is only a function of the channel shape and the depth of flow. A design chart for the purpose of solving the critical depth problem is shown in Fig. 2.4.

Table 2.1.   Semiempirical equations for the estimation of y_c (Straub, 1982).

Channel Type	Equation for y_c in terms of $$\psi = \alpha \frac{Q^2}{g}$$	Comment
Rectangular B	$$\left(\frac{\psi}{b^2}\right)^{1/3}$$	
Trapezoidal B	$$0.81\left(\frac{\psi}{z^{0.75}b^{1.25}}\right)^{0.27} - \frac{b}{30z}$$	Range of applicability $$0.1 < \frac{Q}{b^{2.5}} < 0.4$$ For $$\frac{Q}{b^{2.5}} < 0.1$$ use equation for rectangular channel
Triangular z	$$\left(\frac{2\psi}{z^2}\right)^{0.20}$$	

Channel Type	Equation for y_c in terms of $$\psi = \alpha \frac{Q^2}{g}$$	Comment
Parabolic	$$\left(0.84c\psi\right)^{0.25}$$	Perimeter equation $$y = cx^2$$
Circular	$$\left(\frac{1.01}{d_o^{0.26}}\right)\psi^{0.25}$$	Range of applicability $$0.02 \le \frac{y_c}{d_o} \le 0.85$$
Elliptical	$$0.84b^{0.22}\left(\frac{\psi}{a^2}\right)^{0.25}$$	Range of applicability $$0.05 \le \frac{y_c}{2b} \le 0.85$$ a = major axis b = minor axis
Exponential	$$\left(\frac{m^3\psi c^{1/(2m+1)}}{4}\right)$$	Perimeter equation $$y = cs^{1/(m-1)}$$

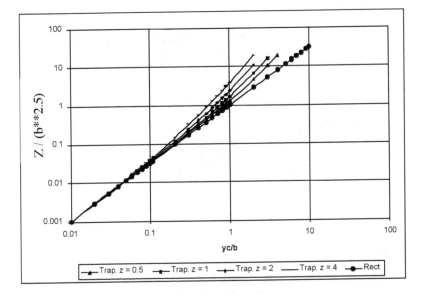

Figure 2.4a.   Curves for estimating critical depth in rectangular and trapezoidal channels ($Z/b^{2.5}$ as a function of y_c/b).

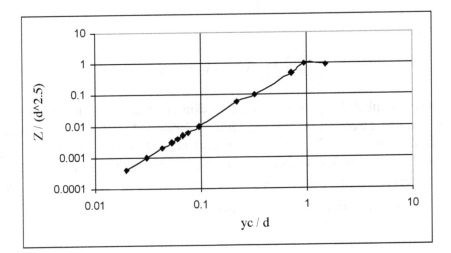

Figure 2.4b.   Curves for estimating critical depth in circular channels ($Z/d^{2.5}$) as a function of y_c/d).

EXAMPLE 2.3

A circular channel 3.0 ft (0.91 m) in diameter conveys a flow of 25 ft³/s (0.71 m³/s); estimate the critical depth of flow.

Solution

From Eq. (2.2.9) with $\alpha = 1$

$$Z = \frac{Q}{\sqrt{\dfrac{g}{\alpha}}} = \frac{25}{\sqrt{32.2}} = 4.41$$

and

$$\frac{d}{d_0^{2.5}} = \frac{4.41}{3^{2.5}} = 0.283$$

From Fig. 2.4

$$\frac{y_c}{d_0} = 0.54$$

or

$$y_c = 0.54(3) = 1.6 \, \text{ft} \, (0.49 \, \text{m})$$

For complex designed sections that cannot be analyzed by either the semi-empirical equations summarized in Table 2.1 or the design chart method presented in Fig. 2.4, a graphical procedure can be used. This method is also applicable to natural channels. In this procedure, a curve of y_c versus Z is constructed such that for a specified value of $Z = Q / \sqrt{g}$ the value of y_c may be estimated.

EXAMPLE 2.4

A trapezoidal channel with $b = 20$ ft (6.0 m) and $z = 1.5$, conveys a flow of 600 ft³/s (17 m³/s); estimate the critical depth of flow.

Solution

The first step in solving this problem is to construct a y versus Z curve (Fig. 2.5). The value of the section factor (Z) is computed from the given data or

$$Z = \frac{Q}{\sqrt{g}} = \frac{600}{\sqrt{32.2}} = 106 = \sqrt{\frac{A^3}{T}}$$

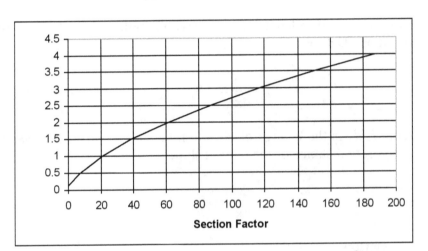

Figure 2.5.   **Critical depth (y_c) versus the section factor**
$(Z = \sqrt{A^3 / T})$ **for a trapezoidal channel with $b = 20$ ft and $z = 1.5$.**

Then, from Fig. 2.5, $y_c = 2.8$ ft (0.85 m).

2.3.   ACCESSIBILITY AND CONTROLS

The introduction of the concepts of specific energy and critical flow makes it possible to discuss the reaction of the flow in a channel to changes in the shape of the channel and hydraulic structures, for different steady-flow regimes,

$$H = \frac{u^2}{2g} + y + z \tag{2.3.1}$$

where y = depth of flow, z = elevation of the channel bottom above a datum, and it is assumed that $\alpha = 1$ and $cos\ \theta = 1$. Differentiating Eq. (2.3.1), with respect to the longitudinal distance x yields,

$$\frac{dH}{dx} = \frac{d\left(\frac{u^2}{2g}\right)}{dx} + \frac{dy}{dx} + \frac{dz}{dx} \tag{2.3.2}$$

The term dH/dx is the change of energy with longitudinal distance or the friction slope. Define

$$\frac{dH}{dx} = -S_f \tag{2.3.3}$$

The term dz/dx is the change of elevation of the bottom of the channel with, respect to distance or the bottom slope. Define

$$\frac{dz}{dx} = -S_0 \tag{2.3.4}$$

For a given flow rate Q, the term

$$\frac{d\left(\frac{u^2}{2g}\right)}{dx}$$

becomes

$$\frac{d\left(\frac{u^2}{2g}\right)}{dx} = -\frac{Q^2}{gA^3}\frac{dA}{dy}\frac{dy}{dx} = -\frac{Q^2 T}{gA^3}\frac{dy}{dx} = -F^2\frac{dy}{dx} \tag{2.3.5}$$

Substituting Eqs. (2.3.3) to (2.3.5) in Eq. (2.3.2) and simplifying, yields

$$\frac{dy}{dx} = \frac{S_0 - S_f}{1 - F^2} \tag{2.3.6}$$

which describes the variation of the depth of flow in a channel of arbitrary shape as a function of S_0, S_f, and F^2. In this chapter, only solutions of Eq. (2.3.6), where $S_f = 0$ (frictionless flow), will be considered. Solutions for the case where $S_f \neq 0$ will be considered in Chapter 6.

In the remainder of this chapter, solutions to Eq. (2.3.6) will be sought for the special case of $S_f = 0$, but first the behavior of the depth of flow in response to changes in cross-sectional shape will be examined from a qualitative viewpoint. For this purpose, a rectangular channel is assumed.

2.3.1.   Case 1:   Constant-Width Channel

In the case of a constant-width rectangular channel, the flow per unit width q is constant, and Eq. (2.3.6) can be rearranged to yield

$$\left(1 - F^2\right)\frac{dy}{dx} + \frac{dz}{dx} \qquad (2.3.7)$$

At this point, a number of subcases can be considered.

1.   If $dz/dx > 0$ and $F < 1$, then $(1 - F^2) > 0$ and dy/dx must be less than zero. Thus, the depth of flow decreases as x increases.

2.   If $dz/dx > 0$ and $F > 1$, then $(1 - F^2) < 0$ and dy/dx must be greater than zero. Thus, under these conditions, the depth of flow increases as x increases.

3.   If $dz/dx < 0$ and $F < 1$, then $(1 - F^2) > 0$ and dy/dx must be greater than zero. Thus, under these conditions, the depth of flow increases as x increases.

4.   If $dz/dx < 0$ and $F > 1$, then $(1 - F^2) < 0$ and dy/dx must be less than zero. Under these conditions, the depth of flow decreases as x increases.

When $dz/dx = 0$, an interesting and very useful case presents itself. Under this condition, Eq. (2.3.7) becomes

$$\left(1 - F^2\right)\frac{dy}{dx} = 0$$

Then, either $dy/dx = 0$ or $\mathbf{F}^2 = 1$. This type of situation can occur at both spillways and broad-crested weirs. By physical observation, it is known that in these situations, $dy/dx \neq 0$, and therefore $\mathbf{F}^2 = 1$. A judicious application of these results can yield an effective flow measurement method. In such cases, critical depth does not occur exactly at the free overfall but slightly before this point.

EXAMPLE 2.5

A broad-crested weir is placed in a channel of width b. If the upstream depth of flow is y_1, and the upstream velocity head of the flow and frictional losses can be neglected, develop a theoretical equation for the discharge in terms of the upstream depth of flow.

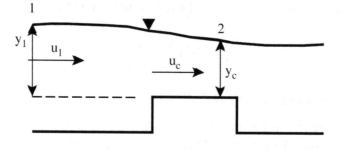

Figure 2.6. Broad-crested weir.

Solution

Given the situation described above and in Fig. 2.6, apply the Bernoulli energy equation between points 1 and 2 or

$$\frac{u_1^2}{2g} + y_1 = \frac{u_c^2}{2g} + y_c = E_c$$

For a deep, relatively slowly moving flow in the upstream section, $(u_1)^2/2g \ll y_1$, and hence

$$y_1 = E_c$$

Then, by Eq. (2.2.8)

$$E = \frac{3}{2}y_c = y_1$$

or

$$y_c = \frac{2}{3}y_1$$

Rearranging Eq. (2.2.6) yields

$$y_c^3 = \left(\frac{2}{3}y_1\right)^3 = \frac{q^2}{g}$$

or

$$\frac{q^2}{g} = \left(\frac{2}{3}\right)^3 y_1^3$$

and therefore,

$$Q = b\sqrt{g}\left(\frac{2}{3}\right)^{3/2} y_1^{3/2}$$

For $g = 32.2$ ft/s^2 (9.8 m/s^2)

$$Q = 3.09 b y_1^{3/2}$$

A comparison of the coefficient in the equation above with coefficients derived empirically (see, King and Brater, 1963), for this type of weir, demonstrates excellent agreement and confirms that, when frictional losses are negligible and $(u_1)^2/2g \ll y_1$, this type of approach is valid.

The foregoing example demonstrates an important practical application of the concepts of specific energy and critical depth; i.e., many structural means of flow measurement are based on the flow passing through critical depth, since for this depth of flow, there is an explicit relationship between the depth of flow and the flow rate.

2.3.2.   Case II: Channel of Variable Width

In this case, the flow per unit is not constant but $S_0 = 0$. Beginning with the total energy equation,

$$H = y + z + \frac{\left[q(x)\right]^2}{2gy^2}$$

where the symbolism $q(x)$ indicates that q is a function of x since the width of the channel is a function of x. Then

$$\frac{dH}{dx} = 0 = \frac{dy}{dx} + \frac{dz}{dx} - \frac{\left[q(x)\right]^2}{gy^3}\frac{dy}{dx} + \frac{q(x)}{gy^2}\frac{d\left[q(x)\right]}{dx} \qquad (2.3.8)$$

Since $Q = qb = $ constant

$$\frac{dQ}{dx} = 0 = b\frac{d\left[q(x)\right]}{dx} + q(x)\frac{db}{dx}$$

or

$$b\frac{d\left[q(x)\right]}{dx} = -q(x)\frac{db}{dx} \qquad (2.3.9)$$

Substitution of Eq. (2.3.9) into Eq. (2.3.8) yields

$$\left(1 - F^2\right)\frac{dy}{dx} - F^2\frac{y}{b}\frac{db}{dx} = 0 \qquad (2.3.10)$$

At this point four subcases can be considered.

1. If $db/dx > 0$ and $F < 1$, then $(1 - F^2) > 0$ and dy/dx must be greater than zero. Thus the depth of flow increases as x increases.

2. If $db/dx > 0$ and $F > 1$, then $(1 - F^2) < 0$ and dy/dx must be less than zero. Thus the depth of flow decreases as x increases.

3. If $db/dx < 0$ and $F < 1$, then $(1 - F^2) > 0$ and dy/dx must be less than zero. Under these conditions, the depth of flow decreases as x increases.

4. If $db/dx < 0$ and $\mathbf{F} > 1$, then $(1 - \mathbf{F}^2) < 0$ and dy/dx must be greater than zero. Under these conditions, the depth of flow increases as x increases.

The above discussion illustrates the interrelationship between the flow rate q and the depth of flow y. Then, by definition, a control is any feature which determines the depth-discharge relationship. From this definition, it follows that at any feature which acts as a control, the discharge can be calculated, once the depth of flow is known. This fact makes control sections attractive for flow measurement; and a critical control section, i.e., one in which the flow passes through critical depth, is especially attractive from the viewpoint of flow measurement.

A primary application of the concepts of specific energy is the prediction of changes in the depth of flow in response to channel transitions, i.e., changes in the channel width and/or the elevation of the channel bottom. In an examination of these problems, the accessibility of various points on the E-y curve must be considered. Consider a rectangular channel of constant width b which conveys a steady flow per unit width q. In the otherwise horizontal channel bed there is a smooth upward step of height z (Fig. 2.7a). Given this situation, an E-y curve can be constructed (Fig 2.7b). In this figure, the flow upstream of the step is represented by point A on the E-y curve. The point A' has the same specific energy as the point A, and the choice between the equally correct points, A and A', is a function of the upstream Froude number. If the point A is chosen to represent the upstream flow, then this flow is subcritical; i.e., $\mathbf{F} < 1$. Since q is a constant, the point representing the flow downstream of the step must also lie on the same E-y curve. The location of the downstream point on the E axis can be determined by applying the Bernoulli equation between the upstream and downstream points or

$$\frac{u_1^2}{2g} + y_1 = \frac{u_2^2}{2g} + y_2 + \Delta z$$

$$E_2 = E_1 - \Delta z \tag{2.3.11}$$

where it has been tacitly assumed that the energy dissipation between these points is negligible. Having mathematically determined the value of E_2, Eq. (2.3.11) can be solved to determine the corresponding values of y_2. There are three values of y which satisfy Eq. (2.3.11). One of these solutions is negative and has no physical meaning; however, the two remaining solutions, points B and B', are equally valid solutions to the problem.   The selection of the one point which is physically correct is the essence of the accessibility problem.

From the foregoing analysis of the varied-flow equation [Eq. (2.3.6)], it can be concluded that since $dz/dx > 0$ and $\mathbf{F} < 1$, the depth of flow decreases from point 1 to 2; however, since both B and B' represent depths of flow which are less than y_1, this conclusion is not useful. The E-y curve itself provides the solution to the problem. Since the width of the channel does not change, q is constant and the "flow point" can move only along the E-y curve defined by this value of q; i.e., the flow point cannot jump across the space which separates the points B and B'. Thus, the flow point must pass through C if it is to move to point B'; however, movement to point C is possible only if the increase in elevation of the channel is greater than the specified change Δz. This situation is represented by the dashed line in Fig. 2.7a. Thus, it is concluded that for the specified situation only point B is accessible from point A.

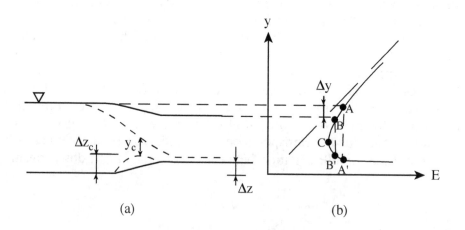

Figure 2.7. The accessibility problem.

The above discussion assumes that a solution to the specified problem exists.   In fact, it is quite easy to specify a problem to which there is no solution.   For example, in Fig. 2.7a and b if the step height exceeds Δz_c, there is no solution.   In essence, the three prescribed values, q, E, and Δz, cannot exist simultaneously.   A physical interpretation of this situation is that the flow area has been sufficiently obstructed that the flow is choked. The flow will back up behind this obstruction, q will decrease because the depth of flow increases, and a new steady flow will be established on an E-y curve to the left of the one shown in Fig 2.7b.

An additional observation is that near the critical condition, point C in Fig. 2.7b, large changes in the water surface can be effected by small changes in the bed level. Thus, flows which occur at near critical depth are inherently unstable and should be avoided.

2.4.   APPLICATION OF THE ENERGY PRINCIPLE TO PRACTICE

2.4.1.   Transition Problem

The primary application of the energy principle, in practice, is the solution of channel transition problems. In general, the solution of these problems can be effected by either algebraic and graphical methods. The following examples illustrate the basic solution techniques.

EXAMPLE 2.6

A rectangular channel expands smoothly from a width of 1.5 m (4.9 ft) to 3.0 m (9.8 ft). Upstream of the expansion, the depth of flow is 1.5 m (4.9ft) and the velocity of flow is 2.0m/s (6.6 ft/s). Estimate the depth of flow after the expansion.

Algebraic Solution

Since there is no change in the elevation of the channel bed, the upstream specific energy E_1 is equal to the downstream specific energy E_2, or

$$E_1 = E_2$$

where

$$E_1 = y_1 + \frac{u_1^2}{2g} = 1.5 + \frac{4}{2(9.8)} = 1.7\,\mathrm{m}\left(5.6\,\mathrm{ft}\right)$$

The velocity at the downstream station is

$$u_2 = \frac{Q}{A_2} = \frac{2(1.5)(1.5)}{3y_2} = \frac{1.5}{y_2}$$

and therefore,

$$E_2 = y_2 + \frac{u_2^2}{2g} = y_2 + \frac{0.11}{y_2^2} = 1.7\,\text{m}\left(5.6\,\text{ft}\right)$$

or

$$y_2^3 - 1.7y_2^2 + 0.11 = 0$$

Solving this equation yields

$$y_2 = 1.6\,\text{m}\left(5.2\,\text{ft}\right) \text{ or } 0.28\,\text{m}\left(0.92\,\text{ft}\right)$$

A consideration of the concepts of accessibility indicates that only the subcritical depth of flow is physically possible, and hence, the correct answer is

$$y_2 = 1.6\,\text{m}\left(5.2\,\text{ft}\right)$$

Graphical Solution

The graphical solution of this problem requires that an appropriate *E-y* curve be constructed for the downstream station.   The governing equation for this curve is

$$E_2 = y_2 + \frac{0.11}{y_2^2}$$

A graph of this equation is shown in Fig. 2.8.   From this graph the *y* values corresponding to $E_2 = 1.7$ m (5.6ft) can be found or

$$y_2 = 1.6\,\text{m}\left(5.2\,\text{ft}\right) \text{ or } 0.28\,\text{m}\left(0.92\,\text{ft}\right)$$

Again, an examination of the problem statement demonstrates that the supercritical depth of flow, i.e., $y_2 = 0.28$ m (0.92ft), is not accessible, and thus $y_2 = 1.6$ m (5.2ft).

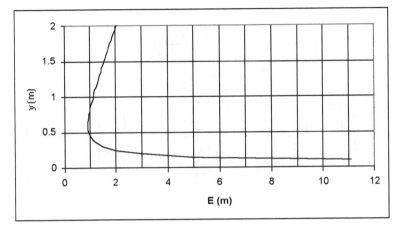

Figure 2.8. Graphical solution of Example 2.6.

Although the above methods provide satisfactory answers, they require either a solution of a cubic equation, or the construction of an E-y curve for each problem.   For rectangular channels, these computational difficulties can be overcome by constructing a dimensionless E-y curve.   Dividing both sides of the specific energy equation by the critical depth, yields

$$\frac{E}{y_c} = \frac{y}{y_c} + \frac{q^2}{2gy^2y_c} \tag{2.4.1}$$

Defining $E' = E/y_c$ and $y' = y/y_c$ and substituting Eq. (2.2.6) in Eq. (2.4.1) yields

$$E' = y' + \frac{1}{2(y')^2} \tag{2.4.2}$$

which is a dimensionless specific energy equation.   A graph of Eq. (2.4.2) is shown in Fig. 2.9, and it is noted in this figure that the critical point occurs at the coordinates (1.5, 1.0). In practice, this graph cannot be read with sufficient precision; for this reason, Babcock (1959) reduced the dimensionless E-y graph to a tabular form (Table 2.2). All transition problems which occur in rectangular channels can be solved efficiently using either Fig. 2.9 or Table 2.2.

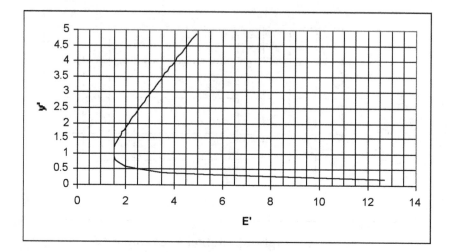

Figure 2.9. Dimensionless *E-y* curve.

EXAMPLE 2.7

Water flows in a rectangular channel 10 ft (3.0 m) wide at a velocity of 10 ft/s (3.0 m/s) and at a depth of 10 ft (3.0 m). There is an upward step of 2.0 ft (0.61 m). What expansion in width must take place simultaneously for this flow to be possible as specified?

Solution

First, examine the upstream flow and the step without considering the expansion. From the problem statement, the following quantities can be calculated:

$$q_1 = \frac{Q}{b_1} = \frac{10(10)(10)}{10} = 100 \text{ ft}^3 / s / \text{ft} \left(9.3 \text{m}^3 / s / m\right)$$

$$y_c = \left(\frac{q^2}{g}\right)^{1/3} = \left(\frac{100^2}{32.2}\right)^{1/3} = 6.76 \text{ft} \left(2.1m\right)$$

$$E_1 = y_1 + \frac{u_1^2}{2g} = 10 + \frac{10^2}{2(32.2)} = 11.6 \text{ft} \left(3.5m\right)$$

and

$$E_1' = \frac{E_1}{y_c} = \frac{11.6}{6.76} = 1.72$$

The downstream dimensionless specific energy, considering only the step, would be

$$E_2' = E_1' - \frac{\Delta z}{y_c} = 1.72 - \frac{2}{6.76} = 1.42$$

From Fig. 2.9, it can be determined that without an expansion, the flow is not possible as specified, since the downstream dimensionless specific energy does not lie on the E'-y' curve. If the downstream flow occurs at critical depth, then the required expansion is a minimum. The downstream condition is thus

$$E_2' = 1.5 = \frac{E_2}{y_{c2}}$$

or

$$y_{c2} = \frac{E_2}{1.5}$$

where $E_2 = E_1 - \Delta z = 9.55$ ft and y_{c2} = downstream critical depth. Then,

$$y_{c2} = \frac{9.55}{1.5} = 6.37 \, \text{ft} \left(1.9 \text{m} \right)$$

and

$$q = \sqrt{gy_{c2}^3} = 91.1 \, \text{ft}^3 / s / \text{ft} \left(8.5 \text{m}^3 / s / m \right)$$

Therefore, the downstream width is

$$b = \frac{Q}{q} = \frac{1000}{91.1} = 11 \, \text{ft} \left(3.4 \text{m} \right)$$

or the minimum expansion required is (11 - 10) = 1.0 ft (0.30 m).

Table 2.2 $E' = E/E_c$ as a function of y' = y/y$_c$ (Babcock, 1959)

y'	0.00	0.01	0.02	0.03	0.04	0.05	0.06	0.07	0.08	0.09
E' as a function of y'
0.0	∞	5000	1250	555.6	312.5	200.1	138.9	102.1	78.21	61.82
0.1	50.10	41.43	34.84	29.71	25.65	22.37	19.69	17.47	15.61	14.04
0.2	12.70	11.55	10.55	9.682	8.920	8.250	7.656	7.129	6.658	6.235
0.3	5.856	5.513	5.203	4.921	4.665	4.432	4.218	4.022	3.843	3.677
0.4	3.525	3.384	3.254	3.134	3.023	2.919	2.823	2.734	2.650	2.572
0.5	2.500	2.432	2.369	2.310	2.255	2.203	2.154	2.109	2.066	2.026
0.6	1.989	1.954	1.921	1.890	1.861	1.833	1.808	1.784	1.761	1.740
0.7	1.720	1.702	1.684	1.668	1.653	1.639	1.626	1.613	1.602	1.591
0.8	1.581	1.572	1.564	1.556	1.549	1.542	1.536	1.531	1.526	1.521
0.9	1.517	1.514	1.511	1.508	1.506	1.504	1.502	1.501	1.501	1.500
1.0	1.500	1.500	1.501	1.501	1.502	1.504	1.505	1.507	1.509	1.511
1.1	1.513	1.516	1.519	1.522	1.525	1.528	1.532	1.535	1.539	1.543
1.2	1.547	1.552	1.556	1.560	1.565	1.570	1.575	1.580	1.585	1.590
1.3	1.596	1.601	1.607	1.613	1.618	1.624	1.630	1.636	1.642	1.649
1.4	1.655	1.662	1.668	1.674	1.681	1.688	1.695	1.701	1.708	1.715
1.5	1.722	1.729	1.736	1.744	1.751	1.758	1.766	1.773	1.780	1.788
1.6	1.795	1.803	1.810	1.818	1.826	1.834	1.841	1.849	1.857	1.865
1.7	1.873	1.881	1.889	1.897	1.905	1.913	1.921	1.930	1.938	1.946
1.8	1.954	1.963	1.971	1.979	1.988	1.996	2.004	2.013	2.022	2.030
1.9	2.038	2.047	2.056	2.064	2.073	2.082	2.090	2.099	2.108	2.116
2.0	2.125	2.134	2.142	2.151	2.160	2.169	2.178	2.187	2.196	2.204
2.1	2.213	2.222	2.231	2.241	2.249	2.258	2.267	2.276	2.285	2.294
2.2	2.303	2.312	2.322	2.330	2.340	2.349	2.358	2.367	2.376	2.385
2.3	2.394	2.404	2.413	2.422	2.431	2.440	2.450	2.459	2.468	2.478
2.4	2.487	2.496	2.505	2.515	2.524	2.533	2.543	2.552	2.561	2.571
2.5	2.580	2.589	2.599	2.608	2.618	2.627	2.636	2.646	2.655	2.665
2.6	2.674	2.683	2.693	2.702	2.712	2.721	2.731	2.740	2.750	2.759
2.7	2.769	2.778	2.788	2.797	2.807	2.816	2.826	2.835	2.845	2.854
2.8	2.864	2.873	2.883	2.892	2.902	2.912	2.921	2.931	2.940	2.950
2.9	2.960	2.969	2.979	2.988	2.998	3.008	3.017	3.027	3.036	3.046
3.0	3.056	3.064	3.075	3.084	3.094	3.104	3.113	3.123	3.133	3.142
3.1	3.152	3.162	3.171	3.181	3.191	3.200	3.210	3.220	3.229	3.239

					E' as a function of y'					
y'	0.00	0.01	0.02	0.03	0.04	0.05	0.06	0.07	0.08	0.09
3.2	3.249	3.258	3.268	3.278	3.288	3.297	3.307	3.317	3.326	3.336
3.3	3.346	3.356	3.365	3.375	3.385	3.395	3.404	3.414	3.424	3.434
3.4	3.443	3.453	3.463	3.472	3.482	3.492	3.502	3.512	3.521	3.531
3.5	3.541	3.551	3.560	3.570	3.580	3.590	3.600	3.609	3.619	3.629
3.6	3.639	3.648	3.658	3.668	3.678	3.688	3.697	3.707	3.171	3.727
3.7	3.737	3.746	3.756	3.766	3.776	3.786	3.795	3.805	3.815	3.825
3.8	3.835	3.844	3.854	3.864	3.874	3.884	3.894	3.903	3.913	3.923
3.9	3.933	3.943	3.953	3.962	3.972	3.982	3.992	4.002	4.012	4.021
4.0	4.031	4.041	4.051	4.061	4.071	4.080	4.090	4.100	4.110	4.120
4.1	4.130	4.140	4.150	4.159	4.169	4.179	4.189	4.199	4.209	4.218
4.2	4.228	4.238	4.248	4.258	4.268	4.278	4.288	4.297	4.307	4.317
4.3	4.327	4.337	4.347	4.357	4.366	4.376	4.386	4.396	4.406	4.416
4.4	4.426	4.436	4.446	4.456	4.465	4.475	4.485	4.495	4.505	4.515
4.5	4.525	4.535	4.544	4.554	4.564	4.574	4.584	4.594	4.604	4.614
4.6	4.624	4.634	4.643	4.653	4.663	4.673	4.683	4.693	4.703	4.713
4.7	4.723	4.732	4.742	4.752	4.762	4.772	4.782	4.792	4.802	4.812
4.8	4.822	4.832	4.842	4.851	4.861	4.871	4.881	4.891	4.901	4.911
4.9	4.921	4.931	4.941	4.951	4.960	4.970	4.980	4.990	5.000	5.010

In the foregoing material, only the specific case of transitions in rectangular channels has been treated. It is necessary to develop an equivalent methodology for non-rectangular shapes. For a channel of arbitrary shape

$$E = y + \frac{Q^2}{2gA^2} \qquad (2.4.3)$$

Dividing both sides of the equation by y_c,

$$\frac{E}{y_c} = \frac{y}{y_c} + \frac{Q^2}{2gA^2 y_c} \qquad (2.4.4)$$

From Eq. (2.2.3)

$$\frac{Q^2}{g} = \frac{A_c^3}{T_c} = \frac{y_c^3 \left(T_c'\right)^3}{T_c} \tag{2.4.5}$$

where A_c and T_c = flow area and channel top width, when the depth of flow is y_c, respectively, and T'_c = top width of an equivalent rectangular channel, i.e., same depth of flow and flow area as the non-rectangular channel, at the critical depth of flow (Fig. 2.10).

Combining Eqs. (2.4.4) and (2.4.5) yields

$$\frac{E}{y_c} = \frac{y}{y_c} + \frac{1}{2}\left(\frac{y}{y_c}\right)^2 \left[\frac{\left(T_c'\right)^3}{T_c \left(T'\right)^2}\right] \tag{2.4.6}$$

For a rectangular channel, that is, $T'_c = T_c = T' = b$, Eq. (2.4.6) reduces to Eq. (2.4.2). The bracketed term in Eq. (2.4.6) is essentially a shape factor, and Silvester (1961) demonstrated that for trapezoidal, triangular,

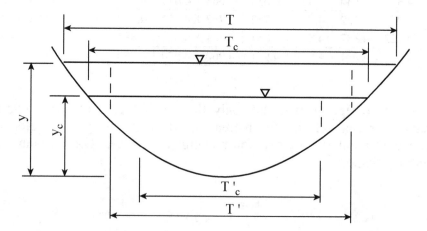

Figure 2.10.   Definition of variables for specific energy in nonrectangular channels.

and parabolic channels this factor can be easily evaluated.   In a trapezoidal channel

$$T = b + 2zy$$

and

$$T' = b + 2zy$$

A triangular channel is a special case of a trapezoidal channel with $b = 0$. For a parabolic channel with $T = ay^{1/2}$, where a is a coefficient, the shape factor is $(2/3)y_c/y$. The use of this technique is best illustrated with an example involving a trapezoidal channel.

EXAMPLE 2.8

A trapezoidal channel with $b = 20$ ft (6.1 m) and $z = 2$, carries a discharge of 4000 ft³/s (110 m³/s) at a depth of 12 ft (3.7 m). If a bridge spans the channel on two piers each 3.0 ft wide (0.91 m), determine the depth of flow beneath the bridge (Fig. 2.11).

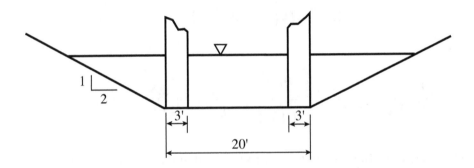

Figure 2.11. Schematic for Example 2.8.

Solution

From Table 2.1, for the section upstream of the bridge

$$y_c = 0.81\left(\frac{\dfrac{Q^2}{32.2}}{z^{0.75}b^{1.25}}\right)^{0.27} - \frac{b}{30z} = 8.5\,\text{ft}\,(2.6\,\text{m})$$

with

$$\frac{Q}{b^{2.5}} = 2.2$$

Then

$$T_c' = b + zy_c = 20 + 2(8.5) = 37\,\text{ft}\,(11\text{m})$$

$$T_c = b + 2zy_c = 20 + 2(2)(8.5) = 54\,\text{ft}\,(16\text{m})$$

$$T' = b + zy = 20 + 2(12) = 44\,\text{ft}\,(13\text{m})$$

The shape factor is

$$\left[\frac{(T_c')^3}{T_c(T_c')^2}\right] = \left[\frac{37^3}{54(44)^2}\right] = 0.48$$

and

$$\frac{E}{y_c} = \frac{y}{y_c} + \frac{1}{2}\left(\frac{y}{y_c}\right)^2\left[\frac{(T_c')^3}{T_c(T')^2}\right] = \frac{12}{8.5} + \frac{1}{2}\left(\frac{8.5}{12}\right)^2(0.48) = 1.5$$

The section of the channel beneath the bridge can be approximated as a trapezoidal channel with $b = 14\,\text{ft}\,(4.3\text{ m})$. In this case

$$\frac{Q}{b^{2.5}} = 5.4$$

and, therefore, the equations in Table 2.1 cannot be used to estimate y_c. The section factor is

$$Z = \frac{Q}{\sqrt{g}} = \frac{4000}{\sqrt{32.2}} = 705$$

and

$$\frac{Z}{b^{2.5}} = \frac{705}{14^{2.5}} = 0.96$$

From Fig. 2.4

$$\frac{y_c}{b} = 0.66$$

$$y_c = 9.2\,\text{ft}\,(2.8\text{m})$$

Then

$$T_c' = b + zy_c = 14 + 2(9.2) = 32.4\,\text{ft}\,(9.9\,\text{m})$$

$$T_c = b + 2zy_c = 14 + 2(2)(9.2) = 50.8\,\text{ft}\,(15\text{m})$$

$$T' = b + zy = 14 + 2(12) = 38\,\text{ft}\,(12\text{m})$$

where the computation of T' is based on an estimate of the depth of flow, at the bridge, to avoid an implicit solution. This assumption will be subsequently checked. The shape factor is

$$\left[\frac{(T_c')}{T_c(T')^2}\right] = \left[\frac{32.4^3}{50.8(38^2)}\right] = 0.46$$

Assuming no loss of energy as the flow passes through the contraction formed by the bridge piers, the dimensionless specific energy at the bridge is

$$\frac{E_2}{y_{c2}} = E_1'\frac{y_{c1}}{y_{c2}} = 1.5\frac{8.5}{9.2} = 1.39$$

Then

$$\frac{E_2}{y_{c2}} = \frac{y_2}{y_{c2}} + \frac{1}{2}\left(\frac{y_{c2}}{y_2}\right)^2 (0.46) = 1.39$$

This equation must be solved for y_2/y_{c2} by trial and error or by some other method. By trial and error, the solution is

$$\frac{y_2}{y_{c2}} = 1.24$$

$$y_2 = 9.2(1.24) = 11.4\,\text{ft}\,(3.5\,\text{m})$$

Then, checking the previously stated approximate computation for T',

$$T' = 14 + 2(11.4) = 36.8\,\text{ft}\,(11\,\text{m})$$

and the effect of this correction on the shape factor

$$\left[\frac{(T'_c)^3}{T_c(T')^2}\right] = \left[\frac{32.4^3}{50.8(36.8)^2}\right] = 0.49$$

which is negligible modification.

2.4.2.   Specific Energy in Channels of Compound Section

Throughout this chapter, the use of the kinetic energy correction factor has been indicated to account for the nonuniformity of the velocity distribution in open-channel flow.   Recall Eq. (2.1.3), which defines specific energy in a one-dimensional channel; i.e.,

$$E = y + \frac{\alpha Q^2}{2gA^2}$$

where, from Eq. (1.3.41), α is

$$\alpha = \frac{\displaystyle\iint_A u^3\,dA}{u^3 A}$$

α becomes especially significant in channels of compound section. Traditionally (Chow, 1959, and Henderson, 1966), this factor has been computed by

$$\alpha = \frac{\displaystyle\sum_{i=1}^{N} \frac{K_i^3}{A_i^2}}{\dfrac{K^3}{A^2}} \tag{2.4.7}$$

where K_i =   conveyance of ith channel subsection
       A_i =   area of ith channel subsection
       K =   conveyance of total section = ΣK_i
       A =   total area of cross section
       N =   number of subsections in given section

As will be discussed in Chapter 5, the subsection conveyance is computed from the Manning equation as

$$K_i = \frac{\phi}{n_i} A_i R_i^{2/3} \qquad (2.4.8)$$

where n_i =   Manning resistance coefficient for ith subsection
       ϕ =   factor correcting for system of units used ($\phi =$ 1.49 for English units and $\phi = 1$ for SI units)
       R_i =   hydraulic radius of ith subsection.

2.5.   BIBLIOGRAPHY

Babcock, H.A., "Tabular Solution of Open Channel Flow Equations," *Proceedings of the American Society of Civil Engineers, Journal of the Hydraulics Division*, vol. 85, no. HY3, March 1959, pp. 17-23.

Chow, V.T., *Open Channel Hydraulics*, McGraw-Hill Book Company, New York, 1959.

Henderson, F.M., *Open Channel Flow*, The Macmillan Company, New York, 1966.

King, H.W., and Brater, E.F., *Handbook of Hydraulics*, 5th ed., McGraw-Hill Book Company, New York, 1963, pp. 5.1-5.51.

Silvester, R., "Specific-Energy and Force Equations in Open-Channel Flow," *Water Power*, March 1961.

Straub, W.O., Personal Communication, Civil Engineering Associate, Department of Water and Power, City of Los Angeles, Jan. 13, 1982.

Streeter, V.L., and Wylie, E.B. *Fluid Mechanics*, McGraw-Hill Book Company, New York, 1975.

2.6.   PROBLEMS

1.   For $q = 100$ ft^3/s (2.8 m^3/s) flow in a rectangular channel, plot specific energy as a function of the depth of flow for channel widths of 10, 15, 20, and 25 ft (3.0, 4.6, 6.1 and 7.6 m, respectively).

2.   Calculate the critical depth for a discharge of 10 m^3/s (350 ft^3/s) in the following channels:
    a.   Rectangular channel; b = 3.0 m (9.8 ft)
    b.   Triangular channel; z = 0.5 m
    c.   Trapezoidal channel; b = 3.0 m (9.8 ft), z = 1.5
    d.   Circular channel; $D = 3.0$ m (9.8 ft).

3.   Calculate the bottom width of channel that is required to convey a discharge of 20 m^3/s (710 ft^3/s), as a critical flow at a depth of 1.5 m (4.9 ft), if the channel section is
    a.   Rectangular
    b.   Trapezoidal, $z = 1.5$.

4.   Water is flowing at a velocity of 4.6 m/s (15 ft/s) and at a depth of 0.75 m (2.5 ft) in a channel of rectangular section. Find the change in the absolute depth of flow produced by
    a.   A smooth upward step of 0.15 m (0.49 ft)
    b.   A smooth downward step of 0.15 m (0.49 ft)
    c.   The maximum allowable size of an upward step for the flow to be possible upstream as described.

5.   Water is flowing at a velocity of 11 ft/s (3.4 m/s) and at a depth of 11 ft (3.4 m) in a channel of rectangular cross section with a width of 11 ft (3.4 m). Find the changes in depth and absolute water level produced by
    a.   A smooth contraction in width of 10 ft (3.0 m)
    b.   A smooth expansion in width of 14 ft (4.3 m)
    c.   The greatest allowable contraction for the flow to be possible upstream as described.

6.   Water is flowing at a velocity of 10 ft/s (3.0 m/s) and a depth of 10 ft (3.0 m) in a channel of rectangular section 10 (f.00 m) wide. There is a smooth upward step of 4 ft (1.2 m) in the channel bed.

What expansion in width must take place for the upstream flow to be possible as specified?

7. Water flows from a lake into a steep rectangular channel 5 m (16 ft) wide. If the lake level is 8 m (26 ft) above the channel bed (channel invert) at the outfall, estimate the discharge from the lake to the channel.

8. A trapezoidal channel with a base width of 20 ft (6.1 m) and side slopes of 2 (horizontal): 1 vertical carries a flow of 2,000 ft³/s (57 m³/s) at a depth of 8 ft (2.4 m). There is a smooth transition to a rectangular section 20 ft (6.1 m) wide accompanied by a gradual lowering of the channel bed by 2 ft (0.61 m)
   a. Find the depth of water within the rectangular section and the change in water surface elevation.
   b. What is the minimum amount by which the bed must be lowered for the upstream flow to be possible as specified?

9. For the situation shown in the figure below, adapt the viewpoint of an observer who is moving with the elementary surge. Neglecting all squares of Δy and Δv, use the continuity and momentum equations to show that

$$c^2 = gy$$

Then, using the continuity and energy equations, show that the same result can be achieved. In both cases, list all important assumptions and justify them. Is on of these two methods of proof preferable? Why?

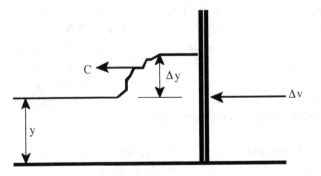

10. For the situation shown in the figure below, prove

$$\Gamma_L = \left(\frac{b_2}{b_1}\right)^2 = \frac{27F_1^2}{\left(2 + F_1^2\right)^3}$$

assuming that $E_1 = E_2$. Γ_L is known as the limiting contraction ratio and can be used to distinguish Yarnell's class A or unchoked flow and class B choked flow.

PLAN

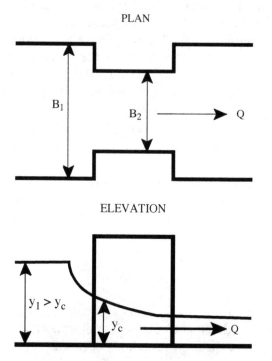

ELEVATION

11.   A rectangular channel 14 ft (4.3 m) wide conveys 756 ft³/s (21 m³/s) at a uniform depth of flow of 9 ft (2.7 m). This flow must pass under a road in a single barrel box culvert without creating a backwater effect and with the water surface a minimum of 1.5 ft (0.46 m), below the top of the culvert. Assuming that the culvert slope is sufficient to convey the flow uniformly, determine the minimum required culvert size.

12.   Flow at the rate of 4,800 ft³/s (136 m³/s) occurs in a rectangular channel 40 ft (12 m) wide and at a depth of 12 ft (3.7 m). Estimate the maximum width a single bridge pier may be if it is placed at midstream and does not cause an increase in the depth

of flow upstream. Compare this result with that obtained from the Yarnell equation in Problem 10.

13. For the upstream data specified in Problem 12, assume that it was decided not to put the bridge pier in mid-channel but to protrude the abutments into the channel. How far could each abutment protrude into the channel if no losses are incurred and no increase in the upstream depth of flow is caused?

14. For the upstream flow specified in Problem 12, assume that the bridge abutments are placed protruding into the channel such that the span between their faces is 24 ft (7.3 m). Also assume that the energy loss through this section is 1.0 ft-lb/lb of flow (0.30 m-N/N of flow).
    a. Will the flow pass through these abutments without causing an increase in the upstream depth of flow?
    b. If the flow cannot pass through the abutments without causing an increase in the upstream depth of flow, estimate the height of the water surface immediately upstream of the bridge.

15. A lake discharges freely into a steep channel. Develop a theoretical equation to estimate the discharge and list all assumptions.

16. The effects of the channel slope have been ignored in this chapter under the assumption that the channel slope is small.
    a. Show, schematically, total energy in a situation where the slope of the channel is not small.
    b. Specific energy for a channel where the slope is small is given by Eq. (2.1.3). Using the results from Part (a), derive a specific energy equation for the situation where the slope of the channel cannot be ignored.
    c. If you are in a boat and measured the depth of flow with a lead line, what depth, with reference to Part (a) of this problem, do you measure?

17. Consider a channel of irregular and varying cross section but with the low point of the section remaining at the same level along the channel. Neglect all losses and show that

$$\frac{dy}{dx}\left(1 - F^2\right) - F^2\left(\frac{y}{B}\right)\left(\frac{db}{dx}\right) = 0$$

The channel is shown in the figure below

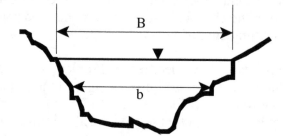

CHAPTER 3

THE MOMENTUM PRINCIPLE

3.1. DEFINITION OF SPECIFIC MOMENTUM

In examining the application of Newton's second law of motion to the basic problems of steady open-channel flow, it is convenient to begin with a general-case problem such as the one shown schematically in Fig. 3.1. Within the control volume defined in this figure, there is an unknown energy loss and/or a force acting on the flow between sections 1 and 2; the result is a change in the linear momentum of the flow. In many cases, this change in momentum is accompanied by a change in the depth of flow. The application of Newton's second law, in a one-dimensional form to this control, yields

$$F_1' + F_3' - F_2' - \sum f_f' - P_f' = \frac{\gamma}{g} Q \left(\beta_2 \bar{u}_2' - \beta_1 \bar{u}_1' \right) \qquad (3.1.1)$$

where F_1' and F_2' = horizontal components of pressures acting at sections 1 and 2, respectively

F_3'	=	horizontal component of $W \sin \theta$
W	=	weight of fluid between sections 1 and 2
γ	=	specific weight of fluid
θ	=	channel slope angle
$\sum f_i'$	=	sum of horizontal components of average velocities of flow at sections 1 and 2, respectively
P_f'	=	horizontal component of unknown force acting between sections 1 and 2
β_1 and β_2	=	momentum correction coefficients

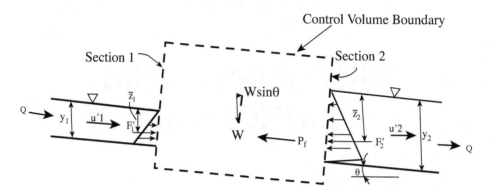

Figure 3.1. Schematic definition for specific momentum.

With the assumptions that, first, θ is small and hence $sin\ \theta = 0$ and cos $\theta = 1$; second, β_1 and $\beta_2 = 1$; and third, $\sum f_i' = 0$, Eq. (3.1.1) becomes

$$\gamma \overline{z_1} A_1 - \gamma \overline{z_2} - P_f = \frac{\gamma}{g} Q\left(\overline{u_2} - \overline{u_1}\right) \tag{3.1.2}$$

where $\overline{z_1}$ and $\overline{z_2}$ = distances to centroids of respective flow areas A_1 and A_2, from free surface

$$F_1 = \gamma \overline{z_1} A_1$$

$$F_2 = \gamma \overline{z_2} A_2$$

Substituting $\overline{u_1} = Q/A_1$, and $\overline{u_2} = Q/A_2$ into Eq. (3.1.2), and rearranging them, yields

$$\frac{P_f}{\gamma} = \left(\frac{Q^2}{gA_1} + \overline{z_1}A_1\right) - \left(\frac{Q^2}{gA_2} + \overline{z_2}A_2\right) \tag{3.1.3}$$

or

$$\frac{P_f}{\gamma} = M_1 - M_2 \tag{3.1.4}$$

where

$$M = \frac{Q^2}{gA} + \overline{z}A \tag{3.1.5}$$

and M is known as the specific momentum or force function.

Plotting the depth of flow y versus M produces a specific momentum curve which has two branches (Fig. 3.2). The lower limb AC asymptotically approaches the horizontal axis while the upper limb BC rises upward and extends indefinitely to the right. Thus, in analogy with the concept of specific energy, for a given value of M, the M-y curve predicts two possible depths of flow. These depths, shown in Fig. 3.2, are termed the *sequent depths of a hydraulic jump*.

The minimum value of the specific momentum function can be found under the assumptions of parallel flow and a uniform velocity distribution by taking the first derivative of M with, respect to y, and setting the resulting expression equal to zero or

$$\frac{dM}{dy} = -\frac{Q^2}{gA^2} + \frac{d\left(\bar{z}A\right)}{dy} = 0 \tag{3.1.6}$$

and

$$-\frac{Q^2}{gA^2} + A = 0 \tag{3.1.7}$$

where

$$d\left(\bar{z}A\right) = \left[A\left(\bar{z} + dy\right) + \frac{T\left(dy\right)^2}{2}\right] - \bar{z}A \approx Ady$$

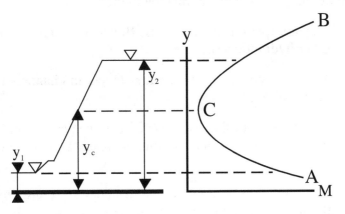

Figure 3.2.   Specific momentum curve and sequent depths y_1 and y_2 of a hydraulic jump.

where it is assumed that $(dy)^2 \approx 0$. Then substituting $dA/dy = T$, $u = Q/A$, and $D = A/T$ in Eq. (3.1.7),

$$\frac{\overline{u}^2}{2g} = \frac{D}{2} \tag{3.1.8}$$

which is the same criterion developed for the minimum value of specific energy. Therefore, for a specified discharge, minimum specific momentum occurs at minimum specific energy or critical depth.

3.2.   THE HYDRAULIC JUMP

Hydraulic jumps result when there is a conflict between upstream and downstream controls which influence the same reach of channel. For example, if the upstream control causes supercritical flow while the downstream control dictates subcritical flow, then there is a conflict. This conflict can only be resolved if there is some means for the flow to pass from one flow regime to the other. Experimental evidence suggests that flow changes from a supercritical to a subcritical state can occur very abruptly through the phenomenon known as a hydraulic jump. A hydraulic jump can take place either on the free surface of a homogeneous flow or at a density interface in a stratified flow. In either case, the hydraulic jump is accompanied by significant turbulence and energy dissipation. In the field of open channel flow, the applications of the hydraulic jump are many and include the following:

1. The dissipation of energy in flows over dams, weirs, and other hydraulic structures.

2. The maintenance of high water levels in channels for water distribution purposes.

3. The increase of the discharge of a sluice gate by repelling the downstream tailwater and thus increasing the effective head across the gate.

4. The reduction of uplift pressure under structures by raising the water depth on the apron of the structure.

5. The mixing of chemicals used for water purification or wastewater treatment.

6. The aeration of flows and the dechlorination of wastewaters.

7. The removal of air pockets from open channel flows in circular channels.

8. The identification of special flow conditions such as the existence of supercritical flow or the presence of a control section for the cost-effective measurement of flow.

3.2.1. Sequent Depths

The computation of the hydraulic jump always begins with Eq. (3.1.3). If the jump occurs in a channel with a horizontal bed and $P_f = 0$, that is, the jump is not assisted by a hydraulic structure, then Eq. (3.1.4) requires

$$M_1 = M_2 \tag{3.2.1}$$

or

$$\frac{Q^2}{gA_1} + \overline{z}_1 A_1 = \frac{Q^2}{gA_2} + \overline{z}_2 A_2 \tag{3.2.2}$$

Rectangular Sections: In the case of a rectangular channel of width b, substitution of $Q = \overline{u}_1 A_1 = \overline{u}_2 A_2$, $A_1 = by_1$, $A_2 = by_2$, $\overline{z}_1 = 0.5y_1$, $\overline{z}_2 = 0.5y_2$ into Eq. (3.2.2) yields

$$\frac{q^2}{g}\left(\frac{1}{y_1} - \frac{1}{y_2}\right) = \frac{1}{2}\left(y_2^2 - y_1^2\right) \tag{3.2.3}$$

where $q = Q/b$ is the flow per unit width. Equation (3.2.3) has the following solutions

$$\frac{y_2}{y_1} = \frac{1}{2}\left(\sqrt{1 + 8F_1^2} - 1\right) \tag{3.2.4}$$

and

$$\frac{y_1}{y_2} = \frac{1}{2}\left(\sqrt{1 + 8F_2^2} - 1\right) \tag{3.2.5}$$

Equations (3.2.4) and (3.2.5) each contain three independent variables, two variables must be known before a value of the third can be calculated. It must be emphasized that the downstream depth y_2 is not the result of upstream conditions, but is the result of a downstream control; i.e., if the downstream control produces a depth y_2 then a jump will form.

The use of Eqs. (3.2.4) and (3.2.5) to solve hydraulic jump problems in rectangular channels is rather obvious; however, in the case of Eq. (3.2.5) significant computational difficulties can arise. In this equation, F_2^2 is usually small, and the term $\sqrt{1 + 8F_2^2}$ is close to 1; hence, the term

$$\sqrt{1 + 8F_2^2} - 1$$

is near zero. The difficulty occurs in retaining computational precision while computing a small difference between two relatively large numbers. This difficulty can be avoided by expressing the term $\sqrt{1 + 8F_2^2}$ by a binomial series expansion or

$$\sqrt{1 + F_2^2} = 1 + 4F_2^2 - 8F_2^4 + 32F_2^6 - \ldots \tag{3.2.6}$$

Then, substituting Eq. (3.2.6) into Eq. (3.2.5)

$$\frac{y_2}{y_1} = 2F_2^2 - 4F_2^4 + 16F_2^6 + \ldots \tag{3.2.7}$$

Equation (3.2.7) should be used when F_2^2 is very small; for example, $F_2^2 \leq 0.05$.

Nonrectangular Sections: In analyzing the occurrence of hydraulic jumps in nonrectangular but prismatic channels, there are no equations analogous to Eqs. (3.2.4) and (3.2.5). In such cases, Eq. (3.2.2) could be solved by trial and error, but semiempirical approximations, and other analytic techniques are available. A few comments, regarding various semiempirical and analytic solutions of Eq. (3.2.2) for circular and other common channel sections, are appropriate at this point.

As noted in Chapter 2, for circular channels, a log-log plot of y_c/d versus $(Q\alpha^{0.5}/g^{0.5})^{2.5}$ yields a straight line in the interval $0.02 < y_c/d < 0.85$ (Fig. 2.4). A regression analysis of this straight line yields the equation

$$y_c = \frac{1.01}{d_0^{0.264}} \left(\frac{Q\sqrt{\alpha}}{\sqrt{g}} \right)^{0.506} \tag{3.2.8}$$

where d_0 = pipe diameter. It should be noted that the ratio y_c/d only rarely exceeds 0.85 in practice, since it is virtually impossible to maintain critical flow near the top of a circular channel. Thus, Eq. (3.2.8) applies to the region of general interest.

Straub (1978) noted that in circular conduits the upstream Froude number F_1 can be approximated by

$$F_1 = \left(\frac{y_c}{y_1} \right)^{1.93} \tag{3.2.9}$$

where y_1 = upstream depth of flow and y_c is estimated by Eq. (3.2.8). Straub (1978) also noted that for $F < 1.7$, the sequent depth, y_2, can be estimated by

$$y_2 = \frac{y_c^2}{y_1} \tag{3.2.10}$$

and for $F > 1.7$

$$y_2 = \frac{y_c^{1.8}}{y_1^{0.73}} \tag{3.2.11}$$

These equations provide a practical basis for estimating hydraulic jump parameters in a channel of circular section.

EXAMPLE 3.1

A flow of 100 ft³/s (2.8 m³/s) occurs in a circular channel 6.0 ft (1.8 m) in diameter. If the upstream depth of flow is 2.0 ft (0.61 m), determine the downstream depth of flow which will cause a hydraulic jump (Fig. 3.3)

Solution by Trial and Error
At the upstream station (Station 1)

(Table 1.1)

$$A_1 = \frac{1}{8}(\theta - \sin\theta)d_0^2 = \frac{1}{8}[2.46 - \sin(2.46)](6)^2 = 8.25\,\text{ft}^2\left(0.77\,\text{m}^2\right)$$

(Table 1.1)

$$D_1 = \frac{d_0}{8}\frac{(\theta - \sin\theta)}{\sin\dfrac{\theta}{2}} = \frac{6}{8}\left[\frac{2.46 - \sin(2.46)}{\sin(1.23)}\right] = 1.46\,\text{ft}\left(0.44\,\text{m}\right)$$

$$\overline{u}_1 = \frac{Q}{A_1} = \frac{100}{8.25} = 12.1\,\text{ft}/\text{s}\left(3.7\,\text{m}/\text{s}\right)$$

$$F_1 = \frac{\overline{u}_1}{\sqrt{gD_1}} = \frac{12.1}{\sqrt{32.2(1.46)}} = 1.76$$

For a partially full circular open channel flow, the distance from the water surface to the centroid of the flow are must be developed using basic principles. With reference to Fig. 3.3, the flow area is given by

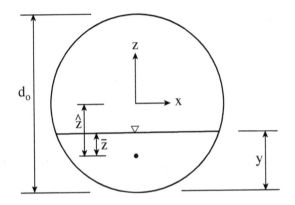

Figure 3.3. Circular conduit flowing partially full.

$$A = 2\int_{-r}^{z} \sqrt{r^2 - z^2}\,dz = \left[z\sqrt{r^2 - z^2} + r^2 \sin^{-1}\left(\frac{z}{r}\right) \right]\Bigg|_{-r}^{z}$$

or

$$A = \frac{\pi r^2}{2} + z\sqrt{r^2 - z^2} + r^2 \sin\left(\frac{z}{r}\right)$$

where r = radius of the circular section and $r^2 = x^2 + z^2$. The distance from the origin of the $z-x$ coordinate system to the centroid, the distance z' in Fig. 3.3, is by definition

$$z'A = \int_{-r}^{z} z\left(2\sqrt{r^2 - z^2}\right)dz = \frac{2}{3}\left(r^2 - z^2\right)^{3/2}\Bigg|_{-r}^{z}$$

and

$$z' = -\frac{2\left(r^2 - z^2\right)^{3/2}}{3A}$$

The distance from the water surface to the centroid of the flow area

$$\bar{z} = y - \left(r + z'\right)$$

Then, for this example

$$z_1' = \cfrac{-\dfrac{2}{3}\left[\left(r^2 - z_1^2\right)^3\right]^{1/2}}{A_1} = \cfrac{\dfrac{2}{3}\left\{\left[3^2 - \left(-1\right)^2\right]^3\right\}^{1/2}}{8.25} = -1.83\,\text{ft}\left(0.56\,\text{m}\right)$$

$$\bar{z} = y_1 - \left(r + z_1'\right) = 2 - \left(3 - 1.83\right) = 0.83\,\text{ft}\left(0.25\,\text{m}\right)$$

$$M_1 = \frac{Q^2}{gA_1} + \bar{z}A_1 = \frac{100^2}{32.2\left(8.25\right)} + 0.83\left(8.25\right) = 44.5\,\text{ft}^3\left(1.26\,\text{m}^3\right)$$

At the downstream station (Station 2)

$$M_2 = \frac{Q^2}{gA_2} + \bar{z}_2 A_2$$

The condition which must be satisfied if a hydraulic jump is to occur between stations 1 and 2 is

$$M_1 = M_2$$

The value $\bar{y}_2$ which satisfies this condition must be found by trial and error.

Trial y_2 ft (m)	$z_2 = y_2 - r$ ft (m)	A_2 ft^2 (m^2)	$\bar{z}_2$ ft (m)	$\bar{y}_2 = y_2 - r - z_2$ ft (m)	M_2 ft^3 (m^3)
3.40	0.40	16.5	-1.06	1.46	42.9
(1.04)	(0.12)	(1.53)	(-0.32)	(0.445)	(1.21)
3.58	0.58	17.5	-0.97	1.55	44.9
(1.09)	(0.18)	(1.64)	(-0.30)	(0.472)	(1.27)
3.65	0.65	18.0	-0.93	1.58	(45.7
(1.11)	(0.20)	(1.67)	(-0.28)	(0.482)	(1.29)

Therefore, the condition for a hydraulic jump is the downstream depth must be 3.58 ft (1.1 m).

Solution by Method of Straub

The upstream critical depth of flow estimated for $\alpha = 1$ is

$$y_c = \frac{1.01}{d_0^{0.264}}\left(\frac{Q}{\sqrt{g}}\right)^{0.506} = \frac{1.01}{6^{0.264}}\left(\frac{100}{\sqrt{32.2}}\right)^{0.506} = 2.69 \, \text{ft} \left(0.820 \, \text{m}\right)$$

Since $y_c/d_0 = 0.45$, Straub's approximations can be used to estimated F_1 and y_2 or

$$F_1 = \left(\frac{y_c}{y}\right)^{1.93} = \left(\frac{2.69}{2}\right)^{1.93} = 1.77$$

Then, for $F_1 > 1.7$

$$y_2 = \frac{y_c^{1.8}}{y^{0.73}} = \frac{2.69^{1.8}}{2^{0.73}} = 3.58 \, \text{ft} \left(1.09 \, \text{m}\right)$$

For other prismatic channel sections such as triangular, parabolic, and trapezoidal, either a graphical or trial-and-error solution of Eq. (3.2.2) is generally required. Silvester (1964, 1965) noted that for any prismatic channel the distance to the centroids of the flow areas $\overline{z}_i$ could be expressed as

$$\overline{z}_i = k_i' y_i \tag{3.2.12}$$

where the subscript i indicates the section, y is the greatest depth of flow at the specified section, and $k' = $ a coefficient. The equation of a hydraulic jump in a horizontal channel then becomes

$$A_1 k_1' y_1 - A_2 k_2' y_2 = \frac{Q^2}{g}\left(\frac{1}{A_2} - \frac{1}{A_1}\right) \tag{3.2.13}$$

Rearrangement of Eq. (3.2.13) yields

$$k_2' \frac{A_2}{A_1} \frac{y_2}{y_1} - k_1' = F_1^2 \frac{A_1}{y_1 T_1}\left(1 - \frac{A_1}{A_2}\right) \qquad (3.2.14)$$

where the parameter F_1^2 was first defined by Silvester (1964) to be $F_1^2 = Q^2/gA_1^2y_1$ and subsequently, (Silvester, 1965), to be $F_1^2 = Q^2/gA_1^2D_1$. This is the standard definition of the Froude number for nonrectangular channels.

For rectangular channels. $k_1' = k_2' = 1/2$, $A_1/A_2 = y_{1/2}$, and $A_1/T_1 = D_1 = y_1$. With these definitions, Eq. (3.2.14) becomes

$$\left(\frac{y_2}{y_1}\right)^2 - 1 = 2F_1^2\left(1 - \frac{y_1}{y_2}\right) \qquad (3.2.15)$$

Although this equation can be reduced to Eq. (3.2.4), for obtaining general graphical solutions of the sequent depth problems, the form given above is satisfactory. For comparison, the solution of this equation is plotted with the other solutions in Fig.3.4.

For triangular channels, $k_1' = k_2' = 1/3$, $A_1/A_2 = y_1^2/y_2^2$, and $A_1/T_1 = D_1 = y_1/2$. With these definitions, Eq. (3.2.14) becomes

$$\left(\frac{y_2}{y_1}\right)^3 - 1 = 1.5F_1^2\left[1 - \left(\frac{y_1}{y_2}\right)^2\right] \qquad (3.2.16)$$

For parabolic channels, whose perimeters can be defined by $y = aT^2/2$, where a is a constant $k_1' = k_2' = 2/5$, $A_1/A_2 = (y_1/y_2)^{1.5}$, and $A_1/T_1 = D_1 = 2y_1/3$. With these definitions, Eq. (3.2.14) becomes

$$\left(\frac{y_2}{y_1}\right)^{2.5} - 1 = 1.67F_1^2\left[1 - \left(\frac{y_1}{y_2}\right)^{1.5}\right] \qquad (3.2.17)$$

This equation is plotted in Fig. 3.4.

For trapezoidal channels, Eq. (3.2.14) must be solved directly. Silvester (1964) defined a shape factor

$$k = \frac{b}{zy_1} \qquad (3.2.18)$$

where b = bottom width of the trapezoidal, and z = trapezoid side slope. Using this shape factor, Eq. (3.2.14) can be solved for this type of channel, and a family of solution curves results (Fig. 3.4).

Silvester (1964, 1965) verified the theoretical results plotted in Fig. 3.4 with laboratory data from the work of Argyropoulos (1957, 1961), Hsing (1937), Sandover and Holmes (1962), and Press (1961). In laboratory results for triangular channels, the value of F_1 did not exceed 4, and some scatter of the data was noted. For this case, all the experimental data were gathered in a single channel which had an included angle of 47.3 degrees and hence, was rather narrow and deep. For parabolic channels, the agreement between the experimental and laboratory results was generally excellent. In this case, the value of F also never exceeded 4. For the trapezoidal channels, there was a fair degree of scatter in the data, but Silvester (1964) ascribed the deviation of laboratory results from theory to various experimental problems and limitations. In general, the curves plotted in Fig. 3.4 provide an adequate method for estimating the value of y_2/y_1 given F_1^2.

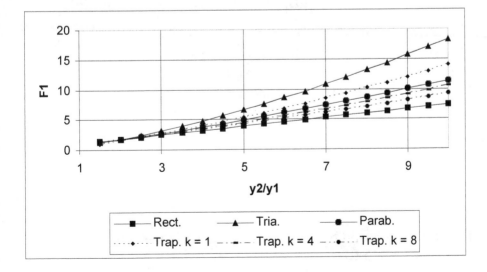

Figure 3.4. Analytical curves for y_2/y_1 versus F_1.

EXAMPLE 3.2

A flow of 100 m³/s (3530 ft³/s) occurs in a trapezoidal channel with side slopes of 2:1, and a base width of 5 m (16 ft). If the upstream depth of flow is 1.0 (3.3 ft), determine the downstream depth of flow which will cause a hydraulic jump.

Solution

At the upstream station

$$A_1 = (b + zy_1)y_1 = [5 + 2(1)]1 = 7.0\,\text{m}^2 \left(75\,\text{ft}^2\right)$$

$$T_1 = b + 2zy_1 = 5 + 2(2)(1) = 9.0\,\text{m} \left(30\,\text{ft}\right)$$

$$D_1 = \frac{A_1}{T_1} = \frac{7}{9} = 0.78\,\text{m} \left(2.6\,\text{ft}\right)$$

$$\bar{u}_1 = \frac{Q}{A_1} = \frac{100}{7} = 14.3\,\text{m/s} \left(47\,\text{ft/s}\right)$$

$$F_1 = \frac{u_1}{\sqrt{gD_1}} = \frac{14.3}{\sqrt{9.8(0.78)}} = 5.2$$

$$k = \frac{b}{zy_1} = \frac{5}{2(1)} = 2.5$$

Then, from Fig. 3.4 for $F_1 = 5.2$ and $k = 2.5$

$$\frac{y_2}{y_1} = 4.7$$

$$y_2 = 4.7\,\text{m} \left(15\,\text{ft}\right)$$

Therefore, the condition for a hydraulic jump is the downstream depth must be 4.7 m (15 ft).

3.2.2. Submerged Jump

In the previous section, methodologies for determining the downstream depth, which must prevail if a hydraulic jump is to form, with a given specified supercritical upstream condition, were discussed.   If the downstream depth is less than the sequent depth y_2, a submerged jump is formed (Fig. 3.5). Submerged jumps commonly form downstream of gates in irrigation systems, and the crucial unknown in such situations is the depth of submergence y_3.

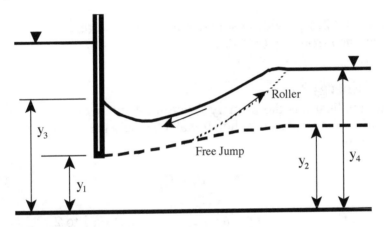

Figure 3.5. Definition sketch for submerged jump.

Using the principles of conservation of momentum and mass, Govinda Rao (1963) demonstrated that in horizontal, rectangular channels

$$\frac{y_3}{y_1} = \left[\left(1+S\right)^2 \varphi^2 - 2F_1^2 + \frac{2F_1^2}{\left(1+S\right)\varphi} \right]^{1/2} \tag{3.2.19}$$

where

$$S = \frac{y_4 - y_2}{y_2} \tag{3.2.20}$$

$$\varphi = \frac{y_2}{y_1} = \frac{1}{2}\left(\sqrt{1 + 8F_1^2} - 1\right) \tag{3.2.21}$$

and y_2 = subcritical sequent depth of the free jump corresponding to y_1 and F_1. Govinda Rao verified Eq. (3.2.19) with data resulting from a number of laboratory experiments. For this same situation, Chow (1959) gave the following equation:

$$\frac{y_3}{y_4} = \left[1 + 2F_4^2\left(1 - \frac{y_4}{y_1}\right)\right]^{1/2} \tag{3.2.22}$$

Equation (3.2.22) provides estimates of y_3 which are comparable with those obtained from Eq. (3.2.19).

3.2.3.   Energy Loss
In many applications the primary function of the hydraulic jump is the dissipation of energy. In a horizontal channel, the change in energy across the jump is

$$\Delta E = E_1 - E_2 \tag{3.2.23}$$

where ΔE =   change in energy from section 1 to 2
            E_1 =   specific energy at section 1
            E_2 =   specific energy at section 2

Energy loss is commonly expressed as either a relative loss $\Delta E/E_1$ or as an efficiency E_2/E_1.

Rectangular Sections
In the case of a horizontal, rectangular channel, it can be demonstrated that

$$\Delta E = \frac{(y_2 - y_1)^3}{4y_1 y_2} \tag{3.2.24}$$

that

$$\frac{\Delta E}{E_1} = \frac{2 - 2\left(\dfrac{y_2}{y_1}\right) + F_1^2\left[1 - \left(\dfrac{y_1}{y_2}\right)^2\right]}{2 + F_1^2} \tag{3.2.25}$$

and that

$$\frac{E_2}{E_1} = \frac{\left(8F_1^2 + 1\right)^{3/2} - 4F_1^2 + 1}{8F_1^2\left(2 + F_1^2\right)} \qquad (3.2.26)$$

Nonrectangular Sections

For other prismatic channel sections, either the general specific energy equation must be solved on a case-by-case basis or a general graphical solution of this equation must be obtained. In terms of relative energy loss, the general specific energy equation is

$$\frac{\Delta E}{E_1} = \frac{y_1 + \dfrac{\overline{u_1^2}}{2g} - y_2 - \dfrac{\overline{u_2^2}}{2g}}{y_1 + \dfrac{\overline{u_1^2}}{2g}}$$

$$= \frac{y_1 - y_2 + \dfrac{Q^2}{2g}\left(\dfrac{1}{A_1^2} - \dfrac{1}{A_2^2}\right)}{y_1 + \dfrac{Q^2}{2gA_1^2}} \qquad (3.2.27)$$

Silvester (1964, 1965) rearranged Eq. (3.2.27) to yield

$$\frac{\Delta E}{E_1} = \frac{\dfrac{2y_1}{D_1}\left(1 - \dfrac{y_2}{y_1}\right) + F_1^2\left[1 - \left(\dfrac{A_1}{A_2}\right)^2\right]}{\dfrac{2y_1}{D_1} + F_1^2} \qquad (3.2.28)$$

The values of y_2/y_1 for any specified F_1 can be obtained from Fig. 3.4, the values of the parameter $[1 - (A_1/A_2)^2]$ from Table 3.1, and the values of the hydraulic depth D from Table 1.1. With these values, a generalized graphical solution of Eq. (3.2.28) can be found (Fig. 3.6). In this figure, the relative energy loss for a hydraulic jump in a rectangular channel is plotted for comparison. Although a general graphical solution of Eq. (3.2.28) can be obtained for a circular channel, in most cases it is much simpler to solve for the energy loss on a case-by-case basis.

Table 3.1   **Formulations of the term [1 - $(A_1/A_2)^2$] for use in Eq. (3.2.28).**

Prismatic channel section	Depth-dependent formulation of $[1 - (A_1/A_2)^2]$
Triangular	$1 - \left(\dfrac{y_1}{y_2}\right)^4$
Parabolic	$1 - \left(\dfrac{y_1}{y_2}\right)^3$
Trapezoidal*	$1 - \left(\dfrac{k+1}{k+\dfrac{y_2}{y_1}}\right)\left(\dfrac{y_1}{y_2}\right)^2$

*$k = b/zy_1$

With regard to Fig. 3.6, it is noted that all channel shapes yield a greater relative energy loss than the rectangular channel for a specified

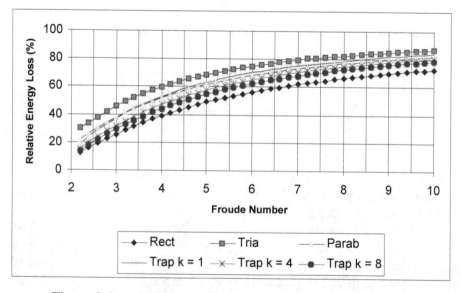

Figure 3.6.   **Relative energy loss in a hydraulic jump for various cross sections.**

Froude number. This result is not unexpected since channels with sloping sides provide for a degree of secondary circulation and, hence, increased energy dissipation. Silvester (1965) noted that for a given flow rate with upstream velocities and depths being equal, the triangular channel provides the greatest energy dissipation.

3.2.4.   Hydraulic Jump Length
Although the length of a hydraulic jump is a crucial design parameter, in general it cannot be derived from theoretical considerations; the results of several experimental investigations. For example, Bakhmeteff and Matzke (1936), Bradley and Peterka (1957a), Peterka (1963), and Rajaratnam (1965) have yielded results, which are, to some degree, contradictory. In this section, the length of the hydraulic jump L_j is defined to be the distance from the front face of the jump to a point on the surface of the flow, immediately downstream of the roller associated with the jump (Fig. 3.7). In estimating this length, a number of special cases must be considered.

Rectangular Sections
In the case of a classic hydraulic jump occurring in a horizontal rectangular channel, the distance L_j is usually estimated from the Bradley and Peterka (1957a) curve, which is a plot of F_1 versus L_j/y_2 (Fig. 3.8). This curve has a flat section in the range of Froude numbers yielding the best performance, and its validity is supported by the data of Rajaratnam (1965). It is noted that there is a marked difference between this curve and the one derived from the Bakhmeteff and Matzke (1936) data. Both Bradley and Peterka (1957a) and Chow (1959) ascribed this difference to the scale effect involved in the Bakhmeteff and Matzke (1936) data; i.e., the prototype behavior was not faithfully reproduced by the model used by these investigators.

The data for the length of the roller L_r result in three curves which are significantly different from each other (Fig. 3.8). From this figure, it is concluded that L_r is always less than L_j, and the curve derived from the Rajaratnam (1965) data lies between the curves defined by the data of Rouse *et al.* (1959) and Safranez (1934).

Silvester (1964) has demonstrated that for horizontal, rectangular channels the ratio L_j/y_1 is a function of the upstream supercritical Froude number.

$$\frac{L_j}{y_1} = 9.75\left(F_1 - 1\right)^{1.01} \tag{3.2.29}$$

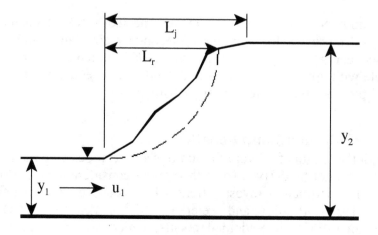

Figure 3.7. Schematic definitions of jump and roller length.

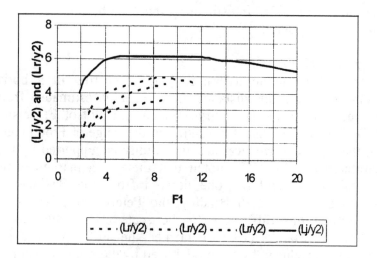

Figure 3.8. Hydraulic jump length as a function of sequent depth, y_2 (y2) and upstream Froude number, F_1 (F1). L_j (Lj) = jump length; L_r (Lr) = roller length. (Bradley and Peterka, 1957a; Rouse *et al*, 1959; Rajaratnam, 1967; Safranez, 1934).

Although this functional relationship was noted in the original work of Bradley and Peterka (1957a), it was apparently not widely used because of the convenience of Fig. 3.8.

Nonrectangular Sections: Silvester (1964) has hypothesized that a functional relationship exists between the ratio L_j/y_1 for prismatic channels of any shape, or

$$\frac{L_j}{y_1} = \sigma\left(F_1 - 1\right)^{\Gamma} \tag{3.2.30}$$

where σ and Γ are both shape factors. Although the relations presented in the material which follows have not been verified by extensive laboratory and field measurements, they provide an approximate method of estimating L_j for various channel shapes.

For triangular channels, the relationship developed by Silvester (1964) is

$$\frac{L_j}{y_1} = 4.26\left(F_1 - 1\right)^{0.695} \tag{3.2.31}$$

This equation is based on data from a single channel which had an included angle of 47.3 degrees (Argyropoulos, 1957), and probably cannot be expected to be valid in triangular channels whose side slopes are either steeper or milder.

For parabolic channels, Silvester's analysis of the Argyropoulos (1957) data yielded

$$\frac{L_j}{y_1} = 11.7\left(F_1 - 1\right)^{0.832} \tag{3.2.32}$$

This equation provided a very poor fit of the available data when F_1 exceeded 3.

In the case of trapezoidal sections, a number of different data sets were available to Silvester, e.g., Hsing (1937), Sandover and Holmes (1962), and Press (1961). In a previous section of this chapter, it was demonstrated that for trapezoidal channels, the ratio y_2/y_1 is dependent not only on F_1 but also on a shape factor k. Thus, it might be expected that for each value of k there is a unique set of values of σ and Γ. In the case of Press (1961) data, this was true, and the results for this data set are summarized in Table 3.2. In the case of Hsing (1937) data, the value of σ had no apparent effect, and these data appeared to be affected by the value of z. The Hsing data also exhibited variations in the value of F_1 for constant values of L_j, which was probably the result of experimental difficulties in measuring the distance L_j, under laboratory conditions.   Sandover and Holmes (1962) found no correlation in their data between L_j and F_1. Thus, the values of σ and Γ used to estimate L_j in the case of trapezoidal channels, are based entirely on the data of Press (1961).

Table 3.2.   Definition of hydraulic jump length parameters σ and Γ for trapezoidal channels (after Press, 1961).

Side slope z	Shape factor k^*	σ ft	Γ
2	16	17.6	0.905
1	8	23.0	0.885
0.5	4	35.0	0.836

$^*k = b/zy$

A salient feature of hydraulic jumps, noted by both Hsing (1937) and Press (1961), was that the reverse flow of the roller caused a depth increase at the sides of the channel and a corresponding decrease in the depth of flow along the centerline of the jump. This traverse variation in the depth of flow could result in problems in determining accurately the length of the jump.

The results available for circular channels from Kindsvater (1936) were limited and not sufficient to draw useful conclusions.

Submerged Jump Length

In the case of a submerged hydraulic jump (Fig. 3.5), the distance L_j is estimated by the empirical equation

$$\frac{L_j}{y_2} = 4.9S + 6.1 \tag{3.2.33}$$

where S = submergence factor defined by Eq. (3.2.20) (Govinda Rao, 1963). Equation (3.2.33) demonstrates that the length of a submerged jump exceeds the length of the corresponding free jump by the term 4.9 S, and further, that L_j is directly proportional to S. Stepanov (1959) found that the length of the roller, in the case of a submerged jump, can be estimated by

$$\frac{L_r}{y_c} = \frac{3.31}{\left(\dfrac{y_4 - y_3}{y_3 F_1}\right)^{0.885}} \tag{3.2.34}$$

This equation has been shown to be valid for $S \leq 2$ and $1 \leq F_1 \leq 8$ (Rajaratnam, 1967).

3.2.5.   Hydraulic Jumps in Sloping Channels

As noted in an earlier section of this chapter, hydraulic jumps can and do occur in channels of such a slope, that the gravitational forces acting on the flow must be considered. Although Eq. (3.1.1) is theoretically applicable to such problems, in practice the number of solutions available is limited. The primary difficulties in obtaining a useful solution to this problem are: (1) The term $W\,sin\,\theta$ is generally poorly quantified because the length and shape of the jump are not well defined, (2) the specific weight of the fluid, in the control volume, may change significantly owing to air entrainment, and (3) the pressure terms cannot be accurately quantified.

According to Rajaratnam (1967), the earliest experiments done by Bidone on the hydraulic jump were performed in what would now be termed a sloping channel. Bazin in 1865 and Beebe and Riegel in 1917 also addressed this problem. In 1934, Yarnell began an extensive study of the hydraulic jump in sloping channels, which was not completed because of his untimely death in 1937. Kindsvater (1944), using the unpublished Yarnell data, was the first investigator to develop a rational solution to the problem. Extensive studies have also been conducted by Bradley and Peterka (1957a) and Argyropoulos (1962).

In discussing the equations and relationships available for hydraulic jumps in sloping channels, it is convenient to consider a number of cases (Fig. 3.9). It is also noted that, in the material which follows, *the*

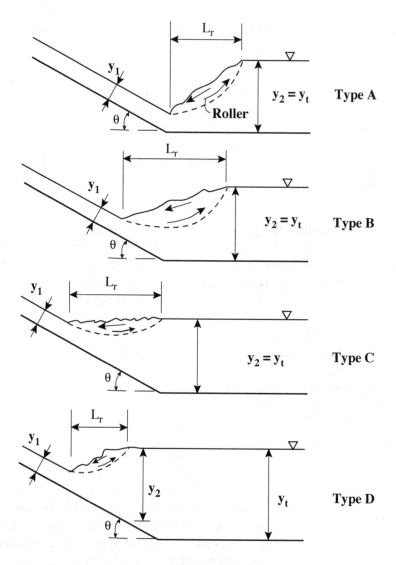

Figure 3.9.   **Definition of type of hydraulic jumps that occur in sloping channels.**

end of the jump is, by definition, at the end of the surface roller. The reader is cautioned that this definition differs from that used in the previous sections of this chapter. Then, with regard to the cases defined in Fig. 3.9, let

y_t =   tailwater depth,

L_r = length of the jump measured horizontally,

y_1 = the supercritical depth of flow on the slope which is assumed to be constant,

y_2 = subcritical sequent depth corresponding to y_1, and

y_2^* = subcritical depth given by Eq. (3.2.4))(see Fig. 3.10).

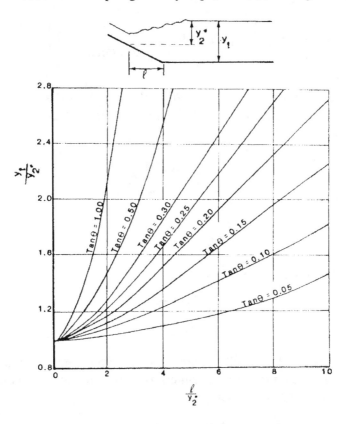

Figure 3.10.   Solution for B jumps (Peterka, 1963; Rajaratnam, 1967).

If the jump begins at the end of the sloping section, then $y_2^* = y_t$ and a type A jump, governed by Eq. (3.2.4), occurs. If the end of the jump coincides with the intersection of the sloping and horizontal beds, a type C jump occurs. If y_t is less than that required for a type C jump but greater than y_2^*, the toe of the jump is on the slope and the end on the horizontal bed. This situation is termed a type B jump. If y_t is greater than that required for a type C jump, then a type D jump occurs completely on the sloping section. Type E jumps occur on sloping beds which have no break in slope, and the rare type F jump occurs only in stilling basins normally found below drop structures. Of the six types of jumps defined in Fig. 3.9, types A to D are the most common and will be discussed first.

For the type C jump, Kindsvater (1944) developed the following equation for the sequent depth:

$$\frac{y_2}{y_1} = \frac{1}{2\cos\theta}\left[\sqrt{1 + 8F_1^2\left(\frac{\cos^3\theta}{1 - 2N\tan\theta}\right)} - 1\right] \tag{3.2.35}$$

where N = an empirical factor related to the length of the jump, and θ = the longitudinal slope angle of the channel. If the following parameters are defined

$$\Gamma_1^2 = \frac{\cos^3\theta}{1 - 2N\tan\theta} \tag{3.2.36}$$

and

$$G_1^2 = \Gamma_1^2 F_1^2 \tag{3.2.37}$$

Eq. (3.2.37) can be written as

$$\frac{y_2}{y_1'} = \frac{1}{2}\left(\sqrt{1 + 8G_1^2} - 1\right) \tag{3.2.38}$$

where $y_1' = y_1/\cos\theta$. Bradley and Peterka (1957a) and Peterka (1963) found that N was primarily a function of θ, and Rajaratnam (1967) stated the following relationship

$$\Gamma_1 = 10^{0.027\theta} \tag{3.2.39}$$

where θ is in degrees. In addition to the type C jump, Bradley and Peterka (1957a) and Peterka (1963) also found that Eqs. (3.2.37) to (3.2.39) could be applied to the type D jump.

Although an analytic solution for the type B jump has not yet been developed, Bradley and Peterka (1957a) and Peterka (1963) have developed a graphical solution for this type of jump based on laboratory experiments. These data are summarized in Fig. 3.10. In this figure, the original data have been supplemented with two additional curves, *tan*θ = 0.50 and 1.0, abstracted from Rajaratnam (1967).

The initial step in solving jump problems, where the slope of the channel must be considered, is to classify the jump which occurs given the slope, initial supercritical flow depth, and the tailwater condition. If it is assumed that the depth of the supercritical stream is constant on the sloping bed. This assumption closely approximates conditions that have been found to exist at the base of high dams and below sluices acting under high heads. Then, a relatively simple jump classification procedure can be developed as shown in Fig. 3.11.

Bradley and Peterka (1957a) and Peterka (1963) also presented a number of results regarding the length of the type D jump, these results are summarized in Fig. 3.12. This graph can also be used to estimate the lengths of type B and C jumps.

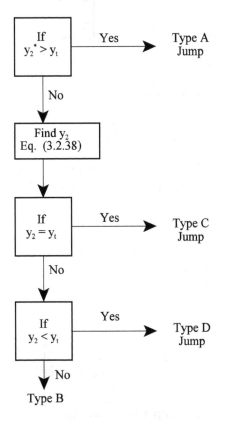

Figure 3.11.   Determination of hydraulic jump type in sloping channels.

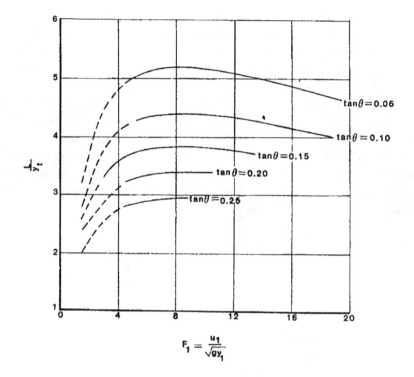

Figure 3.12.   **Hydraulic jump length in sloping channels for jump types B, C, and D (Peterka, 1963).**

The energy loss for the type A jump can be estimated from the equations derived for a horizontal channel, e.g., Eq. (3.2.24). For jump types C and D, an analytic expression can be derived. In Fig. 3.9, if the bed level at the end of the jump is taken as the datum, then the upstream energy is

$$E_1 = L_r \tan\theta + \frac{y_1}{\cos\theta} + \frac{\overline{u}_1^2}{2g} \qquad (3.2.40)$$

where L_r = length of the jump and $\overline{u}_1$ = supercritical upstream velocity. The energy at the end of the jump is

$$E_2 = y_2 + \frac{\overline{u}_2^2}{2g} \qquad (3.2.41)$$

where $\overline{u_2}$ = subcritical velocity at the downstream section. Then the relative energy loss is given by

$$\frac{\Delta E}{E_1} = \frac{\left(1 - \dfrac{y_2}{y_1}\right) + \dfrac{F_1^2}{2}\left[1 - \left(\dfrac{y_1}{y_2}\right)^2\right] + \dfrac{L_r y_t}{y_t y_1}\tan\theta}{1 + \dfrac{F_1^2}{2} + \dfrac{L_r y_t}{y_t y_r}\tan\theta} \qquad (3.2.42)$$

where it has been assumed that $y_1 \fallingdotseq y_1 / \cos\theta$. In general, Eq. (3.2.42) should not be used in situations where F_1 is less than 4, since in this range, there is only very limited information regarding the ratio L_r / y_2, also in this range, $\Delta E/E_1$ is sensitive to this ratio.

The computation of the energy loss for the type B jump is much more involved, and it is best illustrated with an example.

EXAMPLE 3.3

Given a rectangular channel 4.0 ft (1.2 m) wide and inclined at an angle of 3 degrees with the horizontal, determine the jump type for $Q = 5.0$ ft³/s (0.14 m³/s), $y_1 = 0.060$ ft (0.018 m) and $y_t = 1.4$ ft (0.43 m).

Solution

At section 1

$$A_1 = by_1 = 4(0.06) = 0.24\,\text{ft}^2\ \left(0.022\,\text{m}^2\right)$$

$$\overline{u_1} = \frac{Q}{A_1} = \frac{5}{0.24} = 20.8\,\text{ft / s}\ \left(6.3\,\text{m / s}\right)$$

$$F_1 = \frac{\overline{u_1}}{\sqrt{gy_1}} = \frac{20.8}{\sqrt{32.2(0.06)}} = 15$$

The sequent depth in a horizontal channel y_2^* can be computed from Eq. (3.2.4)

$$y_2^* = \frac{y_1}{2}\left(\sqrt{1 + 8F_1^2} - 1\right) = \frac{0.06}{2}\left[\sqrt{1 + 8(15)^2} - 1\right] = 1.2 \, \text{ft} \, (0.37 \, \text{m})$$

The jump can then be classified by the following scheme defined in Fig. 3.11. Since $y_t > y_2^*$, the depth of y_2 must be calculated from Eqs. (3.2.37) to (3.2.39).

$$\Gamma^2 = 10^{0.027\theta} = 10^{0.027(3)} = 1.2$$

$$G_1^2 = \Gamma_1^2 F_1^2 = (1.2)^2 (15)^2 = 324$$

$$y_2 = \frac{y_1}{2\cos\theta}\left(\sqrt{1 + 8G_1^2} - 1\right) = \frac{0.06}{2\cos 3}\left[\sqrt{1 + 8(324)} - 1\right] = 1.5 \, \text{ft} \, (0.46 \, \text{m})$$

Since $y_2 > y_t$, the jump is classified as a type B jump. From Fig. 3.10, the distance can be found for $\tan\theta = 0.05$ and

$$\frac{y_t}{y_2^*} = \frac{1.4}{1.2} = 1.2$$

$$\frac{\ell}{y_2^*} = 6.5$$

and

$$1 = 1.2(6.5) = 7.8 \, \text{ft} \, (2.4 \, \text{m})$$

From Fig. 3.12, the length of the jump is estimated

$$\frac{L_r}{y_t} = 4.9$$

and

$$L_r = 4.9(1.4) = 6.9 \, \text{ft} \, (2.1 \, \text{m})$$

On the basis of these calculations, Fig. 3.13 can be constructed. The energy loss can then be estimated by establishing the horizontal bed as the datum and by computing the total energies at sections 1 and 2 or,

$$E_1 = l\tan\theta + \frac{y_1}{\cos\theta} + \frac{\overline{u_1^2}}{2g} = 7.8\tan(3) + \frac{0.06}{\cos(3)} + \frac{20.8^2}{2(32.2)} = 7.2\,\text{ft}\,(2.2\,\text{m})$$

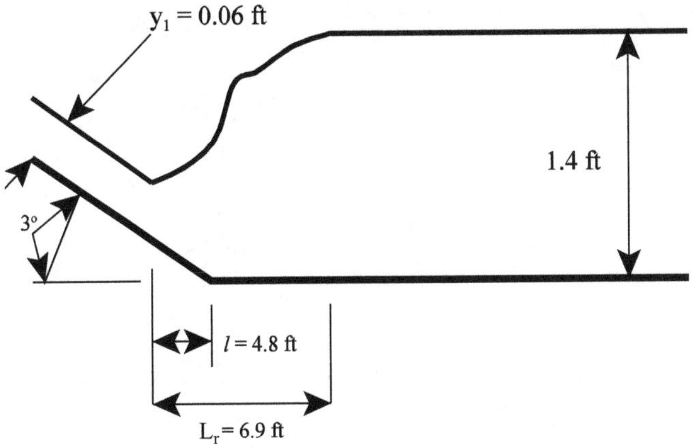

Figure 3.13. Results of Type B hydraulic jump analysis.

Then, at section 2

$$\overline{u_2} = \frac{Q}{A_2} = \frac{5}{1.4(4)} = 0.89\,\text{ft}\,/\,\text{s}\,(0.27\,\text{m}\,/\,\text{s})$$

$$E_2 = y_2 + \frac{\overline{u_2^2}}{2g} = 1.4 + \frac{0.89^2}{2(32.2)} = 1.4\,\text{ft}\,(0.43\,\text{m})$$

and

$$\Delta E = E_1 - E_2 = 7.2 - 1.4 = 5.8\ \text{ft}\ (1.8\ \text{m})$$

or

$$\frac{\Delta E}{E_1} = \frac{5.8}{7.2} = 0.81$$

These calculations indicate that the energy losses in this jump are quite high; however, these results should be used cautiously since they are subject to a number of assumptions.

EXAMPLE 3.4

Given a rectangular channel 6.1 m (20 ft) wide and inclined at an angle of 3 degrees with the horizontal, determine the jump type if $Q = 9.0$ m^3/s(320 ft^3/s), $y_t = 2.6$ m (8.5 ft), and $y_l = 0.09$ m (0.30 ft).

Solution

At section 1

$$A_1 = by_1 = 6.1(0.09) = 0.55 \text{m}^2 \left(5.9 \text{ft}^2\right)$$

$$\overline{u}_1 = \frac{Q}{A_1} = \frac{9.0}{0.55} = 16 \text{m/s} \left(52 \text{ft/s}\right)$$

$$F_1 = \frac{\overline{u}_1}{\sqrt{gy_1}} = \frac{16}{\sqrt{9.8(0.09)}} = 17$$

The depth y_2* can be computed from Eq. (3.2.4).

$$y_2^* = \frac{y_1}{2}\left(\sqrt{1 + F_1^2} - 1\right) = \frac{0.09}{2}\left[\sqrt{1 + 8(17)} - 1\right] = 2.1 \text{m} \left(6.9 \text{ft}\right)$$

The jump can then be classified by following the scheme defined in Fig. 3.11. Since $y_t > y_2$*, the depth y_2 must be calculated from Eqs. (3.2.37) and (3.2.39).

$$\Gamma_1 = 10^{0.027\theta} = 10^{0.027(3)} = 1.2$$

$$G_1^2 = \Gamma_1 F_1^2 = 1.2^2 (17)^2 = 416$$

$$y_2 = \frac{y_1}{2\cos\theta}\left(\sqrt{1+8G_1^2}-1\right) = \frac{0.09}{2\cos(3°)}\left[\sqrt{1+8(445)}-1\right] = 2.6\,m\,(8.5\,ft)$$

Then, since $y_2 > y_t$, the jump is classified as the type C jump. From Fig. 3.12, the distance L_r can be found for $\tan\theta = 0.05$ and $\mathbf{F}_1 = 17$

$$\frac{L_r}{y_t} = 4.8$$

and

$$L_r = 4.8(2.6) = 12\,m\,(39\,ft)$$

The relative energy loss is found from Eq. (3.2.42)

$$\frac{\Delta E}{E} = \frac{\left(1-\dfrac{y_2}{y_1}\right) + \dfrac{F_1^2}{2}\left[1-\left(\dfrac{y_1}{y_2}\right)^2\right] + \tan\theta\left(\dfrac{L_r y_t}{y_t y_1}\right)}{1 + \dfrac{F_1^2}{2} + \dfrac{L_r y_t}{y_t y_1}\tan\theta}$$

with

$$\frac{y_2}{y_1} = \frac{2.6}{0.09} = 29$$

and

$$\frac{L}{y_t} = 4.8$$

$$\frac{y_t}{y_1} = 29$$

Then

$$\frac{\Delta E}{E} = \frac{(1-29) + \dfrac{17^2}{2}\left(1-\dfrac{1}{29^2}\right) + \tan(3°)\left[4.8(29)\right]}{1 + \dfrac{17^2}{2} + 4.8(29)\tan(3°)} = 0.76$$

The analysis of the jump designated as type E (Fig. 3.9) begins with Eq. (3.1.1). For a rectangular channel, if it is assumed that (first) $P_f = 0$, (second) $\beta_1 = \beta_2 = 1$, (third) $\Sigma f_f' = 0$, and (fourth) the weight of the water between sections 1 and 2 is given by

$$W = \frac{A_1 + A_2}{2} \frac{L_2}{\cos\theta} \gamma k'$$

where $L_r/cos2$ can be approximated as

$$\frac{L_r}{\cos\theta} = L' = X(y_2 - y_1) \tag{3.2.43}$$

X = a slope factor and k' = a correction factor, which results from the assumption that the jump profile is linear. Equation (3.1.1) can then be solved to yield

$$\frac{y_2}{y_1} = \frac{1}{2}\left(\sqrt{1 + 8M^2} - 1\right) \tag{3.2.44}$$

where

$$M = \frac{F_1}{\sqrt{1 - \dfrac{k'L'\sin\theta}{y_2 - y_1}}}$$

Equation (3.2.44) has been derived independently by Kennison (1944), Chow (1959), and Argyropoulos (1962). For simplicity the value of k' can be assumed to be unity, the variation of M with X cannot be ignored, except for small values of θ (Rajaratnam, 1967). Kennison (1944) suggested that X could be assumed to have a value of 3. Argyropoulos (1962) and Chow (1959) noted that X was a function of F_1, but neither of these investigators suggests an appropriate value.

As noted previously, the type F jump is very rare, and there are few data available for this type of jump. Rajaratnam (1967) noted that in practice it is almost impossible to keep this jump completely on the reverse slope. If this type of jump is to be used in a design, its characteristics should be determined from model tests.

3.3   HYDRAULIC JUMPS AT DENSITY INTERFACES

It has been demonstrated in laboratory experiments (see, for example, Yih and Guha, 1955, Hayakawa, 1970, Stefan and Hayakawa, 1972, and Stefan *et al.*, 1971) that hydraulic jumps can also occur at density interfaces within stratified flows. Although this phenomenon - termed an internal hydraulic jump - has not been widely observed in nature except in the atmosphere (Turner, 1973) and in tidal inlets (Gardner *et al.*, 1980), this lack of field observations may result from the fact that this is an internal phenomenon which does not necessarily produce a noticeable effect at the air-water interface.

The types of internal hydraulic jumps which can occur are defined schematically in Fig. 3.14. In the first case (Fig. 3.14a), Layer 1 passes from an internally supercritical state to a subcritical state by means of an internal hydraulic jump. In the second case (Fig. 3.14b), Layer 2 passes from an internally supercritical state to an internal subcritical state by means of an internal hydraulic jump. In the discussion of internal hydraulic jumps which follows, the following assumptions are required:

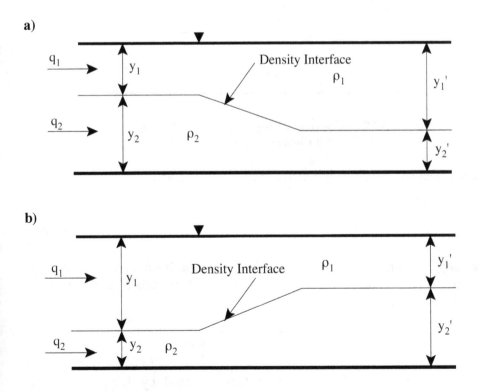

Figure 3.14. Internal hydraulic jump definition.

(1) The channel in which the jump occurs is rectangular and has a width b, (2) the bed of the channel is horizontal, (3) the interfacial shear between the layers is ignored, (4) there is no mixing between the layers, (5) all pressure distributions are hydrostatic, and (6) β_1 and $\beta_2 = 1$. Although assumption (3) through (5) are crucial to the analysis which follows, the results of the analysis are only qualitatively correct because of these assumptions.

In Fig. 3.15, control volumes of unit width are defined for Layers 1 and 2. Considering only Layer 2, the upstream hydrostatic force is

$$F_1 = \left[\frac{\gamma_1 y_1 + (\gamma_1 y_1 + \gamma_2 y_2)}{2} \right] = \gamma_1 y_1 y_2 + \frac{\gamma_2 y_2^2}{2} \qquad (3.3.1)$$

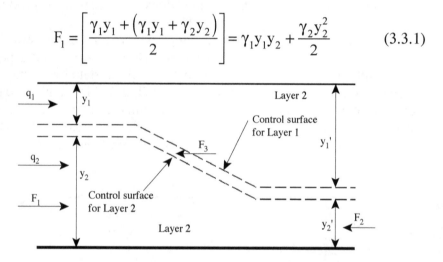

Figure 3.15.   Control volume and definition of variables for an internal hydraulic jump.

and the corresponding downstream hydrostatic force is

$$F_2 = \gamma_1 y_1' y_2' - \frac{\gamma_2 (y_2')^2}{2} \qquad (3.3.2)$$

where γ_1 and γ_2 = specific weights of fluids in Layers 1 and 2, respectively. The force exerted by Layer 1 on the Layer 2 control volume is given by

$$F_3 = \frac{(y_1 + y_1')(y_2 - y_2')\gamma_1}{2} \qquad (3.3.3)$$

where $\gamma_1 (y_1 + y_2')/2$ = mean pressure exerted by Layer 1 on the sloping surface separating Layers 1 and 2 and $y_2 - y_2'$ = projected vertical unit area of the jump. Application of the one-dimensional form of the conservation of momentum equation to the Layer 2 control volume yields

$$F_1 - F_2 - F_3 = \rho_2 q_2 \left(\overline{u_2} - \overline{u_1} \right) \tag{3.3.4}$$

where

$$q_2 = \text{flow per unit width in Layer 2}$$

$$\overline{u_2} = \frac{q_2}{y_2'}$$

$$\overline{u_1} = \frac{q_2}{y_2}$$

Rearranging Eq. (3.3.4) and defining $a_2 = q_2^2/g$ and $r = \rho_1 / \rho_2$

$$2a_2 \left(y_2 - y_1' \right) = y_2 y_2' \left(y_2 + y_2' \right) \left[r \left(y_1 - y_1' \right) + \left(y_2 - y_2' \right) \right] \tag{3.3.5}$$

In an analogous fashion, the application of the one-dimensional form of the conservation of momentum equation to the Layer 1 control volume yields

$$2a_1 \left(y_1 - y_1' \right) = y_2 y_2' \left(y_1 + y_1' \right) \left[r \left(y_1 - y_1' \right) + \left(y_2 - y_2' \right) \right] \tag{3.2.6}$$

If y_1, y_2, ρ_1, and ρ_2 are specified, then Eqs. (3.3.5) and (3.3.6) can be solved simultaneously to determine y_1' and y_2' there are at most nine solutions to this set of equations. One valid solution is $y_1' = y_1$ and $y_2' = y_2$, and of the eight remaining solutions which are conjugate to the specified state, only those which are real and positive must be considered. Yih and Guha (1955) demonstrated that there can be at most three conjugate states which have a real physical interpretation, and under certain conditions a unique conjugate state exists. The conditions for uniqueness are either $F_1^2 = a_1/y_1^3$ or $F_2^2 = a_2/y_2^3$ must be large. The three unique solutions cited by Yih and Guha (1955) are:

1.   If the downstream velocities are the same for both layers, then

$$\left(1 + r\lambda \right) \left(\frac{y_2'}{y_2} \right)^3 + r \left(\lambda - \frac{y_1}{y_2} \right) \left(\frac{y_2'}{y_2} \right) - \left(1 + r \frac{y_1}{y_2} + 2F_2^2 \right) \left(\frac{y_2'}{y_2} \right) + 2F_2^2 = 0 \tag{3.3.7}$$

where $q_1 / q_2 = y_1' / y_2' = \lambda$. Equation (3.3.7) has two positive roots: $y_2' = y_2$ and another which corresponds to the conjugate state.

2.   If $a_1 = 0$, Layer 1 is at rest; and Eqs. (3.3.5) and (3.3.6) can be solved simultaneously to yield

$$\frac{y_2'}{y_2} = \frac{1}{2}\left(\sqrt{1 + 8(F_2')^2} - 1 \right) \tag{3.3.8}$$

where

$$\left(F_2'\right)^2 = F_2^2 \left(1 - r\right)^{-1}$$

The similarity of Eq. (3.3.8) and (3.2.5) should be noted.

3.   If $a_2 = 0$, Layer 2 is at rest, then the simultaneous solution of Eqs. (3.3.5) and (3.3.6) yields

$$\frac{y_1'}{y_1} = \frac{1}{2}\left(\sqrt{1 + 8(F_1')^2} - 1 \right) \tag{3.3.9}$$

where

$$\left(F_1'\right)^2 = F_1^2 \left(1 - r\right)^{-1}$$

With regard to these solutions, in the case $a_1 = 0$ and $a_2 \neq 0$, Eq. (3.3.6) reduces to

$$y_1 + y_2 = y_1' + y_2'$$

This result implies that the free surface is level, and there is no evidence of the jump at the free surface. If $a_2 = 0$ and $a_1 \neq 0$, then from Eq. (3.3.6)

$$r(y_1' - y_1) = y_2 - y_2'$$

and the free surface does not remain level.

In the foregoing material it was explicitly assumed that the fluid composing Layers 1 and 2 did not mix. Many investigators including Stefan and Hayakawa (1972) and Stefan *et al.* (1971) have considered both theoretically and experimentally the case in which mixing occurs.

Figure 3.16 defines the situation considered by these authors and is analogous to cooling water from a thermal power plant entering a reservoir.

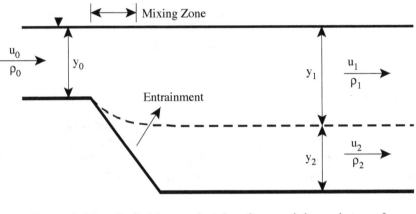

Figure 3.16.   Definition   sketch   for   mixing   internal hydraulic jump.

3.4.   APPLICATION OF THE MOMENTUM PRINCIPLE TO PRACTICE

3.4.1   Pre- and Post-Jump Depths of Flow When the Energy Loss is Specified

The hydraulic jump is one of the most efficient methods of dissipating the energy passing over a weir or a dam spillway. In many design situations, only the flow rate (Q) and the energy to be dissipated in the jump (ΔE) are known, which leaves the designer to determine the upstream and downstream depths of flow. In the past, the solution of this problem has relied on either graphs or trial and error calculations. Chaurasia (2001) using multiple non-linear regression developed equations for estimating the pre- and post-jump depths of flow in a rectangular channel when the flow rate and required energy loss are specified. The following equations are valid for $1.01 \leq F \leq 20$ and $17.7 \leq q \leq 3{,}530$ ft^3/s/ft ($0.50 \leq q \leq$ m^3/s/m). The upstream depth of flow is given by

$$y_1 = \frac{0.196q^{0.98}}{\Delta E^{0.47}\left(1 + \dfrac{1.262q}{\Delta E^{1.50}}\right)^{0.206}} \qquad (3.4.1)$$

and the downstream depth of flow

$$y_2 = \frac{1.0518q^{0.518}\Delta E^{0.223}}{\left(1 + \frac{0.0476q^{2.667}}{\Delta E^{4.0}}\right)^{0.0221}} \tag{3.4.2}$$

EXAMPLE

Flow occurs in a rectangular channel 5 m (16.4 ft) at a rate of 50 m³/s (1,766 ft³/s). If a hydraulic jump occurs with a headloss of 1.63 m (5.35 ft), what are the pre- and post-jump depths of flow and the height of the jump? Compare these results with that from Eq. (3.2.4).

Solution

The flow per unit width is

$$q = \frac{Q}{b} = \frac{50}{5} = 10\,\text{m}^3/\text{s}/\text{m}\left(353\,\text{ft}^3/\text{s}\right)\text{ft}$$

The upstream depth of flow is given by Eq. (3.4.1)

$$y_1 = \frac{0.196q^{0.98}}{\Delta E^{0.47}\left(1 + \frac{1.262q}{\Delta E^{1.50}}\right)^{0.206}} = \frac{0.196(10)^{0.98}}{1.63^{0.47}\left[1 + \frac{1.262(10)}{1.63^{1.50}}\right]^{0.206}} = 0.99\,\text{m}\,(3.02\,\text{ft})$$

The downstream depth of flow is given by Eq. (3.4.2)

$$y_2 = \frac{1.0518q^{0.518}\Delta E^{0.223}}{\left(1 + \frac{0.0476q^{2.667}}{\Delta E^{4.0}}\right)^{0.0221}} = \frac{1.0518(10)^{0.518}(1.63)^{0.223}}{\left[1 + \frac{0.0476(10)^{2.667}}{1.63^{4.0}}\right]^{0.0221}} = 3.75\,\text{m}\,(12.30\,\text{ft})$$

Comparing these results with Eq. (3.2.4)

$$u_1 = \frac{Q}{A} = \frac{Q}{by_1} = \frac{50}{5(0.99)} = 10.1 \text{m/s}$$

$$F_1 = \frac{u_1}{\sqrt{gy_1}} = \frac{10.1}{\sqrt{9.8(0.99)}} = 3.24$$

Then,

$$\frac{y_2}{y_1} = \frac{1}{2}\left(\sqrt{1+8F_1^2} - 1\right) = \frac{1}{2}\left[\sqrt{1+8(3.24)^2} - 1\right] = 4.10$$

and from the results above

$$\frac{y_2}{y_1} = \frac{3.75}{0.99} = 3.79$$

Therefore, the error in estimative equation provided by Chaurasia (2001) is

$$\text{Error} = 100\frac{4.10 - 3.79}{3.79} = 8.2\%$$

1.   The height of the jump is

$$H_j = y_2 - y_1 = 3.75 - 0.99 = 2.76 \text{m} \left(9.06 \text{ ft}\right)$$

3.4.2   Hydrodynamic and Impact Forces
When flood waters flow around a structure, forces are imposed on the structure as shown in Figure 3.17; and these forces include the frontal impact by the mass of the moving water against the projected area of the structure, the drag effect along the sides of the structure, and the eddies (negative) pressure on the downstream side of the structure.

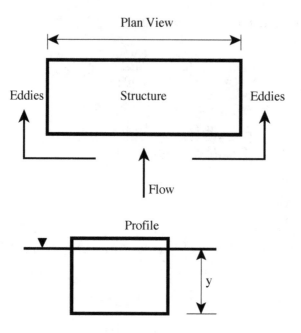

Figure 3.17. Flow around a structure.

The hydrodynamic force (F_D) on a structure is estimated using the standard equation from fluid mechanics for drag on immersed bodies or

$$F_D = C_D A \frac{\rho \overline{u}^2}{2}$$ (3.4.3)

where C_D = drag coefficient (Table 3.3), A = projected area of the structure onto a plane normal to the flow, ρ = fluid density, and $\overline{u}$ = average velocity of flow. Note, the hydrodynamic force on the structure is in addition to the hydrostatic force; and both the hydrodynamic and hydrostatic forces are dependent on the density of the fluid, which in turn, is a function of the sediment concentration. The additional load due to sediment should be considered in all calculations when the volumetric concentration of sediment (C_v) is 5% or greater (FEMA, 1995). The density of a water-sediment mixture $\left(\rho_S\right)$ is given by

$$\rho_s = \left(1 - C_v\right)\rho + C_v S_g \rho$$ (3.4.4)

where ρ = density of clear water and S_g = specific gravity of the sediment.

Table 3.3.   Drag coefficients for structures as a function of the width (b) to height (h) ratio.

Width to height ratio (b/h)	Drag coefficient C_D
$1 \le b/h \le 12$	1.25
$13 \le b/h \le 20$	1.30
$21 \le b/h \le 32$	1.40
$33 \le b/h \le 40$	1.50
$41 \le b/h \le 80$	1.75
$81 \le b/h \le 120$	1.80
$160 \le b/h$	2.00

Impact loads imposed on a structure by objects (floating debris, rocks, *etc.*) carried by moving water are among the most difficult loads to define and predict. By definition, the average impact force acting on a structure is the change in momentum divided by the elapsed time or

$$\overline{F}_I = \frac{\Delta M}{\Delta t} \tag{3.4.5}$$

where $\overline{F}_I$ = the average impact force, ΔM = the change in momentum, and Δt = elapsed time. Therefore, if an object of weight, W_0, is moving a velocity $\overline{u}$ impacts a structure the change in momentum is

$$\Delta M = \frac{W_0}{g}\overline{u} \tag{3.4.6}$$

since the terminal velocity in the direction of flow is zero. Therefore, the average impact (impulse) force of an object striking a structure is

$$\overline{F}_I = \frac{W_0\overline{u}}{g\Delta t} \tag{3.4.7}$$

With regard to Eq. (3.4.7), the following observations are pertinent. First, $\overline{F}_1$ is the average impact force. When an object strikes a structure, the force on the structure increases from zero to a maximum force and then back to zero in time Δt. Thus, the maximum force on the structure will be greater than $\overline{F}_1$. Second, what is an appropriate and reasonable value of W_0? Third, what is an appropriate and reasonable value of Δt? One approach to estimating the average impact force from floating or transport objects is given in FEMA (1995). In this approach, a single object weighing 1,000 lbs (4,448 N) and traveling at the velocity of the water strikes structure at the water surface or any point below it. The average impact force, under this assumption, is in English units,

$$\overline{F}_I = \frac{W_0 \overline{u}}{g\Delta t} = \frac{1000\overline{u}}{32.2\Delta t} = \frac{31.1\overline{u}}{\Delta t} \tag{3.4.8}$$

where Δt is generally taken as one second or less.

3.5.   BIBLIOGRAPHY

Argyropoulos, P.A. "Theoretical and Experimental Analysis of the Hydraulic Jump in a Parabolic Flume," *Proceedings of the 7th Conference, International Association for Hydraulic Research*, vol. 2, 1957, pp. D12.1-D12.20.

Argyropoulos, P.A. "The Hydraulic Jump and the Effect of Turbulence on Hydraulic Structures: Contribution of the research of the Phenomena," *Proceedings of the 9th Conference, International Association for Hydraulic Research*, 1961, p. 173-183.

Argyropoulos, P.A., "A General Solution of the Hydraulic Jump in Sloping Channels," *Proceedings of the American Society of Civil Engineers, Journal of the Hydraulics Division*, vol. 88, no. HY4, July 1962, pp. 61-77.

Bakhmeteff, B.A., and Matzke, A.E., "The Hydraulic Jump in Terms of Dynamic Similarity," *Transactions of the American Society of Civil Engineers*, vol. 101, 1936, pp. 630-647.

Bradley, J.N. and Peterka, A.J., "The Hydraulic Design of Stilling Basins: Hydraulic Jumps on a Horizontal Apron (Basin I)," *Proceedings of the American Society of Civil Engineers, Journal of the Hydraulics Division*, vol. 83, no. HY5, 1957a, pp. 1-24.

Bradley, J.N., and Peterka, A.J., "The Hydraulic Design of Stilling Basins: Stilling Basin with Sloping Apron (Basin V), " *Proceedings of the American Society of Civil Engineers, Journal of the Hydraulics Division*. vol 83, no. HY5, 1957b, p. 1-32.

Chaurasia, S.R., "Direct Equations for Hydraulic Jump Elements in a Rectangular Horizontal Channel," manuscript submitted to *Journal of Irrigation and Drainage Engineering*, American Society of Civil Engineers, 2001.

Chow, V.T. *Open Channel Hydraulics*, McGraw-Hill Book Company, New York, 1959.

FEMA, "Engineering Principles and Practices of Retrofitting Flood-Prone Residential Structures," Federal Emergency Management Agency, Washington, D.C., 1995.

Gardner, G.B., Nowell, A.R.M., and Smith, J.D., "Turbulent Processes in Estuaries," *Estuarine and Wetland Processes*, P. Hamilton and K. McDonald (Eds.), Plenum Press, New York, 1980, pp. 1-34.

Govinda Rao, N.S., and Rajaratnam, N., "The Submerged Hydraulic Jump," *Proceedings of the American Society of Civil Engineers, Journal of the Hydraulics Division*, vol. 89, no. HY1, January 1963, pp. 139-162.

Hayakawa, N., "Internal Hydraulic Jump in Co-current Stratified Flow," *Proceedings of the American Society of Civil Engineers, Engineering Mechanics Division*, vol. 96, no. EM5, 1970, pp. 797-800.

Hsing, P.S., "The Hydraulic Jump in a Trapezoidal Channel," thesis presented to the State University of Iowa, Iowa City, 1937, in partial fulfillment of the requirements for the degree of Doctor of Philosophy.

Kennison, K.R., "Discussion of the Hydraulic Jump in Sloping Channels," *Transactions of the American Society of Civil Engineers*, vol. 109, 1944, pp. 1123-1125.

Kindsvater, C.E., "Hydraulic Jump in Enclosed Conduits," thesis presented to the State University of Iowa, Iowa City, 1936, in partial fulfillment of the requirements for the degree of Master of Science.

Kindsvater, C.E., "The Hydraulic Jump in Sloping Channels," *Transactions of the American Society of Civil Engineers*, vol. 109, 1944, pp 1107-1154.

Peterka, A.J., "Hydraulic Design of Stilling Basins and Energy Dissipators," Engineering Monograph No. 25, U.S. Bureau of Reclamation, Denver, 1963.

Press, M.J., "The Hydraulic Jump," engineering honors thesis presented to the University of Western Australia, Nedlands, Australia, 1961.

Rajaratnam, N. "The Hydraulic Jump as a Wall Jet," *Proceedings of the American Society of Civil Engineers, Journal of the Hydraulics Division*, vol. 91, no. HY5, September 1965, pp. 107-132.

Rajaratnam, N., "Hydraulic Jumps," *Advances in Hydroscience*, vol. 4, Academic Press, New York, 1967, pp. 197-280.

Rouse, H., Siao, T.T., and Nagaratnam, S. "Turbulence Characteristics of the Hydraulic Jump," *Transactions of the American Society of Civil Engineers*, vol. 124, 1959, pp. 926-966.

Safranez, K., "Untersuchungen uber den Wechselsprung," *Der Baningenieur*, vol. 10, no. 37, 1919, pp. 649-651. A brief summary is given in Donald P. Barnes, "Length of Hydraulic Jump Investigated at Berlin," *Civil Engineering*, vol. 4, no. 5, May 1934, pp. 262-263.

Sandover, J.A., and Holmes, P., "The Hydraulic Jump in Trapezoidal Channels," *Water Power*, vol. 14, 1962, p. 445.

Silvester, R., "Hydraulic Jump in All Shapes or Horizontal Channels," *Proceedings of the American Society of Civil Engineers, Journal of the Hydraulics Division*, vol. 90, no. HY1, January 1964, pp. 23-55.

Silvester, R., "Theory and Experiment on the Hydraulic Jump," *2nd Australiasian Conference on Hydraulics and Fluid Mechanics*, 1965, pp. A25-A39.

Stefan, H., Hayakawa, N., and Schiebe, F.R., "Surface Discharge of Heated Water," 16130 FSU 12/71, U.S. Environmental Protection Agency, Washington, D.C., December 1971.

Stefan, H., and Hayakawa, N., "Mixing Induced by an Internal Hydraulic Jump," *Water Resources Bulletin*, American Water Resources Association, Vol. 8, no. 3, June 1972, pp. 531-545.

Stepanov, P.M.., "The Submerged Hydraulic Jump," *Gidrotekhn i Melioratsiya*, Moscow, vol. 10, 1958. English translation by Israel Programme for Scientific Translations, 1959.

Straub, W. O., "A Quick and Easy Way to Calculate Critical and Conjugate Depths in Circular Open Channels," *Civil Engineering*, December 1978, pp. 70-71.

Turner, J.S., "*Buoyancy Effects in Fluids*," Cambridge University Press, Cambridge, England, 1973.

Yih, C.S. and Guha, C.R., "Hydraulic Jump in a Fluid System of Two Layers," *Tellus*, vol. 7, no. 3, 1955, pp. 358-366.

3.6.   PROBLEMS

1. If the channel in the figure below is rectangular in shape and 10 ft (3 m) wide, calculate
    a. The force on the gate.
    b. The Froude number at Sections 1 and 2.
    c. The condition required downstream at Section 3 for a hydraulic jump to form.
    d. Assuming a hydraulic jump occurs downstream of the gate, estimate the efficiency and length of the jump.

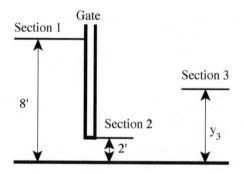

2. Apply the momentum and continuity equations to the submerged hydraulic jump that occurs at a sluice gate outlet in a rectangular channel (see the figure below) to prove

$$\frac{y_s}{y_2} = \left[1 + 2F_2^2\left(1 - \frac{y_2}{y_1}\right)\right]^{05}$$

where all variables are defined in the figure below. Neglect the channel bed friction F_f.

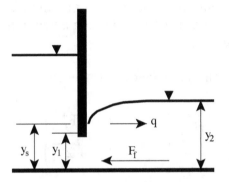

3.   Prove that the energy loss in a hydraulic jump occurring in a horizontal, rectangular channel is given by Eq. (3.2.24) or

$$\Delta E = \frac{(y_2 - y_1)^3}{4y_1y_2}$$

4.   A rectangular channel 8 ft (2.4 m) wide carries 100 ft³/s (2.8 m³/s) at a depth of 0.5 ft (0.15 m). This channel is connected by a 50 ft (15 m) long straight wall transition to a rectangular channel 10 ft (3.0 m) wide. Determine whether a hydraulic jump occurs in this transition, if the headloss within the transition is h_f ft. (Assume that h_f is uniformly distributed through the transition.) If a hydraulic jump occurs, what is the approximate location of the jump relative to the beginning of the transition section?
   a.   Downstream depth 4 ft (1.2 m) and $h_f = 0$ ft.
   b.   Downstream depth 4 ft (1.2 m) and $h_f = 1$ ft (0.30 m).
   c.   Compare the answers to Parts (a) and (b) and discuss any differences.
   d.   Downstream depth 3.6 ft (1.1 m) and $h_f = 1$ ft (0.30 m).

5.   Water flows at a velocity of 25 ft/s (7.6 m/s) and a depth of 3.5 ft (1.1 m), in a rectangular channel 16 ft (4.9 m) wide. Calculate the

downstream depth necessary to form a jump, the head loss, and the horsepower dissipated in the jump.

6.   A hydraulic jump is to be caused in a circular culvert 3 m (9.8 ft) in diameter and conveying a flow of 12 m³/s (420 ft³/s). If the depth of flow downstream of the jump cannot exceed 60 percent of the diameter of the pipe, what is the limiting upstream depth of flow? Is the limiting value found a minimum or maximum value? If assumptions are made in obtaining a solution, then confirm that they are valid.

7.   For the situation shown below, prove that

$$\Gamma_L = \left(2 + \frac{1}{\Gamma_L}\right)^3 \frac{F_3^4}{\left(1 + 2F_3^2\right)^3}$$

assuming that $M_2 = M_3$ and that the channel is rectangular in shape. The limiting contraction ratio, Γ_L, and can be used to distinguish between Yarnell's class A (unchoked) and class B (choked) flow.

Plan View

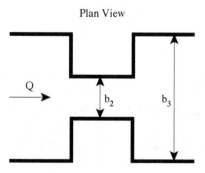

8.   A bridge has piers 1 m (3.3 ft) wide at 7 m (23 ft) centers. Some distance upstream of the bridge, the depth and velocity of flow are 1.5 m (4.9 ft) and 1 m/ss (3.3 ft/s), respectively. Neglecting the slope of the channel and bed friction, estimate the downstream depth of flow if the drag of the piers P_f is given by

$$P_f = \frac{1}{2}C_D A\rho u^2$$

where C_D = drag coefficient, A = area of pier projected on a plane normal to the flow, ρ = fluid density, and u = upstream velocity. For this problem, take C_D = 1.5. Use the results from Prob. 7 to determine if the flow is choked.

9.   A spillway discharges a flow of 8 m³/s/m (86 ft³/s/ft). A model study demonstrated that on the downstream horizontal apron the depth of flow was 0.6 m (2.0 ft). Estimate the tail water depth that is required for a hydraulic jump to form. Assuming that a hydraulic jump is formed,
    a.   Classify the jump type.
    b.   Estimate the jump length.
    c.   Estimate the head loss incurred through the jump.

10.   A trapezoidal channel is 20 ft (6.1 m) wide at the bottom and has side slopes of 1.5 (horizontal to 1 (vertical). For a flow of 500 ft³/s (14 m³/s), calculate the depth sequent to a supercritical depth of 2.0 ft (0.61 m).

11.   A rectangular channel has a longitudinal slope of 0.15 ft/ft (0.15 m/m). This slope channel ends in a horizontal channel section, Fig. 3.9. When a discharge of 120 ft³/s/ft (11 m³/s/m) occurs in the channel at a depth of 2.0 ft (0. 61 m), a hydraulic jump occurs when the sloping channel meets the horizontal channel. Calculate
    a.   The length of the jump.
    b.   The energy lost in the jump.

12.   A sluice gate similar to that shown in Fig. 3.5 is used to control flow in an irrigation system, and the flow from this gate discharges into a rectangular channel of the same width as the gate. Under normal circumstances, the flow immediately downstream of the gate has a depth of 4 ft (1.2 m) and a velocity of 60 ft/s (18 m/s). Given these circumstances, estimate
    a.   The sequent depth of the tail water.
    b.   The length of the lined channel required to protect the natural soil from the supercritical flow and the hydraulic jump.
    c.   The efficiency of the jump with the downstream sequent depth conditions.
    d.   The type of jump that is to be expected if the sequent depth conditions are met.

13. For the data provided in Prob. 10, describe qualitatively the result of having a tail water depth of 32 ft (9.8 m). Can such a situation be described quantitatively? If it can, list all assumptions.

14. A wave is moving down a wide rectangular channel in the direction of flow. Upstream of the wave, the water velocity is 5 ft/s (1.5 m/s) with a depth of 5 ft (1.5 m). Downstream of the wave, the water velocity is 4 ft/s (1.2 m/s) with a depth of 7 ft (2.1 m). What is the velocity of the wave?

15. A rectangular irrigation channel 10 ft (3.0 m) wide conveys a flow of 500 ft³/s (14 m³/s) at a depth of 10 ft (3.0 m). If the upstream discharge is suddenly increased to 700 ft³/s (20 m³/s), estimate the height and velocity of the resulting surge wave.

16. A river is flowing at a depth of 10 m (33 ft) and a velocity of 3 m/s (9.8 ft/s). A tidal bore is moving up the river, and at the point where the river meets the bore, the depth abruptly increases to 15 m (49 ft). Estimate the velocity of the tidal bore up the river. What is the velocity and direction of flow behind the bore?

CHAPTER 4

DEVELOPMENT OF UNIFORM FLOW CONCEPTS

4.1. ESTABLISHMENT OF UNIFORM FLOW

Although the definition of uniform flow and the assumptions required to develop the governing equations are only rarely satisfied in practice, the concept of uniform flow is central to the understanding and solution of most problems in open channel hydraulics. By definition, uniform flow occurs when:

1. The depth, flow area, and velocity at every cross section are constant.

2. The energy grade line, water surface, and channel bottom are all parallel; that is, $S_f = S_w = S_0$, where S_f = slope of the energy grade line, S_w = slope of the water surface, and S_0 = slope of the channel bed (Fig. 4.1).

In general, uniform flow can occur only in a very long, straight, prismatic channel where a terminal velocity of flow can be achieved; i.e., the head loss due to turbulent flow is exactly balanced by the reduction in potential energy due to the uniform decrease in the elevation of the bottom of the channel. In addition, although unsteady uniform flow is theoretically possible, it rarely occurs in open channels; only the case of steady, uniform flow will be treated here.

4.2.   THE CHEZY AND MANNING EQUATIONS

For computational purposes, the average velocity of a uniform flow can be computed approximately by any one of a number of semi-empirical uniform flow equations. All of these equations have the form

$$\overline{u} = CR^x S^y \tag{4.2.1}$$

where
$\overline{u}$	=	average velocity
R	=	hydraulic radius
S	=	longitudinal slope
C	=	resistance coefficient
x and y	=	coefficients.

At some point, during 1768 - 1775 (Levi, 1995), Antoine Chezy, designing an improvement for the water system in Paris, France derived an equation relating the uniform velocity of flow to the hydraulic radius and longitudinal slope of the channel. In 1889, Robert Manning, a professor at the Royal College of Dublin (Levi, 1995) proposed what has become know as Manning's equation. In this book, only the Chezy and Manning equations are considered.

The Chezy equation can be derived from the definition of uniform flow with an assumption regarding the form of the flow resistance coefficient. With reference to Fig. 4.1, the definition of uniform flow requires that the forces resisting motion exactly equal the forces causing motion. The force causing motion is

$$F_m = W \sin\theta = \gamma AL \sin\theta \tag{4.2.2}$$

where
W	=	weight of fluid within the control volume
γ	=	fluid specific weight
A	=	flow area
L	=	control volume length
θ	=	longitudinal slope angle of the channel.

If θ is small, which is usually the case, then $sin(\theta) \approx S_O$. It is assumed that the force per unit area of the channel perimeter resisting motion, F_R, is proportional to the square of the average velocity or

$$F_R \sim \overline{u}^2$$

Then, for a reach of length L with a wetted perimeter P, the force of resistance is

$$F_R = LPk\overline{u}^2 \qquad (4.2.3)$$

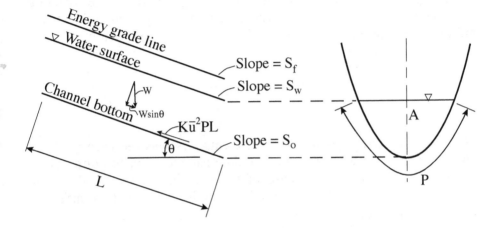

Figure 4.1. Schematic definition of variables for the derivation of the Chezy equation.

where k = a constant of proportionality. Setting the force causing motion, Eq. (4.2.2), equal to the force resisting the motion, Eq. (4.2.3)

$$\gamma ALS_0 = LPk\overline{u}^2$$

or

$$\overline{u} = \left(\frac{\gamma}{k}\right)^{0.5}\sqrt{RS} \qquad (4.2.4)$$

where the subscript associated with S has been dropped. For convenience, define,

$$C = \left(\frac{\gamma}{k}\right)^{0.5} \qquad (4.2.5)$$

The resistance coefficient, C, defined by Eq. (4.2.5) is commonly known as the Chezy C and, in practice, is determined by either measurement or estimate. The coefficient defined by Eq. (4.2.5) is not dimensionless, but has the dimensions of acceleration; i.e., length/time² or LT^{-2}.

The Manning equation is the result of a curve-fitting process and is thus completely empirical in nature. In the application of the Manning equation, it is essential that the system of units being used be identified and that the appropriate coefficient be used. In the SI system of units, the Manning equation is

$$\overline{u} = \frac{1}{n}R^{2/3}\sqrt{S}$$

where n = Manning resistance coefficient. As was the case with the Chezy resistance coefficient, n is not dimensionless but has dimensions of $TL^{-1/3}$, or in the specific case of the equation above, $s/m^{1/3}$. The Manning equation can be converted to English units without affecting the numerical value of n by noting

$$n \text{ in}\left[\frac{s}{m^{1/3}}\right] = 0.6730\, n \text{ in}\left[\frac{s}{ft^{1/3}}\right]$$

Therefore, in English units, the Manning equation is

$$\overline{u} = \frac{1}{0.6730n}R^{2/3}\sqrt{S} = \frac{1.49}{n}R^{2/3}\sqrt{S}$$

For the sake of generality, in this book, the Manning equation will be written

$$\overline{u} = \frac{\phi}{n}R^{2/3}\sqrt{S} \qquad (4.2.6)$$

where $\phi = 1.00$ if SI units are used, and $\phi = 1.49$, if English units are used.

From the viewpoint of modern fluid mechanics, the dimensions of the resistance coefficients C and n are a matter of major importance, see, for example, Simons and Senturk (1976). However, the dimensions of these coefficients appear to have been a problem of no significance to early hydraulic engineers. The reader is cautioned to be aware of the dimensions of C and n and to take these dimensions into account where appropriate.

Since the Chezy and Manning equations describe the same phenomena, the coefficients C and n must be related. Setting Eq. (4.2.4) equal to Eq. (4.2.6) yields

$$C\sqrt{RS} = \frac{\varphi}{n} R^{2/3} \sqrt{S}$$

$$C = \frac{\varphi}{n} R^{1/6} \tag{4.2.7}$$

In this book, the Manning equation will be used almost exclusively, even though the Chezy equation is equally appropriate from a theoretical point of view. The reader is cautioned that the above discussion applies only to channels with plane beds. In alluvial channels where bed forms are present, the analysis which applies is much more complex; see for example, Simons and Senturk (1976) or Garde and Raju (1978).

4.3.   RESISTANCE COEFFICIENT ESTIMATION

4.3.1.   Basic Theoretical Concepts

The primary difficulty in using either the Manning or Chezy equation, in practice, is accurately estimating an appropriate value of the resistance coefficient. In general, it is expected that n and C should depend on the Reynolds number of the flow, the boundary roughness, and the shape of the channel cross section. This is equivalent to hypothesizing that n and C behave in a manner analogous to the Darcy-Weisbach friction factor f, used to assess flow resistance in pipe flow; see for example Streeter and Wylie (1975), or

$$S = \frac{f}{4R} \frac{\overline{u^2}}{2g} \tag{4.3.1}$$

Then

$$n = \varphi R^{1/6} \sqrt{\frac{f}{8g}} \tag{4.3.2}$$

and

$$C = \sqrt{\frac{8g}{f}} \tag{4.3.3}$$

If the behavior of n and C are to be investigated from this viewpoint, then a number of basic definitions regarding the types of turbulent flow must be recalled. *Hydraulically smooth turbulent flow* refers to a flow in which the perimeter roughness elements are completely covered by the viscous sublayer. *Fully rough flow* occurs when the perimeter roughness elements project through the laminar sublayer and dominate the flow behavior. In the latter case, flow resistance is entirely due to form drag, and the resistance coefficient is independent of the Reynolds number, where $Re = 4R\bar{u}/v$ = Reynolds number. Between the two extremes, there is a transitional region. In the analysis which follows, k_s is a length parameter which characterizes the perimeter roughness and is defined as the sand grain diameter for a sand-coated surface which has the same limiting value of f. The three types of turbulent flow - smooth, transitional, and fully rough - are differentiated from each other by a Reynolds number based on k_s and the shear velocity

$$u_* = \sqrt{\tau_0 / \rho} = \sqrt{gRS}$$

or

$$Re_* = \frac{u_* k_s}{v} \tag{4.3.4}$$

The transition region is defined by the limits

$$4 \le Re_* \le 100 \tag{4.3.5}$$

With the lower limit defining the end of the smooth region and the upper limit defining the beginning of the fully rough region. As will be shown in a subsequent section of this chapter, the point at which fully rough flow can be defined in terms of Manning's n or

$$n^6 \sqrt{RS} = 1.9 \times 10^{-13} \tag{4.3.6}$$

When the boundaries are hydraulically smooth and Re < 100,000, the friction factor is given by the Blasius formula developed for pipe flow or

$$f = \frac{0.316}{Re^{0.25}} \qquad (4.3.7)$$

When the Reynolds number exceeds 100,000, but the flow is still hydraulically smooth, the friction factor is given by

$$\frac{1}{\sqrt{f}} = 2.0 \log\left(\frac{Re\sqrt{f}}{2.51}\right) \qquad (4.3.8)$$

For hydraulically rough flow, the friction factor can be estimated by

$$\frac{1}{\sqrt{f}} = 2.0 \log\left(\frac{12R}{k_s}\right) \qquad (4.3.9)$$

where typical values for k_s, the equivalent roughness parameter, are summarized in Table 4.1. It is pertinent to observe that the equivalent roughness parameter, k_s, is sometimes called the "roughness height," and is a measure of the linear dimension of roughness elements, but is not necessarily equal to the actual or even average height of these elements. In the transition region between hydraulically smooth and rough flows, f can be estimated by the modified Colebrook formula or

$$\frac{1}{\sqrt{f}} = -2 \log\left(\frac{k_s}{12R} + \frac{2.5}{Re\sqrt{f}}\right) \qquad (4.3.10)$$

The relationship between n and k_s for hydraulically rough channels is

$$n = \frac{\phi R^{1/6}}{\Gamma \log\left(12.2\dfrac{R}{k_s}\right)} \qquad (4.3.11)$$

where $\Gamma = 32.6$ for English units, and 18.0 for SI units, and $\phi = 1.49$ for English units, and 1.00 for SI units. The advantage of using k_s rather than n is that k_s accounts for changes in the resistance factor due to stage, whereas n does not.

Table 4.1.   Values of k_s for various bed materials (Ackers, 1958, Chow, 1959, and Zegzhda, 1938).

Material	k_s (ft)	k_s (m)
Brass, copper, lead, glass	0.0001-0.0030	0.00003048-0.0009
Wrought iron, steel	0.0002-0.0080	0.0001-0.0024
Asphalted cast iron	0.0004-0.0070	0.0001-0.0021
Galvanized iron	0.0005-0.0150	0.0002-0.0046
Cast iron	0.0008-0.0180	0.0002-0.0055
Wood stave	0.0006-0.0030	0.0002-0.0009
Cement	0.0013-0.0040	0.0004-0.0012
Concrete	0.0015-0.0100	0.0005-0.0030
Untreated gunite	0.01-0.033	0.0030-0.0101
Drain tile	0.0020-0.0100	0.0006-0.0030
Riveted steel	0.0030-0.0300	0.0009-0.0091
Rubble masonry	0.02	0.0061
Straight, uniform earth channels	0.01	0.0030
Natural stream bed	0.1000-3.0000	0.0305-0.9144

4.3.2.   Applied Concepts

For the case of high Reynolds numbers and roughnesses of the type usually found in unlined channels, the friction factor is independent of the Reynolds number and nearly proportional to $R^{-1/3}$ (Anonymous, 1963a). If f is assumed to be given by $R^{-1/3}$ and this value is substituted in Eq. (4.3.2), it is concluded that for this situation, Manning's n is a constant.   Thus, Manning's equation with a constant value of n is applicable only to fully rough, turbulent flow as defined by Eq. (4.3.5) or (4.3.6).   In estimating an appropriate value of n for this case, a qualitative knowledge of the factors on which n depends, is necessary since in most applied situations, the value of n is a function of many variables.

Surface Roughness: The surface roughness of a channel perimeter provides a critical point of reference in estimating n.   When the perimeter material is fine, the value of n is low and is relatively unaffected by changes in the depth of flow; however, when the perimeter is composed of gravels and/or boulders, the value of n is larger and may vary significantly with the depth of flow. For example, large boulders usually collect at the bottom of such channels and result in a high value of n at low stages, and relatively low values of n at high stages.

Vegetation: The estimated value of n should take into account the effect of vegetation in retarding flow and increasing n. In general, the relative importance of vegetation on n is a strong function of the depth of flow and height, density, distribution, and type of vegetation. This consideration is crucial in the design of small drainage channels, since they usually do not receive regular maintenance. Chow (1959) noted that trees from 6 to 8 in (0.15 to 0.20 m) in diameter growing on the side slopes do not impede the flow as much as brushy growths, if the overhanging limbs of the trees are kept trimmed.

Channel Irregularity: This terminology refers to variations in the channel cross section, shape, and wetted perimeter along the longitudinal axis of the channel. In natural channels, such irregularities are usually the result of deposition or scour. In general, gradual variations have a rather insignificant effect on n, while abrupt changes can result in a much higher value of n than would be expected from a consideration of only the surface roughness of the channel perimeter. Greater resistance is associated with alternating large and small cross sections, sharp bends, constrictions, and side to side shifting of the low flow channel. The significance of the effect of changes in channel size depends primarily on the number of alternations of large and small sections, and secondarily, on the magnitude of the changes.   It is pertinent to observe that the effects of abrupt changes may extend downstream several hundred feet (several tens of meters), and the value of n, below a disturbance, may also require adjustment (Arcement and Schneider, 1989).

Where the ratio of channel width to flow depth is small, resistance caused by eroded and scalloped banks, projecting points, and exposed tree roots along the banks must be taken into account. For example,

Benson and Dalrymple (1967) showed that severely eroded and scalloped banks can increase n values by as much as 0.02. Larger adjustments may be required for very large, irregular banks that have projecting points.

Obstruction: The presence of obstructions such as fallen trees, debris flows, and log or debris jams can have a significant impact on the value of n. The degree of effect of such obstructions depends on the number and size of the obstructions. The effect of obstructions on the resistance coefficient is a function of the velocity of flow. When the velocity is high an obstruction exerts a sphere of influence that is much larger than the obstruction. In channels having gentle to moderately steep slopes, the sphere of influence is approximately three to five times the width of the obstruction (Arcement and Schneider, 1989). In addition, obstructions can create overlapping disturbances that increase the value of n even though the obstructions occupy only a small part of the cross section.

Channel Alignment: While curves of large radius without frequent changes in the direction of curvature offer relatively little resistance, severe meandering with curves of small radius will significantly increase the value of n. Meandering is usually discussed in terms of the ratio of the total length of the meandering to the straight length. The meandering is considered minor for ratios of 1.0 to 1.2, moderate for ratios of 1.2 to 1.5, and severe for ratios of 1.5 and greater. Meanders can increase n but as much as 30 percent when the flow is confined within the channel (Chow, 1959); and any adjustment of n should only be considered when the flow is confined to the channel.

Sedimentation and Scouring: In general, active sedimentation and scouring yield channel variation, which results in an increased value of n. Urquhart (1975) noted that it is important to consider whether both of these processes are active and whether they are likely to remain active in the future.

Stage and Discharge: The n value for most channels tends to decrease with an increase in the stage and discharge. This is the result of irregularities which have a crucial impact on the value of n, at low stages, when they are uncovered. However, the n value may increase with increasing stage and discharge if the banks of the channel are rough, grassy, or brush-covered, or if the stage increases sufficiently to cover the flood plain. In such a case, it is necessary to compute a composite value of n - a procedure which is discussed in the next chapter.

Given these qualitative remarks regarding the variation of n with a number of factors, it is now possible to discuss several methods of estimating a value of n given a specific situation.

4.3.3.   Approaches to Estimating n

Soil Conservation Service Method: The Soil Conservation Service (SCS) method for estimating n involves the selection of a basic n value for a uniform, straight, and regular channel, in a native material, and then modifying this value by adding correction factors determined by a critical consideration of some of the factors enumerated above (Urquhart, 1975).   In this process, it is critical that each factor be considered and evaluated independently.   The SCS suggests that the turbulence of a flow can be used as a measure or indicator of the degree of retardance; i.e., factors which induce a greater degree of turbulence should also result in an increase in n.

Step 1:   *Selection of a Basic n*: In this step, a basic value for a straight, uniform, smooth channel in the native material is selected. The channel must be visualized without vegetation, obstructions, changes in shape, and changes in alignment. The basic n values suggested by the SCS are summarized in Table 4.2.

Table 4.2.   **Basic n values suggested by the Soil Conservation Service (Anon., 1963b).**

Channel character	Basic n
Channels in earth	0.02
Channels cut into rock	0.025
Channels in fine gravel	0.024
Channels in coarse gravel	0.028

Table 4.3. Modifying factors for vegetation (Anonymous, 1963b).

Vegetation and flow conditions comparable with:	Degree of effect on n	Range of modifying values
Dense growth of flexible turf grasses or weeds, of which Bermuda grass and blue grass are examples, where the average depth of flow is 2 to 3 times the height of the vegetation. Supple seedling tree switches such as willow, cottonwood, or salt cedar where the average depth of flow is 3 to 4 time the height of the vegetation	Low	0.005 - 0.010
Turf grasses where the average depth of flow is 1 to 2 times the height of the vegetation Stemmy grasses, weeds, or tree seedlings with moderate cover where the average depth of flow is 2 to 3 times the height of the vegetation Brush growths, moderately dense, similar to willows 1 to 2 years old, dormant season along side slopes of channel with no significant vegetation along the channel bottom where the hydraulic radius is greater than 2 ft (0.6 m)	Medium	0.010 - 0.025
Dormant season, willow or cottonwood trees 8 to 10 years old, intergrown with some weeds and brush, none of the vegetation in foliage, where the hydraulic radius is greater than 2 ft (0.6m) Growing season, bushy willows about 1 year old intergrown with some weeds in full foliage along side slopes, no significant vegetation along channel bottom, where the hydraulic radius is greater than 2 ft (0.6 m)	High	0.025 - 0.050
Turf grasses where the average depth of flow is less than one-half the height of vegetation Growing season, bushy willows about 1 year old, intergrown with weeds in full foliage along side slopes, dense growth of cattails along channel bottom, any value of hydraulic radius up to 10 or 15 ft (3 to 4.6 m) Growing season, trees intergrown with weeds and brush, all in full foliage, any value of hydraulic radius up to 10 or 15 ft (3 to 4.6 m)	Very high	0.050 -0.100

Step 2:   *Modification for Vegetation:* The retardance due to vegetation is primarily due to the flow of water around stems, trunks, limbs and branches and only secondarily to the reduction of the flow area.   In assessing the effect of vegetation on retardance, consideration must be given to the height of the vegetation in relation to the depth of flow, the capacity of the vegetation to resist bending, the degree to which the flow is obstructed, the transverse and longitudinal distribution of vegetation of various types, the densities and heights of vegetation in the reach being considered, and the critical season; i.e., is the vegetation dormant or growing? The SCS results regarding vegetation are summarized in Table 4.3.

Step 3:   *Modification for Channel Irregularity:* In determining the modification required for channel irregularity, both changes in flow area and changes in cross-sectional shape must be considered.   The effects of changes in flow area should be examined from the viewpoint of comparing the magnitude of the change with the average area.   While large changes in area, if they are gradual and uniform, result in small modifying values, abrupt changes yield large modifying values.   In the case of changes in channel shape, the degree to which the change causes the greatest depth of flow to migrate from side to side, is critical.   Shape changes, which yield the largest modifying values, are those which shift the main flow from side to side in distances short enough to produce eddies and upstream currents in the shallow area.   The SCS recommendations for the modifying values for this effect are summarized in Table 4.4.

The second consideration in this step is the degree of roughness or irregularity of the surface of the channel perimeter. The existing surface should be compared with the surface smoothness which can, under ideal conditions, be obtained with the native materials and with the specified depth of flow. The results for this are summarized in Table 4.5.

Table 4.4. Modifying factors for changes in cross section size and shape (Anonymous, 1963b).

Character of variations in size and shape of cross sections	Modifying value
Changes in size or shape occurring gradually	0.000
Large and small sections alternating occasionally or shape changes causing occasional shifting of main flow from side to side	0.005
Large and small sections alternating frequently or shape changes causing frequent shifting of main flow from side to side	0.010 - 0.015

Table 4.5. Modifying factors for channel surface irregularity (Anonymous, 1963b).

Degree of irregularity	Surfaces comparable with	Modifying value
Smooth	The best obtainable for the materials involved	0.000
Minor	Good dredged channels; slightly eroded or scoured side slopes of canals or drainage channels	0.005
Moderate	Fair to poor dredged channels; moderately sloughed or eroded side slopes of canals or drainage channels	0.010
Severe	Badly sloughed banks of natural channels; badly eroded or sloughed sides of canals or drainage channels; unshaped, jagged and irregular surfaces of channels excavated in rock	0.020

Step 4:   Modification for Obstruction: The selection of the modifying value for this factor is based on the number and characteristics of the obstructions. Obstructions considered by the SCS included debris deposits, stumps, exposed roots, boulders, and fallen and lodged logs. In assessing the relative effect of obstructions, consideration must given to the following: (a) the degree to which the obstructions reduce the flow area at various depths of flow, (b) the shape of the obstructions (recall that angular objects produce greater turbulence than rounded objects), and (c) the position and spacing of the obstructions in both the transverse and

longitudinal directions. The SCS recommendations for this modification are summarized in Table 4.6.

Table 4.6. Modifying factors for obstruction (Anonymous, 1963b).

Relative effect of obstructions	Modifying value
Negligible	0.000
Minor	0.010-0.015
Appreciable	0.020-0.030
Severe	0.040-0.060

Step 5: *Modification for Channel Alignment:* The modifying value for channel alignment is found by adding the modifying values in Steps 2 to 4 to the basic value of n, Step 1, to form the subtotal n'. Define l_s = straight length of the reach under consideration, and l_m = meander length of the channel in the reach.   The modifying value for alignment can then be estimated from Table 4.7 for various values of the ratio l_m/l_s.

Table 4.7. Modifying values for channel alignment (SCS, 1963b).

l_m / l_s	Degree of meandering	Modifying value
1.0-2.0	Minor	0.00
1.2-1.5	Appreciable	$0.15n'$
>1.5	Severe	$0.30n'$

Step 6: *Estimate of n:* A value of n can then be estimated by summing the results of Steps 1 to 5.

The use of the SCS method for estimating n in a natural channel is best demonstrated by an example.

EXAMPLE 4.1

A dredged channel has a cross section which can be approximated by a trapezoid with a bottom width of 10 ft (3.0 m) and side slopes of 1.5:1. At the maximum stage, the average depth of flow is 8.5 ft (2.6 m) and the top width is 35.5 ft (10.8 m). The alignment of the channel has a moderate degree of meander. While the side slopes are fairly regular, the bottom is uneven and irregular. The variation of the cross sectional area with longitudinal distance is moderate. The material through which the channel is cut is characterized as gray clay with the upper part of the channel being in silty clay loam. The side slopes of the channel are covered with a heavy growth of poplar trees 2 to 3 in (0.05 to 0.08 m) in diameter, large willows, and climbing vines. There is also a thick growth of water weed on the bottom of the channel. Given this description, estimate a value of n for this channel for summer conditions when the vegetation is in full foliage.

Solution

Step	Comment	Modifying value
1	Estimate basic value of n, Table 4.2	0.02
2	Description of vegetation indicates a high degree of retardance, Table 4.3	0.08
3	Description suggests an insignificant change in both channel size and shape, Table 4.4	0.00
4	Description indicates a moderate degree of irregularity, Table 4.5	0.01
5	No obstructions are indicated, Table 4.6	0.00
6	A moderate degree of alignment change is indicated; therefore, $n' = 0.02 + 0.08 + 0.01 = 0.11$ Modifying value = 0.15(0.11) = 0.02 Table 4.7	0.02
Total estimated $n = 0.13$		

Table Method of n Estimation: A second method of estimating n for a channel involves the use of tables of values. Chow (1959) presented an extensive table of n values for various types of channels, and this table has been expanded as additional data have become available, Table 4.8.

In Table 4.8, a minimum, normal, and maximum value of n are stated for each type of channel. The values in bold are those recommended for design. It is noted that the normal values assume that the channel receives regular maintenance.

Table 4.8. Values of the roughness coefficient n from Chow (1959), Richardson *et al.* (1987), and SLA (1982) and other sources. The values in bold are recommended for design.

Closed Conduits flowing partly full	Minimum	Normal	Maximum
Metal			
Brass, smooth	0.009	**0.010**	0.013
Steel			
1. Lockbar and welded	0.010	0.012	0.014
2. Riveted and spiral	0.013	0.016	0.017
Cast iron			
1. Coated	0.010	0.013	0.014
2. Uncoated	0.011	0.014	0.016
Wrought iron			
1. Black	0.012	0.014	0.015
2. Galvanized	0.013	0.016	0.017
Corrugated metal			
1. Subdrain	0.017	0.019	0.021
2. Storm drain	0.021	**0.024**	0.030
Non-metal			
Lucite	0.008	0.009	0.010
Glass	0.009	**0.010**	0.013
Cement			
1. Neat Surface	0.010	0.011	0.013
2. Mortar	0.011	0.013	0.015
Concrete			
1. Culvert, straight and free of debris	0.010	0.011	0.013
2. Culvert, with bends, connections, and some debris	0.011	**0.013**	0.014
3. Finished	0.011	0.012	0.014
4. Sewer and manholes, inlets, etc., straight	0.013	0.015	0.017
5. Unfinished, steel form	0.012	0.013	0.014
6. Unfinished, smooth wood form	0.012	**0.014**	0.016
7. Unfinished, rough wood form	0.015	0.017	0.020
Wood			
1. Stave	0.010	0.012	0.014
2. Laminated, treated	0.015	0.017	0.020
Clay			
1. Common drainage tile	0.011	**0.013**	0.017
2. Vitrified sewer	0.011	0.014	0.017

Closed Conduits flowing partly full	Minimum	Normal	Maximum
3. Vitrified sewer with manholes, inlets, etc.	0.013	0.015	0.017
4. Vitrified subdrain with open joint	0.014	0.016	0.018
Brickwork			
1. Glazed	0.011	0.013	0.015
2. Lined with cement mortar	0.012	0.015	0.017
Sanitary sewers coated with sewage slimes with bends and connections	0.012	0.013	0.016
Paved invert, sewer, smooth bottom	0.016	0.019	0.020
Rubble masonry, cemented	0.018	0.025	0.030
Polyethylene pipe	0.009	-----	-----
PVC	0.010	-----	-----

Lined or built up channels

Metal

	Minimum	Normal	Maximum
Smooth steel surface			
1. Unpainted	0.011	**0.012**	0.014
2. Painted	0.012	0.013	0.017
Corrugated	0.021	0.025	0.030

Nonmetal

	Minimum	Normal	Maximum
Cement			
1. Neat surface	0.010	0.011	0.013
2. Mortar	0.011	0.013	0.015
Wood			
1. Planed, untreated	0.010	0.012	0.014
2. Planed, creosoted	0.011	0.012	0.015
3. Unplaned	0.011	0.013	0.015
4. Plank with battens	0.012	0.015	0.018
5. Lined with roofing paper	0.010	0.014	0.017
Concrete			
1. Trowel, finish	0.011	**0.013**	0.015
2. Float finish	0.013	0.015	0.016
3. Finished, with gravel on the bottom	0.015	0.017	0.020
4. Unfinished	0.014	0.017	0.020
5. Gunite, good section	0.016	0.019	0.023
6. Gunite, wavy section	0.018	0.022	0.025
7. On good excavated rock	0.017	0.020	-----
8. On irregular excavated rock	0.022	0.027	-----
Concrete bottom float with sides of			
1. Dressed stone in mortar	0.015	0.017	0.020
2. Random stone in mortar	0.017	0.020	0.024
3. Cement, rubble masonry, plastered	0.016	0.020	0.024
4. Cement rubble masonry	0.020	0.025	0.030
5. Dry rubble or riprap	0.020	0.030	0.035

Gravel bottom with sides of

Closed Conduits flowing partly full	Minimum	Normal	Maximum
1. Formed concrete	0.017	0.020	0.025
2. Random stone in mortar	0.020	0.023	0.026
3. Dry rubble or riprap	0.023	0.033	0.036
Brick			
1. Glazed	0.011	**0.013**	0.015
2. In cement mortar	0.012	**0.015**	0.018
Masonry			
1. Cemented rubble	0.017	0.025	0.030
2. Dry rubble	0.023	0.032	0.035
Dressed ashlar	0.013	0.015	0.017
Asphalt			
1. Smooth	0.013	0.013	-----
2. Rough	0.016	0.016	-----
Vegetal cover	0.030	-----	0.500
Excavated or dredged - General			
Earth, straight and uniform			
1. Clean and recently completed	0.016	0.018	0.020
2. Clean, after weathering	0.018	**0.022**	0.025
3. Gravel, uniform section, clean	0.022	0.025	0.030
4. With short grass, few weeds	0.022	0.027	0.033
Earth, winding and sluggish			
1. No vegetation	0.023	0.025	0.030
2. Grass, some weeds	0.025	0.030	0.033
3. Dense weeds or aquatic plants in deep channels	0.030	0.035	0.040
4. Earth bottom and rubble sides	0.028	0.030	0.035
5. Stony bottom and weedy banks	0.025	0.035	0.040
6. Cobble bottom and clean sides	0.030	0.040	0.050
Dragline-excavated or dredged			
1. No vegetation	0.025	0.028	0.033
2. Light brush on banks	0.035	0.050	0.060
Rock cuts			
1. Smooth and uniform	0.025	0.035	0.040
2. Jagged and irregular	0.035	0.040	0.050
Channels not maintained, weeds and brush uncut			
1. Dense weeds, high as flow depth	0.050	0.080	0.120
2. Clean bottom, brush on sides	0.040	0.050	0.080
3. Same, highest stage of flow	0.045	0.070	0.110
4. Dense brush, high stage	0.080	0.100	0.140
Channels with maintained vegetation and velocities of 2 to 6 ft/s			
Depth of flow up to 0.7 ft			
1. Bermuda grass, Kentucky bluegrass, buffalo grass			
Mowed to 2 in	0.070		0.045
Length 4 to 6 inches	0.090		0.050

Closed Conduits flowing partly full	Minimum	Normal	Maximum
2. Good stand, any grass			
Length approx. 12 in	0.18		0.09
Length approx. 24 in	0.30		0.15
3. Fair stand, any grass			
Length approx. 12 in	0.14		0.08
Length approx. 24 in	0.25		0.13
Depth of flow up to 0.7 - 1.5 ft			
1. Bermuda grass, Kentucky bluegrass, buffalo grass			
Mowed to 2 in	0.05		0.035
Length 4 to 6 in	0.06		0.04
2. Good stand, any grass			
Length approx. 12 in	0.12		0.07
Length approx. 24 in	0.20		0.10
3. Fair stand, any grass			
Length approx. 12 in	0.10		0.16
Length approx. 24 in	0.17		0.09

Natural Streams

Minor streams (top width at flood stage < 100 ft)

Streams on plain

	Minimum	Normal	Maximum
1. Clean, straight, full stage no riffles or deep pools	0.025	**0.030**	0.033
2. Same as above, but with more stones and weeds	0.030	0.035	0.040
3. Clean, winding, some pools and shoals	0.033	0.040	0.045
4. Same as above, but with some weeds and stones	0.035	0.045	0.050
5. Same as above, but lower stages, more ineffective slopes and sections	0.040	0.048	0.055
6. Same as No. 4, more stones	0.045	0.050	0.060
7. Sluggish reaches, weedy, deep pools	0.050	0.070	0.080
8. Very weedy reaches, deep pools or floodways with heavy stand of timber and underbrush	0.075	0.100	0.150

Mountain streams, no vegetation in channel, banks usually steep, trees and brush along banks submerged at high stages

	Minimum	Normal	Maximum
1. Bottom: gravels, cobbles and few boulders	0.030	0.040	0.050
2. Bottom: cobbles with large boulders	0.040	0.050	0.070

Flood plains

Pasture, no brush

	Minimum	Normal	Maximum
1. Short grass	0.025	0.030	0.035
2. High grass	0.030	0.035	0.050

Cultivated areas

Closed Conduits flowing partly full	Minimum	Normal	Maximum
1. No crop	0.020	0.030	0.040
2. Mature row crops	0.025	0.035	0.045
3. Mature field crops	0.030	0.040	0.050
Brush			
1. Scattered brush, heavy weeds	0.035	0.050	0.070
2. Light brush and trees in winter	0.035	0.050	0.060
3. Light brush and trees in summer	0.040	0.060	0.080
4. Medium to dense brush in winter	0.045	0.070	0.110
5. Medium to dense brush in summer	0.070	0.100	0.160
Trees			
1. Dense willows, summer, straight	0.110	0.150	0.200
2. Cleared land with tree stumps, no sprouts	0.030	0.040	0.050
3. Same as above but with a heavy growth of sprouts	0.050	0.060	0.080
4. Heavy stand of timber, a few down trees, little undergrowth, flood stage below branches	0.080	0.100	0.120
5. Same as above, but with flood stage reaching branches	0.100	0.120	0.160
Major streams (top width at flood stage > 100 ft); the n value is less than that for minor streams of similar description because banks offer less effective resistance			
Regular section with no boulders or brush	0.025	-----	0.060
Irregular and rough section	0.035	-----	0.100
Alluvial sandbed channels (no vegetation and data is limited to channels with $D_{50} < 1.0$ mm)			
Tranquil flow, Fr < 1			
1. Plane bed	0.014	-----	0.020
2. Ripples	0.018	-----	0.030
3. Dunes	0.020	-----	0.040
4. Washed out dunes or transition	0.014	-----	0.025
Rapid flow, Fr > 1			
1. Standing waves	0.010	-----	0.015
2. Antidunes	0.012	-----	0.020

Photographic Method: It is the belief of the U.S. Geological Survey that photographs of channels of known resistance together with a summary of the geometric and hydraulic parameters, which define the channel for a specified flow rate, can be useful in estimating the resistance coefficient (Barnes, 1967). It should also be noted that the U.S. Geological Survey maintains a program which trains engineers in the estimation of channel resistance coefficients. The results of this program indicate that trained engineers can estimate resistance coefficients with an accuracy of $\pm$ 15 per cent under most conditions (Barnes, 1967).

Figures 4.2 to 4.14 are a set of photographs and tables from Barnes (1967) for channels with a wide range of resistance coefficients. Note: Most field offices of the U.S. Geological Survey have three-dimensional viewers and slides from Barnes (1967) which give a three-dimensional view of these channels. All the channels described in these figure are considered stable and meet the following criteria:

1. Sites were studied only after a major flood had occurred. Thus, the photographs represent conditions in a channel reach immediately after a flood.

2. The peak discharge in the channel reach specified was determined either by a current meter survey or from an accurate stage-discharge relation.

3. Within the reach, good high-water marks were available to define the water surface profile at peak discharge.

4. In the vicinity of the gauging station at which the peak discharge was determined, the channel was uniform.

5. The peak discharge was confined within the banks of the channel; i.e., flow did not take place in the floodplains.

Drainage Area:	70.3 mi^2 (182 km^2)
Flood Date:	May 11, 1948
Peak Discharge:	768 ft^3/s (21.7 m^3/s)
Estimate Roughness Coefficient:	$n = 0.026$
Channel Description:	Bed and banks composed of clay. Banks clear except for short grass and exposed tree roots in some places.

Section	Area ft^2 (m^2)	Top Width ft (m)	Mean Depth ft (m)	Hydraulic Radius ft (m)	Average Velocity ft/s (m/s)	Distance Between Sections ft (m)	Fall Between Sections ft (m)
1	280 (26.0)	52 (16)	5.4 (1.6)	4.87 (1.48)	2.74 (0.835)	---	---
2	273 (25.4)	51 (16)	5.4 (1.6)	4.82 (1.47)	2.82 (0.860)	257 (78.3)	0.08 (0.02)
3	279 (25.9)	52 (16)	5.4 (1.6)	4.97 (1.51)	2.76 (0.841)	202 (61.6)	0.05 (0.02)

Figure 4.2. (a) Indian Fork below Atwood Dam near New Cumberland, Ohio (Barnes, 1967).

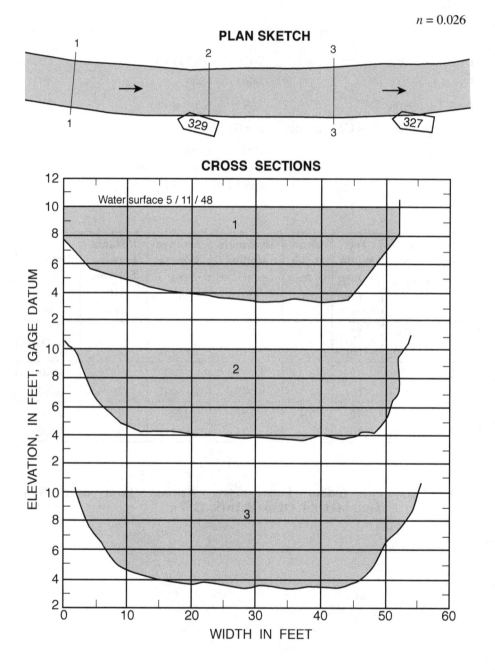

Figure 4.2.   (b) Plan sketch and cross sections, Indian Fork below Atwood Dam, near New Cumberland, Ohio (Barnes, 1967).

Figure 4.2.   (c) Section 329 right bank, Indian Fork below Atwood Dam, near New Cumberland, Ohio (Barnes, 1967). (d) Section 327 from right bank, Indian Fork below Atwood Dam, near New Cumberland, Ohio (Barnes, 1967).

Drainage Area:	174 mi^2 (451 km^2)
Flood Date:	May 2, 1954
Peak Discharge:	1860 ft^3/s (53 m^3/s)
Estimate Roughness Coefficient:	$n = 0.030$
Channel Description:	Bed consists of sand and clay. Banks are generally smooth and free of vegetal growth during floods.

Section	Area ft^2 (m^2)	Top Width ft (m)	Mean Depth ft (m)	Hydraulic Radius ft (m)	Average Velocity ft/s (m/s)	Distance Between Sections ft (m)	Fall Between Sections ft (m)
1	528 (49.1)	72 (22)	7.4 (2.3)	6.7 (2.0)	3.46 (1.05)	---	---
2	502 (46.6)	80 (24)	6.3 (1.9)	5.92 (1.8)	3.71 (1.13)	113 (34.4)	0.09 (0.03)
3	497 (46.2)	69 (21)	7.2 (2.2)	6.6 (2.0)	3.74 (1.14)	110 (33.5)	0.06 (0.02)
4	497 (46.2)	78 (24)	6.4 (2.0)	5.9 (1.8)	3.74 (1.14)	134 (40.8)	0.05 (0.02)

Figure 4.3.   (a) Salt Creek at Roca, Nebraska Ohio (Barnes, 1967).

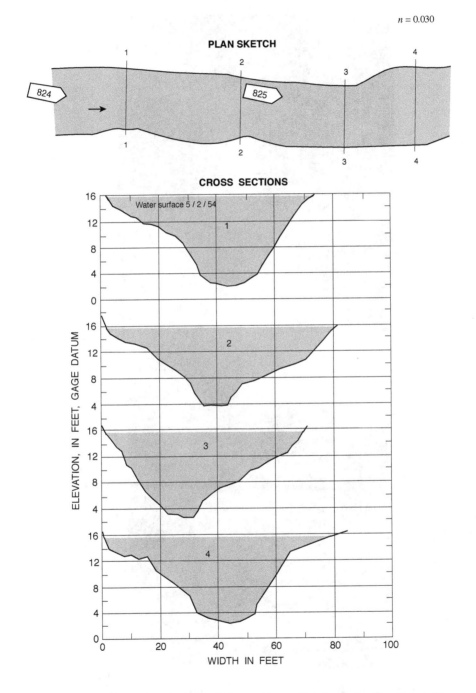

$n = 0.030$

PLAN SKETCH

CROSS SECTIONS

Water surface 5 / 2 / 54

ELEVATION, IN FEET, GAGE DATUM

WIDTH IN FEET

Figure 4.3. **(b) Plan sketch and cross sections, Salt Creek at Roca, Nebraska (Barnes, 1967).**

(c)

(d)

Figure 4.3.   (c) Section 824 downstream, Salt Creek, at Roca, Nebraska (Barnes, 1967). (d) Section 825 downstream, Salt Creek at Roca, Nebraska (Barnes, 1967).

Drainage Area: 3140 mi^2 (8100 km^2)

Flood Date: March 24, 1950; April 3, 1950

Peak Discharge: $1060 \text{ ft}^3/\text{s}$ ($30 \text{ m}^3/\text{s}$); $684 \text{ ft}^3/\text{s}$ ($19 \text{ m}^3/\text{s}$)

Estimate Roughness Coefficient: $n = 0.032$; $s = 0.036$

Channel Description: The bed consists of sand and gravel. The left bank is rock and the right bank is mostly gravel.

Section	Area ft² (m²)	Top Width ft (m)	Mean Depth ft (m)	Hydraulic Radius ft (m)	Average Velocity ft/s (m/s)	Distance Between Sections ft (m)	Fall Between Sections ft (m)
March 24, 1950							
1	312 (29.0)	95 (29)	3.3 (1.0)	3.12 (0.951)	3.40 (1.04)	---	---
2	297 (27.6)	84 (26)	3.5 (1.1)	3.45 (1.05)	3.57 (1.09)	202 (61.6)	0.24 (0.073)
April 3, 1950							
1	235 (21.8)	92 (28)	2.6 (0.79)	2.52 (0.768)	2.91 (0.887)	---	---
2	249 (23.1)	81 (25)	3.1 (0.94)	2.95 (0.899)	2.75 (0.838)	202 (61.6)	0.23 (0.070)

Figure 4.4.   **(a) Rio Chama near Chamita, New Mexico (Barnes, 1967).**

$n = 0.032; 0.036$

PLAN SKETCH

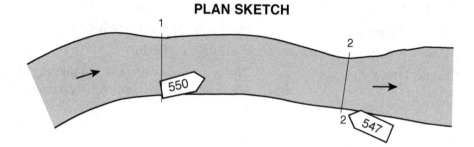

CROSS SECTIONS

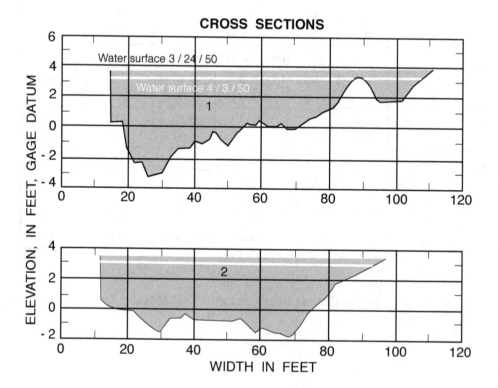

Water surface 3 / 24 / 50

Water surface 4 / 3 / 50

1

2

ELEVATION, IN FEET, GAGE DATUM

WIDTH IN FEET

Figure 4.4.   (b) Plan sketch and cross sections, Rio Chama near Chamita, New Mexico (Barnes, 1967).

(c)

(d)

Figure 4.4.   (c) Section 550 downstream from right bank, Rio Chama near Chamita, New Mexico (Barnes, 1967). (d) Section 547 upstream along right bank, Rio Chama near Chamita, New Mexico (Barnes, 1967).

Drainage Area:	6230 mi² (16,000 km²)
Flood Date:	March 24, 1950
Peak Discharge:	1280 ft³/s (36.2 m³/s)
Estimate Roughness Coefficient:	$n = 0.032$
Channel Description:	Bed and banks consist of smooth cobbles 4-10 in (10-25 cm) in diameter, average diameter 6 in (15 cm). A few boulder are as much as 18 in (15 cm) in diameter.

Section	Area ft² (m²)	Top Width ft (m)	Mean Depth ft (m)	Hydraulic Radius ft (m)	Average Velocity ft/s (m/s)	Distance Between Sections ft (m)	Fall Between Sections ft (m)
1	484 (45.0)	189 (57.6)	2.6 (0.79)	2.55 (0.777)	2.65 (0.808)	---	---
2	408 (37.9)	192 (58.5)	2.1 (0.64)	2.12 (0.646)	3.14 (0.957)	258 (78.6)	0.36 (0.11)
3	384 (35.7)	154 (46.9)	2.6 (0.79)	2.49 (0.759)	3.34 (1.02)	317 (96.6)	0.50 (0.15)
4	449 (41.7)	194 (59.1)	2.3 (0.70)	2.30 (0.701)	2.86 (0.872)	294 (89.6)	0.31 (0.094)
5	420 (39.0)	204 (62.2)	2.1 (0.64)	2.06 (0.628)	3.05 (0.930)	370 (113)	0.63 (0.19)
6	381 (35.4)	207 (63.1)	1.8 (0.55)	1.84 (0.561)	3.36 (1.02)	333 (101)	0.72 (0.22)
7	308 (28.6)	191 (58.2)	1.6 (0.49)	1.61 (0.491)	4.16 (1.27)	314 (95.7)	1.06 (0.32)

Figure 4.5.   (a) Salt River below Stewart Mountain Dam, Arizona (Barnes, 1967).

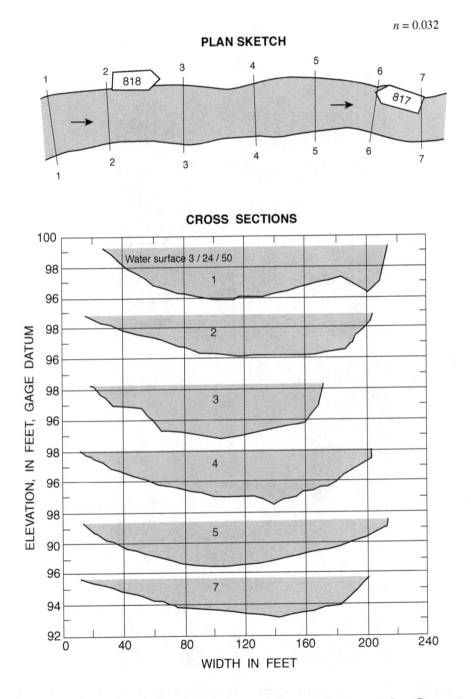

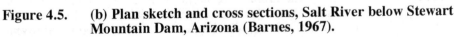

Figure 4.5.   (b) Plan sketch and cross sections, Salt River below Stewart Mountain Dam, Arizona (Barnes, 1967).

(c)

(d)

Figure 4.5.   (c) Upstream along left bank, Salt River below Stewart Mountain Dam, Arizona (Barnes, 1967).   (d) Section 818 downstream along left bank, Salt River below Steward Dam, Arizona (Barnes, 1967).

Drainage Area:	103 mi^2 (267 km^2)
Flood Date:	January 22, February 13, February 14, 1959
Peak Discharge:	2260 ft^3/s (64.0 m^3/s); 1850 ft^3/s (52.4 m^3/s); 515 ft^3/s (14.6 m^3/s)
Estimate Roughness Coefficient:	$n = 0.032$; $n = 0.039$; $n = 0.035$
Channel Description:	Bed is sand and gravel with several fallen trees in the reach. Banks are lined with overhanging trees and underbrush.

Section	Area	Top Width	Mean Depth	Hydraulic Radius	Average Velocity	Distance Between Sections	Fall Between Sections
	ft^2 (m^2)	ft (m)	ft (m)	ft (m)	ft/s (m/s)	ft (m)	ft (m)
January 22, 1959							
7	618 (57.4)	60 (18)	10.3 (3.14)	8.15 (2.48)	3.66 (1.12)	---	---
8	621 (57.7)	66 (20)	9.4 (2.9)	7.77 (2.37)	3.64 (1.11)	316 (96.3)	0.22 (0.067)
9	619 (57.5)	63 (19)	9.8 (3.0)	8.18 (2.49)	3.65 (1.11)	286 (87.2)	0.18 (0.055)
10	610 (56.7)	63 (19)	9.7 (3.0)	7.77 (2.37)	3.70 (1.13)	293 (89.3)	0.14 (0.043)
11	622 (57.8)	66 (20)	9.4 (2.9)	7.89 (2.40)	3.63 (1.11)	203 (61.9	0.17 (0.052)
February 13, 1959							
7	528 (49.1)	57 (17)	9.3 (2.8)	7.37 (2.25)	3.50 (1.07)	---	---
8	531 (49.3)	64 (20)	8.3 (2.5)	6.95 (2.12)	3.48 (1.06)	316 (96.3)	0.19 (0.058
9	532 (49.4)	59 (18)	9.0 (2.7)	7.56 (2.30)	3.48 (1.06)	286 (87.2)	0.20 (0.061)
10	525 (48.8)	61 (19)	8.6 (2.6)	7.00 (2.13)	3.52 (1.07)	293 (89.3)	0.14 (0.043)
11	530	62	8.6	7.10	3.49	203	0.14

Section	Area	Top Width	Mean Depth	Hydraulic Radius	Average Velocity	Distance Between Sections	Fall Between Sections
	ft² (m²)	ft (m)	ft (m)	ft (m)	ft/s (m/s)	ft (m)	ft (m)
January 22, 1959							
	(49.2)	(19)	(2.6)	(2.16)	(1.06)	(61.9)	(0.043)
February 14, 1959							
7	240 (22.3)	51 (16)	4.7 (1.4)	4.04 (1.23)	2.15 (0.655)	---	—
8	209 (19.4)	59 (18)	3.5 (1.1)	3.27 (0.997)	2.46 (0.750)	316 (96.3)	0.16 (0.049)
9	235 (18.6)	53 (16)	4.4 (1.3)	4.04 (1.23)	2.19 (0.668)	286 (87.2)	0.17 (0.052)
10	203 (18.6)	59 (18)	3.4 (1.0)	3.19 (0.972)	2.54 (0.774)	293 (89.3)	0.19 (0.058)
11	219 (20.3)	55 (17)	4.0 (1.2)	3.59 (1.09)	2.35 (0.716)	203 (61.9)	0.15 (0.046)

Figure 4.6.   (a) Etowah River near Dawsonville, Georgia (Barnes, 1967).

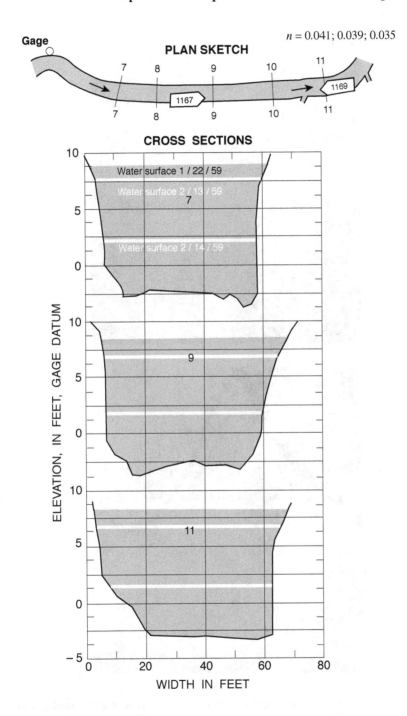

Figure 4.6. (b) Plan sketch and cross sections, Etowah River near Dawsonville, Georgia (Barnes, 1967).

(c)

(d)

Figure 4.6.   (c) Section 1167 downstream from right bank, Etowah River near Dawsonville, Georgia (Barnes, 1967). (d) Section 1168 upstream from right bank, Etowah River near Dawsonville, Georgia (Barnes, 1967).

Drainage Area: 388 mi^2 (1000 km^2)

Flood Date: June 1, 1948

Peak Discharge: 3220 ft^3/s (91.2 m^3/s)

Estimate Roughness Coefficient: $n = 0.041$

Channel Description: The bed is composed of sand, gravel, and angular rocks. The banks are irregular, eroded and have sparse cover of grass and scattered small trees. Although the channel between Sections 1 and 2 is straight, the channel curves sharply above the reach and moderately below it.

Section	Area ft^2 (m^2)	Top Width ft (m)	Mean Depth ft (m)	Hydraulic Radius ft (m)	Average Velocity ft/s (m/s)	Distance Between Sections ft (m)	Fall Between Sections ft (m)
1	818 (76.0)	114 (34.7)	7.2 (2.2)	6.79 (2.07)	3.94 (1.20)	---	---
2	735 (68.3)	102 (31.1)	7.2 (2.2)	6.87 (2.09)	4.38 (1.34)	315 (96.0)	0.38 (0.12)

Figure 4.7.   (a) Bull Creek near Ira, Texas (Barnes, 1967).

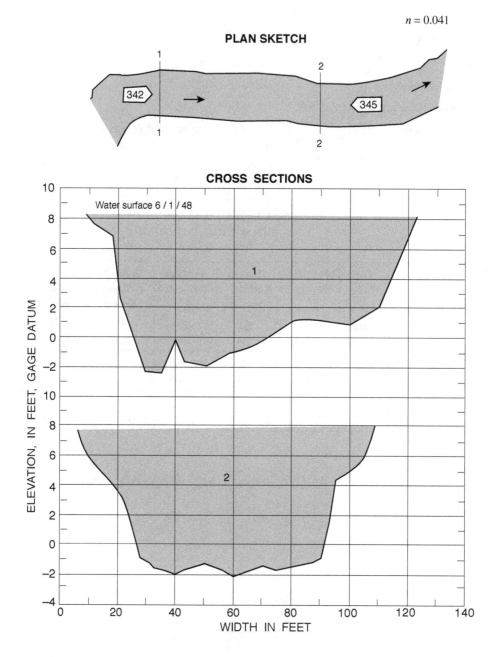

Figure 4.7.   **(b) Plan sketch and cross sections, Bull Creek near Ira, Texas (Barnes, 1967).**

(c)

(d)

Figure 4.7.   (c) Section 342 downstream, Bull Creek near Ira, Texas
Georgia (Barnes, 1967). (d) Section 345 upstream, Bull
Creek near Ira, Texas (Barnes, 1967).

Drainage Area:	24 mi^2 (1000 km^2)
Flood Date:	February 7, 1958
Peak Discharge:	840 ft^3/s (23.8 m^3/s)
Estimate Roughness Coefficient:	$n = 0.045$
Channel Description:	Bed consists of sand and gravel. Both banks are lined with trees above and below the waterline.

Section	Area ft^2 (m^2)	Top Width ft (m)	Mean Depth ft (m)	Hydraulic Radius ft (m)	Average Velocity ft/s (m/s)	Distance Between Sections ft (m)	Fall Between Sections ft (m)
2	215 (20.0)	45 (14)	4.8 (1.5)	4.21 (1.28)	3.93 (1.20)	---	---
3	212 (19.7)	42 (13)	5.0 (1.5)	4.20 (1.28)	3.99 (1.22)	78 (24)	0.21 (0.064)
4	227 (21.1)	50 (15)	4.5 (1.4)	4.02 (1.23)	3.73 (1.14)	124 (37.8)	0.35 (0.11)
5	202 (18.8)	43 (13)	4.7 (1.4)	4.02 (1.23)	4.19 (1.28)	92 (28)	0.21 (0.064)
6	187 (17.4)	41 (12)	4.6 (1.4)	3.74 (1.14)	4.52 (1.38)	71 (22)	0.35 (0.11)
7	200 (18.6)	48 (15)	4.2 (1.3)	3.64 (1.11)	4.23 (1.29)	88 (27)	0.11 (0.033)
8	198 (18.4)	45 (14)	4.4 (1.3)	3.94 (1.20)	4.27 (1.30)	149 (45.4)	0.38 (0.12)

Figure 4.8. (a) Murder Creek near Monticello, Georgia (Barnes, 1967).

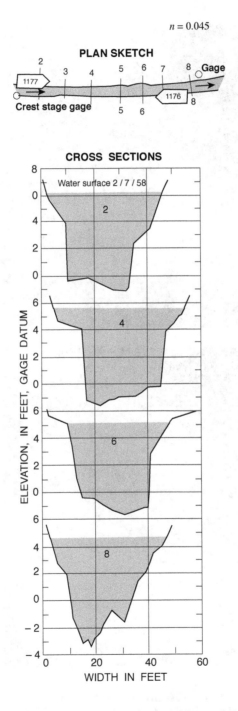

Figure 4.8.   (b) Plan sketch and cross sections, Murder Creek near Monticello, Georgia (Barnes, 1967).

(c)

(d)

Figure 4.8.   (c) Section 1176 upstream from right bank, Murder Creek near Monticello, Georgia (Barnes, 1967). (d) Section 1177 downstream from left bank, Murder Creek near Monticello, Georgia (Barnes, 1967).

Drainage Area:	233 mi^2 (604 km^2)
Flood Date:	June 13; October 7, 1952
Peak Discharge:	1200 ft^3/s (34.0 m^3/s); 64.8 ft^3/s (1.83 m^3/s)
Estimate Roughness Coefficient:	$n = 0.045$; $n = 0.073$
Channel Description:	Bed and banks consist of smooth, rounded rocks as much as 1 ft (0.0305 m) in diameter. Some undergrowth is below the water surface elevations of June 13, 1952.

Section	Area ft^2 (m^2)	Top Width ft (m)	Mean Depth ft (m)	Hydraulic Radius ft (m)	Average Velocity ft/s (m/s)	Distance Between Sections ft (m)	Fall Between Sections ft (m)
			June 13, 1952				
1	184 (17.1)	47 (14)	3.9 (1.2)	3.70 (1.13)	6.52 (1.99)	---	---
3	171 (15.9)	49 (15)	3.5 (1.1)	3.33 (1.02)	7.02 (2.14)	88 (26.8)	0.67 (0.20)
5	173 (16.1)	55 (17)	3.1 (0.94)	3.02 (0.920)	6.95 (2.12)	109 (33.2)	1.04 (0.317)
7	173 (16.1)	48 (15)	3.6 (1.1)	3.43 (1.05)	6.95 (2.12)	117 (35.7)	1.10 (0.335)
9	183 (17.0)	55 (17)	3.3 (1.0)	3.22 (0.982	6.56 (2.00)	116 (35.4)	1.04 (0.317)
			October 7, 1952				
1	36 (3.3)	38 (12)	1.0 (0.30)	0.95 (0.29)	1.79 (0.546)	---	---
3	38 (3.5)	34 (10)	1.1 (0.34)	1.10 (0.335)	1.70 (0.518)	88 (26.8)	0.32 (0.098)
5	34 (3.2)	32 (9.8)	1.1 (0.34)	0.82 (0.25)	1.90 (0.579)	109 (33.2)	0.84 (0.26)
7	31 (3.2)	39 (12)	0.9 (0.27)	0.86 (0.26)	1.91 (0.582)	117 (35.7)	1.28 (0.390)
9	31 (2.9)	41 (12)	0.8 (0.24)	0.76 (0.23)	2.08 (0.634)	116 (35.4)	1.12 (0.341)

Figure 4.9.   (a) Provo River near Hailstone, Utah (Barnes, 1967).

$n = 0.045; 0.073$

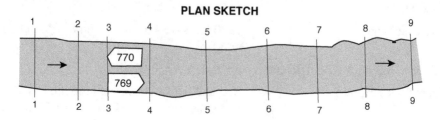

PLAN SKETCH

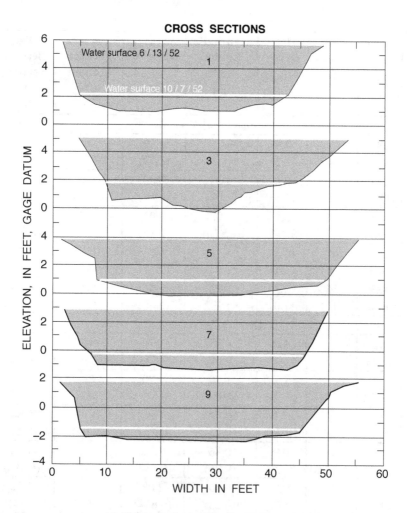

Figure 4.9.   (b) Plan sketch and cross sections, Provo River near Hailstone, Utah (Barnes, 1967).

(c)

(d)

Figure 4.9.   (c) Section 1176 upstream from right bank, Murder Creek near Monticello, Georgia (Barnes, 1967). (d) Section 769 downstream, Provo River near Hailstone, Utah (Barnes, 1967).

Drainage Area:	399 mi² (1030 km²)
Flood Date:	May 26, 1958
Peak Discharge:	1380 ft³/s (39.1 m³/s)
Estimate Roughness Coefficient:	$n = 0.050$
Channel Description:	Bed and banks are composed of angular boulders as much as 2.0 ft (0.61 m) in diameter.

Section	Area ft² (m²)	Top Width ft (m)	Mean Depth ft (m)	Hydraulic Radius ft (m)	Average Velocity ft/s (m/s)	Distance Between Sections ft (m)	Fall Between Sections ft (m)
12	200 (18.6)	43 (13)	4.6 (1.4)	4.22 (1.29)	6.90 (2.10)	---	---
13	206 (19.1)	50 (15)	4.1 (1.2)	3.83 (1.17)	6.70 (2.04)	47 (14)	0.85 (0.26)
14	183 (17.0)	52 (16)	3.5 (1.1)	3.29 (1.00)	7.54 (2.30)	39 (12)	0.65 (0.20)
15	184 (17.1)	51 (16)	3.6 (1.1)	3.36 (1.02)	7.50 (2.29)	46 (14)	0.65 (0.20)
16	184 (17.1)	55 (17)	3.3 (1.0)	2.69 (0.820)	7.50 (2.29)	32 (9.8)	0.60 (0.18)

Figure 4.10.   (a) Clear Creek near Golden, Colorado (Barnes, 1967).

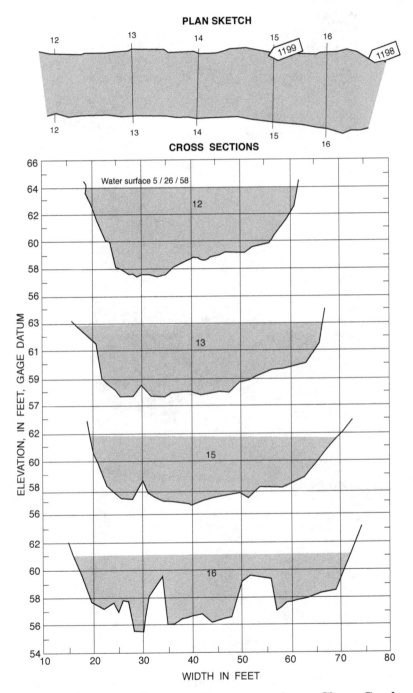

Figure 4.10.   (b) Plan sketch and cross sections, Clear Creek near Golden, Colorado (Barnes, 1967).

(c)

(d)

Figure 4.10.   (c) Section 1199 upstream from left bank, Clear Creek near Golden, Colorado (Barnes, 1967). (d) Section 1198 upstream, Clear Creek near Golden Colorado Utah (Barnes, 1967).

Drainage Area:	528 mi^2 (1370 km^2)
Flood Date:	January 24, 1951; January 25, 1951
Peak Discharge:	3840 ft^3/s (109 m^3/s); 1830 ft^3/s (51.8 m^3/s);
Estimate Roughness Coefficient:	$n = 0.053$; $n = 0.079$
Channel Description:	Bed is composed of large, angular boulders. Banks consist of exposed rock, boulders, and some trees.

Section	Area ft^2 (m^2)	Top Width ft (m)	Mean Depth ft (m)	Hydraulic Radius ft (m)	Average Velocity ft/s (m/s)	Distance Between Sections ft (m)	Fall Between Sections ft (m)
January 24, 1951							
1	401 (37.2)	81 (25)	5.0 (1.5)	4.27 (1.30)	9.59 (2.92)	---	---
2	384 (35.7)	65 (20)	5.9 (1.8)	5.02 (1.53)	10.00 (3.05)	102 (31.1)	1.80 (0.549)
3	295 (27.4)	45 (14)	6.6 (2.0)	5.27 (1.61)	13.02 (3.97)	62 (18.9)	3.10 (0.945)
January 25, 1951							
1	271 (25.2)	76 (23)	3.6 (1.1)	3.09 (0.942)	6.75 (2.06)	---	---
2	236 (21.9)	51 (16)	4.6 (1.4)	3.88 (1.18)	7.75 (2.36)	102 (31.1)	2.75 (0.838)
3	211 (19.6)	38 (12)	5.6 (1.7)	4.45 (1.36)	8.67 (2.64)	62 (18.9)	2.55 (0.777)

Figure 4.11.   (a) Cache Creek near Lower Lake, California (Barnes, 1967).

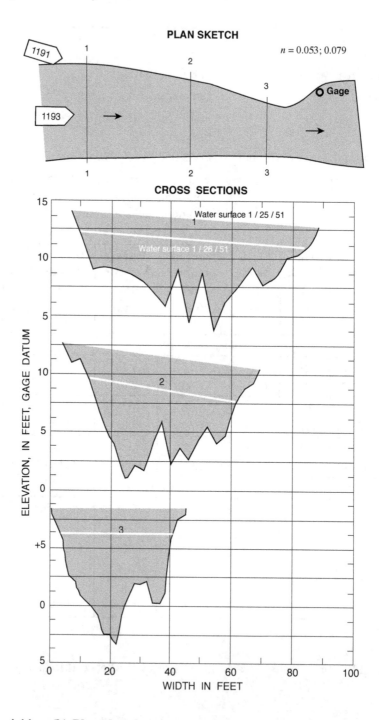

Figure 4.11. (b) **Plan sketch and cross sections, Cache Creek near Lower Lake, California (Barnes, 1967).**

(c)

(d)

Figure 4.11.   **(c) Section 1191 downstream from left bank, Cache Creek near Lower Lake, California (Barnes, 1967). (d) Section 1193 downstream, Cache Creek near Lower Lake, California (Barnes, 1967).**

Drainage Area:	89,700 mi² (232,000 km²)
Flood Date:	May 22, 1949
Peak Discharge:	406,000 ft³/s (11,500 m³/s)
Estimate Roughness Coefficient:	$n = 0.024$
Channel Description:	Bed consists of slime-covered cobbles and gravels. The straight and steep left bank is composed of cemented cobbles and gravel. The gently sloping right bank consists of cobbles set in gravel and is free of vegetation.

Section	Area ft² (m²)	Top Width ft (m)	Mean Depth ft (m)	Hydraulic Radius ft (m)	Average Velocity ft/s (m/s)	Distance Between Sections ft (m)	Fall Between Sections ft (m)
1	47,100 (4380)	1800 (549)	26.2 (7.99)	26.16 (7.97)	8.65 (2.64)	---	---
2	49,000 (4550)	1650 (503)	29.7 (9.05)	29.56 (9.01)	8.28 (2.52)	2500 (762)	0.48 (0.15)
3	49,600 (4610)	1760 (536)	28.2 (8.60)	28.10 (8.56)	8.17 (2.49)	2500 (762	0.49 (0.15)

Figure 4.12.   (a) Columbia River at Vernita Washington (Barnes, 1967).

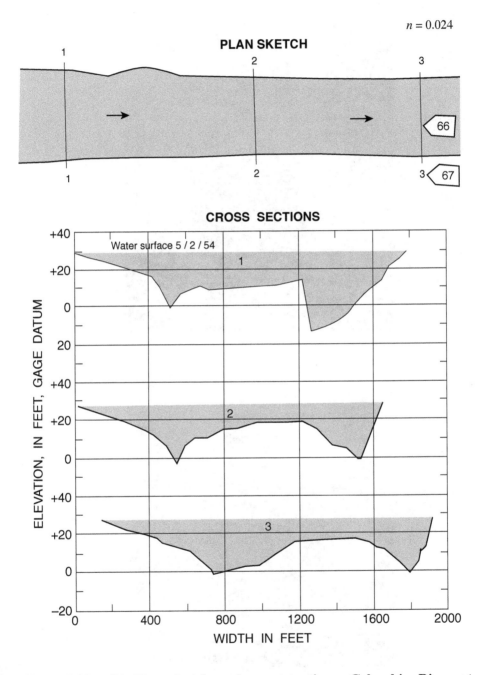

Figure 4.12.   (b) Plan sketch and cross sections, Columbia River at Vernita, Washington (Barnes, 1967).

(c)

(d)

Figure 4.12.   **(c) Section 66 upstream along right bank, Columbia River at Vernita, Washington (Barnes, 1967).   (d) Section 67 upstream from top of bank, Columbia River at Vernita, Washington (Barnes, 1967).**

Drainage Area:	181 mi^2 (469 km^2)
Flood Date:	May 17, 1950
Peak Discharge:	1950 ft^3/s (55.2 m^3/s)
Estimate Roughness Coefficient:	$n = 0.065$
Channel Description:	Fairly straight channel is composed of boulders with trees along top of banks.

Section	Area ft^2 (m^2)	Top Width ft (m)	Mean Depth ft (m)	Hydraulic Radius ft (m)	Average Velocity ft/s (m/s)	Distance Between Sections ft (m)	Fall Between Sections ft (m)
1	308 (28.6)	78 (24)	4.0 (1.2)	3.90 (1.19)	6.33 (1.93)	---	---
2	263 (24.4)	64 (20)	4.1 (1.2)	3.98 (1.21)	7.41 (2.26)	200 (61)	3.40 (1.04)
3	327 (30.4)	63 (19)	4.92 (1.5)	4.68 (1.43)	6.31 (1.92)	40 (12)	0.50 (0.15)
4	327 (30.4)	78 (24)	4.2 (1.3)	4.09 (1.25)	5.96 (1.82)	180 (55)	1.55 (0.47)

Figure 4.13. (a) **Merced River at Happy Isles Bridge, near Yosemite, California (Barnes, 1967).**

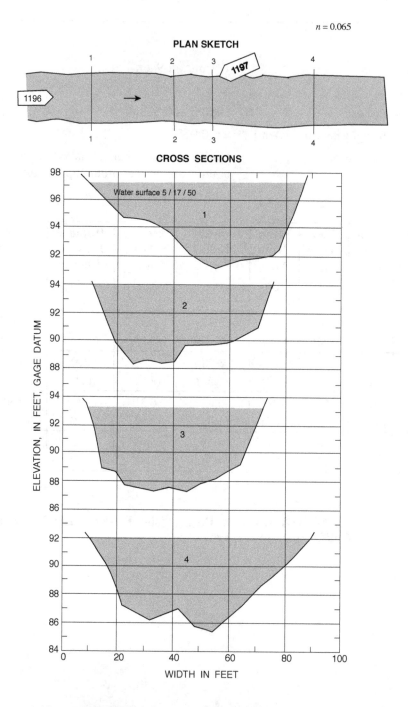

Figure 4.13.   (b) Plan sketch and cross sections, Merced River at Happy Isles Bridge, near Yosemite, California (Barnes, 1967).

(c)

(d)

Figure 4.13.   (c) Section 1196 downstream, Merced River at Happy Isles Bridge, near Yosemite, California (Barnes, 1967).   (d) Section 1197 upstream from right bank, Merced River at Happy Isles Bridge, near Yosemite, California (Barnes, 1967).

Drainage Area: 64.0 mi^2 (469 km^2)

Flood Date: February 14, 1950

Peak Discharge: 1480 ft^3/s (41.9 m^3/s)

Estimate Roughness Coefficient: $n = 0.070$

Channel Description: Bed is fine sand and silt.   Banks are
                                        irregular with fairly heavy growth of 2 to
                                        8 in (5 to 20 cm) trees on the banks above
                                        low water particularly on the left bank.
                                        Reach sections 1, 2, 5, 6, and 7 used to
                                        determine   the   roughness   coefficient.
                                        Bridge abutments form the constriction at
                                        Section 3.

Section	Area ft^2 (m^2)	Top Width ft (m)	Mean Depth ft (m)	Hydraulic Radius ft (m)	Average Velocity ft/s (m/s)	Distance Between Sections ft (m)	Fall Between Sections ft (m)
1	888 (82.5)	115 (35.0)	7.7 (2.3)	7.10 (2.16)	1.67 (0.509)	---	---
2	830 (77.1)	122 (37.2)	6.8 (2.1)	6.48 (1.98)	1.78 (0.543)	90 (27)	0.05 (0.015)
3	591 (54.9)	52 (16)	11.4 (3.47)	8.23 (2.51)	2.50 (0.762)	95 (29)	0.10 (0.031)
4	837 (77.8)	116 (35.4)	7.2 (2.2)	6.80 (2.07)	1.77 (0.540)	45 (14)	0.02 (0.0061)
5	818 (76.0)	94 (29)	8.7 (2.7)	8.06 (2.46)	1.81 (0.552)	79 (24)	0.05 (0.015)
6	791 (73.5)	107 (32.6)	7.4 (2.3)	7.00 (2.13)	1.87 (0.570)	156 (47.5)	0.10 (0.031)
7	854 (79.3)	115 (35.1)	7.4 (2.3)	7.00 (2.13)	1.73 (0.527)	147 (44.8)	0.07 (0.021)

Figure 4.14.   (a) Pond Creek near Louisville, Kentucky (Barnes, 1967).

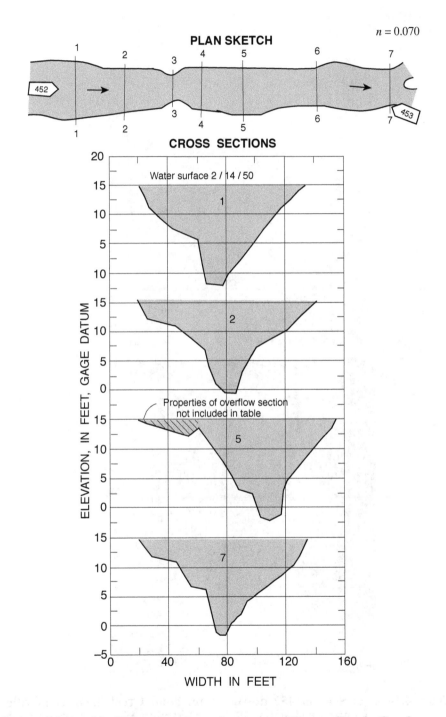

Figure 4.14.   (b) Plan sketch and cross sections, Pond Creek near Louisville, Kentucky (Barnes, 1967).

(c)

(d)

Figure 4.14.   (c) Section 452 downstream, Pond Creek near Louisville, Kentucky (Barnes, 1967). (d) Section 453 downstream from right bank, Pond Creek near Louisville, Kentucky (Barnes, 1967).

The Manning resistance coefficients presented in Figs. 4.2 to 4.14 were estimated from the measured discharges, water surface profiles, and reach properties, as defined by more than two cross sections in each case. For a case in which there were N cross sections

$$n = \frac{\varphi}{Q} \left[\frac{\left(h + h_u\right)_1 - \left(h + h_u\right)_N - \sum_{j=2}^{N}\left(k\,\Delta h_u\right)_{j-1,j}}{\sum_{j=2}^{N}\dfrac{L_{j-1,j}}{\left(AR^{2/3}\right)_{j-1}\left(AR^{2/3}\right)_{j}}} \right]^{1/2} \qquad (4.3.12)$$

where with reference to Fig. 4.15, h = elevation of the water surface at the section with respect to a datum common to all sections: $(\Delta h_u)_{j-1,j}$ = change in velocity head between sections j-1 and j; $h_u = \alpha u^2 / 2g =$

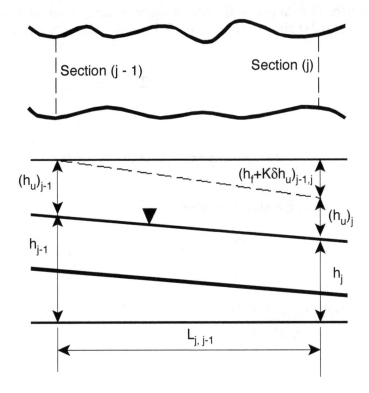

Figure 4.15. Definition sketch.

velocity head at a section; k = a coefficient accounting for the nonuniformity of the channel ($k = 0$ for a uniform reach; $k = 0.5$ for a nonuniform reach); h_f = energy loss in a reach due to boundary friction; ϕ = a coefficient which accounts for the system of units used ($\phi = 1.49$ for the English system and $\phi = 1.00$ for SI system); and $L_{j-1,j}$ = distance between sections j - 1 and j.

Velocity Measurement: From a theoretical viewpoint, a value of the resistance coefficient can be estimated from velocity measurements since the velocity profile depends on the perimeter roughness height. In hydraulically rough flows, the vertical velocity distribution can be approximated by

$$u = 5.75u_* \log \frac{30y}{k_s} \qquad (4.3.13)$$

where y = distance from the bottom boundary and k_s = roughness height. Let $u_{0.2}$ = velocity at two-tenths of the depth; i.e., at $0.8D$ above the bottom where D = depth of flow, and $u_{0.8}$ = velocity at eight-tenths the depth; i.e., at $0.2D$ above the bottom. Substituting these variables in Eq. (4.3.13)

$$u_{0.2} = 5.75u_* \log \frac{24D}{k_s}$$

and

$$u_{0.8} = 5.75u_* \log \frac{6D}{k_s}$$

Estimating u_* from the above equations

$$\log \frac{D}{k_s} = \frac{0.778\psi - 1.381}{1 - \psi} \qquad (4.3.14)$$

where $\psi = u_{0.2}/u_{0.8}$. Substituting Eq. (4.3.14) in Eq. (1.4.33) where it is assumed that $R \cong D$ yields

$$\frac{u}{u_*} = \frac{1.78(\psi + 0.95)}{\psi - 1} \qquad (4.3.15)$$

Recall that

$$\frac{u}{u_*} = \frac{C}{\sqrt{g}}$$

and using this relation and Eq. (4.2.7) yields

$$\frac{u}{u_*} = \frac{D^{1/6}}{3.81} \qquad (4.3.16)$$

Equating the right-hand sides of Eqs. (4.3.15) and (4.3.16) and solving for n

$$n = \frac{(\psi - 1)D^{1/6}}{6.78(\psi + 0.95)} \qquad (4.3.17)$$

Equation (4.3.17) estimates values of n for fully rough flows in a wide channel which have a logarithmic vertical velocity distribution. In the case of actual stream, D may be taken as the mean depth. Chow (1959) hypothesized this methodology and noted that it should be used with caution since it had not been verified. Although French and McCutcheon (1977) found that Eq. (4.3.17) provided reasonable estimates of n in a study on the Cumberland River, Tennessee, additional verification of the methodology is still required.

Empirical Methods: A number of empirical methods have been advanced for estimating n. Perhaps, the best known of these methods is the one proposed by Strickler in 1923; see, for example, Simons and Senturk (1976), which hypothesizes that

$$n = 0.047d^{1/6} \qquad (4.3.18)$$

where d = diameter in millimeters of the uniform sand particles, pasted to the sides and bottom of the flume used by Strickler in his experiments. Simons and Senturk (1976) asserted that, because of the experimental design used by Strickler, Eq. (4.3.18) does not apply to flows over mobile beds. Note, there is significant disagreement among authors regarding the circumstances of the original Strickler experiments, the value of the coefficient in Eq. (4.3.18), the definition of the variable d, and the units associated with d. For example,

1.   Henderson (1966) claimed that Strickler's research was based on streams with gravel beds, not a flume, and that d was the median size of the bed material. The equation given by Henderson (1966) and attributed to Strickler was

$$n = 0.034d^{1/6} \qquad (4.3.19)$$

where the units of d were not defined.

2.   Raudkivi (1976) stated that the Strickler equation was

$$n = 0.042d^{1/6} \qquad (4.3.20)$$

where d is measured in meters or

$$n = 0.013d_{65}^{1/6} \qquad (4.3.21)$$

where d_{65} = diameter of the bed material in millimeters, such that 65 percent of the material by weight is smaller.   If the dimensions of Eq. (4.3.21) are given in feet, then

$$n = 0.034d_{65}^{1/6} \qquad (4.3.22)$$

where the coefficient has the same numerical value as that given by Henderson, Eq. (4.3.19).   Raudkivi (1976) further noted that Eqs. (4.3.20) to (4.3.22) are useful in selecting roughness heights in fixed bed hydraulic models.

3.   Garde and Raju (1978) state that Strickler analyzed data from various streams in Switzerland which had beds formed of coarse materials and which were free from bed undulations.   The equation given by these authors and attributed to Strickler was

$$n = 0.039d_{50}^{1/6} \qquad (4.3.23)$$

where d_{50} = diameter of the bed material, in feet, such that 50 percent of the material by weight is smaller.

4.   Subramanya (1982) gave the Strickler equation as

$$n = 0.047d_{50}^{1/6}$$ (4.3.24)

where d_{50} = diameter of the bed material, in meters, such that 50 percent of the material by weight is smaller. It can be easily shown that Eqs. (4.3.23) and (4.3.24) are equivalent; and therefore, there is no difference between the equations given by Garde and Raju (1978) and Subramanya (1982).

The above discussion is presented to highlight that there are several different interpretations of an important concept published in what is now a rather obscure publication.

For mixtures of bed materials with a significant proportion of coarse grained sizes, Meyer-Peter and Muller (1948) suggested the following equation

$$n = 0.038d_{90}^{1/6}$$ (4.3.25)

where d_{90} = bed size in meters such that 90 percent of material by weight is smaller. In field experiments involving canals paved with cobbles, Lane and Carlson (1953) determined that

$$n = 0.026d_{75}^{1/6}$$ (4.3.26)

where d_{75} = diameter of the bed material in inches such that 75 percent of the material is smaller by weight.

Limerinos (1970) related n values to the hydraulic radius and bed particle size, based on samples from eleven stream channels having bed materials ranging from small gravel to medium sized boulders. The Limerinos equation is

$$n = \frac{0.0926R^{1/6}}{1.16 + 2.0\log\left(\dfrac{R}{d_{84}}\right)}$$ (4.3.27)

where R = hydraulic radius in feet [data range was 1.0 to 6.0 ft (0.30 to 1.8 m)], and d_{84} = particle diameter in feet that equals or exceeds 84 percent of the particles (data range was 1.5 mm to 250 mm). Limerinos selected reaches that had a minimum amount of roughness, other than that resulting from the bed material. Thus, Eq. (4.3.27) provides an estimate of the base value of n.

Jarrett (1984) developed an equation for high gradient stream ($S > 0.002$) by regressing 75 data sets derived from 21 different streams. The Jarrett equation is

$$n = \frac{0.39S^{0.38}}{R^{0.16}} \tag{4.3.28}$$

where S = friction slope and R = hydraulic radius in feet. Jarrett (1984) noted the following limitations for Eq. (4.3.28)

1.   The equation is applicable to natural main channels, having stable bed and bank materials (gravels, cobbles, and boulders), without backwater.

2.   The equation can be used for slopes from 0.002 to 0.04 and for hydraulic radii from 0.5 to 7.0 ft (0.15 to 2.1 m). Results of the regression analysis indicated that for hydraulic radii greater than 7.0 ft (2.1 m) n did not vary significantly with depth; and therefore, extrapolating to larger flows, should not result in too much error if the bed and bank material remain stable.

3.   The hydraulic radius does not include the wetter perimeter of the bed particles.

4.   The equation is applicable to streams having relatively small amounts of suspended sediment.

EXAMPLE 4.2

For the channel defined in Example 4.1, determine the minimum slope which is required if Manning's n is to be assumed constant.

Solution

From Example 4.1

$$n = 0.13$$

Depth of flow $y_N = 8.5$ ft (2.6 m)

Bottom width $b = 10$ ft (3.0 m)

Side slope 1.5:1

Then, from Table 1.1

$$A = (b + zy)y = [10 + 1.5(8.5)] = 194\,\text{ft}^2 \left(18.0\,\text{m}^2\right)$$

$$P = b + 2y\sqrt{1 + z^2} = 10 + 2(8.5)\sqrt{3.25} = 40.6\,\text{ft} \left(12.4\,\text{m}\right)$$

$$R = \frac{A}{P} = \frac{194}{40.6} = 4.8\,\text{ft} \left(1.5\,\text{m}\right)$$

From Eq. (4.3.6)

$$S = \left(\frac{1.9 \times 10^{-13}}{n^6 \sqrt{R}}\right)^2 = \left(\frac{1.9 \times 10^{-13}}{0.13^6 \sqrt{4.8}}\right)^2 = 3.2 \times 10^{-16}$$

Therefore, if the slope of the channel described in Example 4.1 exceeds 3.2×10^{-16}, the assumption of a constant value of n is valid.

Vegetated Channels and Flood Plains: Vegetated channels and flood plains present unique challenges from the viewpoint of estimating roughness. For example, in grass-lined channels and overbank regions, the traditional approach assumed that n was a function of vegetal retardance and VR (Coyle, 1975). However, there are approaches more firmly based on the principles of fluid mechanics and the mechanics of materials (Kouwen and Li, 1980 and Kouwen 1988). Data also exist that suggest that in such channels, flow duration is not a factor as long as the vegetal elements are not destroyed or removed. Further, inundation times, and/or hydraulic stresses, or both, that are sufficient, to damage vegetation have been found to reduce the resistance to flow (Temple, 1991).

Roughness values for channels and flood plains should be determined separately since the composition, physical shape, and vegetation of a flood plain is generally quite different from those of a channel. One approach to estimating a value of n for the flood plain results from modifying the procedure developed by Cowan, (1956) for estimating n for channels or

$$n = \left(n_b + n_1 + n_2 + n_3 + n_4\right)m \qquad (4.3.29)$$

where

n_b	=	a base value of n for the natural bare soil of the flood plain,
n_1	=	a correction factor for the effect of surface irregularities of the flood plain,
n_2	=	a value for variations in shape and size of the flood plain cross section, assumed equal to 0.0,
n_3	=	a value for obstructions on the flood plain,
n_4	=	a value for vegetation on the flood plain, and
m	=	a correction factor for sinuosity of the flood plain, assumed equal to 1.0.

In using Eq. (4.3.29), a base value of n_b is selected for the natural bare soil of the flood plain and adding adjustment factors to account for surface irregularity, obstructions, and vegetation. The justification, for assuming n_2 is 0.0, is that the cross section of the flood plain is subdivided where abrupt changes occur in the shape of the flood plain. The adjustment for meandering, m, is assumed 1.0 because there will be very little flow in the meandering channel when there is flow in the flood plain.

Irregularity in the surface of a flood plain causes an increase in the resistance to flow. Adjustment values for surface irregularities are summarized in Table 4.9. In this table, the terminology slough refers to a stagnant swamp, marsh, bog, or pond; and a hummock is a low mound or ridge of earth above the level of an adjacent depression. Shallow depths of water, accompanied by an irregular ground surface in pastures or brush areas accompanied by deep furrows perpendicular to the flow in cultivated fields, can increase the value of n by as much as 0.02 (Arcement and Schneider, 1989).

Obstructions in the flood plain such as debris deposits, stumps, exposed roots, logs, or isolated boulders increase the resistance to flow, but their effect cannot be measure directly. In Table 4.9, adjustment values for different percentages of obstruction are summarized.

Visual observation, judgement, and experience are used in selecting an adjustment factor for the effects of vegetation, Table 4.9. Although

Table 4.9.   Adjustment values for factors that affect flow resistance in flood plains, adapted from Aldridge and Garrett (1973).

Adjustment	Flood plain conditions	n value adjustment	Example
Degree of irregularity (n_1)	Smooth	0.000	Compares to the smoothest, flattest, flood plain attainable in a given bed material
	Minor	0.001 - 0.005	Is a flood plain slightly irregular in shape. A few rises and dips or sloughs may be visible on the flood plain.
	Moderate	0.006 - 0.010	Has more rises and dips. Sloughs and hummocks may occur.
	Severe	0.011 - 0.020	Flood plain very irregular in shape. Many rises and dips or sloughs are visible. Irregular ground surfaces in pastureland and furrows perpendicular to the flow are also included.
Effect of obstructions (n_3)	Negligible	0.000 - 0.004	Few scattered obstructions, which include debris deposits, stumps, exposed roots, logs, or isolated boulders, occupy less than 5 percent of the cross sectional area.
	Minor	0.005 - 0.019	Obstructions occupy less than 15 percent of the cross sectional area.
	Appreciable	0.020 - 0.030	Obstructions occupy from 15 to 50 percent of the cross sectional area.
Amount of vegetation (n_4)	Small	0.001 - 0.010	Dense growth of flexible turf grass, such as Bermuda, or weeds growing where the average depth of flow is at least two times the height of the vegetation, or supple tree seedlings such as willow, cottonwood, arrowweed, or saltcedar growing where the average depth of flow is at least three time the height of the vegetation.
	Medium	0.011 - 0.025	Turf grass growing where the average depth of flow is from one to two times the height of the vegetation, or moderately dense stemmy grass, weeds, or tree seedlings growing where the average depth of flow is from two to three times the height of the vegetation; brushy moderately dense vegetation, similar to 1- ro 2-year old willow trees in the dormant season.
	Large	0.025 - 0.050	Turf grass growing where the average depth of flow is about equal to the height of the vegetation, 8- to 10-year old willow or cottonwood trees intergrown with some weeds and brush (none of the vegetation is in foliage)

Adjustment	Flood plain conditions	*n* value adjustment	Example
			where the hydraulic radius exceeds 2 ft, or mature row crops such as small vegetables, or mature field crops where depth of flow is at least twice the height of the vegetation.
	Very large	0.050 - 0.100	Turf grass growing where the average depth of flow is less than half the height of the vegetation, or moderate to dense brush, or heavy stand of timber with few down trees and little undergrowth where the depth of flow is below the branches, or mature filed crops where the depth of flow is less than the height of the vegetation.
	Extreme	0.100 - 0.200	Dense bushy willow, mesquite, and saltcedar (all vegetation in full foliage), or heavy stand of timber, few down trees, depth of flow reaching branches.

measuring the area occupied by tree trunks and large diameter vegetation is relatively easy, measuring the area occupied by low vines, briars, or crops is more difficult. Petryk and Bosmajian (1975) presented a relation for *n* in vegetated channels base on a balance of drag and gravitational forces or

$$n = \varphi R^{2/3} \left[\frac{C_d (Veg)_d}{2g} \right]^{1/2} \qquad (4.3.30)$$

where C_d = a coefficient accounting for the drag characteristics of the vegetation and $(Veg)_d$ = vegetation density. Flippin-Dudley (1997) developed a rapid and objective procedure using a horizontal point frame to measure $(Veg)_d$. Equation (4.3.30) has limited utility because there is limited information regarding C_d (Flippin-Dudley *et al.*, 1997).

There are other approaches to estimating *n* in flood plains. For example, Arcement and Schneider (1989) provide a series of heavily vegetated flood plain photographs with computed values of *n*. Ree and Crow (1977) conducted experiments to determine flow resistance in earthen channels planted with certain crops and grasses. These authors provide photographs and brief descriptions of the vegetation and a tabulation of the hydraulic elements. Aldridge and Garrett (1973) provide photographs of selected channels and flood plains in Arizona, having known roughness coefficients.

Kouwen and Fathi-Moghadam (2000) noted that while the reported Manning's n values cover a reasonable range of application, there is no quantitative and reliable guideline to assist the engineer in selecting an appropriate Manning's n for a particular condition of flow and vegetation. Based on previous work and experiments with coniferous trees, Kouwen and Fathi-Moghadam demonstrated that the Darcy-Weisbach friction factor, f, can be estimated for flow inside a stand of coniferous trees, on a flood plain or in vegetated zones of rivers, by the following equations derived by linear regression of the experimental data

$$f = 4.06 \left(\frac{u}{\sqrt{\dfrac{\varsigma E}{\rho}}} \right)^{-0.46} \tag{4.3.31}$$

when the tree is just submerged and

$$f = 4.06 \left(\frac{u}{\sqrt{\dfrac{\varepsilon E}{\rho}}} \right)^{-0.46} \left(\frac{y_N}{y_T} \right) \tag{4.3.32}$$

when the tree is partially submerged.; that is, $y_N \leq y_T$. In the above equations,

f	=	Darcy-Weisbach friction factor,
u	=	mean flow velocity,
$\in E$	=	vegetation index (Table 4.10),
ρ	=	fluid density,
y_N	=	normal depth of flow; and
y_T	=	average height of vegetation in the canopy.

Since linear regression was used to derive Eqs. (4.3.31) and (4.3.32), it is pertinent to note that the height of the trees tested ranged from approximately 1 ft (0.3 m) to 13 ft (4 m) and the velocities used in testing ranged up to 91 ft/s (100 km/hr) - small tree sizes were tested in flume while the larger sizes were tested using a specially instrumented truck. Using the relationship between n and f, Kouwen and Fathi-Moghadam (2000) then demonstrated that

$$n = 0.228 \left(\frac{u}{\sqrt{\frac{\xi E}{\rho}}} \right)^{-0.23} \left(\frac{y_N}{y_T} \right)^{0.5} \quad \text{for } y_N \leq y_T \qquad (4.3.33)$$

Equation (4.3.33) shows that the value of Manning's n increases with the square root of the relative depth of submergence for the non-submerged condition; and therefore, for a constant velocity, there is an approximate twofold increase of Manning's n from a shallow flow ($y_N/y_T = 0.25$) to the nearly submerged condition ($y_N/y_T = 1.0$).

Equation (4.3.33) was developed based on the just covered density and for the situation where the trees just cover the ground in plan view. Trees can be spaced so not all the ground is either covered by foliage or crowded because of overlapping foliage from adjoining trees. Raupach *et al* (1980) noted the pattern or distribution of the trees does not have a significant effect on the friction factor and this was confirmed by Fathi-Moghadam (1996). Therefore, the application of Eq. (4.3.33) to a real canopy with a different size, maturity, density and species of trees requires only a correction for vegetation densities either lower or higher than the just-covered density. The correction equation proposed by Kouwen and Fathi-Moghadam (2000) is

$$n_c = n \left(\frac{a}{a_T} \right)^{0.5} \qquad (4.3.34)$$

where n_c = corrected Manning's n, n = estimated Manning's n, a = total plan (top) view area covered by trees, and a_T = total plan (top) view area. For a canopy with a density higher than the just-covered density, the covered ratio is greater than one.

The capabilities and limitations of Eqs. (4.3.33) and (4.3.34), noted by Kouwen and Fathi-Moghadam (2000), are as follows. First, these equations are capable of estimating the resistance for any canopy height, density, and any relative depth of submergence ($y_N/y_T \leq 1.0$), where the resistance is dominated by the coniferous vegetation rather than the channel surface friction. Kouwen and Fathi-Moghadam (2000) did note that the trees tested represented a wide range of tree flexibility ranging from cedar with the lowest stiffness up to Austrian pine, with the highest. Thus, the application of Eqs. (4.3.33) and (4.3.34), for other tree species, may have some error; such applications would be valid. Second, application of Eqs. (4.3.33) and (4.3.34) is limited to subcritical, turbulent flow, through foliage and the flexible parts of woody species of trees. Fourth, trees with bare trunks were not included in the study; however, the drag on trunks that support branches of a tree were implicitly included, because all trees tested consisted of trunks and branches.

EXAMPLE 4.3

A wide, flat floodplain is 80% covered by equal amounts of cedar and spruce trees having an average height of 3 m (9.8 ft). If the depth of flow is 2 m and the velocity of flow is 0.8 m/s, estimate the value of Manning's n.

Solution

If the floodplain were totally covered in cedar trees and $\rho = 998.2$ kg/m^3, then

$$n = 0.228 \left(\frac{u}{\sqrt{\dfrac{\xi E}{\rho}}} \right)^{-0.23} = 0.228 \left(\frac{0.8}{\sqrt{\dfrac{2.07}{998.2}}} \right)^{-0.23} = 0.118$$

If the floodplain were totally cover in spruce trees, then

$$n = 0.228 \left(\frac{u}{\sqrt{\dfrac{\xi E}{\rho}}} \right)^{-0.23} = 0.228 \left(\frac{0.8}{\sqrt{\dfrac{3.36}{998.2}}} \right)^{-0.23} = 0.125$$

Therefore, for a just-covered, and just-submerged floodplain covered with and equal mixture of cedar and spruce trees, the value of Manning's *n* is

$$n = \frac{0.118 + 0.125}{2} = 0.122$$

The specified floodplain is neither fully covered nor just-submerged, and therefore, the estimated *n* must be corrected or

$$n_c = n \left(\frac{y_N}{y_T} \frac{a}{a_T} \right)^{0.5} = 0.122 \left(\frac{2}{3} 0.8 \right)^{0.5} = 0.089$$

It is pertinent to note that in an actual situation, the floodplain cover and flow rate would be known; and therefore, the determination of Manning's *n*, for the overbank region, would be an iterative process.

Table 4.10. Estimated vegetation indices for coniferous trees (after Kouwen and Fathi-Moghadam, 2000).

Species	$\in E$ (N/m^2)
Cedar (*Thuja occidentalis*)	2.07
Spruce (*Picea glauca*)	3.36
White Pine (*Pinus Strobus*)	2.99
Austrian Pine (*Pinus palustris*)	4.54

4.4.   BIBLIOGRAPHY

Arcement, G.J., Jr. and Schneider, V.R., "Guide for Selecting Manning's Roughness Coefficients for Natural Channels and Flood Plains," U.S. Geological Survey Water-Supply Paper 2339, U.S. Geological Survey, Washington, 1989.

Ackers, P., Resistance to Fluids Flowing in Channels and Pipes," Hydraulic Research Paper No. 1, H.M.S.O., London, 1958.

Aldridge, B.N. and Garrett, J.M., "Roughness Coefficients for Stream Channels in Arizona," U.S. Geological Survey Open File Report, U.S. Geological Survey, Washington, 1973.

Anonymous, "Report of the American Society of Civil Engineers' Task Force on Friction Factors in Open Channels," Proceedings of the American Society of Civil Engineers, Journal of the Hydraulics Division, vol. 89, No. HY2, March 1963a, pp. 97-143.

Anonymous, "Guide for Selecting Roughness Coefficient 'n' Values for Channels," U.S. Department of Agriculture, Soil Conservation Service, Washington, December 1963b.

Barnes, H.H., Jr., "Roughness Characteristics of Natural Channels," U.S. Geological Survey Water-Supply Paper 1849, U.S. Geological Survey, Washington, 1967.

Benson, M.A. and Dalrymple, T., "General Field and Office Procedures for Indirect Discharge Measurements," Techniques of Water-Resources Investigations of the United States Geological Survey, book 3, Chapter A1, U.S. Geological Survey, Washington, 1967.

Chow, V.T., Open Channel Hydraulics, McGraw-Hill Book Company, New York, 1959.

Cowan, W.L., "Estimating Hydraulic Roughness Coefficients," Agricultural Engineering, vol. 37, no. 7, 1956, pp. 473-475.

Coyle, J.J., "Grassed Waterways and Outlets," Engineering Field Manual, U.S. Department of Agriculture, U.S. Soil Conservation Service, Washington, DC, April, 1975, pp. 7-1 - 7-43.

Fathi-Moghadam, M., "Momentum Absorption in Non-Rigid, Non-Submerged, Tall Vegetation Along Rivers." Thesis submitted in partial fulfillment of requirement for Ph.D. in Civil Engineering, University of Waterloo, Waterloo, Ontario, Canada, 1996.

Flippin-Dudley, S.J., "Vegetation Measurements for Estimating Flow Resistance," Doctoral Dissertation, Colorado State University, Fort Collins, CO,1997.

Flippin-Dudley, S.J., Abt, S.R., Bonham, C.D., Watson, C.C., and Fischenich, J.C., "A Point Quadrant Method of Vegetation Measurement for Estimating Flow Resistance," Technical Report EL-97-XX, U.S. Army Engineers Waterways Experiment Station, Vicksburg, MS, 1997.

French, R.H., and McCutcheon, S.C., "The Stability of a Two Layer Flow without Shear in the Presence of Boundary Generated Turbulence: Field Verification," Technical Report No. 39, Environmental and Water Resources Engineering, Vanderbilt University, Nashville, Tenn., October 1977.

Garde, R.J. and Ranga Raju, K.G., Mechanics of Sediment Transportation and Alluvial Stream Problems, Wiley Eastern, New Delhi, 1978.

Henderson, F.M., Open Channel Flow, The Macmillan Company, New York, 1966.

Jarrett, R.D., "Hydraulics of High-Gradient Streams," Journal of Hydraulic Engineering, American Society of Civil Engineers, vol. 110, No. 11, November, 1984, pp. 1519-1539.

Kouwen, N., "Field Estimation of the Biomechanical Properties of Grass," Journal of Hydraulic Research, International Association for Hydraulic Research, vol. 26, no. 5, 1988, pp. 559-568.

Kouwen, N. and M. Fathi-Moghadam, "Friction Factors for Coniferous Trees Along Rivers," Journal of Hydraulic Engineering, American Society of Civil Engineers, vol. 126, no. 10, October, 2000, pp. 732-740.

Kouwen, N. and Li, R., "Biomechanics of Vegetative Channel Linings," Journal of the Hydraulics Division, American Society of Civil Engineers, vol. 106, no. HY6, June, 1980, pp. 1085-1103.

Lane, E.W., and Carlson, E.J., "Some Factors Affecting the Stability of Canals Constructed in Coarse Granular Materials," Proceedings of the Minnesota International Hydraulics Convention, September 1953.

Levi, E., The Science of Water: The Foundation of Modern Hydraulics, Translated from Spanish by D.E. Medina, ASCE Press, New York, 1995.

Limerinos, J.T., "Determination of the Manning Coefficient from Measured Bed Roughness in Natural Channels,"U.S. Geological Survey Water Supply Paper 1898-B, U.S. Geological Survey, Washington, 1970.

Meyer-Peter, E., and Muller, R., "Formulas for Bed-Load Transport," Proceedings of the 3rd Meeting of IAHR, Stockholm, pp. 39-64, 1948.

Petryk, S. and Bosmajian, G., "Analysis of Flow Through Vegetation," Journal of the Hydraulics Division, American Society of Civil Engineers, vol. 101, no. HY7, July, 1975, pp. 871-884.

Raudkivi, A.J., Loose Boundary Hydraulics, 2nd ed., Pergamon Press, New York, 1976.

Raupach, M.R., A.S. Thom, and I. Edwards, "A Wind-Tunnel Study of Turbulent Flow Close to Regularly Arrayed Rough Surfaces," Boundary Layer Meteorology, Vol. 18, pp. 373-397, 1980.

Ree, W.O. and Crow, F.R., "Friction Factors for Vegetated Waterways of Small Slope," ARS-S-151, Agricultural Research Service, U.S. Department of Agriculture, Washington, 1977.

Richardson, E.V., D.B. Simons, and P.Y. Julien, Highways in the River Environment, U.S. Department of Transportation, Federal Highway Administration, Washington, DC, 1987.

SLA, Engineering Analysis of Fluvial Systems, Simons, Li & Associates, Fort Collins, CO, 1982.

Simons, D.B., and Senturk, F., Sediment Transport Technology, Water Resources Publications, Fort Collins, CO, 1976/1992.

Streeter, V. L., and Wylie, E.B., Fluid Mechanics, McGraw Hill Book Company, New York, 1975.

Subramanya, K., Flow in Open Channels, vol. 1, Tata McGraw-Hill Publishing Company, New Delhi, 1982.

Temple, D.M., "Changes in Vegetal Resistance During Long-Duration Flows," Transactions of the ASAE, vol. 34, 1991, pp. 1769-1774.

Urquhart, W.J., "Hydraulics," Engineering Field Manual, U.S. Department of Agriculture, Soil Conservation Service, Washington, 1975.

Zegzhda, A.P., "Teoriia podobiia I metodika rascheta gidrotekhnicheskikh modelei" ("Theory of Similarity and Methods of Design of Models for Hydraulic Engineering"), Gosstroiisdat, Leningrad, 1938.

CHAPTER 5

COMPUTATION OF UNIFORM FLOW

5.1. CALCULATION OF NORMAL DEPTH AND VELOCITY

In the previous chapter, uniform flow, its governing equations, and various methodologies for estimating the resistance coefficient were described. Recall from that chapter that the Manning equation asserts that

$$\bar{u} = \frac{\phi}{n} R^{2/3} \sqrt{S} \qquad (5.1.1)$$

and by the law of conservation of mass this equation, when multiplied by the flow area, yields an equation for the uniform flow rate or

$$Q = \bar{u}A = \frac{\phi}{n} AR^{2/3} \sqrt{S} \qquad (5.1.2)$$

In Eq. (5.1.2), the parameter $AR^{2/3}$ is termed the *section factor* and

$$K = \frac{\phi}{n} AR^{2/3} \qquad (5.1.3)$$

is, by definition, the conveyance of the channel. For a given channel where $AR^{2/3}$ always increases with increasing depth, each discharge has a corresponding unique depth at which uniform flow occurs.

An examination of Eqs. (5.1.1) and (5.1.2) demonstrates that the average uniform velocity of flow or the flow rate is a function of (1) the

221

channel shape, (2) the resistance coefficient, (3) the longitudinal slope of the channel, and (4) the depth of flow or

$$Q = f\left(\Gamma, n, S, y_N\right) \qquad (5.1.4)$$

where Γ = a channel shape factor and y_N = normal depth of flow. If four of the five variables in Eq. (5.1.4) are known, then the fifth variable can be determined.

EXAMPLE 5.1

Given a trapezoidal channel with a bottom width of 3 m (10ft), side slopes of 1.5:1, a longitudinal slope of 0.0016, and a resistance coefficient of $n = 0.013$, determine the normal discharge if the normal depth of flow is 2.6m (8.5ft).

Solution

From Table 1.1

$$A = \left(b + zy\right)y = \left[3.0 + 1.5(2.6)\right]2.6 = 18\,m^2\left(190\,ft^2\right)$$

$$P = b + 2y\sqrt{1 + z^2} = 3.0 + 2(2.6)\sqrt{3.25} = 12\,m\left(39\,ft\right)$$

$$R = \frac{A}{P} = \frac{18}{12} = 1.5\,m\left(4.9\,ft\right)$$

From Eq. (5.1.12)

$$Q = \frac{\phi}{n}AR^{2/3}\sqrt{S} = \frac{1.0}{0.013}(18)(1.5)^{2/3}\sqrt{0.0016} = 73\,m^3\,/\,s\left(2600\,ft^3\,/\,s\right)$$

In general, the most difficult and tedious normal flow calculation occurs when Q, Γ, S, and n are known, and y_N must be estimated. In such a case, an explicit solution of Eq. (5.1.2) is not possible and the problem must be solved by trial and error, design charts, or numerical methods. The initial section of this chapter will examine three methodologies which are useful in solving this type of problem.

Trial-and-Error and Graphical Solutions

If Q, Γ, S, and n are known, then Eq. (5.1.2) may be rearranged to yield

$$AR^{2/3} = \frac{nQ}{\phi\sqrt{S}} \qquad (5.1.5)$$

where the right-hand side of the equation is a constant and the left-hand side is the section factor, which is a function of the channel shape and the depth of flow. The most direct approach to finding y_N is a trial-and-error solution of Eq. (5.1.5).

EXAMPLE 5.2

Given a trapezoidal channel with a bottom width of 10 ft (3.0 m), side slopes of 1.5:1, a longitudinal slope of 0.00016, and estimated n of 0.13, find the normal depth of flow for a discharge of 250 ft³/s (7.1 m³/s).

Solution

$$AR^{2/3} = \frac{nQ}{\phi\sqrt{S}} = \frac{0.13(250)}{1.49\sqrt{0.0016}} = 545$$

with

$$A = \left(b + zy\right)y = \left(10 + 1.5y\right)y$$

$$P = b + 2y\sqrt{1 + z^2} = 10 + 2y\sqrt{3.25} = 10 + 3.6y$$

$$R = \frac{A}{P} = \frac{\left(10 + 1.5y\right)y}{10 + 3.6y}$$

Then, the following table (Table 5.1) is constructed by assuming values of y and computing a corresponding value of $AR^{2/3}$. When the computed value of $AR^{2/3}$ matches the value computed from the problem statement, the correct value of y_N has been determined.

From the information contained in this table, it is concluded that y_N = 8.5 ft (2.6m), which is the normal depth of flow for the channel and specified flow rate.

Table 5.1.

Trial y ft	A ft²	P ft	R ft	AR²ᐟ³
8.0	176	38.8	4.54	482
9.0	212	42.4	5.00	620
8.5	193	40.6	4.75	546

A second method of solving this problem is by the construction of a graph of the section factor versus the depth of flow.

EXAMPLE 5.3

Given a circular culvert 3.0 ft (0.91 m) in diameter with $S = 0.0016$ and $n = 0.015$, find the normal depth of flow for a discharge of 15 ft³/s (0.42 m³/s).

Solution

$$AR^{2/3} = \frac{nQ}{\phi\sqrt{S}} = \frac{0.015(15)}{1.49\sqrt{0.0016}} = 3.78$$

A graph of $AR^{2/3}$ versus y is constructed in Fig. 5.1, for $AR^{2/3} = 3.78$, $y_N = 1.7$ ft. With regard to Fig. 5.1, it is noted that the section factor first increases with depth, and then as the full depth is approached it decreases with increasing depth. Therefore, it is possible to have two values of y_N for some values of $AR^{2/3}$.

The methods discussed above have the advantage of directness and the disadvantage of being cumbersome if a number of problems must be solved. When a number of such problems must be solved, a different approach is required.

General Design Charts

In order to simplify the computation of the normal depth for common channel shapes, dimensionless curves for the section factor as a function of the depth have been prepared for rectangular, circular, and trapezoidal channels (Fig. 5.2). Although these curves provide solutions to the problems of normal depth computation for these channel shapes, in a manner similar to that used in Example 5.3, they do not provide a general method of solution.

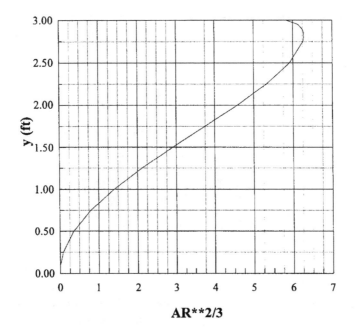

Figure 5.1. $AR^{2/3}$ versus y for a circular conduit 3.0 ft (0.91 m) in diameter.

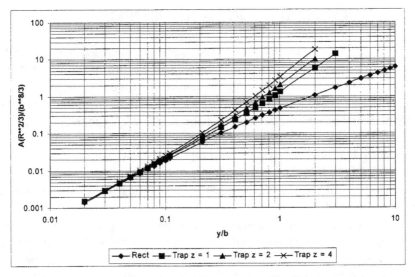

Figure 5.2a.   Dimensionless curves $AR^{2/3}/b^{8/3}$ versus y/b for determining the normal depth of flow in rectangular and trapezoidal channels where b is the bottom width.

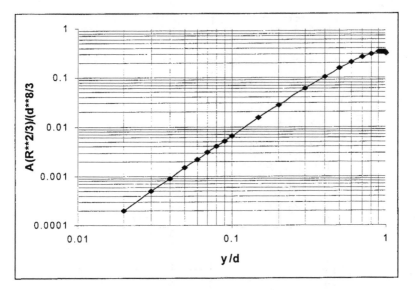

Figure 5.2b. Dimensionless curves $AR^{2/3}/d^{8/3}$ versus y/d for determining the normal depth of flow in circular channels where d is the diameter.

5.2.   NORMAL AND CRITICAL SLOPES

If Q, n, y_N, and the channel section are defined, then Eq. (5.1.2) can be solved explicitly for the slope which allows the flow to occur as specified. By definition, this slope is a normal slope.

If the slope of a channel is varied while the discharge and roughness are held constant, it is possible to find a value of the slope such that normal flow occurs in a critical state, i.e., a slope such that normal flow occurs with $\mathbf{F} = 1$. The slope obtained in this fashion is, by definition, the critical slope, but also a normal slope. The smallest critical slope for a specified channel shape, discharge and roughness is termed the *limiting critical slope.*

On the basis of these definitions, it must be possible to define a critical slope and discharge, which correspond to a specified normal depth in a given channel. This slope is known as the critical slope at the specified normal depth. The following two examples illustrate the crucial points of this discussion.

EXAMPLE 5.4

A rectangular channel has a bottom width of 6.0 m (19.7 ft) and $n = 0.020$

a.   For $y_N = 1.0$ m (3.3 ft) and $Q = 11$ m³/s (388 ft³/s), find the normal slope.

b.   Find the limiting critical slope and normal depth of flow for $Q = 11$ m³/s (388 ft³/s).

c.   Find the critical slope for $y_N = 1.0$ m (3.3 ft) and determine the discharge which corresponds to this depth of flow and slope.

Solution

a.   For the given data

$$A = by = 6.0(1.0) = 6.0\,\text{m}^2 \left(65\,\text{ft}^2\right)$$

$$P = b + 2y = 6 + 2(1.0) = 8\,\text{m}\,(26\,\text{ft})$$

$$R = \frac{A}{P} = \frac{6}{8} = 0.75\,\text{m}\,(26\,\text{ft})$$

Rearranging Eq. (5.1.2),

$$S = \left(\frac{nQ}{\phi AR^{2/3}}\right)^2 = 0.002$$

Thus, $S = 0.002$ will maintain a uniform flow of 11 m³/s (388 ft³/s) in this channel, at a depth of 1.0 m (3.3 ft).

b.   The critical depth of flow is found by Eq. (2.2.3) or

$$\frac{\overline{u}^2}{2g} = \frac{D}{2}$$

where

$$\overline{u} = \frac{Q}{by} = \frac{11}{6y_N}$$

and since this is a rectangular channel

$$D = y_c = \frac{Q^2}{g(by_N)^2}$$

$$y_N = \left(\frac{Q^2}{gb^2}\right)^{1/3} = \left[\frac{11^2}{9.8(6)^2}\right] = 0.70\,\text{m}\,(2.3\,\text{ft})$$

Then

$$A = by = 6(0.70) = 4.2\,\text{m}^2\,(45\,\text{ft}^2)$$

$$P = b + 2y = 6 + 2(0.70)\,7.4\,\text{m}\,(24\,\text{ft})$$

$$R = \frac{4.2}{7.4} = 0.57\,\text{m}\,(1.9\,\text{ft})$$

Rearranging Eq. 5.1.2),

$$S_c = \left(\frac{nQ}{\phi AR^{2/3}}\right)^2 = \left[\frac{0.02(11)}{1.0(4.2)(0.57)^{2/3}}\right]^2 = 0.0058$$

where S_c = a critical slope. This is the slope which will maintain a critical and uniform flow of 11 m³/s (388 ft³/s) in the specified channel.

c.   From part *a*, for a normal depth of 1.0 m (3.3 ft)

$$A = 6.0\,\text{m}^2\,(65\,\text{ft}^2)$$

$$P = 8.0\,\text{m}\,(26\,\text{ft})$$

$$R = 0.75\,\text{m}\,(2.5\,\text{ft})$$

The velocity may then be found from the definition of critical flow or

$$F = \frac{\bar{u}}{\sqrt{gy_N}}$$

$$\bar{u} = \sqrt{9.8(1)}$$

Then, rearranging Eq. (5.1.1)

$$S_{cN} = \left(\frac{n\bar{u}}{\phi R^{2/3}}\right)^2 = \left[\frac{0.02(3.1)}{1.0(0.75)^{2/3}}\right]^2 = 0.0056$$

where S_{cN} = slope, which will maintain a critical and uniform flow at a depth of 1.0 m (3.3 ft), in the specified channel. The flow rate is given by

$$Q = \bar{u}A = 3.1(6) = 19\,\text{m}^3/\text{s}\left(670\,\text{ft}^3/\text{s}\right)$$

EXAMPLE 5.5

For a trapezoidal channel with a bottom width of 20.4 ft (6.2 m), side slopes of 0.5:1, and n = 0.02, develop a graph of Q versus S_c, the limiting critical slope.

Solution

From Eq. (2.2.3), the condition for critical flow in a trapezoidal channel is

$$\frac{\bar{u}^2}{g} = D$$

where from Table 1.1

$$A = (b + zy)y = (20.4 + 0.5y)y$$

$$T = b + 2zy = 20.4 + y$$

$$D = \frac{A}{T} = \frac{(20.4 + 0.5y)y}{20.4 + y}$$

and

$$\bar{u} = \frac{Q}{A} = \frac{Q}{(20.4 + 0.5y)y}$$

Combining these relations yields an implicit equation for the critical depth or

$$\frac{Q^2}{g}(20.4 + y_c) = \left[(20.4 + 0.5y_c)y_c\right]^3$$

This equation, when solved, yields values of y_c, which correspond to a specified flow rate. Then y_c is used in the Manning equation, in a manner similar to that demonstrated in Example 5.5b to estimate S_c. The results of this computation are summarized in Fig. 5.3. The curve in this figure separates the regions of subcritical and supercritical normal flow for the specified values of Q and S.

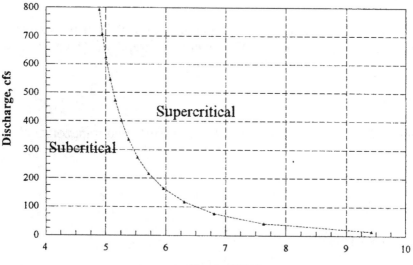

Figure 5.3.   Curve defining subcritical and supercritical areas of normal flow for a given channel as a function of discharge and slope.

5.3.   CHANNELS OF COMPOSITE ROUGHNESS

In many designed channels and most natural channels, the roughness varies along the perimeter of the channel, and in such cases it is sometimes necessary to calculate an equivalent value to the roughness coefficient for the entire perimeter. This effective roughness coefficient is then used in the normal flow calculations for the entire channel. In this section, a number of different methods of estimating the equivalent roughness coefficient for natural, designed, and laboratory channels are considered.

In the natural channel, the flow area is divided into N parts, each with an associated wetted perimeter P_i and roughness coefficient n_i, which are known. In these methods, the wetted perimeters, i.e., the P_i's, do not include the imaginary boundaries between the subsections. The methods of computing the equivalent roughness coefficient for this type of channel are:

1.   Horton (1933) and Einstein and Banks (1950) each developed a method in which it is assumed that each of the subdivisions of the flow area is assumed to have the average velocity of the total section or $\bar{u} = \bar{u}_1 = \bar{u}_2 = \cdots = \bar{u}_N$. Where $\bar{u}_i$ = the average velocity in the ith subsection. Then

$$n_e = \left[\frac{\sum_{i=1}^{N}\left(P_i n_i^{3/2}\right)}{P} \right]^{2/3} \tag{5.3.1}$$

where n_e = equivalent Manning roughness coefficient
   P = wetted perimeter of the complete section
   N = number of subsections

2.   If it is assumed that the total force-resisting motion is equal to the sum of the subsection-resisting forces, then

$$n_e = \left[\frac{\sum_{i=1}^{N}\left(P_i n_i^{2}\right)}{P} \right]^{1/2} \tag{5.3.2}$$

3.   If it is assumed that the total discharge of the section is equal to the sum of the subsection discharges, then

$$n_e = \frac{PR^{5/3}}{\displaystyle\sum_{i=1}^{N} \frac{P_i R_i^{5/3}}{n_i}} \tag{5.3.3}$$

where R_i = the hydraulic radius of the ith subsection, and R = hydraulic radius of the complete section.

In a designed channel or a laboratory flume, the boundary angles are bisected and the subdivision is composed of the channel perimeter, the boundary angle bisectors, and the water surface. Although the two methods of computing n_e, described below, could be used for natural channels, the required geometric calculations can be quite cumbersome in such a situation.

1.   In the method attributed by Cox (1973) to the Los Angeles U.S. Army Corp of Engineers District, n_e is computed by

$$n_e = \frac{\displaystyle\sum_{i=1}^{N} n_i A_i}{A} \tag{5.3.4}$$

where A_i = area of the ith subsection and A = total flow area.

2.   The Colebatch method also described by Cox (1973) uses the following equation:

$$n_e = \left(\frac{\displaystyle\sum_{i=1}^{N} A_i n_i^{3/2}}{A} \right)^{2/3} \tag{5.3.5}$$

In general, any of the five methods discussed above is satisfactory for estimating equivalent values of n for natural and designed channels; however, the appropriateness and accuracy of the resulting estimates are unknown.

EXAMPLE 5.6

Given the channel described schematically in Fig. 5.4, estimate the value of n_e by the methods described above.

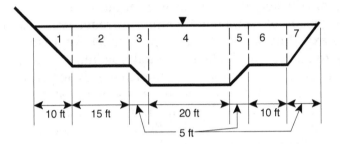

Figure 5.4. Channel definition for Example 5.6.

Solution

The data required for the solution of this problem are summarized in the following table (Table 5.2):

Table 5.2.

Sub-Division	Eqs. (5.3.1), (5.3.2), and (5.3.3)				Eqs. (5.3.4) and (5.3.5)
	A ft^2	P ft	R ft	n	A ft^2
1	25.0	11.2	2.23	0.01	27.9
2	75.0	15.0	5.00	0.015	77.2
3	37.5	7.07	5.30	0.020	53.0
4	200	20.0	10.0	0.030	159
5	37.5	7.07	5.30	0.020	53.0
6	50.0	10.0	5.00	0.015	50.0
7	12.5	7.07	1.77	0.010	17.7
Total Section	437	77.4	5.65		437

Then, using

Eq. (5.3.1), $n_e = 0.019$
Eq. (5.3.2), $n_e = 0.020$
Eq. (5.3.3), $n_e = 0.019$

$$\text{Eq. (5.3.4)}, \, n_e = 0.022$$
$$\text{Eq. (5.3.5)}, \, n_e = 0.023$$

It is noted that each of the methods of estimating an equivalent value of n yields approximately the same answer for this example.

The problem of compound perimeter roughness also occurs in laboratory experiments; i.e., the bottom boundary of a laboratory flume is usually roughened to duplicate natural conditions, but the walls are usually not roughened. In this case, the friction factor of the composite section is estimated from measured values of the slope of the water surface and the hydraulic radius; the problem is to estimate the friction factor or shear velocity associated with the bottom boundary roughness. The methodology described below was developed by Vanoni and Brooks (1957). In this discussion, the variables subscripted b are associated with the bottom of the channel, the variables subscripted w are associated with the walls, and the variables without a subscript are associated with the composite flow.

The following assumptions are required:

1. The channel flow area can be divided into two parts: one section in which the flow produces shear on the bottom boundary and a second section in which the flow produces shear on the walls (Fig. 5.5). The boundaries between the bottom and wall sections are assumed to be zero shear surfaces and are not included in P_b and P_w.

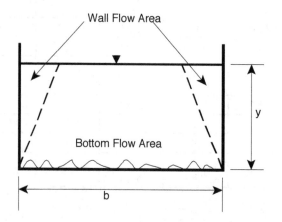

Figure 5.5. **Schematic definition of the Vanoni and Brooks (1957) composite roughness calculation.**

2.   The average velocity in the bottom section $\overline{u}_b$ equals the average velocity in the wall section $\overline{u}_w$.

3.   The bottom and wall flow sections act as independent channels.

4.   The bottom and wall roughnesses are homogeneous, although different.

In this analysis, it is assumed that the following fundamental quantities are known:

$$R,\ u_*,\ \nu,\ P_w,\ P_b,\ A,\ f,\ u$$

where

$$f = 8\left(\frac{u_*}{\overline{u}}\right)^2$$

= Darcy-Weisbach friction factor, and ν = kinematic viscosity. The primary unknowns are

$$u_{*b},\ f_b,\ R_b$$

It is crucial to the solution of this problem that the wall friction factor be estimated. Although the problem can be solved if the walls are not hydraulically smooth, the assumption of smooth side walls provides for a much simpler solution, since under such an assumption f_w is only a function of the Reynolds number of the wall $\mathbf{Re}_w$ where

$$\mathbf{Re}_w = \frac{\overline{u}_w R_w}{\upsilon} \qquad (5.3.6)$$

In this equation, $\mathbf{Re}_w$ is an unknown. Equation (5.3.6) may also be written as

$$\mathbf{Re}_w = \frac{4\overline{u}R}{\upsilon}\frac{R_w}{R} = \mathbf{Re}\frac{R_w}{R} \qquad (5.3.7)$$

The friction factor for the wall is given by

$$gR_w S = \frac{f_w \overline{u}^2}{8} \qquad (5.3.8)$$

and the friction factor for the composite flow is

$$gRS = \frac{f\overline{u^2}}{8}$$

(5.3.9)

Dividing Eq. (5.3.8) by Eq. (5.3.9),

$$\frac{R_w}{R} = \frac{f_w}{f}$$

(5.3.10)

Substitution of this equation in Eq. (5.3.7) yields

$$\frac{Re_w}{f_w} = \frac{Re}{f}$$

(5.3.11)

Although the parameter Re_w/f_w can be calculated from this equation, neither Re_w nor f_w can be estimated alone. An auxiliary equation relating f_w and Re_w is required. For hydraulically smooth boundaries with $Re_w <$ 100,000, Eq. (4.3.7) can be used as the auxiliary equation. For $Re_w \geq$ 100,000, Eq. (4.3.8) can be used. For use with Eq. (5.3.11), Eqs. (4.3.7) and (4.3.8) are plotted in Fig. 5.6. With this figure f_w can be estimated when Re_w/f_w is known.

Then from geometrical considerations

$$A = A_b + A_w$$

(5.3.12)

where, with a slight rearrangement, Eq. (5.3.9) yields

$$A = \frac{Pf\overline{u^2}}{8gS}$$

(5.3.13)

$$A_b = \frac{P_b f_b \overline{u^2}}{8gS}$$

(5.3.14)

and

$$A_w = \frac{P_w f_w \overline{u^2}}{8gS}$$

(5.3.15)

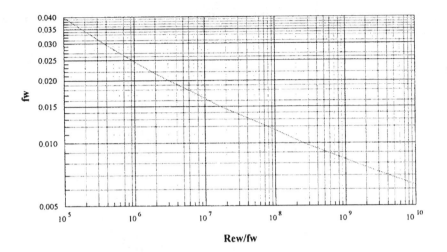

Figure 5.6. f_w **as a function of** Re_w/f_w.

Substituting Eqs. (5.3.13) to (5.3.15) in Eq. (5.3.12) and rearranging them,

$$f_b = \frac{P}{P_b} f - \frac{P_w}{P_b} f_w \qquad (5.3.16)$$

For a rectangular channel, Eq. (5.3.16) becomes

$$f_b = f + \frac{2y}{b}\left(f - f_w\right) \qquad (5.3.17)$$

where y = depth of flow. Also for a rectangular channel

$$R_b = R\frac{f_b}{f} \qquad (5.3.18)$$

and

$$u_{*b} = \sqrt{gR_bS} \qquad (5.3.19)$$

EXAMPLE 5.7

A laboratory experiment is performed in a rectangular flume 2.75 ft (0.84 m) wide with hydraulically smooth side walls and an artificially roughened bottom. If $Q = 1.00$ ft^3/s (0.028 m^3 /s), $S = 0.00278$, and $y =$ 0.30 ft (0.091 m), estimate the values of f_b and u_{*b}

Solution

From the given data, the following variable values can be computed:

$$A = by = 2.75(0.30) = 0.825\,\text{ft}^2 \ (0.25\,\text{m}^2)$$

$$P = b + 2y = 2.75 + 2(0.30) = 3.35\,\text{ft} \ (1.0\,\text{m})$$

$$R = \frac{A}{P} = \frac{0.825}{3.35} = 0.246\,\text{ft} \ (0.075\,\text{m})$$

$$u_* = \sqrt{gRS} = \sqrt{32.2(0.246)(0.00278)} = 0.148\,\text{ft}/\text{s} \ (0.045\,\text{m}/\text{s})$$

$$\bar{u} = \frac{Q}{A} = \frac{1.00}{0.825} = 1.21\,\text{ft}/\text{s} \ (0.37\,\text{m}/\text{s})$$

$$f = 8\left(\frac{u_*}{\bar{u}}\right)^2 = 8\left(\frac{0.148}{1.21}\right)^2 = 0.120$$

Assume that $v = 9.55 \times 10^{-6}$ ft^2/s (8.87 x 10^{-7} m^2/s); then

$$Re = \frac{4\bar{u}R}{\upsilon} = \frac{4(1.21)(0.246)}{9.55 \times 10^{-6}} = 1.25 \times 10^6$$

and from Eq. (5.3.11)

$$\frac{Re_w}{f_w} = \frac{Re}{f} = \frac{1.25 \times 10^5}{0.120} = 1.04 \times 10^6$$

For this value of the parameter Re_w/f_w Fig. 5.6 yields a value of

$$f_w = 0.028$$

and

$$f_b = f + \frac{2y}{b}\left(f - f_w\right) = 0.120 + \frac{2\left(0.30\right)}{2.75}\left(0.120 - 0.028\right) = 0.140$$

Using Eqs. (5.3.18) and (5.3.19),

$$R_b = R\frac{f_b}{f} = 0.246\frac{0.140}{0.120} = 0.287$$

and

$$u_{*b} = \sqrt{gR_b S} = \sqrt{32.2\left(0.287\right)\left(0.00278\right)} = 0.160\,\text{ft} / \text{s}\left(0.049\,\text{m} / \text{s}\right)$$

5.4.   APPLICATION OF UNIFORM FLOW CONCEPTS TO PRACTICE

5.4.1.   Introduction

In this section a number of specific applications of the principles of uniform flow are considered. Although space limits the number of special cases that can be treated, it is hoped that the cases considered will indicate the manner in which the theoretical principles of uniform flow can be applied.

5.4.2.   Slope-Area Peak Flood Flow Computations

Although flood flows are both spatially varied and unsteady, in some cases it is possible and/or necessary to analyze such flows with steady, uniform flow concepts. From a theoretical viewpoint, the use of such an approach can be justified only when the changes in flood stage and discharge are sufficiently gradual, so that the friction slope reflects only the losses due to boundary friction. Also the slope-area approach is justified if the change in conveyance in the reach is less than 30 percent. From a practical viewpoint, slope-area methods can and are used when a peak flow discharge estimate is required and the data available are not sufficient to justify the use of more sophisticated techniques. For example, slope-area techniques have been used to estimate unsteady events such as flash floods in arid regions, where both stream flow and precipitation records are essentially nonexistent (Glancy and Harmsen, 1975).

In applying the slope-area technique, the following data are required: (1) a measurement of the average flow area in the reach, (2) the change of elevation of the water surface through the reach, (3) the length of the reach, and (4) an estimate of the average resistance coefficient for the reach. Dalrymple and Benson (1976) provided the following guidelines for selecting a suitable reach:

1. A primary consideration in selecting a reach is the availability of high-water marks within the reach. For example, a steep-sided rock channel might have nearly perfect hydraulic characteristics, but if no high-water marks are available, it cannot be used. Benson and Dalrymple (1976) provide a number of techniques for identifying reliable high-water marks.

2. A reach with significant changes in the channel shape should be avoided, if possible, because of the uncertainties regarding the head loss in these sections. Although a straight, uniform reach is preferred, a contracting reach should be chosen over an expanding reach, if there is a choice.

3. The slope-area method assumes that the total cross-sectional area of the channel is effective in transporting the flow. Therefore, conditions either upstream or downstream of the reach which cause an unbalanced flow distribution, should be avoided, e.g., bridges and channel bends.

4. The slope-area technique is not applicable to reaches which include free-fall situations, e.g., waterfall.

5. The reach should be long enough to develop a fall in the water surface, whose magnitude can be determined accurately. In general, the accuracy of the slope-area method improves as the length of the reach is increased. Dalrymple and Benson (1976) recommended that one or more of the following criteria should be met in determining the reach length: (a) the length should be greater than or equal to 75 times the mean depth of flow, (b) the fall of the water surface should be equal to or greater than the velocity head, and (c) the fall should be equal to or greater than 0.50 ft (0.15 m).

It must be emphasized that the selection of a suitable reach is crucial to obtaining accurate estimates of the peak flow from the slope-area method.

Given the required information, the slope-area computation proceeds as follows:

1. For the specified cross-sectional areas, estimate R. Recall that the conveyance of a channel is by definition

$$K = \frac{\phi}{n} AR^{2/3} \qquad (5.1.3)$$

Compute the upstream, K_u and downstream, K_d conveyances.

2. Compute the geometric mean conveyance of the reach or

$$\overline{K} = \sqrt{K_u K_d} \qquad (5.3.1)$$

3. As a zero-order approximation of the true energy slope, let

$$S^0 = \frac{F}{L} \qquad (5.3.2)$$

where S^0 = zero-order approximation of S
       F = change of water surface elevation in reach
       L = length of reach.

The zero-order estimate of the peak flood flow discharge is

$$Q^0 = \overline{K}\sqrt{S^0} \qquad (5.3.3)$$

Compute the first-order approximation of the discharge by refining the energy slope estimate; i.e.,

$$S^1 = S^0 + k\left(\frac{\alpha_u \dfrac{\overline{u_u^2}}{2g} - \alpha_d \dfrac{\overline{u_d^2}}{2g}}{L} \right) \qquad (5.4.4)$$

where α = kinetic energy correction factor

       $\overline{u_u}$ and $\overline{u_d}$ = upstream and downstream average velocities of flow calculated from Q^0

and the cross-sectional areas, respectively

k = constriction/expansion correction factor

If the reach is expanding, that is, $\overline{u_u} > \overline{u_d}$, then $k = 0.5$. If the reach is contracting, that is, $\overline{u_d} > \overline{u_u}$, then $k = 1.0$.

Then

$$Q^l = \overline{K}\sqrt{S^l} \tag{5.4.5}$$

6.   Step 5 is repeated until

$$Q^{n-1} \cong Q^n$$

7.   It is considered appropriate to average the discharges estimated for several reaches.

At best, the slope-area method provides a crude estimate of the peak flood flow, but in many cases the estimate is obtained by this technique may be very cost-effective.

EXAMPLE 5.8

Estimate the flood discharge in a reach 4300 ft (1300 m) long if $F = 7.3$ ft (2.2 m), $\alpha_u = \alpha_d = 1$, $n = 0.035$, $A_u = 1{,}189$ ft^2 (110 m^2), $P_u = 248$ ft (76 m), $A_d = 1{,}428$ ft^2 (133 m^2), and $P_d = 298$ ft (91 m).

Solution

Upstream Hydraulic Radius

$$R_u = \frac{A_u}{P_u} = \frac{1189}{248} = 4.79\,\text{ft}\left(1.46\,\text{m}\right)$$

Upstream Conveyance

$$K_u = \frac{\phi}{n}A_u R_u^{2/3} = \frac{1.49}{0.035}\left(1189\right)\left(4.79\right)^{2/3} = 144{,}000$$

Downstream Hydraulic Radius

$$R_d = \frac{A_d}{P_d} = \frac{1428}{298} = 4.79 \, \text{ft} \, (1.46 \, \text{m})$$

Downstream Conveyance

$$K_d = \frac{\phi}{n} A_d R_d^{2/3} = \frac{1.49}{0.035}(1428)(4.79)^{2/3} = 173,000$$

The percentage change in conveyance through the reach is

$$100 \frac{K_d - K_u}{K_d} = 100 \frac{173,000 - 144,000}{173,000} = 17\% < 30\%$$

The geometric average conveyance for the reach is

$$\overline{K} = \sqrt{K_u K_d} = \sqrt{144,000(173,000)} = 158,000$$

The zero-order flow estimate is

$$S^0 = \frac{F}{L} = \frac{7.3}{4300} = 0.0017$$

$$Q^1 = \overline{K}\sqrt{S^1} = 158,000\sqrt{0.0017} = 6514 \, \text{ft}^3 / s \, (184 \, \text{m}^3 / s)$$

Then, since $A_u < A_d$, $k = 0.50$, and the following data apply:

Order of Estimate n	Q^{n-1} ft³/s	$\overline{u}_u$ ft/s	$\overline{u}_d$ ft/s	S^n	Q^n ft³/s
1	6510	5.50	4.56	0.0017	6540
2	6540	5.50	4.58	0.0017	6540

Therefore, the discharge is estimated to be 6500 ft³/s (184 m³/s).

5.4.3.   Ice-Covered Channels

Before a discussion of ice-covered channels can be initiated, it is essential that the reader have some knowledge of the material and the manner in which it is formed. First, as a material, ice is complex and exhibits the following characteristics:

1. Ice is nonhomogeneous and nonisotropic.

2. Under conditions of rapid loading, ice behaves elastically; however, if the load is applied slowly, viscoelastic deformation occurs.

3. Unlike most materials, ice is less dense than the water from which it is formed.   Therefore, ice floats on its own melt water, at a temperature near its melting point.

4. Although the upper surface temperature of an ice cover is variable, the temperature of the lower surface, which is in contact with water, is always equal to the melting point temperature.

5. Within an ice cover, the ratio of the temperature of the ice to the melting point temperature, both in kelvins, is always greater than 0.80 (Sharp, 1981).

6. Although all ice crystals have a hexagonal shape, the grain structure of ice varies as a function of the mechanism of formation.

Second, two basic methods of ice formation in open channels have been identified:

1. In slow-moving or static water, nucleation occurs when the water becomes supercooled, so the result is a thin layer of skim ice on the top surface. In this stage of ice development, the crystals formed are usually needle-shaped. If the growth process of these crystals is slow, they may grow into large grains with dimensions measured in feet. However, if the growth process is rapid, the needle-shaped crystals interlock and a complex structure is formed. Since in open channels it is common for the midchannel temperature to exceed the boundary temperature, the growth of ice by this process is usually inhibited by the higher midchannel temperatures or turbulence.

    After the development of an initial surface ice sheet, subsequent growth usually takes place in a direction parallel to the heat flow, i.e., in the vertical coordinate direction. This type of growth results in a columnar structure (Fig. 5.7). Although a columnar structure may be present over the complete ice column, in general most vertical ice column structures include some ice which is termed *snow ice*. Snow ice, which is opaque and white, is the result of snow falling on an ice cover, being wetted, and then freezing.

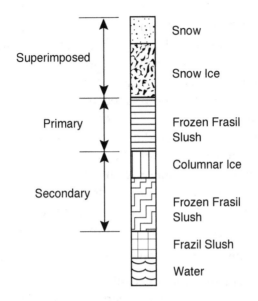

Figure 5.7. Typical ice profile (Michel and Ramsier, 1971).

2. Under turbulent flow conditions, the formation of surface ice is inhibited and nucleation occurs within the water column rather than at the surface. For this process to take place, a few hundredths of a degree of supercooling is still required. Ice formed under these conditions is termed *frazil ice* and can take place throughout the entire turbulent, supercooled vertical column. Frazil ice forms by the development of small discoids which are held in suspension below the surface. Under ideal conditions, e.g., an open surface cooled by high winds, the production of frazil ice can be enormous.

Frazil ice crystals are initially very angular and adhere to one another forming floc particles. These floc particles float to the surface where they form what is termed *frazil slush*. Frazil ice floc particles also tend to adhere to any suitable object, e.g., other ice, rocks, or structures. As the frazil ice collects, a complete surface cover may be formed even under turbulent flow conditions. An excellent review of frazil ice formation was given by Osterkamp (1978).

When a channel is covered with floating ice, the wetted perimeter associated with the flow is increased; thus, for a constant bed slope, the uniform discharge of the channel must decrease. Given the significant impact that ice cover can have on channels in cold climates, it is not surprising that this problem has received attention since the early 1930s.

Most of the investigations of this problem have sought to develop modifications of either the Manning or Chezy resistance coefficients, which would reflect the composite nature of the channel roughness. That is, the channel is divided into two parts: one with area A_1 and hydraulic radius R_1 which is dominated by the bed roughness and the balance of the section $A_2 = A_0 - A_1$ and hydraulic radius R_2 which is dominated by the ice cover roughness, where A_0 is the total flow area. Uzuner (1975) reviewed most of the hypotheses developed between 1931 and 1969, and some of these results are summarized in Table 5.3. In this table, the subscripts 0, 1, and 2 designate parameters associated with the composite, the bed, and the ice-covered sections of the channel, respectively; $\phi = P_2/P_1$, y_j = distances from the two types of boundary to the point of maximum velocity and $\overline{u}_j$ = mean section velocities.

The limitations of the equations summarized in Table 5.3, which are primarily for use with floating ice covers, are: (1) With the exception of Hancu's method, all the equations are based on the Manning equation and thus are applicable only to fully turbulent flow; (2) all these methods assume that the cross section can be divided into two parts with the upper part being solely affected by the ice cover and the lower part solely by the bed; (3) reliable estimates of n_2 for various types of ice cover are not available; and (4) the estimated resistance coefficient is a constant and cannot reflect the change of the ice roughness with the progression of the winter season.

Given the limitations of the equations in Table 5.3, Tsang (1982) began his derivation of uniform flow in an ice-covered channel with a force balance equation analogous to the one used in the previous chapter to develop the Chezy equation or

$$\rho g A_w \Delta h = \tau_b A_b + \tau_i A_i \qquad (5.4.6)$$

where ρ = density of water

A_w = cross-sectional area of channel

A_b = surface area of channel bed

A_i = undersurface area of ice

Δh = head loss across section

τ_b = shear stress on channel bed

τ_i = shear stress on ice cover

Table 5.3. Composite resistance relationships for ice-covered channels.

Developer	Equation	Comment
Pavlovskiy in Uzuner (1975)	$$\frac{n_0}{n_1} = \left[\frac{1 + \varphi\left(\dfrac{n_2}{n_1}\right)^2}{1 + a} \right]^{1/2}$$	
Lotter in Uzuner (1975)	$$\frac{n_0}{n_1} = \frac{\varphi + 1}{1 + \varphi\dfrac{n_1}{n_2}}$$	
Belokon in Uzuner (1975)	$$\frac{n_0}{n_1} = \left[1 + \varphi\left(\frac{n_2}{n_1}\right)^{3/2} \right]^{2/3}$$	
Sabaneev in Uzuner (1975)	$$\frac{n_0}{n_1} = \frac{\left[1 + \left(\dfrac{n_2}{n_1}\right)^{2/(2r+1)} \right]^{(2r+1)/2}}{1 + \phi}$$	
Chow (1959)	$$\frac{n_0}{n_1} = \frac{1}{\sqrt{1 + \varphi}}\left(\frac{n_2}{n_1}\right)\left[\varphi^{3/4} + \left(\frac{n_1}{n_2}\right)^{3/2} \right]^{2/3}$$	$r = 1/6$
Hancu in Uzuner (1975)	$$\frac{n_0}{n_1} = \frac{1}{\sqrt{2}}\left(\frac{R_0}{R_1}\right)^{1/6}\left[\left(\frac{\overline{u_1}}{\overline{u_0}}\right)^2 + \left(\frac{\overline{u_2}}{\overline{u_0}}\right)^2 \times \left(\frac{n_2}{n_1}\right)^2\left(\frac{R_1}{R_2}\right)^{1/3} \right]^{1/2}$$	

Developer	Equation	Comment
Yu-Graf-Levine in Uzuner (1975)	$$\frac{n_0}{n_1} = \left[\frac{1 + \varphi^z \left(\dfrac{n_2}{n_1} \right)^{3/2}}{\left(1 + \varphi \right)^z} \right]^{2/3}$$	$z = \left(\dfrac{n_2}{n_1} \right)^{1/6}$
Larsen (1969)	$$\frac{n_0}{n} = \frac{0.63 \left(\dfrac{y_2}{y_1} + 1 \right)^{5/3}}{\dfrac{n_1}{n_2} \left(\dfrac{y_2}{y_1} \right)^{5/3} + 1}$$	

Simplification of (5.4.6) yields

$$\rho g A_w S = \tau_b P_b \left(1 + \frac{\tau_i A_i}{\tau_b A_b} \right) \tag{5.4.7}$$

where P_b = wetted perimeter of the cross section, excluding the ice cover. Let r_s represent the shear stress ratio or

$$r_s = \frac{\tau_i}{\tau_b}$$

and C_i represents the fractional surface ice coverage of the channel surface or

$$C_i = \frac{A_i}{A_b}$$

With these definitions, Eq. (5.4.7) becomes

$$\rho g R S = \tau_b \left(1 + r_s C_i \right) \tag{5.4.8}$$

It can be hypothesized that

$$\bar{u} = \varphi C \sqrt{RS} \qquad (5.4.9)$$

where

$$\varphi = \left[\frac{1}{\psi(1 + rC_i)} \right]^{1/2}$$

where ψ designates an unknown function, ϕ is less than 1, and $C =$ Chezy roughness coefficient of the channel with no ice cover. It is noted that Eq. (5.4.9) is analogous to the Chezy equation [Eq. (4.2.4)], and is applicable to any type of flow whether fully turbulent or not.

In examining winter conditions in the Beauharnois Canal between Lakes St. Francis and St. Louis on the St. Lawrence River, Tsang (1982) found that ϕ was a strong function of time measured from the day of ice formation (Fig. 5.8). With regard to this figure, which is based on 8 years of data, the following should be noted. First, following the first appearance of ice in the canal, the value of ϕ decreases significantly. Although most of the decrease takes place in the 2 days after the appearance of ice, the minimum value of ϕ is reached, on the average, in a week. Second, after the minimum value of ϕ is achieved, the value of this parameter slowly increases until it reaches a local maximum about three-fifths of the way through the ice season. Third, after reaching the local maximum, the value of ϕ decreases slowly until the day of ice breakup is reached. Fourth, following the breakup of ice, the value of ϕ either rapidly increases or rapidly decreases and then increases, depending on how the ice breaks up. This is not shown in Fig. 5.8.

It is necessary that the observations made above be correlated with a physical description of the situation. Tsang (1982) attributed the rapid decrease in the value of ϕ after the formation of ice to the presence of large amounts of frazil ice in the flow. He noted that when the concentration of frazil ice was the highest, the resistance of the canal to flow was also a maximum. The reduced velocity of flow under these conditions resulted in a consolidation of the frazil ice into a complete ice cover. As the underside of the ice cover was smoothed by the flow, the value of ϕ increased. Toward the end of winter, ϕ reaches a local maximum and begins to decrease, a phenomenon attributed to a very slight warming of the water flowing below the ice. Ashton and Kennedy (1972) have noted that an underflow, only fractions of a degree above freezing, can produce undulations in the underside of an ice cover, which increase the roughness of the channel. Once formed, ice ripples tend to persist until the end of the winter season. Then, when the ice breaks up, φ should either rapidly increase to a value of 1, if the ice melts, or rapidly decreases, if blocks of ice are broken off and an ice jam develops.

It must be noted that in many ways these results represent a special case. For example, the flow in this channel is regulated, and during the winter months the channel is completely ice covered. In a natural river with a significant velocity of flow, the channel is seldom completely covered with ice. Thus, the results presented in Fig. 5.8 must be considered to be indicative of a pattern which may be encountered in channels which develop ice covers. The salient point is that φ, and therefore C or *n*, are a function of time or the development of the ice cover through the winter season. Thus, the assumption of a resistance coefficient, whose value is stationary in time, is probably inappropriate.

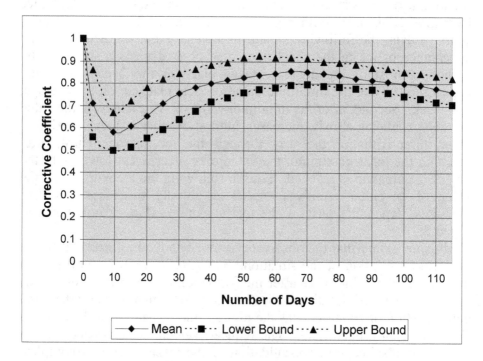

Figure 5.8.   Variation of φ (corrective coefficient) with time. Note, time _0_ is the first day of ice.

In a subsequent discussion of this work, Calkins and Ashton (1983) criticized the work of Tsang (1982), because it did not explicitly separate the large effect of the approximate doubling of the wetted perimeter from secondary effects, such as the variation of the ice cover roughness coefficient, flow blockages due to frazil ice, and changes in the head difference associated with changes in the gradually varied flow profiles. Calkins and Ashton (1983) concluded that the methodology advocated by Tsang (1982) would appear to be inappropriate except in site-specific cases, where there are extensive historical records for estimating ϕ. In reply to these comments, Tsang (1983) noted that his paper clearly demonstrated that in the Beauharnois Canal, the factor which had the greatest influence on the conveyance capacity of the channel, was not a solid ice cover but the presence of frazil ice in the flow. In such a case, where increased flow resistance is the result of a nonsolid boundary, the methods described previously in this section are not applicable. However, until additional results are available, the methodology advocated by Tsang (1982) must be considered innovative but site-specific. *Note*: the time-dependent model of Tsang (1982) is conceptually similar to the hypothesis of Nezhikhovskiy (1964), which involved a time-dependent value of *n*; see, for example, Shen and Yapa (1982). Further, Shen (1983) provides a brief review regarding flow resistance, due to other forms of ice such as ice jams, hanging ice dams, frazil suspensions, and moving ice covers.

5.4.4.   Normal Discharge in Channels of Compound Section

In Chapter 2, the problem of channels composed of several distinct subsections, with each subsection having its own roughness coefficient, was discussed from the viewpoint of specific energy. This type of channel also requires special consideration from the viewpoint of the normal discharge computation. As the flow begins to cover the flood plain, the wetted perimeter of the section increases rapidly, while the flow area increases slowly. Thus, in such a situation, the hydraulic radius, velocity, and discharge decrease with increasing flow depth, a situation which, although computationally correct, is physically incorrect. A number of methods have been suggested to circumvent this computational problem:

1.   The whole cross-sectional area *A(abcdefgh)*, Fig. 5.9, is divided by the total wetted perimeter *P(abcdefgh)*, to obtain a mean or composite hydraulic radius. A mean velocity and discharge may then be computed. As noted above, this method yields erroneous results for shallow depths of flow, in the over-bank sections.

2.   The total channel is divided into a main section and two over-
        bank sections by the vertical lines *ic* and *jf*. The area of the main
        channel is *A(icdefj)*, with wetted perimeter *P(icdefj)*. The over-
        bank sections have the following areas and wetted perimeters:
        A(abci), *P(abc)*, *A(jfgh)*, and *P(fgh)*. In this method, the lines *ic*
        and *jf* are included in the wetted perimeter of the main channel,
        because the shear on these lines is nonzero (see, for example,
        Myers (1978); however, it is assumed that the shear on these
        lines does not affect the overbank flow.

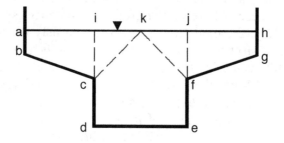

Figure 5.9. Cross section definition.

3.   The channel is divided as in method 2, but the lengths, *ic* and *jf*,
        are not considered part of the wetted perimeter.

4.   The cross section is divided into sections by the lines *ck* and *fk*,
        which bisect the reentrant angles. The wetted perimeter of each
        of the three subsections is computed including only the solid
        boundaries, and a weighted hydraulic radius for the whole
        section is computed by

$$R = \frac{\dfrac{\left[A\left(abck\right)\right]^2}{P\left(abc\right)} + \dfrac{\left[A\left(kcdef\right)\right]^2}{P\left(cdef\right)} + \dfrac{\left[A\left(kfggh\right)\right]^2}{P\left(cgh\right)}}{A\left(abck\right) + A\left(kcdef\right) + A\left(kfgh\right)}$$

Posey (1967) evaluated the accuracy of the above methods with a set of
data, from flume experiments, performed in a flume with the
dimensions shown in Fig. 5.9. In these experiments, the roughness of
the main channel and the over-bank sections was uniform, and thus this

evaluation primarily examined the ability of the above methods to account for a complex channel geometry. Posey (1967) concluded that: (1) method 2 is superior when the depth of flow in the over-bank sections is shallow, and (2) method 1 is superior, when the depth of flow in the over-bank sections is at least half as deep as the bank full main channel. It is noted that the data available by Posey were limited and additional work in this area is required.

5.5.   BIBLIOGRAPHY

Ashton, G.D., and Kennedy, J.F., "Ripples on Underside of River Ice Covers," *Proceedings of the American Society of Civil Engineers, Journal of the Hydraulics Division*, vol. 98, no. HY9, September 1972, pp.1603-1624.

Benson, M.A., and Dalrymple, T., "General Field and Office Procedures for Indirect Discharge Measurements," *Techniques of Water-Resources Investigations of the United States Geological Survey*, book 3, chapter Al, U.S. Geological Survey, Washington, 1976.

Calkins, D.J., and Ashton, G.D., Discussion of "Resistance of Beauharmois Canal in Winter," by G. Tsang, *Journal of Hydraulic Engineering, American Society of Civil Engineers*, vol. 109, no. 4, April 1983, pp. 641-642.

Chow, V.T., *Open Channel Hydraulics*, McGraw-Hill Book Company, New York, 1959.

Cox, R.G., "Effective Hydraulic Roughness for Channels Having Bed Roughness Different from Bank Roughness,: Miscellaneous paper H-73-2, U.S. Army Engineers Waterways Experiment station, Vicksburg, MS, February 1973.

Dalrymple, T., and Benson, M.A., "Measurement of Peak Discharge by the Slope-Area Method," *Techniques of Water-Resources Investigations of the United States Geological Survey*, book 3, chapter A2, U.S. Geological Survey, Washington, 1976.

Dawdy, D.R., "Flood Frequency Estimates on Alluvial Fans," *Journal of Hydraulics Division, American Society of Civil Engineers*, vol. 105, no. 11, November 1979, pp. 1407-1413.

Einstein, H.A., and Banks, R.B., "Fluid Resistance of Composite Roughness," *Transactions of the American Geophysical Union*, vol. 31, no. 4, August 1950, pp. 603-610.

French, R.H., 1987. *Hydraulic Processes on Alluvial Fans*, Elsevier Science Publishers, Amsterdam, The Netherlands.

Glancy, P.A., and Harmsen, L., "A Hydrologic Assessment of the September 14, 1974 Flood in Eldorado Canyon, Nevada," Geological Survey Professional Paper 930, U.S. Geological Survey, Washington, 1975.

Horton, R.E., "Separate Roughness Coefficients for Channel Bottom and Sides," *Engineering News Record*, Vol. III, no. 22, Nov. 30, 1933, pp. 652-653.

Jarrett, R.D., "Hydraulics of High Gradient Streams," *Journal of Hydraulic Engineering, American Society of Civil Engineers*, vol. 110, no. 11, November 1984, pp. 1519-1539.

Larsen, P.A., "Head Losses Caused by an Ice Cover on Open Channels," *Journal of the Boston Society of Civil Engineers*, vol. 56, no. 1, 1969, pp. 45-67.

Michel, B., and Ramseier, R.O., "Classification of River and Lake Ice," *Canadian Geotechnical Journal*, vol. 8, no. 36, 1971, pp. 36-45.

Myers, W.R.C., "Momentum Transfer in A Compound Channel," *International Association for Hydraulic Research, Journal of Hydraulic Research*, vol. 16, no. 2, 1978, pp. 139-150.

Nezhikhovskiy, R.A., "Coefficients of Roughness of Bottom Surface of Slush Ice Cover," *Soviet Hydrology: Selected Papers*, no. 2, 1964, pp. 127-149.

Osterkamp, T.E., "Frazil Ice Formation: A Review," *Proceedings of the American Society of Civil Engineers, Journal of the Hydraulics Division*, vol. 104, no. HY9, September 1978, pp. 1239-1255.

Pine, L.S., and Allen, R.C., *Numerical Computing: Introduction*, Saunders, Philadelphia, 1973.

Posey, C.J., "Computation of Discharge Including Over-Bank Flow," *Civil Engineering, American Society of Civil Engineers*, April 1967, pp. 62-63.

Sharp, J.J., *Hydraulic Modelling*, London, Butterworth, 1981.

Shen, H.T., "Hydraulic Resistance of River Ice," ASCE, *Proceedings of the Conference on Frontiers in Hydraulic Engineering*, Massachusetts Institute of Technology, Cambridge, MA., 1983, pp. 224-229.

Shen, H.T., and Yapa, P.N.D.D., "Simulation of Undersurface Roughness Coefficient of River Ice Cover," Report No. 82-6, Department of Civil and Environmental Engineering, Clarkston College, Potsdam, NY, July, 1982.

Tsang, G., "Resistance of Beauharnois Canal in Winter," *Proceedings of the American Society of Civil Engineers, Journal of the Hydraulics Division*, vol. 108, no. HY2, February 1982, pp. 167-186.

Tsang, G., Closure to "Resistance of Beauharnois Canal in Winter," *Journal of Hydraulic Engineering, American Society of Civil Engineers*, vol. 109, no. 4, April 1983, pp. 642-643.

Uzuner, M.S., "The Composite Roughness of Ice Covered Streams," *International Association for Hydraulic Research, Journal of Hydraulic Research*, vol. 13, no. 1, 1975, pp. 79-102.

Vanoni, V., and Brooks, N.H., "Laboratory Studies of the Roughness and Suspended Load of Alluvial Streams," Report No. 11, California Institute of Technology, Pasadena, CA, 1957.

5.6.   PROBLEMS

1.   Manning's n for a channel is estimated as $n = 0.015$. If the channel is rectangular in shape with a bottom width of 10 ft (3.0 m) and a longitudinal slope of 0.001 ft/ft,

a.   Estimate the discharge of this channel by both the Manning and Chezy equations, when the depth of flow is 2 ft (0.61 m).

b.   Estimate the discharge of this channel, by both the Manning and Chezy equations, when the depth of flow is 20 ft (6.1 m).

c.   Discuss the results of parts (a) and (b).

2.   The following field notes describe a channel reach. Using these field notes, estimate a value of the Manning n, for the channel reach.

Course:	straight, 661 ft (200 m) long
Cross Section:	very little variation in shape
Side Slopes:	fairly regular
Soil:	lower part, yellowish gray clay
	upper part, light gray silty clay loan
Condition:	side slopes covered with heavy growth of poplar trees 2 to 3 inches (5.1 to 7.6 cm) in diameter, large willows, and climbing vines; thick growth of water weeds on the channel bottom.

3.   The following field notes describe a channel reach. Using these field notes, estimate a value of the Manning n, for the channel reach.

Course:	straight, 1,710 ft (520 m) long
Cross Section:	not much variation in shape
Side Slopes:	somewhat irregular
Bottom:	rather irregular and uneven
Soil:	heavy silty clay
Condition:	practically no vegetation or obstructions, appears to be newly dredged.

4.   Determine the normal discharges for $y_N = 6.0$ ft (1.8 m), $n = 0.015$, and $S = 0.0020$ ft/ft in channels having the following geometric characteristics:

a.   A rectangular section having a width of 20 ft (6.1 m).

b.   A triangular section with an internal bottom angle equal to 60°.

c.   A trapezoidal section with a bottom width of 20 ft (6.1 m) and side slopes of 1 (vertical): 2 (horizontal).

5.   Prove in English units

$$Q = \frac{0.47 y^{8/3} \sqrt{S}\, f(z)}{n}$$

where

$$f(z) = \frac{z^{5/3}}{\left(1 + \sqrt{1 + z^2}\right)^{2/3}}$$

is the equation for discharge in a triangular highway gutter having one vertical side, one side sloped 1 on z, Manning's n, depth of flow y, and longitudinal slope S. The figure below defines the problem; note the break in side slope and possible change in n, when the gutter overflows.

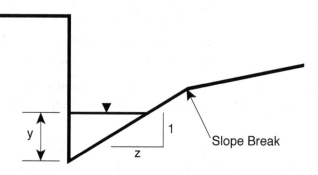

6.   For the previous problem, estimate the flow in the gutter if

$z = 24$
$n = 0.017$
$y = 0.22$ ft (0.07 m)
$S = 0.30$ ft/ft

7.   Prove for a circular pipe of diameter D that $R = D/4$. Show in doing this that the Darcy-Weisbach equation for pipe flow is equivalent to a special case of the Chezy equation provided that

$$C = \left(\frac{8g}{f}\right)^{1/2}$$

8. A canal of rectangular section and $n = 0.014$ is to be laid on a slope of 0.001 ft/ft and is to carry a discharge of 1,000 ft³/s (28 m³/s). Calculate the flow area required for the following values of the ratio b/y: 1, 1.5, 2, 2.5, 3, 3.5 and plot the flow areas as a function of b/y.

9. Estimate the normal depth of flow and velocity for a trapezoidal channel with the following characteristics: $b = 6.1$ m (20 ft), $z = 2$, $Q = 11$ m³/s (390 ft³/s), $S = 0.0016$ m/m, and $n = 0.025$.

10. A critical uniform flow, at a specified normal depth, may be produced in a channel of fixed geometry and roughness, by varying the slope and discharge. For a wide rectangular channel of width b, develop a theoretical expression for the critical slope in terms of n and q.

11. A lined rectangular channel ($n = 0.011$) is to be built to convey flows that range from 400 ft³/s (11 m³/s) to 8,000 ft³/s (230 m³/s). If other considerations dictate that the channel must be 40 ft (12 m) wide, what is the range of slopes for this channel, such that the flow will be subcritical?

12. Estimate the normal discharge in a channels having the following cross sections when $y = 2$ m (6.6 m), $n = 0.015$ and $S = 0.0020$ m/m:
    a. Triangular section with an internal bottom angle of 45°.
    b. Circular section 4.5 m (15 ft) in diameter.
    c. Parabolic section, having a top width of 5 m (16 ft) at a depth of 1.2 m (3.9 ft).

13. Based on field observation of historical floods on alluvial fans, it has been suggested that natural flow cut alluvial channels stabilize at the point where a decrease in the flow would result in a 200-fold increase in the top width or

$$\frac{dy}{dT} = -0.005$$

(see for example, French (1987) or Dawdy (1979)). For a wide rectangular channel and specified values of Q, n, and S, develop equations for
a.   The channel top width.
b.   The depth of flow.
c.   The average velocity.

In doing this, the resulting equations should include a factor that accounts for either English or SI units.

14.   A number of investigators (see for example, Jarrett (1984)) have recognized that the available guidelines for estimating n in high gradient streams ($S > 0.002$) are based on very limited data. An empirical expression for n in such cases is

$$n = 0.39S^{0.38}R^{-0.16}$$

Assume in natural channels, on alluvial fans, that n is given by the above equation and the channel is subject to the condition of Prob. 13. Then for a wide rectangular channel
a.   Estimate T in terms of Q and S.
b.   Estimate the depth of flow in terms of Q and S.
c.   Estimate u in terms of Q and S.

15.   For $Q = 28$ m³/s (990 ft³/s), $n = 0.0115$, and $S = 0.05$ m/m, use the results from both Probs. 13 and 14 to
            a.   Estimate T.
            b.   Estimate y.
            c.   Estimate u.
            d.   Estimate F.
            e.   Discuss and compare the results.

16.   In a laboratory flume experiment, the bed of a rectangular flume is artificially roughened, but the side walls are smooth. If $Q = 0.024$ m³/s (0.85 ft³/s), $y = 0.073$ m (0.24 ft), $b = 0.85$ m (2.8 ft), $S = 0.00278$ m/m, and the water temperature is 25.5° C, estimate f_b and u_{*b}.

17.   A trapezoidal irrigation channel is constructed through alluvial materials ($n = 0.02$) and has the following dimensions: $b = 15$ ft (4.6 m), $z = 1.5$, $S = 0.0003$ ft/ft and a total safe depth of 4 ft (1.2 m). After it is built, it is found that there is excessive seepage and two proposals are advanced to solve this problem: (1) line only the sides (Alternative 1) and (2) line the bottom (Alternative 2). If the

proposed lining material is concrete ($n = 0.012$), determine the following:

a.   The normal discharge in the unlined channel if 1 ft (0.30 m) of freeboard is required.

b.   The equivalent roughness and normal discharge [allowing 1 ft (0.30 m) of freeboard] for Alternative 1.

c.   The equivalent roughness and normal discharge [allowing 1 ft (0.30 m) of freeboard] for Alternative 2.

18.   A circular storm water drainage pipe 6 ft (1.8 m) in diameter is laid on a slope of 0.0004 ft/ft. If $n = 0.015$ and $Q = 75$ ft³/s (2.1 m³/s), what is the depth of flow in the drainage pipe?

19.   In a circular open channel, at what depth does the maximum normal discharge take place? How much larger is the maximum discharge that the just-full pipe discharge?

20.   A trapezoidal channel excavated in earth, with a straight alignment and uniform cross section, has a bottom width of 2 ft (0.61 m), side slopes of 1:1, and a longitudinal slope of 0.003 ft/ft. If the clean channel is 10 years old and the normal depth is 1.0 ft (0.30 m), estimate the velocity of flow and discharge.

21.   For the channel in the following figure with a slope of 0.005 ft/ft, estimate the following:

a.   For $n_A = n_B = 0.02$, what is the discharge? Is there more than one estimate of this discharge?

b.   For $n_A = 0.02$ and $n_B = 0.05$, what is the discharge? Is there more than one estimate of this discharge?

c.   For the situation defined in the figure below, what is the meaning of average discharge?

d.   Compare and comment on the results from Parts (a) and

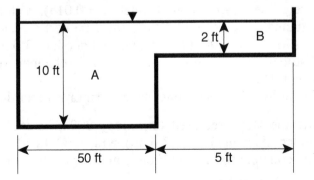

22. A trapezoidal channel parallels a highway and a bottom width of 4 ft (1.2 m), side slopes 3:1, and a grade of 1 percent. At the lower end of the channel, the flow must be turned 90° and passed under the roadway. The designer can line the channel with concrete ($n = 0.04$) or grass ($n = 0.04$). The design flow is 300 ft³/s (8.5 m³s).
    a. From the viewpoint of hydraulic engineering only, which lining would you choose? Why?
    b. Discuss other factors that would affect the choice made in Part (a).

23. A trapezoidal channel has a bottom width of 11 m (36 ft) and $z = 1.5$. If this channel is lined with smooth concrete and $S = 0.0003$ m/m, estimate the mean velocity and discharge for a normal depth of flow of 3.5 m (11 ft). Considering that Manning's n is an approximate coefficient, estimate the maximum and minimum values of discharge. Discuss the results.

24. A trapezoidal channel with a bottom width of 5 m (16 ft), $n = 0.015$, $z = 2$, and $S = 0.0015$ m/m, conveys a flow of 25 m³/s (880 ft³/s). Estimate the normal depth of flow for this situation.

25. A triangular channel, with an apex angle of 50°, conveys a flow of 50 ft³/s (1.4 m³/s), at a depth of 1 ft (0.30 m). If the longitudinal slope of the channel is 0.009 ft/ft, estimate the roughness coefficient of the channel. From the viewpoint of engineering design and construction, is this an attainable roughness coefficient value?

26. A trapezoidal channel has a bottom width of 11 m (36 ft) and $z = 1.5$. If the channel is lined with smooth concrete, estimate the longitudinal channel slope required to convey a flow of 60 m³/s (2,100 ft³/s), at a depth of 3.5 m (11 ft).

27. A concrete lined trapezoidal channel ($n = 0.015$), with $z = 1.0$ and a longitudinal slope of 0.004 m/m must be designed to convey a flow of 150 m³/s (5,300 ft³/s), at a normal depth of 3.0 m (9.8 ft).
    a. Estimate the bottom width of the channel required to accomplish this.
    b. Is the flow in this channel subcritical or supercritical?

28. Estimate the slope required to convey 200 ft³/s (5.7 m³/s) in a rectangular channel 10 ft (3.0 m) wide and lined with poorly troweled concrete so that the Froude number of the flow is 2.

CHAPTER 6

THEORY AND ANALYSIS OF GRADUALLY AND SPATIALLY VARIED FLOW

6.1. BASIC ASSUMPTIONS AND THE EQUATION OF GRADUALLY VARIED FLOW

From Chapter 2, the gradual variation of the depth of flow, in the longitudinal direction in an open channel, is governed by

$$\frac{dy}{dx} = \frac{S_0 - S_f}{1 - Fr^2} \qquad (6.1.1)$$

While Chapter 2 solutions of this equation for the special case of $S_f = 0$ were determined, in this chapter, solutions of this equation for $S_f \neq 0$ will be sought. In general, all the solution techniques, which will be developed for gradually varied flow, depend on the following assumptions:

1. The headloss for a specified reach is equal to the headloss in the reach for a uniform flow having the same hydraulic radius and average velocity, or in terms of the Manning equation

$$S_f = \frac{n^2 \overline{u}^2}{\phi^2 R^{4/3}} \qquad (6.1.2)$$

It is believed that the errors, which arise from this assumption, are small, compared with those incurred in the normal use of the

Manning equation and in the estimation of a resistance coefficient. This assumption is probably more accurate in the case of contracting flow fields than in expanding flow fields since, in the former, the headloss is primarily the result of frictional effects, while in the latter, there may be significant losses attributable to eddies.

2.   The slope of the channel is small; therefore, the depth of flow is the same, whether it is measured vertically or perpendicular, to the bottom.

3.   There is no air entrainment. If there is significant air entrainment, the problem is solved under the assumption of no air entrainment and the resulting profile is modified to account for air entrainment.

4.   The velocity distribution in the channel section is fixed; therefore, α, the kinetic energy correction factor, is constant.

5.   The resistance coefficient is independent of the depth of flow and constant throughout the reach under consideration.

6.2.   CHARACTERISTICS AND CLASSIFICATION OF GRADUALLY VARIED FLOW PROFILES

In examining the computation of gradually varied flow profiles, it is first necessary to develop a systematic method of classifying the profiles, which may occur in a specified channel reach. Recall

$$Fr^2 = \frac{Q^2 T}{g A^3}$$

for a channel of arbitrary shape, and note that Eq. (6.1.2) can be rewritten as

$$S_f = \frac{n^2 Q^2 P^{4/3}}{\phi^2 A^{10/3}}$$

Substitution of the above equations in Eq. (6.1.1) yields

$$\frac{dy}{dx} = \frac{S_0 - \dfrac{n^2 Q^2 P^{4/3}}{\phi^2 A^{10/3}}}{1 - \dfrac{Q^2 T}{gA^3}} \tag{6.2.1}$$

For a specified value of Q, F and S_f are functions of the depth of flow y, and in a wide channel F and S_f will vary in much the same way with y since $P \simeq T$ and both S_f and F have a strong inverse dependence on the flow area. In addition, as y increases, both F and S_f will decrease. By definition, $S_f = S_0$, when $y = y_N$; therefore, the following set of inequalities must apply:

$$S_f \overset{>}{\underset{<}{}} S_0 \text{ according to } y \overset{<}{\underset{>}{}} y_N \tag{6.2.2}$$

and

$$F \overset{>}{\underset{<}{}} 1 \text{ according to } y \overset{<}{\underset{>}{}} y_c \tag{6.2.3}$$

These inequalities divide the channel into three sections in the vertical dimension (Fig. 6.1). By convention, these sections are labeled 1 to 3, starting at the top. For a channel of mild slope (Fig. 6.1), the following results can be stated, where y is the actual depth of flow.

1. Level 1: $y > y_N > y_c$; $S_0 > S_f$; $F < 1$; therefore $dy/dx > 0$.

2. Level 2: $y_N > y > y_c$; $S_0 < S_f$; $F < 1$; therefore $dy/dx < 0$.

3. Level 3: $y_N > y_c > y$; $S_0 < S_f$; $F > 1$; therefore $dy/dx > 0$.

With the sign of dy/dx determined in each region, the behavior of the water surface profile can be predicted. Let us consider each of the above cases in turn.

Level 1:   In this situation, at the upstream boundary, $y \rightarrow y_N$ and, by definition, $S_f \rightarrow S_0$; thus, $dy/dx \rightarrow 0$. At the downstream boundary, $y \rightarrow \infty$ and S_f and F approach zero; thus $dy/dx \rightarrow S_0$ and the water surface asymptotically approaches a horizontal line. With these results, a rather clear picture of the circumstances, which could cause such a profile emerges. This type of situation can occur behind a dam and is designated an M1 backwater curve, where M indicates the channel slope (mild) and 1 identifies the vertical region of the channel in which the profile occurs.

Level 2:   In this situation, at the upstream boundary, $y \rightarrow y_N$ and, by definition, $S_f \rightarrow S_0$; thus, $dy/dx \rightarrow 0$. At the downstream boundary, $y \rightarrow y_c$ and $dy/dx \rightarrow \infty$. The downstream boundary condition can never be exactly met since the water surface can never form a right angle with the bed of the channel. This type of profile can occur at either a free overfall or at a transition between a channel of mild slope and a channel with a steep slope. This water surface profile is termed an M2 drawdown curve.

Level 3:   At the upstream boundary, $y \rightarrow 0$ and both S_f and F approach infinity, with the result being that dy/dx tends to a positive, finite limit. This result is of limited interest since a zero depth of flow can never occur. At the downstream boundary, $y \rightarrow y_c$, the derivative dy/dx is positive, and the depth of flow increases until the sequent depth is attained and a hydraulic jump forms. This profile is termed an M3 profile and can occur downstream of a sluice gate in a channel of mild slope or at the point where a channel of steep slope meets a channel of mild slope.

The information above has presented a rather detailed description of gradually varied flow profiles in channels of mild slope. In Fig. 6.1, the possibly gradually varied flow profiles for channels of various slopes are summarized schematically, and in Table 6.1, the various types of water surface profiles are summarized in a tabular form. The material which follows briefly describes the various possibilities.

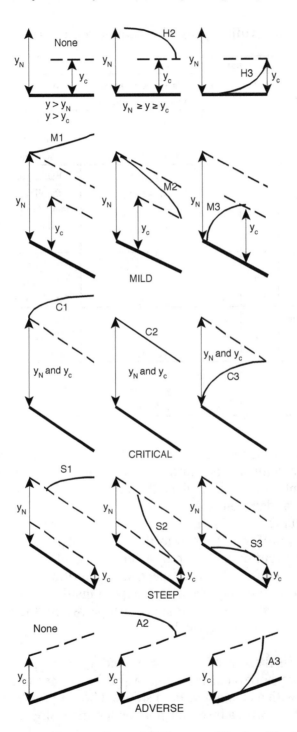

Figure 6.1. Gradually varied flow profile classification.

Table 6.1,   Flow profile types. (* Assumed a positive value)

Channel Slope	Profile Designation			Relative Relationship Among y, y_N, and y_C	Type of Curve	Type of Flow
	Zone 1	Zone 2	Zone 3			
Mild $0 < S_0 < S_C$	M1			$y > y_N > y_C$	Backwater	Subcritical
		M2		$y_N > y > y_C$	Drawdown	Subcritical
			M3	$y_N > y_C > y$	Backwater	Subcritical
Critical $S_0 = S_C > 0$	C1			$y > y_C = y_N$	Backwater	Subcritical
		C2		$y_C = y = y_N$	Parallel to channel bottom	Uniform-critical
			C3	$y_C = y_N > y$	Backwater	Supercritical
Steep $S_0 > S_C > 0$	S1			$y > y_C > y_N$	Backwater	Subcritical
		S2		$y_C > y > y_N$	Drawdown	Supercritical
			S3	$y_C > y_N > y$	Backwater	Supercritical
Horizontal $S_0 = 0$	None					
		H2		$y_N > y > y_C$	Drawdown	Subcritical
			H3	$y_N > y_C > y$	Backwater	Supercritical
Adverse $S_0 < 0$	None					
		A2		$y_N{}^* > y > y_C$	Drawdown	Subcritical
			A3	$y_N{}^* > y_C > y$	Backwater	Supercritical

S Profiles: In these profiles, $S_0 > S_c$ and $y_N < y_c$. The *S1* profile usually begins with a jump at the upstream boundary and terminates with a profile tangent to a horizontal at the downstream boundary, e.g., the flow behind a dam built in a steep channel. The *S2* profile is a drawdown curve and is usually very short. At the downstream boundary, this profile is tangent to the normal depth of flow. This type of profile can occur downstream of an enlargement of a channel section and also transitional in that it connects a supercritical flow with normal depth. Such a profile may occur at a slope transition between steep and mild, or below a sluice gate, on a steep slope, where the water initially flows at a depth less than normal.

C Profiles: In these profiles, $S_0 = S_c$ and $y_N = y_c$. The *C1* profile is asymptotic to a horizontal line, e.g., a profile connecting a channel of critical slope with a channel of mild slope. The *C3* profile can connect a supercritical flow with a reservoir pool on a critical slope.

H Profiles: In this case, $S_0 = 0$, and the H profiles can be considered to be a limiting case of the M profiles. The H2 drawdown profile can be found upstream of a free overfall, while the *H3* profile can connect the supercritical flow below a sluice gate with a reservoir pool.

A Profiles: In these profiles $S_0 < 0$. In general, the *A2* and *A3* profiles occur only infrequently and are similar to the *H2* and *H3* profiles.

In summary, the following principles regarding gradually varied flow profiles can be stated:

1.   The sign of dy/dx can be readily determined from Eqs. (6.2.1) through (6.2.3).

2.   When the water surface profile approaches normal depth, it does so asymptotically.

3.   When the water surface profile approaches critical depth, it crosses this depth at a large but finite angle.

4.   If the flow is subcritical upstream but passes through critical depth, then the feature which produces critical depth determines and locates the whole water surface profile. If the upstream flow is supercritical, e.g., in the case of an *M3* profile, then the control cannot come from downstream.

5.   Every gradually varied flow profile exemplifies the principle that subcritical flows are controlled from downstream, while supercritical flows are controlled from upstream. Gradually varied flow profiles would not exist if it were not for the upstream and downstream controls.

6.   In channels with horizontal and adverse slopes, the terminology *normal depth of flow* has no meaning since the normal depth of flow is either negative or imaginary. However, in these cases, the numerator of Eq. (6.2.1) is negative and the shape of the profile can be deduced.

6.3.   COMPUTATION OF GRADUALLY VARIED FLOW

In this section, a number of methods of integrating Eq. (6.1.1) to determine the variation of the depth of flow with distance are discussed.

Although there are a variety of solution techniques for accomplishing this integration, given a particular situation, one method may be superior to the others. Thus, the reader is cautioned to carefully consider the problem before proceeding to a computational procedure. All solutions of the gradually varied flow equation must begin with the depth of flow at a control and proceed in the direction in which the control operates. At the upstream boundary, the gradually varied flow profile may approach a specified depth of flow asymptotically. At this point, under the condition of asymptotic approach, a reasonable definition of convergence must be developed.

6.3.1.   Uniform Channels

In the case of a uniform channel, i.e., a prismatic channel with a constant slope and resistance coefficient, a tabular solution of the gradually varied flow equation is possible. In such a channel,

$$\frac{dE}{dx} = S_0 - S_f = S_0 - \frac{n^2 \bar{u}^2}{\phi^2 R^{4/3}} \tag{6.3.1}$$

or in finite difference from

$$\frac{\Delta E}{\Delta x} = \frac{\Delta\left(y + \frac{\bar{u}^2}{2g}\right)}{\Delta x} = S_0 - \frac{n^2 \overline{\bar{u}^2}}{\phi^2 R^{4/3}} \tag{6.3.2}$$

In this equation, all variables with the exception of Δx, are functions of y, the depth of flow. Thus, Eq. (6.3.2) can be solved by selecting values of y, computing $\bar{u}$ and R, and solving for the value of Δx corresponding to y or

$$\Delta x = \frac{\Delta\left(y + \frac{\bar{u}^2}{2g}\right)}{\left(S_0 - \frac{n^2 \bar{u}^2}{\phi^2 R^{4/3}}\right)_m} \tag{6.3.3}$$

where the subscript m indicates a mean value over the interval being considered. This is commonly termed the *step method*. The primary difficulty with this technique is that, in most cases, ΔE or

$$\Delta\left(y + \frac{\overline{u^2}}{2g}\right)$$

is small; thus, the difficulty of accurately determining a small difference between two relatively large numbers must be resolved. This computational difficulty can be overcome by using Eq. (2.2.1) in a finite difference form or

$$\Delta E = \Delta y\left(1 - Fr^2\right) \tag{6.3.4}$$

and Eq. (6.3.3) then becomes

$$\Delta x = \frac{\Delta y\left(1 - Fr^2\right)_m}{\left(S_0 - \frac{n^2 u^2}{\phi^2 R^{4/3}}\right)_m} \tag{6.3.5}$$

A secondary difficulty with this solution technique occurs in situations where $S_f \rightarrow S_0$ since, in this case, the difference S_f - S_0 is small. In general, this does not present grave difficulties from a practical viewpoint, since the details of the asymptotic approach of the depth of flow to normal depth is not of interest. The details of this method of solution are best illustrated with an example.

EXAMPLE 6.1

A trapezoidal channel with $b = 20$ ft (6.1m), $n = 0.025$, $z = 2$, and $S_0 = 0.001$ carries a discharge of 1000 ft^3/s (28 m^3/s). If this channel terminates in a free overfall, determine the gradually varied profile by the step method.

Solution

The flow profile will be a drawdown curve, and the first step in solving the problem is to establish the depth of flow at the upstream and downstream boundaries.

1. *Upstream Boundary*: At the upstream boundary, the depth of flow will be y_N. For the given information y_N can be readily determined by trial and error or another appropriate method. For exmaple,

$$AR^{2/3} = \frac{nQ}{\phi\sqrt{S}} = \frac{0.025(1000)}{1.49\sqrt{0.001}} = 531$$

and

$$y_N = 6.25\,\text{ft}\,(1.90\,\text{m})$$

Since at this boundary the depth of flow approaches y_N asymptotically, a reasonable definition of convergence must be established; e.g., computations will proceed until the depth of flow is $0.90 y_N$ or 5.62 ft (1.71m).

2. *Downstream Boundary*: At the downstream boundary, the depth of flow will be y_c. From Eq. (2.2.3)

$$\frac{\overline{u}^2}{2g} = \frac{D}{2} \quad \rightarrow \quad \frac{Q^2}{gA^2} = \frac{A}{T}$$

or

$$\frac{Q^2}{g} = \frac{A^3}{T}$$

where

$$A = (b + zy)y = (20 + 2y)y$$

$$T = b + 2zy = 20 + 4y$$

Then

$$\frac{y_c^3(20 + 2y_c)^3}{20 + 4y_c} = \frac{1000^3}{32.2} = 31{,}060$$

By trial and error

$$y_c = 3.74\,\text{ft}\,(1.14\,\text{m})$$

Thus, the range of depths in this computation is from 3.74 ft (1.14m) at the control to 5.62 ft, at the upstream boundary. The Froude number at the upstream boundary is

$$F = \frac{\bar{u}}{\sqrt{gD}} = 0.41$$

Therefore, the drawdown profile is an *M2* curve. The details of the computation are contained in Table 6.2*a* and *b*, and the results are summarized in Fig. 6.2. In Table 6.2*a*, Eq. (6.3.3) was used, while in Table 6.2b, Eqs. (6.3.4) and (6.3.5) were used. The actual difference in the results at the upstream boundary is less than 2 percent.

This method is useful for artificial channels and has the distinct advantage of predicting a longitudinal distance for a specified depth of flow. However, the method has the disadvantage of not being capable of predicting a depth of flow at a specified longitudinal distance.

A second method of estimating a gradually varied flow profile in uniform channels is by direct integration of Eq. (6.1.1). In this equation, the Froude number for a channel of arbitrary shape is defined as

$$Fr^2 = \frac{\alpha Q^2 T}{gA^3}$$

Recall that in Eq. (2.2.9), the section factor was defined as

$$Z = \sqrt{\frac{A^3}{T}}$$

Assume that a critical flow of discharge Q occurs in the subject channel; then

$$Q = Z_c \sqrt{\frac{g}{\alpha}} \tag{6.3.6}$$

where Z_c = the critical flow section factor. It is crucial that the reader differentiate between Z, which is simply a numerical value computed for a specified Q as if the flow occurred at critical depth. Substitution of the definition of Z and Eq. (6.3.6) into the Froude number definition yields

$$Fr^2 = \frac{Z_c^2}{Z^2} \tag{6.3.7}$$

Table 6.2. Example 6.1.

(a). Solution of Example 6.1 using Eq. (6.3.3)

y	A	P	R	$\overline{u}$	$\dfrac{\overline{u}^2}{2g}$	E	$\dfrac{n^2\overline{u}^2}{\phi^2 R^{4/3}}$	$\left(\dfrac{n^2\overline{u}^2}{\phi^2 R^{4/3}}\right)_m$	$S_0 - \left(\dfrac{n^2\overline{u}^2}{\phi^2 R^{4/3}}\right)_m$	ΔE	Δx	$x = \Sigma\delta x$
(ft)	(ft²)	(ft)	(ft)	(ft/s)	(ft)	(ft)				(ft)	(ft)	(ft)
(1)	(2)	(3)	(4)	(5)	(6)	(7)	(8)	(9)	(10)	(11)	(12)	(13)
3.74	103	36.7	2.81	9.71	1.46	5.20	0.00670					
								0.00601	-0.00501	+0.0400	-8.00	
4.00	112	37.9	2.95	8.93	1.24	5.24	0.00531					-8.0
								0.00478	-0.00378	+0.0700	-18.5	
4.25	121	39.0	3.10	8.26	1.06	5.31	0.00425					-27
								0.00386	-0.00286	+0.110	-38.5	
4.50	130	40.1	3.24	7.69	0.918	5.42	0.00347					-65
								0.00314	-0.00214	+0.120	-56.1	
4.75	140	41.2	3.40	7.14	0.792	5.54	0.00281					-121
								0.00256	-0.00156	+0.150	-96.1	
5.00	150	42.3	3.55	6.67	0.691	5.69	0.00231					-217
								0.00213	-0.00113	+0.170	-150	
5.25	160	43.5	3.68	6.25	0.607	5.86	0.00194					-367
								0.00179	-0.000790	+0.180	-228	
5.50	170	44.6	3.81	5.88	0.537	6.04	0.00164					-595
								0.00156	-0.000560	+0.080	-143	
5.62	176	45.1	3.90	5.68	0.501	6.12	0.00148					-738

Table 6.2. Example 6.1.

(b). Solution of Example 6.1 using Eqs. (6.3.4) and (6.3.5)

y	A	P	T	R	D	$\bar{u}$	$\dfrac{n^2\bar{u}^2}{\phi^2 R^{4/3}}$	F^2	$\left(\dfrac{n^2 u^2}{\phi^2 R^{4/3}}\right)_m$	F^2_m	$S_0 - \left(\dfrac{n^2 u^2}{\phi^2 R^{4/3}}\right)_m$	ΔE	Δx	$\Sigma\delta x$
(ft)	(ft²)	(ft)	(ft)	(ft)	(ft)	(ft/s)						(ft)	(ft)	(ft)
(1)	(2)	(3)	(4)	(5)	(6)	(7)	(8)	(9)	(10)	(11)	(12)	(13)	(14)	(15)
3.74	103	36.7	35.0	2.81	2.94	9.71	0.00670	0.996						
									0.00601	0.896	-0.00501	0.0270	-5.39	
4.00	112	37.9	36.0	2.95	3.11	8.93	0.00531	0.796						-5
									0.00478	0.722	-0.00378	0.0695	-18.4	
4.25	121	39.0	37.0	3.10	3.27	8.26	0.00425	0.648						-24
									0.0386	0.593	-0.00286	0.102	-35.7	
4.50	130	40.1	38.0	3.24	3.42	7.69	0.00347	0.537						-59
									0.00314	0.489	-0.00214	0.128	-59.8	
4.75	140	41.2	39.0	3.40	3.59	7.14	0.00281	0.441						-119
									0.00256	0.404	-0.00156	0.149	-95.5	
5.00	150	42.3	40.0	3.55	3.75	6.67	0.00231	0.368						-215
									0.00213	0.340	-0.00113	0.165	-146	
5.25	160	43.5	41.0	3.68	3.90	6.25	0.00194	0.311						-361
									0.00179	0.288	-0.00079	0.178	-225	
5.50	170	44.6	42.0	3.81	4.05	5.88	0.00164	0.265						-586
									0.00156	0.254	-0.00056	0.0895	-160	
5.62	176	45.1	42.5	3.90	4.14	5.68	0.00148	0.242						-746

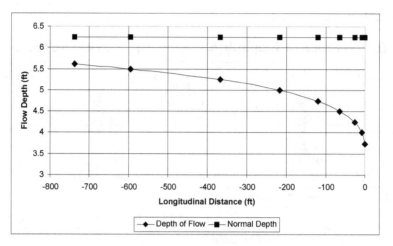

Figure 6.2. Drawdown curve for Example 6.1.

Both the slope of the channel bottom and the energy slope can be represented in terms of the conveyance defined in Eq. (5.1.3) or

$$S_0 = \frac{Q^2}{K_N^2} \tag{6.3.8}$$

and

$$S_f = \frac{Q^2}{K^2} \tag{6.3.9}$$

Again it is necessary to distinguish between K, which is the numerical value of the conveyance, when the flow occurs at depth y and K_N, which is the conveyance computed for a specified Q, as if the flow occurred at normal depth. Rearranging Eq. (6.1.1) and substituting Eqs. (6.3.7) and (6.3.9) yields

$$\frac{dy}{dx} = S_0 \left[\frac{1 - \left(\dfrac{K_N}{K} \right)^2}{1 - \left(\dfrac{Z_c}{Z} \right)^2} \right] \tag{6.3.10}$$

If Eq. (6.3.10) is to be solved by direct integration, it is necessary that the parameters $(K_N/K)^2$ and $(Z_c/Z)^2$ be expressed in terms of the depth of flow. Since the section factor is only a function of the channel shape and y, assume

$$Z^2 = Cy^M \tag{6.3.11}$$

where C = a coefficient and M = hydraulic exponent for critical flow. The functional relationship between M, the depth of flow, and the geometry of the channel can be established by first taking the logarithm of both sides of Eq. (6.3.11), and then, differentiating that equation with respect to y or

$$\frac{d(\ln Z)}{dy} = \frac{M}{2y} \tag{6.3.12}$$

Taking the logarithm and differentiating the definition of the section factor, Eq. (2.2.9), with respect to y yields

$$\frac{d(\ln Z)}{dy} = \frac{3}{2}\frac{T}{A} - \frac{1}{2T}\frac{dT}{dy} \tag{6.3.13}$$

Combining Eqs. (6.3.12) and (6.3.13) and solving for M yields

$$M = \frac{y}{A}\left(3T - \frac{A}{T}\frac{dT}{dy}\right) \tag{6.3.14}$$

With a channel geometry specified, the value of M can be calculated for each depth of flow. For a rectangular channel, M has a value of 3. Expressions for calculating M in channels of various shapes are summarized in Table 6.3.

In an analogous fashion, it is assumed that conveyance is also a function of the depth of flow or

$$K^2 = C'y^N \tag{6.3.15}$$

where C' = a coefficient and N = hydraulic exponent for uniform flow computations. Following a procedure analogous to the technique used for M, yields

$$N = \frac{2y}{3A}\left(5T - 2R\frac{dP}{dy}\right) \tag{6.3.16}$$

Table 6.3. Values of M and N for rectangular, trapezoidal, and triangular channels.

	M	N
Rectangular	3.0	3.33
Trapezoidal	$\dfrac{3\left(1+2z\dfrac{y}{b}\right)^2 - 2z\dfrac{y}{b}\left(1+z\dfrac{y}{b}\right)}{\left(1+2z\dfrac{y}{b}\right)\left(1+z\dfrac{y}{b}\right)}$	$\dfrac{10\left(1+2z\dfrac{y}{b}\right)}{3\left(1+z\dfrac{y}{b}\right)} - \dfrac{8\dfrac{y}{b}\sqrt{1+z^2}}{3\left(1+2\dfrac{y}{b}\sqrt{1+z^2}\right)}$
Triangular	5	5.33

The value of N depends on whether the conveyance is expressed in terms of Manning's n or the Chezy C. Equation (6.3.16) is based on the Manning equation. For example, in a rectangular channel, $N = 3.33$, if the Manning equation is used, but $N = 3$, if the Chezy equation is used. Expressions for computing N in channels of various shapes are summarized in Table 6.3. It should be noted that, in general, N and M are functions of y and are not constants.

Substitution of Eqs.(6.3.11) and (6.3.15) into Eq. (6.3.10) yields

$$\frac{dy}{dx} = S_0 \left[\frac{1 - \left(\dfrac{y_N}{y}\right)^N}{1 - \left(\dfrac{y_C}{y}\right)^M} \right] \qquad (6.3.17)$$

Letting $u = y/y_N$ and rearranging the equation yields

$$dx = \frac{y_N}{S_0}\left[1 - \frac{1}{1-u^N} + \left(\frac{y_C}{y_N}\right)^M \frac{u^{N-M}}{1-u^N}\right] du \qquad (6.3.18)$$

If Eq. (6.3.18) is to be integrated, then N and M must be assumed to be constants. In general, it is possible to divide a channel into subsections so that N and M are essentially constant. Under this assumption

$$x = \frac{y_N}{S_0}\left[u - \int_0^u \frac{du}{1-u^N} + \left(\frac{y_C}{y_N}\right)^M \int_0^u \frac{u^{N-M}}{1-u^N}du\right] + C \qquad (6.3.19)$$

where the integral

$$\int_0^u \frac{du}{1-u^N} = F(u,N)$$

is known as the Bakhmeteff varied flow function and C = a constant of integration. The second integral in Eq. (6.3.19) can also be expressed in the form of the varied-flow function by substituting $v = u^{N/J}$, where $J = N/(N - M + 1)$ (Chow, 1959). Then

$$\int_0^u \frac{u^{N-M}}{1-u^N}du = \frac{J}{N}\int_0^v \frac{dv}{1-v^J} = \frac{J}{N}F(v,J) \qquad (6.3.20)$$

With this definition, Eq. (6.3.19) becomes

$$x = \frac{y_N}{S_0}\left[u - F(u,N) + \frac{J}{N}\left(\frac{y_C}{y_N}\right)^M F(v,J)\right] + C$$

The constant of integration in this equation, C, can be eliminated if Eq. (6.3.17) is integrated between two stations or

$$L = x_2 - x_1 = A\left\{u_2 - u_1 - F(u_2,N) - F(u_1,N) + B\left[F(v_2,J) - F(v_1,J)\right]\right\}(6.3.21)$$

where $A = y_N/S_0$ and $B = (y_c/y_N)^M J/N$. Although tables for the evaluation of $(F(u,N)$ and $F(v,J)$ are available (Appendices 1 and 2), it is usually convenient, when using a computer to estimate gradually varied flow profiles, to express $F(u,N)$ in the form of an infinite series or

1. For $u < 1$

$$\int_0^u \frac{du}{1-u^N} = u + \frac{1}{N+1}u^{N+1} + ... + \frac{1}{(P-1)N+1}u^{(P-1)N+1} + R_P \qquad (6.3.22)$$

where N = hydraulic exponent
P = term number ($P = 1, 2, 3, ... \infty$)
R_P = remainder after (P - 1) terms have been summed.

With regard to Eq. (6.3.22), the following are noted: First,

$$R_P < \frac{u^{PN+1}}{PN+1}\left(\frac{1}{1-u^N}\right) \tag{6.3.23}$$

Second, for $u < 0.7$, the series converges rapidly, but for $u = 1$ the series diverges. Third, the smaller the value of u, the more rapidly the series converges.

2. For $u > 1$

$$\int_0^u \frac{du}{1-u^N} = \frac{1}{(N-1)u^{N-1}} + \frac{1}{(2N-1)u^{2N-1}} + \ldots + \frac{1}{(PN-1)u^{PN-1}} + R_P \tag{6.3.24}$$

For this series expansion, the following are noted: First,

$$R_P < \frac{u^N}{(PN-1)^{PN-1}u^{N-1}} \tag{6.3.25}$$

Second, for $u > 1.5$, the series converges rapidly, but for $u = 1$, the series diverges. Third, the larger the value of u, the more rapid the series convergence.

At this point, it should be noted that Eq. (6.3.17) can be integrated exactly for the special case of $N = M = 3$, that is, a rectangular channel with the conveyance expressed in terms of the Chezy equation. In such a case,

$$x = \frac{1}{S_0}\left\{y - y_N\left[1 - \left(\frac{y_c}{y_N}\right)^3\right]\phi\right\} \tag{6.3.26}$$

where

$$\phi = \int \frac{du}{1-u^3} = \frac{1}{6}\log\left[\frac{u^2+u+1}{(u-1)^2}\right] - \frac{1}{\sqrt{3}}\tan^{-1}\left(\frac{\sqrt{3}}{2u+1}\right) + C$$

which is known as the *Bresse function* and C = constant of integration. For computational purposes, it is usually convenient to express (y_c/y_N) as

$$\left(\frac{y_c}{y_N}\right)^3 = \frac{C^2 S_0}{g}$$

EXAMPLE 6.2

A trapezoidal channel, with b = 20 ft (6.1 m), n = 0.025, z = 2, and S_0 = 0.001, carries a discharge of 1000 ft³/s (28 m³/s). If this channel terminates in a free overfall, determine the gradually varied flow profile by the method of direct integration.

Solution

In Example 6.1, for the same problem statement it was determined that

$$y_N = 6.25\,\text{ft}\,(1.90\,\text{m})$$

and

$$y_c = 3.74\,\text{ft}\,(1.14\,\text{m})$$

A tabular solution of this problem, by the method of direct integration, is summarized in Table 6.4. This method of solution predicts that at 745 ft (227 m), from the free overfall, the depth of flow is 5.62 ft (1.71 m). This result is not significantly different from the results of Example 6.1.

As with the step method discussed previously, the method of direct integration has the disadvantage of not being capable of predicting a depth of flow given specified longitudinal distance. However, this method has the distinct advantage that each step is independent of the previous step; i.e., the total length of the gradually varied flow profile may be estimated with a single computation. For example, in Example 6.2, the distance between the points where y = 3.74 ft (1.14 m) and y = 5.62 ft (1.71 m) is given by

$$L = \frac{6.25}{0.001}\left[\left(0.899 - 0.598\right) - \left(1.12 - 0.620\right) + \frac{2.99}{3.71}\left(\frac{3.74}{6.25}\right)^{3.47}\left(1.13 - 0.552\right)\right]$$

$$L = -754\,ft\,(-230\,m)$$

where the values of M, J, and N used in this computation were obtained from the table accompanying Example 6.2 and averaged over the reach considered. This type of single-step computation is possible with the method of direct integration, but cannot be done with the step method, because of this method each step is dependent on the result obtained in the previous step. Unfortunately, this advantage of the method of direct integration cannot be exploited effectively since the intermediate values of x usually must be known, so that a gradually varied flow profile may be plotted.

In Table 6.4, it is noted that the hydraulic exponents show very little variation throughout the entire length of channel considered. When the hydraulic exponents are a function of longitudinal distance, the Bakhmeteff-Chow methodology can be applied if the channel length is divided into subreaches so that the hydraulic exponents are constant in each subreach. In rectangular, trapezoidal, and triangular channels, the assumption of constant hydraulic exponents is usually satisfactory, while in circular channels or channels which have abrupt cross-sectional changes in shape with elevation, this is not a valid assumption.

For channels with gradually closing crowns, such as circular channels, the hydraulic exponents are variable near the crown. While the methods described above can be applied, more accurate results can be obtained by using a numerical integration procedure such as that developed by Kiefer and Chu (1955). Let Q_0 be the discharge of a circular conduit of diameter d_0, when the open-channel depth of flow is y and the energy slope is equal to the bottom of S_0. Then, by Eq. (5.1.3)

$$Q_0 = K_0 \sqrt{S_0} \qquad (6.3.27)$$

For a uniform flow in this conduit of discharge Q, the same equation yields

$$Q = K_N \sqrt{S_0} \qquad (6.3.28)$$

from Eqs. (6.3.27) and (6.3.28), the following relationship can be developed

$$\left(\frac{K_N}{K}\right)^2 = \left(\frac{K_N}{K_0}\right)^2 \left(\frac{K_0}{K}\right)^2 = \left(\frac{Q}{Q_0}\right)^2 \left(\frac{K_0}{K}\right)^2 \qquad (6.3.29)$$

Table 6.4. Tabular solution of Example 6.2

y ft	$\bar{u}$ ft/s	N	M	J	V	F(u,N)	F(v,J)	$\bar{J}$	$\bar{N}$	$\bar{M}$	Δx ft	ΣΔx ft
3.74	0.598	3.63	3.39	2.93	0.529	0.620	0.552					
								2.930	3.645	3.40	-1.7	
4.00	0.640	3.66	3.41	2.93	0.573	0.670	0.607					-2
								2.950	3.670	3.425	-19	
4.25	0.680	3.68	3.44	2.97	0.620	0.721	0.664					-21
								2.975	3.690	3.450	-35	
4.50	0.720	3.70	3.46	2.98	0.665	0.775	0.725					-56
								2.990	3.710	3.470	-60	
4.75	0.760	3.72	3.48	3.00	0.712	0.834	0.794					-116
								3.010	3.730	3.490	-92	
5.00	0.800	3.74	3.50	3.02	0.759	0.899	0.871					-208
								3.025	3.750	3.510	-147	
5.25	0.840	3.76	3.52	3.03	0.805	0.974	0.958					-355
								3.040	3.770	5.530	-237	
5.50	0.880	3.78	3.54	3.05	0.853	1.066	1.065					-591
								3.055	3.785	3.545	-153	
5.62	0.899	3.79	3.55	3.06	0.876	1.118	1.130					-745

Since the ratio K_0/K is a function of y/d_0, Eq. (6.3.29) can be expressed as

$$\left(\frac{K_N}{K}\right)^2 = \left(\frac{Q}{Q_0}\right)^1 f_1\left(\frac{y}{d_0}\right) \tag{6.3.30}$$

In an analogous fashion, a corresponding expression for the section factor can be developed or

$$\left(\frac{Z_c}{Z}\right)^2 = \frac{\alpha Q^2 T}{gA^3} = \frac{\alpha Q^2}{d_0^5}\left[\frac{\dfrac{T}{d_0}}{g\left(\dfrac{A}{d_0^2}\right)^3}\right] = \frac{\alpha Q^2}{d_0^5} f_2\left(\frac{y}{d_0}\right) \tag{6.3.31}$$

where the expression

$$\frac{\dfrac{T}{d_0}}{g\left(\dfrac{A}{d_0^2}\right)^3}$$

is a function of y/d_0 and represented as $f_2(y/d_0)$. Then, substituting Eqs. (6.3.30) and (6.3.31) into Eq. (6.3.10) and simplifying yields

$$dx = \frac{d_0}{S_0}\left[\frac{1 - \dfrac{\alpha Q^2}{d_0^5}f_2\left(\dfrac{y}{d_0}\right)}{1 - \left(\dfrac{Q}{Q_0}\right)^2 f_1\left(\dfrac{y}{d_0}\right)}\right]d\left(\frac{y}{d_0}\right) \qquad (6.3.32)$$

Integration of Eq. (6.3.32) yields

$$x = \frac{d_0}{S_0}\left[\int_0^{y/d_0}\frac{d\left(\dfrac{y}{d_0}\right)}{1 - \left(\dfrac{Q}{Q_0}\right)^2 f_1\left(\dfrac{y}{d_0}\right)} - \frac{\alpha Q^2}{d_0^5}\int_0^{y/d_0}\frac{f_2\left(\dfrac{y}{d_0}\right)d\left(\dfrac{y}{d_0}\right)}{1 - \left(\dfrac{Q}{Q_0}\right)^2 f_1\left(\dfrac{y}{d_0}\right)}\right] + C \quad (6.3.33)$$

where C = a constant of integration, the integrals must be evaluated by numerical methods, and $f_1(y/d_0)$ and $f_2(y/d_0)$ can be evaluated numerically. The constant of integration can be eliminated by integrating over a designated interval or

$$L = x_2 - x_1 = A\left[\left(X_2 - X_1\right) - B\left(Y_2 - Y_1\right)\right] \qquad (6.3.34)$$

where

$$X = -\int_0^{y/d_0}\frac{d\left(\dfrac{y}{d_0}\right)}{1 - \left(\dfrac{Q}{Q_0}\right)^2 f_1\left(\dfrac{y}{d_0}\right)}$$

$$Y = - \int_{0}^{y/d_0} \frac{f_2\left(\dfrac{y}{d_0}\right) d\left(\dfrac{y}{d_0}\right)}{1 - \left(\dfrac{Q}{Q_0}\right)^2 f_1\left(\dfrac{y}{d_0}\right)}$$

$$A = - \frac{d_0}{S_0}$$

and

$$B = \frac{\alpha Q^2}{d_0^5}$$

Values of the functions X and Y are summarized in Tables 6.5 and 6.6 as functions of y/d_0 and Q/Q_0. In these tables, values between the tabulated values can be estimated by interpolation; however, the interpolation process cannot cross the heavy lines in these tables which identify the location of the normal depth of flow.

EXAMPLE 6.3

A circular conduit, 6.0 ft (1.8 m) in diameter, conveys a flow of 160 ft^3/s (4.5 m^3/s), with $S = 0.001$ and $n = 0.013$. Determine the distance between a critical depth control section and the section at which the conduit flows full.

Solution

Assume $\alpha = 1$, then from Table 2.1

$$\varphi = \frac{\alpha Q^2}{g} = \frac{1(160)}{32.2} = 795$$

and

$$y_c = \frac{1.01}{d_0^{0.26}} \psi^{0.25} = \frac{1.01}{6^{0.26}} (795)^{0.25} = 3.4 \, \text{ft} \, (1.04 \, \text{m})$$

The capacity of the conduit flowing full but as an open channel is

$$Q_0 = \frac{1.49}{n} AR^{2/3} \sqrt{S}$$

Table 6.5. The function Y as a function of y/d_0 and Q/Q_0 for circular channels (Kiefer and Chu, 1955)

Ratio y/d_0	Values of Q/Q_0									
	0.40	0.60	0.70	0.80	0.90	1.00	1.07	1.20	1.40	1.60
1.00	0	0	0	0	0	--	0.0245	0.1128	0.0680	0.0482
0.98	0.0003	0.0003	0.0004	0.0006	0.0009	0.0016	0.0186	0.1120	0.0677	0.0479
0.96	0.0008	0.0010	0.0012	0.0015	0.0024	0.0053	0	0.1105	0.0671	0.0477
0.94	0.0015	0.0019	0.0023	0.0028	0.0043	0.0098	0.0824	0.1079	0.0663	0.0470
0.92	0.0023	0.0029	0.0036	0.0045	0.0067	0.0150	0.1614	0.1051	0.0652	0.0466
0.90	0.0034	0.0042	0.0050	0.0063	0.0096	0.0215	0.2500	0.1028	0.0641	0.0458
0.88	0.0044	0.0055	0.0067	0.0086	0.0132	0.0303	0.1852	0.0983	0.0627	0.0452
0.86	0.0058	0.0072	0.0087	0.0112	0.0175	0.0438	0.1604	0.0950	0.0613	0.0444
0.84	0.0071	0.0090	0.0110	0.0143	0.0230	0.0667	0.1439	0.0913	0.0598	0.0435
0.82	0.0087	0.0112	0.0136	0.0180	0.0299	--	0.1321	0.0879	0.0583	0.0426
0.80	0.0106	0.0136	0.0166	0.0225	0.0389	0.1806	0.1230	0.0847	0.0568	0.0416
0.78	0.0128	0.0164	0.0204	0.0282	0.0524	0.1546	0.1153	0.0816	0.0552	0.0405
0.76	0.0150	0.0196	0.0247	0.0349	0.0768	0.1394	0.1088	0.0784	0.0537	0.0396
0.74	0.0177	0.0326	0.0300	0.0440	1.8457	0.1288	0.1032	0.0755	0.0521	0.0384
0.72	0.0206	0.0278	0.0361	0.0571	0.1950	0.1200	0.0980	0.0727	0.0505	0.0375
0.70	0.0240	0.0333	0.0443	0.0817	0.1650	0.1130	0.0933	0.0699	0.0489	0.0363
0.68	0.0280	0.0395	0.0556	0.1747	0.1465	0.1066	0.0891	0.0673	0.0473	0.0354
0.66	0.0324	0.0475	0.0724	0.2320	0.1352	0.1011	0.0851	0.0648	0.0458	0.0343
0.64	0.0377	0.0576	0.0978	0.1882	0.1263	0.0960	0.0813	0.0622	0.0442	0.0332
0.62	0.0438	0.0721	0.2500	0.1680	0.1189	0.0916	0.0777	0.0599	0.0426	0.0322
0.60	0.0511	0.0915	0.2780	0.1545	0.1120	0.0870	0.0743	0.0575	0.0411	0.0310
0.58	0.0600	0.1290	0.2265	0.1437	0.1054	0.0830	0.0710	0.0550	0.0396	0.0299
0.56	0.0705	0.3909	0.2006	0.1343	0.0997	0.0790	0.0678	0.0529	0.0381	0.0288
0.54	0.0840	0.3080	0.1828	0.1260	0.0944	0.0753	0.0647	0.0506	0.0365	0.0278
0.52	0.1014	0.2580	0.1686	0.1186	0.0895	0.0717	0.0618	0.0485	0.0351	0.0266
0.50	0.1253	0.2293	0.1562	0.1117	0.0850	0.0682	0.0588	0.0463	0.0336	0.0256
0.48	0.1609	0.2085	0.1452	0.1054	0.0803	0.0648	0.0561	0.0441	0.0321	0.0244
0.46	0.2235	0.1912	0.1350	0.0993	0.0761	0.0616	0.0533	0.0420	0.0306	0.0233
0.44	1.4143	0.1770	0.1256	0.0935	0.0718	0.0582	0.0505	0.0399	0.0291	0.0221
0.42	0.0500	0.1643	0.1170	0.0880	0.0680	0.0552	0.0478	0.0378	0.0276	0.0210
0.40	0.4116	0.1528	0.1090	0.0826	0.0641	0.0519	0.0451	0.0357	0.0261	0.0199
0.35	0.3090	0.1271	0.0917	0.0700	0.0546	0.0444	0.0386	0.0306	0.0224	0.0171
0.30	0.2432	0.1038	0.0755	0.0580	0.0454	0.0370	0.0321	0.0255	0.0188	0.0143
0.25	0.1903	0.0825	0.0607	0.0466	0.0366	0.0297	0.0258	0.0205	0.0151	0.0114
0.20	0.1425	0.0627	0.0462	0.0356	0.0281	0.0227	0.0197	0.0157	0.0115	0.0088
0.15	0.1000	0.0442	0.0327	0.0250	0.0197	0.0158	0.0138	0.0110	0.0081	0.0062
0	0	0	0	0	0	0	0	0	0	0

Table 6.6. The function X as a function of y/d_0 and Q/Q_0 for circular channels (Kiefer and Chu, 1955).

Ratio y/d_0	Values of Q/Q_0									
	0.40	0.60	0.70	0.80	0.90	1.00	1.07	1.20	1.40	1.60
1.00	0	0	0	0	0	--	1.1610	0.9166	0.4075	0.2543
0.98	0.0235	0.0299	0.0368	0.0495	0.0817	0.1085	0.7600	0.8567	0.3830	0.2397
0.96	0.0468	0.0592	0.0720	0.0954	0.1517	0.2777	0	0.7822	0.3555	0.2238
0.94	0.0700	0.0883	0.1067	0.1403	0.2188	0.4277	2.8720	0.7016	0.3269	0.2074
0.92	0.0932	0.1173	0.1415	0.1851	0.2857	0.5763	5.0520	0.6203	0.2982	0.1909
0.90	0.1165	0.1465	0.1764	0.2305	0.3541	0.7345	2.7491	0.5427	0.2701	0.1748
0.88	0.1398	0.1759	0.2119	0.2769	0.4257	0.9160	1.4754	0.4714	0.2432	0.1591
0.86	0.1632	0.2057	0.2481	0.3250	0.5023	1.1459	1.0082	0.4076	0.2178	0.1441
0.84	0.1868	0.2360	0.2852	0.3754	0.5868	1.5020	0.7486	0.3515	0.1942	0.1299
0.82	0.2105	0.2668	0.3236	0.4389	0.6835	2.5439	0.5803	0.3025	0.1724	0.1166
0.80	0.2344	0.2985	0.3637	0.4868	0.8002	0.8911	0.4619	0.2600	0.1524	0.1043
0.78	0.2585	0.3311	0.4059	0.5508	0.9552	0.6046	0.3740	0.2233	0.1343	0.0928
0.76	0.2830	0.3649	0.4509	0.6237	1.2090	0.4518	0.3065	0.1916	0.1179	0.0823
0.74	0.3077	0.4002	0.4997	0.7110	1.5967	0.3521	0.2533	0.1643	0.1031	0.0727
0.72	0.3329	0.4374	0.5536	0.8239	0.5844	0.2810	0.2105	0.1407	0.0899	0.0639
0.70	0.3585	0.4770	0.6151	0.9955	0.4016	0.2275	0.1765	0.1202	0.0781	0.0560
0.68	0.3818	0.5199	0.6885	1.5361	0.3001	0.1858	0.1465	0.1026	0.0676	0.0488
0.66	0.4117	0.5674	0.7831	0.5102	0.2329	0.1527	0.1228	0.0873	0.0583	0.0424
0.64	0.4394	0.6215	0.9263	0.3311	0.1845	0.1259	0.1023	0.0741	0.0500	0.0366
0.62	0.4683	0.6860	1.3682	0.2403	0.1479	0.1039	0.0837	0.0627	0.0428	0.0315
0.60	0.4985	0.7699	0.4083	0.1825	0.1196	0.0858	0.0713	0.0529	0.0364	0.0269
0.58	0.5304	0.8986	0.2596	0.1416	0.0970	0.0708	0.0593	0.0444	0.0308	0.0229
0.56	0.5647	--	0.1851	0.1113	0.0789	0.0583	0.0492	0.0371	0.0259	0.0193
0.54	0.6022	0.3114	0.1380	0.0882	0.0640	0.0478	0.0406	0.0309	0.0217	0.0163
0.52	0.6446	0.1971	0.1051	0.0701	0.0512	0.0391	0.0334	0.0255	0.0180	0.0136
0.50	0.6947	0.1385	0.0811	0.0557	0.0410	0.0318	0.0274	0.0210	0.0149	0.0112
0.48	0.7591	0.1014	0.0628	0.0442	0.0331	0.0257	0.0221	0.0171	0.0122	0.0092
0.46	0.8585	0.0756	0.0487	0.0347	0.0265	0.0206	0.0178	0.0138	0.0099	0.0075
0.44	3.3677	0.0569	0.0377	0.0273	0.0210	0.0164	0.0142	0.0110	0.0080	0.0061
0.42	0.1827	0.0430	0.0291	0.0212	0.0164	0.0130	0.0113	0.0088	0.0063	0.0048
0.40	0.1107	0.0324	0.0223	0.0165	0.0127	0.0101	0.0088	0.0069	0.0050	0.0038
0.35	0.0419	0.0155	0.0110	0.0082	0.0064	0.0052	0.0045	0.0035	0.0026	0.0020
0.30	0.0169	0.0068	0.0049	0.0038	0.0029	0.0024	0.0021	0.0016	0.0012	0.0009
0.25	0.0062	0.0027	0.0019	0.0015	0.0012	0.0009	0.0009	0.0006	0.0005	0.0004
0.20	0.0019	0.0008	0.0006	0.0005	0.0004	0.0003	0.0003	0.0002	0.0002	0.0001
0.15	0.0004	0.0002	0.0001	0.0001	0.0001	0.0001	0.0001	--	--	--
0	0	0	0	0	0	0	0	0	0	0

where

$$A = \frac{\pi d_0^2}{4} = \frac{3.14(6.0)^2}{4} = 28.3\,\text{ft}^2 \left(2.6\,\text{m}^2\right)$$

$$R = \frac{A}{P} = \frac{\dfrac{\pi d_0^2}{4}}{\pi d_0} = \frac{28.3}{3.14(6.0)} = 1.5\,\text{ft}\left(0.46\,\text{m}\right)$$

and

$$Q_0 = \frac{1.49}{0.013}(28.3)(1.5)^{2/3}\sqrt{0.001} = 134\,\text{ft}^3\,/\,\text{s}\left(3.79\,\text{m}^3\,/\,\text{s}\right)$$

then, with reference to Eq. (6.3.34)

$$A = -\frac{d_0}{S_0} = -\frac{6}{0.001} = -6000$$

$$B = \frac{\alpha Q^2}{d_0^5} = \frac{1(160)}{6.0^5} = 3.29$$

and

$$\frac{Q}{Q_0} = \frac{160}{134} = 1.19$$

By use of the data summarized in Tables 6.5 and 6.6, the solution proceeds as indicated in Table 6.7. From these computations, it is concluded that the conduit flows full approximately 3900 ft (1200 m) upstream of the control section.

When the Bakhmeteff-Chow methodology is applied to channels with zero and negative (adverse) slopes, the techniques discussed above must be modified. In a horizontal channel, $S_0 = 0$ becomes

$$\frac{dy}{dx} = \frac{\left(\dfrac{Q}{K}\right)^2}{1 - \left(\dfrac{Z_c}{Z}\right)^2} \tag{6.3.35}$$

Table 6.7. Tabular solution of Example 6.3.

y ft	$\dfrac{y}{d_0}$	X_2	X_1	Y_2	Y_1	Δx	$\Sigma \Delta x$
3.4	0.57	---	---	---	---	----	0
3.6	0.60	0.0543	0.0418	0.0588	0.0552	-3.9	-3.9
4.2	0.70	0.1245	0.0543	0.0717	0.0588	-170	-170
4.8	0.80	0.2755	0.1245	0.0876	0.0717	-590	-760
5.4	0.90	0.7124	0.2755	0.1141	0.0876	-2100	-2900
5.9	0.98	0.8493	0.7124	0.1048	0.1141	-1000	-3900

Then, by definition, the critical slope S_c is the slope which will sustain a flow Q at critical depth, y_c; and therefore

$$Q = K_c \sqrt{S_c} \tag{6.3.36}$$

Substituting this into Eq. (6.3.35) and letting

$$\left(\frac{K_c}{K}\right)^2 = \left(\frac{y_c}{y}\right)^N$$

$$\left(\frac{Z_c}{Z}\right)^2 = \left(\frac{y_c}{y}\right)^M$$

and

$$P = \frac{y}{y_c}$$

yields

$$\frac{dy}{dx} = S_c \frac{P^{M-N}}{1 - P^M} \tag{6.3.37}$$

For the case where M and N are constant, Eq. (6.3.37) can be integrated to yield

$$x = \frac{y_c}{S_c}\left(\frac{P^{N-M+1}}{N-M+1} - \frac{P^{N+1}}{N+1}\right) + C \qquad (6.3.38)$$

where C = a constant of integration.

In channels of adverse slope, the slope of the channel bottom is negative, and Eq. (6.1.1) becomes

$$\frac{dy}{dx} = \frac{-S_0 - S_f}{1 + \alpha\dfrac{d\left(\dfrac{u^2}{2g}\right)}{dy}} \qquad (6.3.39)$$

It can then be shown that under these conditions with N and M constant [see, for example, Chow (1959)]

$$x = -\frac{y_N}{S_0}\left[u - \int_0^u \frac{du}{1+u^N} - \left(\frac{y_c}{y_N}\right)\int_0^u \frac{u^{N-M}}{1+u^N}du\right] + C \qquad (6.3.40)$$

where the varied-flow functions for adverse slopes are

$$F(u,N)_{-S_0} = \int_0^u \frac{du}{1+u^N} \qquad (6.3.41)$$

$$F(\upsilon,J)_{-S_0} = \int_0^u \frac{u^{N-M}}{1+u^N}du = \frac{J}{N}\int_0^\upsilon \frac{d\upsilon}{1+\upsilon^J} \qquad (6.3.42)$$

and

$$\upsilon = u^{N/J}$$

The functions $F(u,N)_{-S_0}$ and $F(\upsilon,J)_{-S_0}$ are tabulated in Appendix I.

6.3.2. Nonprismatic Channels

In the foregoing section, two methods of calculating gradually varied flow profiles in prismatic channels were discussed. In general, neither of these techniques is well suited to the computation of water surface profiles in nonprismatic channels, where both the resistance coefficient and the channel shape are functions of longitudinal distance. In addition, because the channel properties are a function of longitudinal distance in nonprismatic channels, the method used to estimate the water surface elevation should predict the depth of flow at a specified longitudinal distance rather than vice versa, which is the case in both of the methods previously discussed.

The application of the energy equation between the two stations shown in Fig. 6.3 yields

$$z_1 + \alpha_1 \frac{\overline{u_1^2}}{2g} = z_2 + \alpha_2 \frac{\overline{u_2^2}}{2g} + h_f + h_e$$

where z_1 and z_2 = elevation of water surface above a datum at stations 1 and 2, respectively

$$h_e = \text{eddy loss incurred in reach}$$

$$h_f = \text{each friction loss}$$

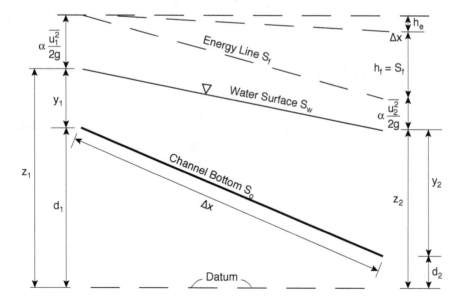

Figure 6.3. Channel reach definition.

The eddy losses, which may be appreciable in nonprismatic channels, are generally expressed in terms of the change in the velocity head in the reach or

$$h_e = k\Delta\left(\alpha\frac{\overline{u^2}}{2g}\right)$$

where k = a coefficient. Unfortunately, there is very little generalized information regarding k, and it is common practice for values of k to be determined on the basis of field measurements. Where this is not possible, k is assumed to have a value of 0.5 for abrupt expansions or contractions and to range between 0 and 0.2, for gradually converging or diverging reaches. The friction loss term is approximated by

$$h_f = S_f\Delta x = \frac{\left(S_{f1} + S_{f2}\right)\Delta x}{2} \tag{6.3.44}$$

where S_{f1} and S_{f2} = friction slopes at stations 1 and 2, respectively. With these approximations, the energy equation becomes

$$z_1 + \alpha_1\frac{\overline{u_1^2}}{2g} = z_2 + \alpha_2\frac{\overline{u_2^2}}{2g} + \frac{\Delta x\left(S_{f1} + S_{f2}\right)}{2} + k\Delta\left(\alpha\frac{\overline{u^2}}{2g}\right) \tag{6.3.45}$$

For convenience, define

$$H_1 = z_1 + \alpha_1\frac{\overline{u_1^2}}{2g} \tag{6.3.46}$$

and

$$H_2 = z_2 + \alpha_2\frac{\overline{u_2^2}}{2g} \tag{6.3.47}$$

With these definitions, Eq. (6.3.45) becomes

$$H_1 = H_2 + h_f + h_e \tag{6.3.48}$$

Equation (6.3.48) is solved by trial and error; i.e., given a longitudinal distance, a water surface elevation at station 1 is assumed which allows the computation of H_1 by Eq. (6.3.46). Then, h_f and h_e are computed, and H_1 is estimated by Eq. (6.3.48). If the two computed values of H_1 agree, then the assumed stage at station 1 is correct. In essence, the trial-and-error procedure attempts to make the difference between H_1 computed by Eq. (6.3.46) and the value estimated by Eq. (6.3.48) zero. The use of this trial-and-error step method of water surface profile computation is best illustrated with an example. Although this method was designed for use in nonprismatic channels, the initial example will be in a prismatic channel with $\alpha_1 = \alpha_2 = 1$, and $h_e = 0$.

EXAMPLE 6.4

A trapezoidal channel with $b = 20$ ft (6.1 m), $n = 0.025$, $z = 2$, and $S_0 = 0.001$ carries a discharge of 1,000 ft³/s (28m³/s). If this channel terminates in a free overfall and there are no eddy losses, determine the gradually varied flow profile by the trial-and-error step method.

Solution

It is noted that this problem statement is identical to that of Examples 6.1 and 6.2. For comparative purposes, the elevation of the water surface will be determined at the longitudinal stations found in Example 6.2. If the elevation at the free overall is 100 ft (30 m), then the elevation of the water surface at this point is 103.74 ft (31.62 m) (see Example 6.1). The solution of this problem is summarized in Table 6.8; with regard to this table, the following points should be noted:

Column 1:   Each longitudinal station is identified by its distance from the control.

Column 2:   With the exception of the first station, this is a trial value of the water surface elevation, which is either verified or rejected on the basis of subsequent computations. When a trial value is verified, it becomes the basis of the verification process for the trial value at the next longitudinal station.

Column 3:   The depth of flow corresponds to the water surface elevation in col. 2. This value is computed by subtracting, from the assumed water surface elevation at the station, the elevation of the control and the product of the longitudinal distance from the control, and the bottom slope of the channel. For example, in the second line of Table 6.8

$$104.62 - 100 - 0.001\,(116) = 4.50 \text{ ft } (1.37 \text{ m})$$

Column 4:   The flow area corresponding to the depth of flow in col. 3.

Table 6.8. Tabular solution of Example 6.4.

Station	z	y	A	$\bar{u}$	$\dfrac{\bar{u}^2}{2g}$	H_1	R	S_f	$\overline{S_f}$	Δx	h_f	h_e	H
(ft)	(ft)	(ft)	(ft²)	(ft/s)	(ft)	(ft)	(ft)			(ft)	(ft)	(ft)	(ft)
(1)	(2)	(3)	(4)	(5)	(6)	(7)	(8)	(9)	(10)	(11)	(12)	(13)	(14)
0	103.74	3.74	103	9.71	1.46	105.20	2.81	0.00670				--	105.20
116	104.62	4.50	130	7.69	0.918	105.54	3.24	0.00347	0.00509	116	0.590	--	105.79
	105.02	4.90	146	6.85	0.729	105.75	3.48	0.00251	0.00461	116	0.535		105.73
355	105.56	5.20	158	6.33	0.622	106.18	3.65	0.00201	0.00226	355	0.802	--	106.55
	105.93	5.57	173	5.78	0.519	106.45	3.85	0.00156	0.00204	355	0.724		106.47
745	106.34	5.60	175	5.71	0.506	106.85	3.89	0.00150	0.00153	745	1.14	--	107.59
	106.96	6.21	201	4.98	0.385	107.34	4.21	0.00103	0.00130	745	0.968		107.42

Column 5:   The average velocity of flow.

Column 6:   The velocity head corresponds to the velocity in col. 5.

Column 7:   The total head computed by Eq. (6.3.46).

Column 8:   The hydraulic radius corresponds to the depth of flow in col. 3.

Column 9:   The friction slope at the station estimated from

$$S_f = \frac{n^2\bar{u}^2}{2.22R^{4/3}}$$

Column 10:   The average friction slope in the reach which is approximated as the average value of the friction slope in col. 9 and the corresponding value from the previous reach. Note: Other methods of computing the average friction slope are available; see, for example, Table 6.11.

Column 11:   The incremental length of the reach.

Column 12:   The friction loss in a reach which is estimated as the product of cols. 10 and 11.

Column 13:   The estimated eddy loss for the reach.

Column 14:   The total head computed from Eq. (6.3.48) or the sum of cols. 12 and 13 and the total energy of the previous reach, col. 14, which should be identical to the value in col. 7. If the value obtained in this computation does not closely agree with the value in col. 7, then a new value of the water surface elevation in col. 2 must be assumed and the calculations repeated.

The crucial step in this analysis is the selection of an appropriate trial value of the water surface elevation. In Table 6.8 it is noted that no more than two trials were required to obtain the correct value of the water surface elevation in col. 7. This is because an error function equation was used to adjust the first trial value. Recall that the trial-and-error solution of Eq. (6.3.48) depends on making the difference in H_l determined by Eq. (6.3.46) and H_l determined by Eq. (6.3.48) zero. Let the difference be designated by e. The value of e can be modified only by changing the depth of flow y_2; therefore, the change in e, with respect to changes in y_2, can be evaluated by

$$\frac{de}{dy_2} = \frac{d}{dy_2}\left(y_2 + \frac{\overline{u_2^2}}{2g} - \frac{1}{2}\Delta x S_{f2} \right)$$

or

$$\frac{de}{dy_2} = 1 - F_2^2 + \frac{3 S_{f2}\Delta x}{2R_2}$$

Then,

$$y_2 = \frac{e}{1 - F_2^2 + \left(\dfrac{3S_{f2}\Delta x}{2R_2}\right)} \qquad (6.3.49)$$

where Δy_2 = amount by which the water level must be adjusted to force e to zero. In Table 6.5, the correction which was applied to the first trial value at $x = 116$ ft (35.4 m) was calculated by

$$D_2 = \frac{A_2}{T_3} = \frac{130}{38} = 3.42\,\text{ft}\left(1.04\,\text{m}\right)$$

$$F = \frac{\overline{u_2}}{\sqrt{gD_2}} = \frac{7.69}{\sqrt{32.2\left(3.42\right)}} = 0.73$$

Then, by Eq. (6.3.49)

$$\Delta y_2 = \frac{105.54 - 105.79}{1 - 0.73^2 + \dfrac{3\left(0.00347\right)\left(116\right)}{2\left(3.24\right)}} = -0.382 \approx 0.40\,\text{ft}\left(0.12\,\text{m}\right)$$

In this case, Δy_2 must be added to the initial trial value. Properly used, Eq. (6.3.49) can significantly reduce the number of trial values needed to estimate the water surface elevation correctly.

   With regard to this trial-and-error step method of calculating gradually varied flow profiles, the following are noted: (1) This method yields answers similar to those estimated in Examples 6.1 and 6.2; and (2) as with all step methods, the calculations are sequential; i.e., each step is dependent on the preceding step.

   In most natural channels, the possibility that the channel is composed of a main section and one or more overbank sections, must be considered. In such a situation, the depth of flow in the overbank section is different than the depth of flow in the main channel, the resistance coefficient of the overbank section may be significantly different from the one which characterizes the main section, and the length between sections on the river may differ. In general, the step method previously discussed with computational modifications may also be applied to this situation.

Since there is a difference in velocity heads between the main and overbank sections, the energy correction factor α must be used to define the velocity head component of the total head. From Eq. (2.4.7), the energy coefficient is given by

$$\alpha = \frac{\sum \overline{u}_i^3 A_i}{u^3 \sum A_i} = \frac{\left(\sum A_i\right)^2}{\left(\sum Q_i\right)^3} \sum \left(\frac{Q_i^2}{A_i^2}\right)$$

where the subscripts indicate distinct flow subsections and the unsubscripted variables are total channel variables. If the definition of conveyance, Eq. (5.1.3), is combined with the above expression for α, then

$$\alpha = \frac{\left(\sum A_i\right)^3}{\left(\sum K_i\right)^3} \sum \frac{K_i^3}{A_i^2} \tag{6.3.50}$$

In addition, the friction slope for the reach under consideration is given by

$$S_f = \frac{Q^2}{\left(\sum K_i\right)^2} \tag{6.3.51}$$

When the reach under consideration is on a curve, then the length of the reach may depend on the subsection being considered. The simplest method of taking this effect into consideration is to vary the value of the resistance coefficient. For example, if the length of the overbank section is less than the main section, then the value of the resistance coefficient must be reduced to maintain S_f constant for the reach. In the case of Manning's n, the n for the overbank section is reduced in proportion to the parameter $(L_0/L_M)^{1/2}$ where L_0 = overbank section length and L_M = main channel length (Henderson, 1966). The U.S. Army Corps of Engineers HEC2 model (Anonymous, 1979) and Shearman (1976) calculate a discharge-weighted reach length by the equation

$$L = \frac{\sum L_i K_i}{\sum K_i} \qquad\qquad (6.3.52)$$

Where L_i = subsection length. The use of the various equations and definitions, in conjunction with the step method to determine a gradually varied flow profile, is best demonstrated with an example.

EXAMPLE 6.5

The gradually varied flow profile for Q = 100,000 ft³/s (2800 m³/s) is to be estimated over a 1 mile reach (1,600 m) whose cross section is defined in Fig. 6.4. In this reach, there is a distinct main channel and an overbank section, both of which are approximately rectangular in section. The properties of the channel are described in Table 6.9. Between river miles 10 and 10.5, the channel is straight, and it is assumed that there are no eddy losses. Between river miles 10.5 and 11.0, the channel is curved in such a manner that the overbank section is on the inside of the curve and has a length of 1,920 ft (585 m). In this curved section, there is an eddy loss equal to 0.2 times the average velocity heads at the ends of the reach. If the stage of flow is 70 ft (21 m) at river mile 10, determine the stage at river mile 11.0.

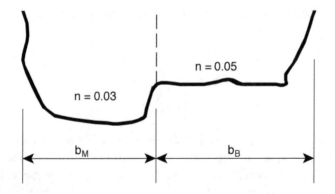

Figure 6.4. Channel definition for Example 6.5.

Table 6.9. Channel properties for Example 6.5.

River Mile	Widths		Bed Elevation	
	b_M (ft)	b_B (ft)	Main (ft)	Berm (ft)
10.00	100	600	50.0	55.0
10.25	150	550	52.5	57.5
10.50	200	500	55.0	60.0
11.00	300	400	60.0	65.0

Solution

Computationally, the solution to this problem proceeds in a manner similar to that used in Example 6.4 with minor modifications to account for the compound section. The solution is summarized in Table 6.10, and with regard to these computations, the following points should be noted:

Column 2:   The notation M in this column refers to the main channel, with OB referring to the overbank section.

Column 8:   This is the resistance coefficient defined in Fig. 6.4 or, in the case of the overbank section, between river miles 10.5 and 11

$$n' = \left(\frac{1920}{2640}\right)^{1/2}(0.05) = 0.04$$

If the Hydrologic Engineering Center (HEC) methodology were used, then the modified Δx between these stations would be

$$L = \frac{18.2\left(10^5\right)(2640) + 11.0\left(10^5\right)(1920)}{(18.2 + 11.0)\left(10^5\right)} = 2370\,\text{ft}\,(722\,\text{m})$$

where the values used in this computation are from the second trial in this reach with $n_{OB} = 0.05$. The result would have been approximately the same as given; e.g., see the line in Table 6.10, designated with an asterisk. In this solution, Eq. (6.3.49) also served as a guide to the

Table 6.10. Tabular solution of Example 6.5.

River Mile (1)	Sub-section (2)	Stage ft (3)	A ft² (4)	P ft (5)	R ft (6)	$R^{2/3}$ (7)	n (8)	$K \times 10^{-5}$ (9)	$(K^3/A^2) \times 10^{-10}$ (10)	α (11)	$\bar{u}$ ft/s (12)	$\alpha\frac{\bar{u}^2}{2g}$ ft (13)	H ft (14)	S_f (15)	$\overline{S_f}$ (16)	Δx ft (17)	h_f ft (18)	h_e ft (19)	H ft (20)
10	M	70.	2,000	125	16.0	6.36	0.03	6.3	6.3	1.24	9.09	1.59	71.6	0.00201					71.6
	OB		9,000	615	14.6	5.99	0.05	16.0	5.1										
	Total		11,000					22.3	11.4										
10.25	M	70	2,625	172	15.2	6.15	0.03	8.0	7.5	1.55	10.5	2.29	79.7	0.00305	0.00253	1,320	3.33	--	74.9
	OB		6,875	562	12.2	5.31	0.05	10.0	2.7										
	Total		9,500					18.0	10.2										
	M	72.1	2,940	175	16.8	6.58	0.03	9.6	10.2	1.33	9.12	1.72	73.8	0.00180	0.00191	1,320	2.52	--	74.1
	OB		8,030	565	14.2	5.88	0.05	14.0	4.3										
	Total		10,970					23.6	14.5										
	M	72.4	2,985	175	17.1	6.64	0.03	9.8	10.6	1.32	8.94	1.64	74.0	0.00169	0.00185	1,320	2.44	--	74.0
	OB		8,195	565	14.5	5.95	0.05	14.5	4.6										
	Total		11,180					24.3	15.2										
10.5	M	74.0	3,800	224	17.0	6.61	0.03	12.0	12.0	1.31	9.26	1.74	75.7	0.00174	0.00171	1,320	2.26	--	76.3
	OB		7,000	514	13.6	5.71	0.05	12.0	3.5										
	Total		10,800					24.0	15.5										
	M	74.5	3,900	224	17.4	6.71	0.03	13.0	14.4	1.35	8.97	1.69	76.2	0.00153	0.00161	1,320	2.13	--	76.1
	OB		7,250	514	14.1	5.84	0.05	12.6	3.8										
	Total		11,150					25.6	18.2										
11	M	78	5,400	323	16.7	6.54	0.03	17.5	18.4	1.17	9.43	1.60	79.6	0.00128	0.00140	2,640	3.70	--	79.8
	OB		5,200	413	12.6	5.42	0.04	10.5	4.3										
	Total		10,600					28.0	22.7										
	M	78.4	5,520	323	17.1	6.64	0.03	18.2	19.8	1.16	9.19	1.52	79.9	0.00117	0.001135	2,640	3.56	0.33	80.0
	OB		5,360	413	13.0	5.53	0.04	11.0	4.6									0.32	
	Total		10,880						24.4										
0	M	78.4	5,520	323	17.1	6.64	0.03	18.2	19.8	1.34	9.19	1.76	80.2	0.00137	0.00145	2,370	3.44	0.32	80.0
	OB		5,360	413	13.0	5.53	0.05	8.8	2.4										
	Total		10,880					27.0	22.2										

selection of the second trial value of the stage at each station. In the reach between river miles 10.0 and 10.25, the first trial was incorrect and a second trial value was calculated from

$$\alpha F_2^2 = \frac{2}{R}\left(\alpha \frac{\overline{u^2}}{2g}\right) = \frac{2}{12.9}(2.29) = 0.36$$

$$\frac{3S_{f2}\Delta x}{2R} = \frac{3(0.00277)1320}{2(12.9)} = 0.36$$

and therefore,

$$\Delta y = \frac{72.7 - 74.9}{1 - 0.36 + 0.43} = -2.1 \text{ft}\,(0.64\,\text{m})$$

Thus, the second trial value was 70 + 2.1 = 72.1 ft (22.0 m). It should be noted that since the first trial was significantly in error, more than two trials were required for this reach to determine the proper stage.

Although any of the methods discussed in the foregoing material can be adapted for solution by a digital computer, some general-case computer models have been developed by agencies of the U.S. government for this purpose. In particular, there is the HEC-2 model developed by the Hydrologic Engineering Center (HEC) of the U.S. Army Corps of Engineers (Anonymous, 1979; Feldman, 1981). For many years, HEC-2 was the industry standard; but as the computational environment changed, HEC realized the need for a new model that could model both steady and unsteady flow, and sediment transport. It is pertinent to note that the new model, HEC-RAS (Anonymous, 2001), is a completely new piece of software. During the development process, HEC greatly improved the computational capabilities of the model and developed a user friendly interface. Although HEC-2 data can be imported into HEC-RAS, the reader should be aware that for the same the problem there may be minor differences in results (Anonymous, 2001). Finally, it is pertinent to observe that HEC-RAS is not the only model available for solving gradually varied flow problems; for example, the U.S. Geological Survey developed a model similar to HEC-2 (Shearman, 1976,1977). The purpose of this section is to briefly summarize the capabilities of Hydrologic Engineering Center HEC-RAS model. For additional information, the reader is referred to the

HEC User's Manual (Anonymous, 2001) and the Hydraulic Reference Manual (Anonymous, 1995), both of which can be down-loaded from the HEC web site.

The HEC-RAS model was developed to calculate water surface profiles for steady, gradually varied flows in both prismatic and nonprismatic channels. Both subcritical and supercritical flow profiles can be estimated, and the effects of various obstructions such as bridges, culverts, weirs, and structures in the overbank region, are considered. Subsequently, the model was updated to include unsteady flow (Barkau, 1992 and Anonymous, 1993); but as of this version (2001), the option to model sediment transport has not been implemented. The model is subject to four crucial assumptions:

1. HEC-RAS can be run as either a steady or unsteady flow model. In the steady flow mode, subcritical, supercritical, or mixed flow conditions can be modeled.

2. The flow must be gradually varied since the model equations assume a hydrostatic distribution of pressure.

3. The flow is one-dimensional.

4. The slope of the channel is small.

The purpose of the HEC-RAS model is to determine water surface elevations for specified discharges in natural and constructed channels to aid in the U.S. Army Corps of Engineers floodplain management program. In this capacity, the model has been used to:

1. Determine areas inundated by various flood discharges for the assessment of damages,

2. Study the effects of land use in floodplains from the viewpoint of flood damages,

3. Study how flood damages can be mitigated by various channel improvements.

The HEC-2 model has a number of options and caveats regarding data which must also be noted as follows:

1. Within the model, there are four methods of estimating the average friction slope for each reach.
    a. Average conveyance

$$\overline{S_f} = \left(\frac{Q_1 + Q_2}{K_1 + K_2} \right)^2 \tag{6.3.53}$$

where K_1 and K_2 are the total conveyances at stations 1 and 2, respectively.

b. Average friction slope

$$\overline{S_f} = \frac{S_{f1} + S_{f2}}{2} \tag{6.3.54}$$

c. Geometric mean

$$\overline{S_f} = \sqrt{S_{f1} S_{f2}} \tag{6.3.55}$$

d. Harmonic mean

$$\overline{S_{f1}} = \frac{S_{f1} S_{f2}}{S_{f1} + S_{f2}} \tag{6.3.56}$$

In addition to the above equations, Reed and Wolfkill (1976), in their study of friction slope models, noted that the following formulations have also been used to estimate the friction slope in a reach.

e.

$$\overline{S_f} = \frac{Q^2 n^2}{\varphi^2 \left[\dfrac{2A_1 A_2}{A_1 + A_2} \right]^2 \left[\dfrac{R_1 + R_2}{2} \right]^{4/3}} \tag{6.3.57}$$

f.

$$\overline{S_f} = \frac{Q^2 n^2}{\varphi^2 \left(\dfrac{A_1 + A_2}{2} \right)^2 \left(\dfrac{A_1 + A_2}{P_1 + P_2} \right)^{4/3}} \tag{6.3.58}$$

g.

$$S_f = \frac{Q^2 n^2}{\varphi^2 \left(\dfrac{A_1 + A_2}{2}\right)^2 \left(\dfrac{R_1 + R_2}{2}\right)^{4/3}} \qquad (6.3.59)$$

h.

$$S_f = \frac{Q^2 n^2}{\varphi^2 \left(\dfrac{A_1 R_1^{2/3} + A_2 R_2^{2/3}}{2}\right)} \qquad (6.3.60)$$

where R_1 and R_2 = hydraulic radii at the beginning and end of the reach, respectively; A_1 and A_2 = flow areas at the beginning and end of the reach, respectively, n = Manning's resistance coefficient for the reach, and Q = the flow in the reach. Although any of the above equations will provide a satisfactory estimate of the friction slope provided that the reach is not too long. The concept of providing alternative methodologies within a model to maximize the reach length is attractive. Equation (6.3.53) is the standard HEC-RAS formulation, and Eq. (6.3.55) is the formulation used by the U.S. Geological Survey in its gradually varied flow model (Shearman, 1976,1977). The results of Reed and Wolfkill (1976) are summarized in Table 6.11 along with the criteria used by the HEC-RAS model in selecting an optimum model of the friction slope.

Table 6.11. Recommended friction slope models.

Profile Type	Friction slope model recommended by Reed and Wolfkill (1976)	Friction slope model recommended for HEC-RAS (Anonymous, 1995)
M1	(6.3.53)	(6.3.54)
M2	(6.3.55)	(6.3.56)
M3	(6.3.56)	(6.3.55)
S1	(6.3.55)	(6.3.54)
S2	(6.3.53)	(6.3.54)
S3	(6.3.54)	(6.3.55)

2.   Adequate data, regarding the channel cross section, are essential to the proper and accurate operation of the model. Cross sections are required along the channel where there are significant changes in either the geometry of the channel or in the hydraulic characteristics. Critical geometric changes usually involve either artificial or natural contractions or expansions, while changes in channel slope, roughness, or discharge can be regarded as significant hydraulic characteristic changes. In general, the cross sections provided should be perpendicular to the direction of flow and extend completely across the channel to the high ground on either side of the channel. In some cases where there are very significant changes in either the channel geometry or its hydraulic characteristics, a number of cross sections will be required. Care should also be exercised to ensure that only the effective flow area of the channel is included.

3.   The distance between cross sections, also known as the reach length, is a function of the required degree of hydraulic detail and the available human, temporal, and financial resources. Small nonuniform, or steep channels may require very short reach lengths, e.g., 200 ft (60 m), while in large uniform channels with small slopes, reach lengths of up to 2 miles (3 km) may be appropriate. In addition, the type of study being performed may dictate the detail required and hence the reach length. For example, a navigation study of the Missouri River required reach lengths of 500 ft (150 m) (Feldman, 1981 and Anonymous, 1995). Reach lengths can be optimized by adapting an appropriate friction slope equation (Table 6.11). The HEC-RAS model requires reach lengths for both the right and left overbank regions of the channel, if such areas exist in the main channel. In the main channel, this distance is measured either along the centerline in an artificial channel or along the thalweg in a natural channel. While in the overbank areas, the reach length should represent the flow path length, of the center of mass, of the water moving in this area. Since for a given length of channel these three lengths may vary, it is sometimes necessary to calculate a hypothetical reach length by Eq. (6.3.52).

6.4.   SPATIALLY VARIED FLOW

Steady, spatially varied flow is defined as flow in which the discharge varies in the direction of flow and occurs in situations involving side-channel spillways, channels with permeable boundaries, gutters for conveying storm water runoff, and drop structures in the bottom of channels. Since the salient principles, governing flow in which the discharge increases with distance are different from those governing a flow in which the discharge decreases with distance, these two types of flow must be considered separately.

6.4.1.   Increasing Discharge
In the case of spatially varied flow, which is exemplified by the side-channel spillway, a significant portion of the energy loss results from the mixing of the water entering the channel and the water flowing in the channel. Since this energy loss cannot be accurately quantified, the momentum principle has been traditionally used to develop governing equations.

Given a channel with a lateral inflow (Fig. 6.5), the continuity equation for the control volume shown is

$$\frac{dQ}{dx} = q^* \tag{6.4.1}$$

where Q = the flow rate in the channel and q^* = the lateral inflow rate per unit length. The left-hand side of Eq. (6.4.1) can be expanded to yield

$$u\frac{dA}{dx} + A\frac{\overline{du}}{dx} = q^* \tag{6.4.2}$$

where A = cross-sectional flow area and $\overline{u}$ = mean velocity in the longitudinal direction. For this same control volume, the momentum flux in the x direction is

$$\Delta M = M_2 - \left(M_1 + M_L\right) \tag{6.4.3}$$

where

$$M_2 - M_1 = \iint_A \rho u^2 dA \qquad (6.4.4)$$

and

$$M_L = \rho q^* \Delta x \, V \cos\phi \qquad (6.4.5)$$

In Eqs. (6.4.4) and (6.4.5), ρ = fluid density, and V = average velocity of lateral inflow (Fig. 6.5).

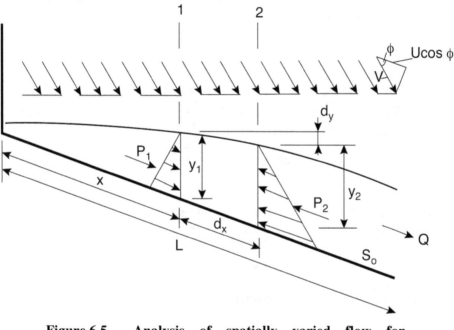

Figure 6.5.   Analysis of spatially varied flow for increasing discharge.

If the effects of surface tension and turbulence are neglected, the forces acting on the control volume are the component of the weight of the fluid, within the control volume, acting in the longitudinal direction, the boundary shear resistance, and the pressure forces at stations 1 and 2. If the total change of momentum flux, Eq. (6.4.3), is equated to the sum of the forces acting, then

$$\frac{d}{dx} \iint \rho u_x^2 \, dA - \rho q^* V \cos\phi = \gamma S_0 A - \tau_b P - \frac{d}{dx} \iint_A p \, dA \qquad (6.4.6)$$

where S_0 = $sin \theta$ = bottom slope
     p =   pressure
     P =   wetted perimeter
     γ =   fluid specific weight
     τ_b =   average boundary shear stress, acting in x direction
              over entire cross section.

Equation (6.4.6) can be simplified by introducing standard expressions for the first term on the left-hand side of the equation and the last term on the right-hand side of the equation. This result is then combined with Eq. (6.4.2), to yield an equation for the slope of the water surface (see, for example, Yen and Wenzel, 1970), or

$$\frac{dy}{dx} = $$

$$\frac{S_0 - S_f + \dfrac{q^*}{gA}\left(V\cos\theta - 2\beta\overline{u}\right) - \dfrac{\overline{u^2}}{g}\dfrac{dB}{dx} - y\dfrac{d\left(\alpha'\cos\theta\right)}{dx}}{\alpha'\cos\theta\left(1 + \dfrac{y}{D}\right) - \dfrac{\beta u^2}{gD}} \qquad (6.4.7)$$

where β =   momentum correction factor
     α' =   pressure correction coefficient
     D =   hydraulic depth
     S_f =   friction slope

If 2 is constant, Eq. (6.4.7) becomes

$$\frac{dy}{dx} = \frac{S_0 - S_f + \dfrac{q^*}{gA}\left(V\cos\theta - 2\beta\overline{u}\right) - \dfrac{\overline{u^2}}{g}\dfrac{d\beta}{dx}}{\cos\theta\left[\alpha'\left(1 + \dfrac{y}{D}\right) + y\dfrac{d\alpha'}{dy}\right] - \dfrac{\beta u^2}{gD}} \qquad (6.4.8)$$

If it is assumed that the pressure distribution is hydrostatic, $V\cos\phi = 0$, $\cos\theta = 1$, and $\beta = 1$, then Eq. (6.4.8) becomes

$$\frac{dy}{dx} = \frac{S_0 - S_f - \dfrac{2q^*V}{gA}}{1 - \dfrac{u^2}{gD}} = \frac{S_0 - S_f - \dfrac{2Q}{gA^2}\dfrac{dQ}{dx}}{1 - \dfrac{Q^2}{gA^2D}} \quad (6.4.9)$$

which is the most common form of the gradually varied flow equation with lateral flow (see, for example, Chow, 1959, or Henderson, 1966).

Equation (6.4.9) can be solved by a number of techniques, but as was the case with the computation of the regular gradually varied flow profile, the solution procedure must begin at a control. Thus, for a given channel with a specified lateral inflow, the situation must be examined to determine if a critical flow section occurs. Following the reasoning of Henderson (1966), critical flow occurs or $dy/dx = 0$, when the numerator of Eq. (6.4.9) is zero or

$$S_0 - S_f - \frac{2Q}{gA}\frac{dQ}{dx} = 0 \quad (6.4.10)$$

If the rate of lateral inflow is constant, as it is in most cases of practical interest,

$$\frac{dQ}{dx} = \frac{Q}{x} \quad (6.4.11)$$

where x = longitudinal distance from the beginning of the channel. Note: This formulation assumes that the flow at the beginning of the channel is zero. The friction slope can be estimated from either the Manning or Chezy equations, and for notational convenience the Chezy equation is used or

$$S_f = \frac{Q^2P}{C^2A^3} \quad (6.4.12)$$

Substitution of Eqs. (6.4.11) and (6.4.12) in Eq. (6.4.10) yields

$$S_0 = \frac{Q^2P}{C^2A^3} + \frac{2Q^2}{gA^2x} = F^2\left(\frac{gP}{C^2T} + \frac{2A}{Tx}\right) \quad (6.4.13)$$

where T = channel top width. As in Chapter 2, the possibility that dy/dx = 0 is now excluded, and the alternative possibility that

$$F^2 = \frac{Q^2 T}{gA^3} = 1$$

is considered. Using this definition to eliminate A in Eq. (6.4.13), after some rearrangement yields,

$$S_0 = \frac{gP}{C^2 T} + \frac{2}{T}\left(\frac{Q_x^2 T}{gx}\right)^{1/3}$$

$$x = \frac{8Q_x^2}{gT^2\left(S_0 - \frac{gP}{C^2 T}\right)^3} \qquad (6.4.14)$$

where $Q_x = dQ/dx$. Equation (6.4.14) may be used to estimate the location of the critical flow section, if one exists. With regard to Eq. (6.4.14), the following observations may be made. First, this type of development was first presented by Keulegan (1952). Second, if the x estimated is greater than the length of the channel, then a critical flow section does not occur. Third, if there is a critical flow section and beyond this section there is a downstream control, then it is possible that the critical section will be drowned if the downstream depth is sufficient. Fourth, as demonstrated by Keulegan (1952), explicit solutions of this equation are possible for special cases, e.g., a wide, rectangular channel. In general, Eq. (6.4.14) must be solved by trial and error since neither T nor P are known and C may not be constant. Fifth, although at this time it is not possible to justify the possibility of a critical section at which dy/dx = 0 and $F \neq 0$, this alternative must remain an item for discussion. Sixth, in the case of a wide rectangular channel with a constant C, Eq. (6.4.14) becomes

$$x = \frac{8q_x^2}{g\left(S_0 - \frac{g}{C^2}\right)^3} \qquad (6.4.15)$$

where $q = Q_x/b$ and b = width of the channel.

Although spatially varied flow profiles can be determined by a number of trial-and-error computational procedures, the most effective solution technique is numerical integration combined with trial and error. For this procedure, Newton's second law of motion can be written as

$$\frac{\gamma}{g}\left[Q\Delta\overline{u} + \left(\overline{u} + \Delta\overline{u}\right)\Delta Q\right] = -\gamma\overline{A}\Delta y + \gamma S_0\overline{A}\Delta x - \gamma S_f\overline{A}\Delta x$$

where $\overline{A}$ = the average flow area between stations 1 and 2 (Fig. 6.5), which are separated by a distance Δx. The average flow area may be expressed as

$$\overline{A} = \frac{Q_1 + Q_2}{\overline{u}_1 + \overline{u}_2}$$

where Q_1 and Q_2 are the flow rates at stations 1 and 2, respectively, and $\overline{u}_1$ and $\overline{u}_2$ are the average velocities of flow at these stations, respectively. Let

$$Q = Q_1$$

and

$$\overline{u}_2 = \overline{u} + \Delta\overline{u}$$

Substitution of these relations in the momentum equation yields, after some rearrangement,

$$\Delta y = -\frac{Q_1\left(\overline{u}_1 + \overline{u}_2\right)}{g\left(Q_1 + Q_2\right)}\left(\Delta\overline{u} + \frac{\overline{u}_2}{Q_1}\Delta Q\right) + S_0\Delta x - S_f\Delta x \qquad (6.4.16)$$

With reference to Fig. 6.5, the drop in the water surface between stations 1 and 2 is given by

$$\Delta y' = -\Delta y + S_0\Delta x \qquad (6.4.17)$$

Substitution of Eq. (6.4.16) in Eq. (6.4.17) yields

$$\Delta y' = \frac{Q_1\left(\overline{u_1} + \overline{u_2}\right)}{g\left(Q_1 + Q_2\right)}\left(\overline{\Delta u} + \frac{\overline{u_2}}{Q_1}\Delta Q\right) + S_f \Delta x \qquad (6.4.18)$$

which can be used to determine the spatially varied water surface profile for the case of increasing discharge.

EXAMPLE 6.10

A trapezoidal lateral spillway channel 400 ft (122 m) long is designed to carry a discharge which increases at a rate of 40 (ft³/s)/ft [3.72 (m³/s)/m]. The cross section has a bottom width of 10 ft (3.0 m) and side slopes of 0.5:1. The longitudinal slope of the channel is 0.015 and begins at an upstream bottom elevation of 73.7 (22.5 m). If $n = 0.1505$ and the velocity distribution is uniform, estimate the water surface profile of the design discharge. Note: This example was first presented by Hinds (1926).

Solution

The first step in the solution of this problem is to determine if a critical flow section exists and, should such a section exist, the second step is to determine its longitudinal position. This determination requires a trial-and-error solution Eq. (6.4.14) (Table 6.16). In this table, the following points should be noted:

Column 1:   An assumed longitudinal position.

Column 2:   The total discharge at the assumed longitudinal position determined as the product of col. 1 and the lateral discharge of 40 (ft³/s)/ft[3.72(m³/s)/m].

Column 3:   The critical depth of flow corresponding to the assumed longitudinal distance in col. 1 and the discharge in col. 2, Eq. (2.2.3), and Table 2.1.

Cols. 4 to 8:   The area, wetted perimeter, top width, hydraulic radius, and   Chezy   resistance   coefficient,   respectively, corresponding to the depth of flow in col. 3 or

$$A = (b + zy) = (10 + 0.5y)y$$

TABLE 6.16. Location of the critical flow section for Example 6.10.

Trial x ft (1)	Q ft³/s (2)	y_c ft (3)	A ft² (4)	P ft (5)	T ft (6)	R ft (7)	C (8)	x ft (9)
200	8000	20.8	425	56.6	30.8	7.51	139	131
131	5240	16.4	299	46.7	26.4	6.40	135	178
178	7120	19.5	385	53.7	29.5	7.17	138	142
142	5680	17.2	320	48.5	27.2	6.59	136	168
168	6720	18.9	368	52.3	28.9	7.03	137	149
149	5960	17.7	333	49.6	27.7	6.71	136	162
162	6480	18.5	356	51.4	28.5	6.92	137	153
153	6120	17.9	340	50.1	27.9	6.79	137	159
159	6360	18.3	351	51.0	28.3	6.88	137	155
155	6200	18.1	345	50.5	28.1	6.83	137	157
157	6280	18.2	348	50.8	28.2	6.85	137	156
156	6240	18.1	345	50.5	28.1	6.83	137	157

$$P = b + 2(1 + z^2)^{0.5} = 10 + 2.24y$$

$$T = b + 2zy = 10 + y$$

and

$$C = \frac{\phi}{n} R^{1/6} = \frac{1.49}{0.015} R^{1/6}$$

Column 9:   The calculated distance to the critical section by Eq. (6.4.14).

When the value in col. 9 of Table 6.16 agrees with the estimated value in col. 1, the calculation is completed. From Table 6.16 it is estimated that a critical flow section occurs approximately 156 ft (48 m) downstream from the beginning of the channel.

With the critical flow section located, the water surface profile both upstream and downstream of this point can be estimated from Eq. (6.4.18). The calculations for the water surface profile upstream of the critical flow section are contained in Table 6.17a, while those for the profile downstream of this section are contained in Table 6.17b. In these tables the columns contain the following data.

Column 1:　This is the longitudinal distance between the point of computation and the beginning of the channel.

Column 2:　This is the incremental longitudinal distance between two adjacent points of computation.

Table 6.17. Data for Example 6.10 (Note, all velocities used in this table are average velocities.)

x ft (1)	Δx ft (2)	z_0 ft (3)	y ft (4)	z ft (5)	$\Delta y'$ ft (6)	A ft² (7)	Q ft³/s (8)	u ft/s (9)	Q_1+Q_2 ft³/s (10)	u_1+u_2 ft/s (11)	ΔQ ft³/s (12)	Δu ft/s (13)	Δy_m ft (14)	R ft (15)	h_f ft (16)	Δy ft (17)

(a) Computation of upstream subcritical water surface profile for Example 6.10

x ft (1)	Δx (2)	z_0 (3)	y (4)	z (5)	$\Delta y'$ (6)	A (7)	Q (8)	u (9)	Q_1+Q_2 (10)	u_1+u_2 (11)	ΔQ (12)	Δu (13)	Δy_m (14)	R (15)	h_f (16)	Δy (17)
156		50.20	18.10	68.30		345	6240	18.09								
			16.00	74.70	6.33	288	4000	13.89	10,240	31.98	2,240	4.20	5.56	6.28	0.09	5.65
			15.50	74.20	5.80	275	4000	14.54	10,240	32.63	2,240	3.55	5.42	6.15	0.11	5.53
100	56	58.70	15.0	73.70	5.33	262	4000	15.24	10,240	33.33	2,240	2.85	5.25	6.02	0.12	5.37
			12.80	79.00	5.33	210	2000	9.52	6,000	24.76	2,000	5.72	5.37	5.43	0.05	5.42
50	50	66.20	12.90	79.10	5.42	212	2000	9.42	6,000	24.66	2,000	5.82	5.38	5.45	0.05	5.43
			11.30	81.20	2.16	177	1000	5.65	3,000	15.07	1,000	3.77	2.06	5.01	0.01	2.07
25	25	69.90	11.20	81.10	2.06	175	1000	5.72	3,000	15.14	1,000	3.70	2.06	4.98	0.01	2.07
			9.90	82.10	0.96	148	400	2.70	1,400	8.42	600	3.02	0.87	4.60	0	0.87
10	15	72.20	9.80	82.00	0.86	146	400	2.74	1,400	8.46	600	2.98	0.87	4.57	0	0.87
0	10	73.70	8.61	82.31												

$\Delta y' = 2u^2/2g$ at x = 10 ft (assumed)

(b) Computation of downstream water surface profile for Example 6.10

x ft (1)	Δx (2)	z_0 (3)	y (4)	z (5)	$\Delta y'$ (6)	A (7)	Q (8)	u (9)	Q_1+Q_2 (10)	u_1+u_2 (11)	ΔQ (12)	Δu (13)	Δy_m (14)	R (15)	h_f (16)	Δy (17)
156		50.20	18.10			345	6,240	18.09								
			18.60	62.20	6.12	359	8,000	22.28	14,240	40.37	1,760	4.19	5.75	6.95	0.17	5.92
			18.50	62.10	6.22	356	8,000	22.46	14,240	40.55	1,760	4.37	5.91	6.92	0.17	6.08
200	44	43.60	18.40	62.00	6.32	353	8,000	22.64	14,240	40.73	1,760	4.55	6.06	6.89	0.17	6.23
			19.70	55.77	6.22	391	10,000	25.57	18,000	48.21	2,000	2.93	6.20	7.22	0.24	6.44
250	50	36.07	19.90	55.97	6.02	397	10,000	25.19	18,000	47.83	2,000	2.55	5.84	7.27	0.23	6.07
300	50	28.55	21.00	49.55	6.42	430	12,000	27.91	22,000	53.10	2,000	2.72	6.22	7.54	0.27	6.49
			21.20	42.22	7.32	437	14,000	32.04	26,000	59.95	2,000	4.13	8.13	7.60	0.35	8.48
350	50	21.02	22.00	43.02	6.53	462	14,000	30.30	26,000	58.21	2,000	2.39	6.21	7.79	0.30	6.51
			23.00	36.50	6.53	495	16,000	32.32	30,000	62.62	2,000	2.02	6.02	8.05	0.33	6.35
400	50	13.50	22.90	36.40	6.63	491	16,000	32.57	30,000	62.87	2,000	2.27	6.31	8.01	0.33	6.64

Column 3:　This is the elevation of the channel bottom and is obtained by substituting the product of col. 1 and the longitudinal slope of the channel from the bottom

elevation at the beginning of the channel; e.g., at the critical flow section.

$$z_0 = 73.7 - 156(0.1505) = 50.2 \text{ ft } (15.3 \text{ m})$$

Column 4: This is the assumed depth of flow.

Column 5: This is the elevation of the water surface and is obtained by adding cols. 3 and 4.

Column 6: This is the change in the water surface elevation and is calculated from

$$\Delta y' = -\Delta y + S_0 \Delta x$$

Cols. 7-13: These are, respectively, the flow area for the assumed depth of flow, the discharge, the average velocity, the sum of the flow rates, the sum of the average velocities, the difference in discharge between two adjacent stations, and the change in average velocity.

Column 14: This is the drop in the water surface due to the impact loss or

$$\Delta y'_m = \frac{Q_1\left(\overline{u_1} + \overline{u_2}\right)}{g\left(Q_1 + Q_2\right)}\left(\overline{\Delta u} + \frac{\overline{u_2}}{Q_1}\Delta Q\right)$$

Column 15: This is the hydraulic radius associated with the assumed depth of flow.

Column 16: This is the head loss due to friction and is computed from

$$h_f = S_f\Delta x = \left(\frac{nQ}{1.49AR^{2/3}}\right)^2 \Delta x$$

Column 17: This is the drop in the water surface between two adjacent stations estimated by Eq. (6.4.18).

At each longitudinal station, a trial-and-error procedure is used until the values in cols. 6 and 17 agree. This procedure is applied to the profiles above and below the critical section (Table 6.17*a* and *b*).

6.4.2.   Decreasing Discharge

In this type of spatially varied flow, which is exemplified by side weirs and bottom racks, there is no significant energy loss and the water surface profile can be estimated from the energy principle. The total energy at a channel section relative to a datum is

$$H = z + y + \frac{Q^2}{2gA^2} \qquad (6.4.19)$$

Differentiating this equation with respect to the longitudinal coordinate x

$$\frac{dH}{dx} = \frac{dz}{dx} + \frac{dy}{dx} + \frac{1}{2g}\left(\frac{2Q}{A^2}\frac{dQ}{dx} - \frac{2Q^2}{A^3}\frac{dA}{dx} \right) \qquad (6.4.20)$$

in this equation,

$$\frac{dH}{dx} = -S_f$$

$$\frac{dz}{dx} = -S_0$$

and

$$\frac{dA}{dy} = \frac{dA}{dy}\frac{dy}{dx} = T\frac{dy}{dx}$$

Substituting these relations in Eq. (6.4.20) and rearranging yields

$$\frac{dy}{dx} = \frac{S_0 - S_f - \dfrac{Q}{gA^2}\dfrac{dQ}{dx}}{1 - \dfrac{Q^2}{gA^2 D}} \qquad (6.4.21)$$

If a nonuniform velocity profile is present, the energy correction factor is used and Eq. (6.4.21) becomes

$$\frac{dy}{dx} = \frac{S_0 - S_f - \dfrac{\alpha Q}{gA^2}\dfrac{dQ}{dx}}{1 - \dfrac{\alpha Q^2}{gA^2 D}} \qquad (6.4.22)$$

Equations (6.4.21) and (6.4.22) are known as the dynamic equations for spatially varied flow with decreasing discharge. It is noted that the only difference between Eqs. (6.4.9) and (6.4.21) is in the coefficient of the third term in the numerator.

In the case of withdrawal of water through a rack in the bottom of a rectangular channel (Fig. 6.6), an equation for the water surface profile can be derived. Assume that $\alpha = 1$ and $\theta = 0$; then the specific energy at any section of the channel is

$$E = y + \frac{Q^2}{2g(by)^2} \qquad (6.4.23)$$

where b = width of the channel. According to Chow (1959), it can be assumed that the specific energy along the channel is constant. Using this assumption and differentiating Eq. (6.4.23) with respect to longitudinal distance yields

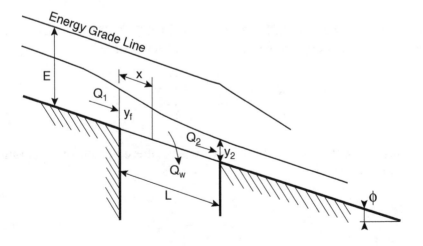

Figure 6.6. Analysis of spatially varied flow with bottom withdrawal.

$$\frac{dy}{dx} = -\frac{Qy\dfrac{dQ}{dx}}{gb^2y^3 - Q^2} \tag{6.4.24}$$

where dQ/dx is the flow withdrawn through a length Δx of the rack. In contrast to the problem of spatially varied flow with increasing discharge, dQ/dx cannot be assumed to be constant since this term depends on the effective head on the rack.

In the case of a rack where the direction of flow through the rack openings is nearly vertical, e.g., in the case of a rack composed of vertical bars, the effective head on the rack is approximately equal to the specific energy, and the energy loss is negligible. In such a situation,

$$-\frac{dQ}{dx} = \varepsilon C_D b\sqrt{2gE} \tag{6.4.25}$$

where ε = ratio of the open area in the rack to the total rack surface area and C_D = the coefficient of discharge. According to Chow (1959), C_D varies for this situation from 0.435 for a grade of 1 on 5 to 0.497 for a horizontal channel. Then from Eq. (6.4.23), the discharge is

$$Q = by\sqrt{2g(E - y)} \tag{6.4.26}$$

Substitution of Eqs. (6.4.25 and 6.4.26) in Eq. (6.4.24) yields a first-order differential equation for the water surface profile or

$$\frac{dy}{dx} = \frac{2\varepsilon C_D \sqrt{E(E - y)}}{3y - 2E}$$

and integrating this equation and evaluating the constant of integration with the condition that at $x = 0$, $y = y_1$, yields

$$x = \frac{E}{\varepsilon C_D}\left(\frac{y_1}{E}\sqrt{1 - \frac{y_1}{E}} - \frac{y_1}{E}\sqrt{1 - \frac{y}{E}}\right) \tag{6.4.27}$$

The length of rack required to remove all the flow in the reach where therack is located is given by

$$L = \frac{E}{\varepsilon C_D}\left(\frac{y_1}{E}\sqrt{1 - \frac{y_1}{E}} \right) \qquad (6.4.28)$$

since at this point, $y = 0$.

In the case where the openings in the rack make an appreciable angle with the vertical, e.g., in the case of a rack composed of a perforated screen, the energy loss through the rack is not negligible. It is assumed that the effective head on the rack is equal to the depth of flow. In this case, the discharge through the rack for an incremental distance dx is

$$-\frac{dQ}{dx} = \varepsilon C_D b \sqrt{2gy} \qquad (6.4.29)$$

Then in a fashion analogous to that described above

$$x = \frac{E}{\varepsilon C_D}\left[0.5\cos^{-1}\sqrt{\frac{y}{E}} - 1.5\sqrt{\frac{y}{E}\left(1 - \frac{y}{E}\right)} \right] + \Gamma \qquad (6.4.30)$$

where

$$\Gamma = \frac{E}{\varepsilon C_D}\left[1.5\sqrt{\frac{y_1}{E}\left(1 - \right)} - 0.5\cos^{-1}\left(\frac{y_1}{E}\right) \right]$$

The length of rack required to complete flow withdrawal is

$$L = \frac{E}{\varepsilon C_D}\left[1.5\sqrt{\frac{y_1}{E}\left(1 - \frac{y_1}{E}\right)} - 0.25\sin^{-1}\left(1 - \frac{2y_1}{E}\right) + \frac{\pi}{8} \right] \qquad (6.4.31)$$

In the case of racks composed of perforated screens, Chow (1959) noted that C_D, the discharge coefficient, ranges from 0.750 for grades of 1:5 to 0.800 for channels built on horizontal slopes.

It should be noted that the concept of flow removal by a vertical drop structure is particularly important because this situation is found in most stormwater channel systems.

6.5.   APPLICATION TO PRACTICE

6.5.1.   Introduction
In the foregoing sections, the subject of gradually varied flow profile estimation has been treated from the abstract viewpoint of discussing computational procedures. In this section, a number of situations in which gradually varied flow profiles interact with other aspects of flow in open channels are considered.

6.5.2.   Profile Behind a Dam
One of the most common gradually varied flow problems involves the estimation of backwater effects behind a dam. Although theoretically the gradually varied flow profile, behind a dam on a mild slope, extends in the upstream direction indefinitely. It is usually assumed to terminate at the point where the elevation of the water surface is 1 percent greater than the elevation under normal flow conditions. In most investigations, a backwater envelope is found; i.e., the terminus of the backwater curve is determined for a number of flow rates.

The backwater envelope begins at the point where the static pool of the reservoir intersects the channel floor - the case when there is zero flow. As the flow into the reservoir increases, the endpoint of the backwater curve may migrate either upstream or downstream depending on many factors. For example:

1.   If the reservoir level is constant and the channel prismatic with a simple cross section, then the terminus of the backwater curve will migrate downstream as the flow rate increases.

2.   An increase in the channel roughness results in the shortening of the length of the gradually varied flow profile and, hence, in a downstream migration of the backwater terminus.

3.   The presence of flood plains also results in a shortening of the profile length and a downstream movement of the terminus.

In many cases, it is convenient to define the endpoint of the backwater curve in terms of the requirements of a specific problem. For example, the endpoint may be defined as the point where the rise in the water surface begins to cause damage.

6.5.3.   Discharge Computation

When an open channel is used to connect two reservoirs, the possibility that the discharge of this channel is affected by the interaction of local features and the gradually varied flow profile must be considered. It is noted that this problem has been previously addressed by Bakhmeteff (1932), Chow (1959), and Henderson (1966), and the classification system developed by Bakhmeteff is used here. The classification system is based on the behavior of three crucial variables: (1) the depth of flow at the upstream entrance to the channel, y_1; (2) the depth of flow at the downstream outlet, y_2; and (3) the channel discharge Q. These variables are described schematically in Fig. 6.7. Based on the behavior of these variables, the following cases can be identified.

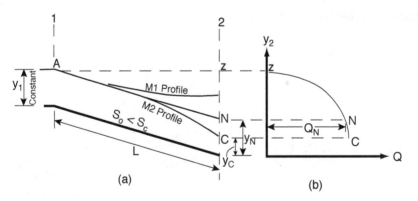

Figure 6.7. Channel discharge for y_1 = constant, mild slope.

Case I (y_1 = constant)

For y_1 constant, several discharge conditions are possible. First, if $y_1 = y_N = y_2$, the flow in the channel is uniform and the discharge can be computed from the Manning equation. In Fig. 6.7a this situation is represented by the line AN, and the corresponding discharge is represented by point N in Fig. 6.7b.

Second, if $y_2 = y_c$, the discharge of the channel is maximum attainable (line AC in Fig. 6.7a and point C in Fig.6.7b). It is noted that if $y_2 < y_c$, then the channel terminates in a free overfall, but the discharge will not exceed the value represented by point C in Fig. 6.7b.

Third, if $y_2 > y_N$, then the flow profile in the channel is a M1 curve. For this situation to exist, the following condition must be satisfied: $y_N < y_2 < y_c$. At this point, the length of the M1 profile becomes a crucial consideration. If y_1 is constant, then the length of the M1 profile, L', cannot exceed L; however, L' may be less than L. In this latter case, the channel is termed long and the M1 curve is said to "run out"; i.e., the discharge at the upstream station is not affected by y_2 and the depth of flow in the initial channel length is normal. If $L' = L$, then the discharge to the canal is less than the normal flow.

Fourth, if $y_c < y_2 < y_N$, then an *M2* gradually varied flow profile occurs in the channel and the discharge relationship can be discussed as was done for the *M1* profile.

Case II (y_2 = constant)

For y_2 constant, the same types of situations discussed for Case I are possible. First, if $y_1 = y_N = y_2$, flow occurs at normal depth and the discharge is the normal discharge (Fig. 6.8a and b).

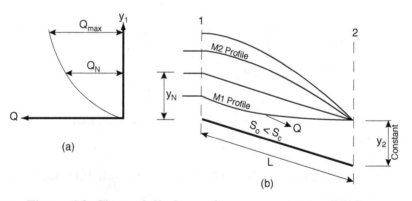

Figure 6.8. Channel discharge for y_2 = constant, mild slope.

Second, maximum discharge occurs when the depth y_1 corresponds to a critical discharge at section 2. It is noted that this value of y_1 is the maximum depth which can occur since any increase in y_1 would necessitate a corresponding increase in y_2.

Third, for $y_1 < y_N$, an *M1* flow profile occurs, and the discharge is less than that which would occur at normal depth.

Fourth, for $y_1 > y_N$ but less than the depth which would cause critical discharge, an *M2* profile occurs. In this case, the discharge would be greater than normal but less than the maximum.

Case III (Q = constant)

For a constant channel discharge, the depths y_1 and y_2 are allowed to fluctuate. This first possibility is $y_2 = y_N = y_1$, and the discharge can be found by the Manning equation, line *AN* in Fig. 6.9.

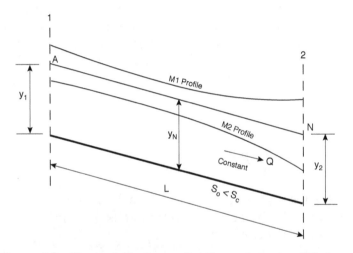

Figure 6.9. Channel discharge for Q = constant, mild slope.

Second, for water surface positions above the line *AN*, an *M1* profile occurs. The obvious upper limit occurs when $y_2 = y_1 = S_0L$. It is noted that as this limiting situation is approached, both the elevation difference between the reservoirs and the velocity of flow decrease, however, since the flow area increases, the discharge can be held constant.

Third, for water surface positions below *AB* an *M2* profile occurs. The minimum depth of flow occurs when $y_2 = y_c$.

The application of these rather abstract concepts to practice is best examined through an example.

EXAMPLE 6.11

A rectangular channel 12.0 m (39 ft) wide conveys water from a reservoir whose surface elevation is 3.0 m (9.8 ft) above the channel bed at the reservoir outlet. At a distance of 1,000 m (3,280 ft) downstream of the reservoir, the channel contracts to a width of 7.3 m (24 ft). If $n = 0.020$ and $S = 0.001$, determine the discharge from the reservoir.

Solution

The solution of this problem requires a number of assumptions which are either accepted or rejected on the basis of calculations. Thus:

1. Assume that the contraction is not severe enough to cause a reduction in the flow. Under this assumption, the flow in the channel is uniform, and the solution is found by calculating uniform flow conditions over a range of depths until the value of the specific energy in the channel matches the value at the beginning of the channel.

    The calculations are summarized in Table 6.18 where

$$\bar{u} = \frac{\phi\sqrt{S}}{n}R^{2/3} = \frac{\sqrt{0.001}}{0.020}R^{2/3} = 1.58R^{2/3}$$

From the results summarized in this table, it is determined that the unrestricted discharge of the reservoir is 75.8 m³/s (2,680 ft³/s). Then, within the constriction

$$q = \frac{Q}{b} = \frac{75.8}{7.3} = 10.4\,\text{m}^3/\text{s}\left(112\,\text{ft}^3/\text{s}/\text{ft}\right)$$

and for this value of q the critical depth is, from Eq. (2.2.6)

$$y_c = \left(\frac{q^2}{g}\right)^{1/3} = \left(\frac{10.4^2}{9.8}\right)^{1/3} = 2.22\,\text{m}\left(7.28\,\text{ft}\right)$$

and the critical specific energy from Eq. (2.2.8) is

$$E_c = 1.5y_c = 1.5\left(2.22\right) = 3.33\,\text{m}\left(10.9\,\text{ft}\right)$$

which is greater than the available upstream specific energy if the flow is uniform. Thus, it is concluded that the contraction acts as a choke, and hence the flow in the channel cannot be uniform. The question now is if the M1 backwater profile caused by the construction influences the flow rate.

2.   If it is assumed that the discharge from the reservoir is not affected by the constriction, then the depth just upstream of the constriction is found from

$$E = y + \frac{q^2}{2gy^2} = y + \frac{\left(\frac{75.8}{12}\right)^2}{2(9.8)y^2} = y + \frac{2.04}{y^2} = 3.33\,\text{m}\left(10.9\,\text{ft}\right)$$

where it is assumed that there is no energy loss at the entrance to the constriction. Solving the above equation by trial and error yields $y = 3.15$ m (10.3 ft). Then, the gradually varied upstream flow profile can be estimated from the initial depth, e.g., Example 6.1. The results of this analysis are summarized in Table 6.18 *a*. This analysis demonstrates that for $Q = 75.8$ m³/s (2,680 ft³/s) the gradually varied flow profile extends more than 1,500 m (4,920 ft) upstream or beyond the entrance to the channel. Therefore, it is concluded that the downstream constriction controls the flow in the channel

3.   Since the constriction 1,000 m (3280 ft) downstream of the reservoir affects the flow rate, a new trial flow rate must be determined; i.e., a value of Q must be determined such that the value of E at the reservoir is 3.0 m (9.8 ft). By interpolation in Table 6.18, it is estimated that the value of E at the reservoir is 3.12 m (10.2 ft) for $Q = 75.8$ m³ /s (2,680 ft³/s); thus, the correct value was missed by 0.12 m (0.39 ft). The obvious second trial value of Q is the one which produces a value of $E = 0.12$ m (0.39 ft), less than the previously estimated flow rate at the entrance to the constriction or

$$E_c = 3.33 - 0.12 = 3.21\,\text{m}\left(10.5\,\text{ft}\right)$$

Table 6.18. Data for and solution of Example 6.11.

y	A	P	R	$\bar{u}$	F	Q	$\dfrac{\bar{u}^2}{2g}$	E
m	m²	m	m	m/s		m³/s	m	m
(1)	(2)	(3)	(4)	(5)	(6)	(7)	(8)	(9)
2	24	16	1.5	2.1	0.47	50.4	0.22	2.22
3	36	18	2.0	2.5	0.46	90.0	0.32	3.32
2.5	30	17	1.8	2.3	0.46	69.0	0.27	2.77
2.7	32	17.4	1.84	2.47	0.46	75.8	0.29	2.99

y	A	P	R	$\bar{u}$	$\dfrac{\bar{u}^2}{2g}$	E	$\dfrac{n^2\bar{u}^2}{\phi^2 R^{4/3}}$	$\left(\dfrac{n^2\bar{u}^2}{\phi^2 R^{4/3}}\right)_m$	$S_0 - \left(\dfrac{n^2\bar{u}^2}{\phi^2 R^{4/3}}\right)_m$	ΔE	Δx	x = $\Sigma\Delta$x
(m)	(m²)	(m)	(m)	(m/s)	(m)	(m)				(m)	(m)	(m)
(1)	(2)	(3)	(4)	(5)	(6)	(7)	(8)	(9)	(10)	(11)	(12)	(13)
(a) Gradually varied flow profile computation, Example 6.11, Q = 75.8 m³/s												
3.15	37.8	18.3	2.07	2.01	0.21	3.36	0.000614					0
								0.000661	0.000339	0.13	383	
3.0	36.0	18.0	2.00	2.11	0.23	3.23	0.000708					383
								0.000807	0.000193	0.21	1090	
2.75	33.0	17.5	1.89	2.30	0.27	3.02	0.000906					1470
(b) Gradually varied flow profile computation, Example 6.11, Q = 72 m³/s												
3.0	36.0	18.0	2.00	2.00	0.20	3.20	0.000635					0
								0.000667	0.000333	0.08	240	
2.9	34.8	17.8	1.96	2.07	0.22	3.12	0.000699					240
								0.000736	0.000264	0.09	341	
2.8	33.6	17.6	1.91	2.14	0.23	3.03	0.000773					581
								0.000794	0.000206	0.04	194	
2.75	33.0	17.5	1.89	2.18	024	2.99	0.000814					775
(c) Gradually varied flow profile computation, Example 6.11, Q = 73.3 m³/s												
3.09	37.1	18.1	1.98	1.98	0.20	3.29	0.000614					0
								0.000670	0.000330	0.16	485	
2.90	34.8	17.8	1.96	2.11	0.23	3.13	0.000726					485
								0.000765	0.000235	0.09	382	
2.80	33.6	17.6	1.91	2.18	0.24	3.04	0.000804					867
								0.000825	0.000175	0.04	224	
2.75	33.0	17.5	1.89	2.22	0.25	3.00	0.000845					1090

With this value of E_c

$$y_c = \frac{2}{3}E_c = \frac{2}{3}(3.21) = 2.14\,\text{m} = (7.02\,\text{ft})$$

and

$$u_c = \sqrt{gy_c} = 4.58\,\text{m/s}\,(15.0\,\text{ft/s})$$

Then

$$Q = u_c y_c b = 4.58(2.14)(7.3) = 71.6 \approx 72\,\text{m}^3/\text{s}\,(2{,}540\,\text{ft}^3/\text{s})$$

The depth of flow immediately upstream of the constriction is estimated by a trial-and-error solution of

$$E = y + \frac{q^2}{2gy^2} = y + \frac{1.84}{y^2} = 3.21\,\text{m}\,(10.5\,\text{ft})$$

or $y = 3.0$ m (9.8 ft). The gradually varied flow profile for $Q = 72.0$ m³/s (2,540 ft³/s) is estimated in Table 6.18b. In this case, the profile terminates or "runs out" before the reservoir is reached. Thus, it is concluded that the correct flow rate for this situation lies between 75.8 m³/s (2,680 ft³/s) and 72.0 m³/s (2,540 ft³/s). Interpolation in Table 6.18a and b yields a new flow estimate of 73.3 m³/s (2,590 ft³/s). Then

$$q = \frac{Q}{b} = \frac{73.3}{7.3} = 10.0\,\text{m}^3/\text{s/m}\,(108\,\text{ft}^3/\text{s/ft})$$

$$y_c = \left(\frac{q^2}{g}\right)^{1/3} = \left(\frac{10.0^2}{9.8}\right)^{1/3} = 2.17\,\text{m}\,(7.12\,\text{ft})$$

$$E_c = 1.5y_c = 1.5(2.17) = 3.26\,\text{m}\,(10.7\,\text{ft})$$

and the depth of flow just upstream of the constriction is estimated from

$$E = y + \frac{q^2}{2gy^2} = y + \frac{\left(\dfrac{73.3}{12}\right)^2}{2(9.8)y^2} = 3.26\,\text{m}\,(10.7\,\text{ft})$$

by trial and error or $y = 3.09$ m (10.0 ft). The validity of this set of parameters is confirmed in Table 6.18*c*.

The discharge problem can also arise in channels of steep slope; i.e., the slope of the channel is greater than the critical slope. If the slope is not so steep that the flow becomes unsteady, then the flow rate is governed by the upstream critical section and can be easily computed. The water surface profile in such a channel is governed by the downstream water surface elevation. First, if the downstream water surface elevation is less than the depth of flow at station 2, then a free overall occurs (Fig. 6.10). In this situation, the flow passes through critical depth at point *C* and approaches normal depth by means of an *S2*

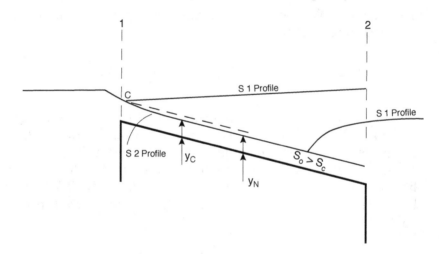

Figure 6.10. Channel discharge for supercritical flow.

drawdown curve. Second, if the downstream water surface elevation is greater than the depth of flow at station 2, then the *S2* curve terminates in a hydraulic jump. The flow upstream of the jump is not affected by the downstream water surface elevation. Third, as the downstream water surface elevation increases, the location of the jump moves upstream until point *C* is reached. If the downstream water surface elevation increases beyond this elevation, the flow is affected by the downstream elevation since a drowned jump would occur.

6.5.4.   Flow Past Islands
In natural channels, there may be reaches in which a long island divides the flow into two channels (Fig. 6.11). If the flow is normal, then Q_1 and Q_2 may be estimated from the following set of equations:

$$Q = Q_1 + Q_2$$

$$Q_1 = K_1\sqrt{S_1}$$

$$Q_2 = K_2\sqrt{S_2}$$

However, if the flow is gradually varied, then the solution procedure is more complex.

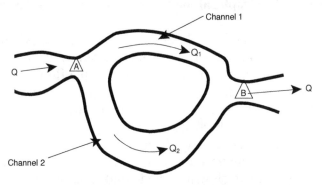

Figure 6.11. Schematic of flow past an island.

If the flow is gradually varied and subcritical, the computation begins at point B. For convenience, define E_1^A = water surface elevation at point A, estimated by beginning at point B and proceeding to point A by way of channel 1 and E_2^A = water surface elevation at point A, estimated be beginning at point B and proceeding to point A by way of channel 2. The discharges Q_1 and Q_2, are assumed such that their total is Q. Then using the methods of Section 6.3, E_1^A and E_2^A are estimated for the flow rates Q_1 and Q_2, respectively. If $E_1^A = E_2^A$, as they must be if the assumed values of Q_1 and Q_2 are correct, then the solution has been determined. In general a number of trials are required. Two graphs may be used in most cases to expedite the solution (Fig. 6.12). In Fig. 6.12a, E_1^A is plotted as a function of E_2^A. On this figure, a 45 degree line is plotted, and the correct water surface elevation at point A is estimated from the intersection of this line and the E_1^A versus E_2^A curve (point C). Then an auxiliary curve (Fig. 6.12b) of E_1^A versus Q is used to estimate Q_1.

If the flow in the channel is supercritical, then the solution proceeds in the same manner but begins at point A. In addition, it is noted that this procedure also serves to extend the gradually varied flow profile past the island.

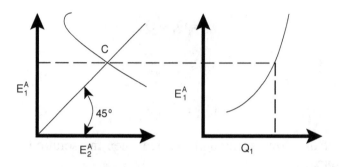

Figure 6.12. Graphical solution of flow passing and island.

Although the procedure described above provides an adequate method for determining the flow around a single island, it is not a technique which can be extended to determine the flow around a group of islands. In addition, the foregoing technique does not account for changes in the shape of the channel as it rounds the island and the minor losses associated with such changes. Wylie (1972) noted that the flow around islands can be efficiently determined by visualizing the system as a series of nodes connected by links. For example, the situation described in Fig. 6.11 is shown in node and link notation in Fig. 6.13. The solution of the node and link system rests on two principles:

1. Each node has a total energy associated with it which is common to each element or link terminating at the node.

2. The equation of continuity must be satisfied at each node.

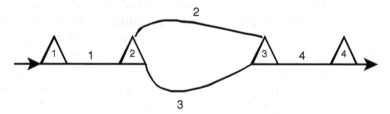

Figure 6.13.   Link and node representation of flow past an island.

The most common type of link is the single channel with no minor losses. In this case, the energy equation for the link k, connecting an upstream node i with a downstream node j, is

$$E_i = E_j + \overline{S}_f L_k$$

where L_k = the length of the link k, $\overline{S}_f$ = average slope of the energy grade line between the nodes i and j,

$$E = z + y + \frac{Q^2}{2gA^2} \tag{6.5.1}$$

and z = elevation of the channel bottom above a datum. Although any of the previously defined expressions for $\overline{S}_f$ [Eqs. (6.3.55) to (6.3.62)] can be used, the average friction slope expression, Eq. (6.3.35), will be used here or

$$E_i = E_j + \left(\frac{S_{ki} + S_{kj}}{2}\right) L_k \tag{6.5.2}$$

The slope of the energy grade line is usually determined from the Manning equation [Eq. (4.2.6)] or

$$S = \frac{Q^2 n^2 P^{4/3}}{\phi^2 A^{10/3}}$$

Combining this expression for S and Eq. (6.5.2) yields an equation for the flow in link k or

$$Q_k = \frac{\phi}{n} \left[\frac{2(E_i - E_j)}{L_k}\right]^{1/2} \left(\frac{P_{ki}^{4/3}}{A_{ki}^{10/3}} + \frac{P_{ji}^{4/3}}{A_{ji}^{10/3}}\right)^{-1/2} \tag{6.5.3}$$

Although Eq. (6.5.3) theoretically applies only to prismatic channels, it can be applied with confidence to channels which are moderately nonprismatic. However, if there is a rapid change in the cross-sectional shape of the channel, then it may be necessary to include a link which accounts for the local or minor losses associated with the change. King and Brater (1963) suggest that the energy equation at a localized loss in subcritical flow is

$$E_i = E_j + \frac{K_k Q_k^2}{2g} \left(\frac{1}{A_{ki}^2} + \frac{1}{A_{kj}^2}\right) \tag{6.5.4}$$

where K = a loss coefficient. Rearrangement of this equation yields

$$Q_k = \left[2g\left(E_i - E_j\right)\right]^{1/2}\left[K_k\left(\frac{1}{A_{ki}^2} - \frac{1}{A_{kj}^2}\right)\right] \tag{6.5.5}$$

When the element equations defined above are combined with a continuity equation written at each node, a system of nonlinear simultaneous equations which must be solved results. The continuity equation at each node is given by

$$F_i = \sum_{k=1}^{M} Q_k + Q_{Ni} = 0 \tag{6.5.6}$$

where M is the number of links connected to the node i, and $Q_{Ni} =$ flow, which is added at a node. Application of Eq. (6.5.6) to node 2 in Fig. 6.13 yields

$$F_2 = Q_1 - Q_2 - Q_3 + Q_{N2} = 0 \tag{6.5.7}$$

Then, since $Q_{N2} = 0$, substitution of the link flow equations in Eq. (6.5.7) yields

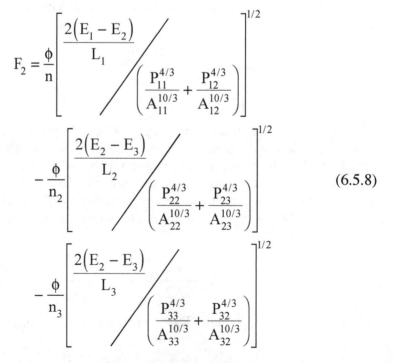

$$F_2 = \frac{\phi}{n}\left[\frac{\dfrac{2\left(E_1 - E_2\right)}{L_1}}{\left(\dfrac{P_{11}^{4/3}}{A_{11}^{10/3}} + \dfrac{P_{12}^{4/3}}{A_{12}^{10/3}}\right)}\right]^{1/2}$$

$$- \frac{\phi}{n_2}\left[\frac{\dfrac{2\left(E_2 - E_3\right)}{L_2}}{\left(\dfrac{P_{22}^{4/3}}{A_{22}^{10/3}} + \dfrac{P_{23}^{4/3}}{A_{23}^{10/3}}\right)}\right]^{1/2}$$

$$- \frac{\phi}{n_3}\left[\frac{\dfrac{2\left(E_2 - E_3\right)}{L_3}}{\left(\dfrac{P_{33}^{4/3}}{A_{33}^{10/3}} + \dfrac{P_{32}^{4/3}}{A_{32}^{10/3}}\right)}\right]^{1/2} \tag{6.5.8}$$

At this point, it should be noted that the subscripts on the variables F and E refer to a node, the subscripts on n and L refer to a link, and on P and A, the first subscript refers to a link while the second refers to a node.   For the other nodes in the system shown in Fig. 6.13, the corresponding equations are

Node 1

$$F_1 = Q_{N1} - Q_1 = 0$$

In this case $Q_{N1} \neq 0$ and substitution of Eq. (6.5.3) yields

$$F_1 = Q_{N1} - \frac{\phi}{n_1} \left[\frac{\dfrac{2(E_1 - E_2)}{L_1}}{\left(\dfrac{P_{11}^{4/3}}{A_{11}^{10/3}} + \dfrac{P_{12}^{4/3}}{A_{12}^{10/3}} \right)} \right]^{1/2} \tag{6.5.9}$$

Node 3

$$F_3 = Q_{N1} - Q_1 = 0$$

and $Q_{N3} = 0$. Therefore,

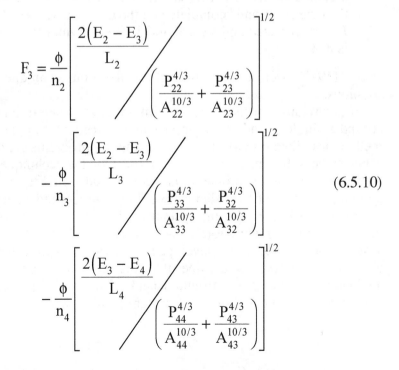

$$F_3 = \frac{\phi}{n_2} \left[\frac{\dfrac{2(E_2 - E_3)}{L_2}}{\left(\dfrac{P_{22}^{4/3}}{A_{22}^{10/3}} + \dfrac{P_{23}^{4/3}}{A_{23}^{10/3}} \right)} \right]^{1/2}$$

$$- \frac{\phi}{n_3} \left[\frac{\dfrac{2(E_2 - E_3)}{L_3}}{\left(\dfrac{P_{33}^{4/3}}{A_{33}^{10/3}} + \dfrac{P_{32}^{4/3}}{A_{32}^{10/3}} \right)} \right]^{1/2} \tag{6.5.10}$$

$$- \frac{\phi}{n_4} \left[\frac{\dfrac{2(E_3 - E_4)}{L_4}}{\left(\dfrac{P_{44}^{4/3}}{A_{44}^{10/3}} + \dfrac{P_{43}^{4/3}}{A_{43}^{10/3}} \right)} \right]^{1/2}$$

In a typical steady-flow problem, the unknowns would likely be Q_1, Q_2, E_1, E_2, and E_3. The depth of flow and flow rate at node 4 would generally be known and, therefore, E_4. Also, since the flow is by definition steady, $Q_{N1} = Q_{N2}$. Since the system of simultaneous equations which govern this problem are nonlinear, a direct solution is not possible; however, a trial-and-error solution can be effected. The steps in obtaining a solution are:

1. Trial values of E_3, E_2, and E_1 are assumed. The speed at which any trial-and-error solution process will converge to the correct values of the variables depends on the accuracy of the values initially assumed. Wylie (1972) began his solutions with the assumption of constant energy throughout the system.

2. With the assumed values of E_3, E_2, and E_1, Eqs. (6.5.1) and (6.5.3) are used to determine the corresponding trial values of the depths of flow and the element flow rates.

3. Equations (6.5.8) and (6.5.10) are then used to compute values of F_3, F_2, and F_1. If the values of these functions are zero or meet a predetermined error criterion, then the solution is complete and the assumed values of E_3, E_2, and E_1 are correct. If this is not the case, then the solution is complete and the assumed values of E_3, E_2, and E_1 are corrected and steps 2 and 3 repeated until the error criterion is met.

Wylie (1972) used a Newton-Raphson iteration technique to achieve solutions.

It is obvious that the node-and-link method of determining the flow around a single island is not an efficient procedure compared with the method described previously, unless computer programs are available to achieve the solution. However, the power of the technique is evident when more complex problems are considered. In Fig. 6.14a a flow system consisting of two islands, two flow sources, and a single outflow is shown. In Fig. 6.14b, the system is viewed as a series of links and nodes. Although many more equations will be required to describe this system, the methodology developed by Wylie is applicable. It would also be a mistake to assume that the node-and-link technique is applicable only to flow around islands. This technique can also be applied to open-channel flow in a system of storm sewers.

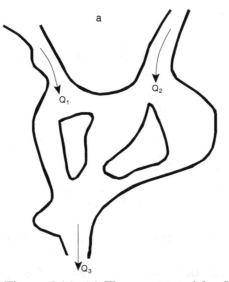

Figure 6.14. (a) Flow past two islands.

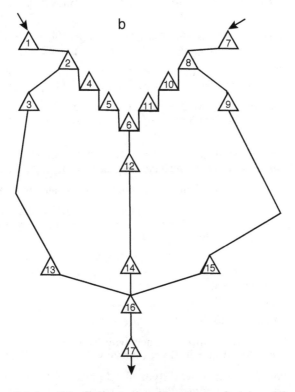

**Figure 6.14.   (b) Schematic diagram of (a) with eddy
losses at junctions included.**

6.6   BIBLIOGRAPHY

Anonymous, "HEC-RAS River Analysis System, Version 3.0, Users Manual," U.S. Army Corps of Engineers, Hydrologic Engineering Center, Davis, CA, January 2001.

Anonymous, "HEC-RAS River Analysis System, Version 3.0, Hydraulic Reference Manual," U.S. Army Corps of Engineers, Hydrologic Engineering Center, Davis, CA., January 2001.

Anonymous, "UNET, One-Dimensional Unsteady Flow Through a Full Network of Open Channels, User's Manual," U.S. Army Corps of Engineers, Hydrologic Engineering Center, Davis, CA, 1993.

Anonymous, "HEC-2, Water Surface Profiles, Computer Program 723-X6-L202A, Users Manual," U.S. Army Corps of Engineers, Hydrologic Engineering Center, Davis, CA, August 1979.

Bakhmeteff, B.A., *Hydraulics for Open Channels*, McGraw-Hill Book Company, New York, 1932.

Barkau, R.L., "UNET, One-Dimensional Unsteady Flow Through a Full Network of Open Channels, Computer Program," St. Louis, MO.

Chow, V.T., *Open-Channel Hydraulics*, McGraw-Hill Book Company, New York, 1959.

Chow, V.T., "Integrating the Equation of Gradually Varied Flow," *Proceedings of the American Society of Civil Engineers*, vol. 81, 1955, pp. 1-32.

Escoffier, F.F., "Graphic Calculation of Backwater Eliminates Solution by Trial," *Engineering News-Record*, vol. 136, no. 26, June 27, 1946, p. 71.

Ezra, A.A., "A Direct Step Method for Computing Water Surface Profiles," *Transactions of the American Society of Civil Engineers*, vol. 119, 1954, pp. 453-462.

Feldman, A.D., "HEC Models for Water Resources Simulation: Theory and Experience," *Advances in Hydroscience*, vol. 12, 1981, pp. 297-423.

Henderson, F.M., *Open Channel Flow*, The Macmillan Company, New York, 1966.

Hinds, J., "Side Channel Spillways: Hydraulic Theory, Economic Factors, and Experimental Determination of Losses," *Transactions of the American Society of Civil Engineers*, vol. 89, 1926, pp. 881-927.

Keulegan, G.H., "Determination of Critical Depth in Spatially Variable Flow," *Proceedings of the 2d Midwestern Conference on Fluid Mechanics*, Ohio State University, Columbus, 1952.

Kiefer, C.J., and Chu, H.H., "Backwater Functions by Numerical Integration," *Transactions, American Society of Civil Engineers*, vol. 120, 1955, pp. 429-442.

King, H.W., and Brater, E.F., "Open Channels with Nonuniform Flow," *Handbook of Hydraulics*, 5th ed., McGraw-Hill Book Company, New York, 1963.

Rajaratnam, N. and Subramanya, K., "Hydraulic Jumps Below Abrupt Symmetrical Expansions," *Proceedings of the American Society of Civil Engineers, Journal of the Hydraulics Division*, vol. 94, no. HY2, February 1968, pp. 481-503.

Reed, J.R., and Wolfkill, A.J., "Evaluation of Friction Slope Models," *Rivers '76 Symposium on Inland Waterways for Navigation, Flood Control, and Water Diversions*, Colorado State University, Ft. Collins, 1976, pp. 1159-1178.

Shearman, J.O., "Computer Applications for Step Backwater and Floodway Analyses," U.S. Geological Survey, Open File Report 76-499, Washington, 1976.

Shearman, J.O., "User's Guide, Step Backwater and Floodway Analyses, Computer Program J635," U.S. Geological Survey, Water Resources Division, Reston, VA, May 1977.

Wylie, E.G., "Water Surface Profiles in Divided Channels," *Journal of Hydraulic Research* (International Association for Hydraulic Research), vol. 10, no. 3, 1972, pp. 325-341.

Yen, B.C., and Wenzel, H.G., Jr., "Dynamic Equations for steady Spatially Varied Flow," *Proceedings of the American Society of Civil Engineers, Journal of the Hydraulics Division*, vol. 96, no. HY3, March 1970, pp. 801-814.

6.7   PROBLEMS

1.   The very wide rectangular channel shown in the accompanying figure has a unit discharge of 25 ft³/s/ft (2.3 m³/s/m) and $n = 0.02$. For these conditions,

   a.   Calculate the normal and critical depths of flow for each slope. Draw lines representing these depths of flow on the figure and label them as either NDL (normal depth of flow) or CDL (critical depth of flow).

   b.   On the figure, draw all possible flow profiles and classify and label them; for example, M1, S3, *etc.*

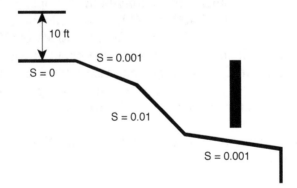

2.   A rectangular channel 2.4 m (7.9 ft) wide with a longitudinal slope of 0.008 m/m conveys a discharge of 4.2 m³/s (150 ft³/s). If $n = 0.014$ and the depth of flow is 1.5 m (4.9 ft), what is the flow profile classification? What is the upstream boundary condition?

3. A rectangular channel 25 ft (7.6 m) wide has a longitudinal slope of 0.002 ft/ft and conveys a discharge of 400 ft³/s (11 m³/s). If $n = 0.03$, $\alpha = 1.15$, and the downstream depth of flow at a control point is 7 ft (2.1 m), classify the flow profile upstream of the control point and estimate the length of the flow profile. What is the upstream boundary condition?

4. A trapezoidal channel, with a longitudinal slope of 0.001 m/m carrying a discharge of 30 m³/s (1,060 ft³/s), a Manning's n of 0.025, a bottom width of 6 m (19.7 ft), and side slopes 1.5 (horizontal): 1 (vertical), terminates in a weir. What is the upstream boundary condition for this problem? Define and estimate the gradually varied flow profile that results from this situation.

5. A trapezoidal channel with a longitudinal slope of 0.001 ft/ft, carrying a discharge of 1,000 ft³/s, a Manning's n of 0.025, a bottom width of 20 ft (6.1 m), and side slope of 1.5(horizontal): 1(vertical) terminates in a sluice gate structure. If the depth of water at the sluice gate is 10 ft (3.0 m), use the step method with a depth increment of 0.5 ft (0.15 m) to estimate the water surface profile behind this sluice gate. What is the upstream boundary condition for this problem?

6. A trapezoidal channel with a longitudinal slope of 0.0169 ft/ft, carrying a discharge of 400 ft³/s (11 m³/s), a Manning's n of 0.025, bottom width of 20 ft (6.1 m), and side slopes of 2(horizontal): 1(vertical) terminates at a dam where the depth of water is 5 ft (1.5 m). If the upstream boundary condition is critical flow, estimate the water surface profile behind the dam using the step method with a depth increment of 0.5 ft (0.15 m).

7. Determine the hydraulic exponents M and N for a trapezoidal channel with bottom width b and side slopes z.

8. A flow of 5,000 ft³/s (140 m³/s) is carried in a rectangular channel 40 ft (12 m) wide. At one point, this channel constricts to a width of 25 ft (7.6 m). It is known from model tests that the energy loss in this constriction is approximately 2.6 ft (0.76 m). This channel has a longitudinal slope of 0.002 ft/ft, and Manning's n has been estimated to be 0.16. Given this information,

    a.   Is the constriction sufficient to cause a backwater curve?

    b.   If the answer to Part (a) is yes, then estimate the depth of flow immediately upstream of the constriction.

    c.   If the answer to Part (a) is yes, what is the depth of flow 2,000 ft (610 m) upstream of the constriction?

9.   A trapezoidal irrigation channel has the following characteristics: $b = 15$ ft (4.6 m), $z = 2.0$, $S = 0.0004$ ft/ft, and a normal depth of 10 ft (3.0 m). If this channel empties into a pond at the downstream end and the pond elevation is 4 ft (1.2 m) higher than the channel elevation at the downstream end, calculate and plot the gradually varied flow profile that results from this situation.

10.   A 6-ft (1.8 m) diameter sewer pipe ($n = 0.015$) has a slope of 0.001 ft/ft and carries a discharge of 100 ft^3/s (2.8 m^3/s). If this pipe ends in a free overfall, estimate the resulting gradually varied flow profile.

11.   A small river has a cross section that can be approximated as a trapezoid. If $n = 0.015$, $Q = 3,500$ ft^3/s (99 m^3/s), and the cross sectional properties of three typical cross sections are given in the following table, find the water surface elevations at Sections II and III if the water surface elevation at Section I is 345 ft (105 m).

Section	Distance from Mouth (mi) (km)	Bed Elevation above MSL (ft) (m)	Bed Width (ft) (m)	Side Slope z
I	100 (161)	328 (99.97)	46 (14)	2
II	101.5 (163)	331 (100.9)	41 (12)	1.5
III	102.5 (165)	333 (101.5)	33 (10)	3

12.   In the accompanying table, values of the wetted perimeter, P, and the cross-sectional area, A, are given for three channel sections as a function of the water surface elevation or stage. The distance from Section 1 to Section 2 is 600 m (1,970 ft), and the distance from Section 2 to Section 3 is 700 m (2,300 ft). If $n = 0.035$, $Q = 200$ m^3/s (7,060 ft^3/s), and the stage at Section 1 corresponding to this flow rate is 508 m (1,670 ft), estimate the water surface elevations at Sections 2 and 3.

Stage	Section 1		Section 2		Section 3	
(m)	P	A	P	A	P	A
(ft)	(m)	(m²)	(m)	(m²)	(m)	(m²)
	(ft)	(ft²)	(ft)	(ft²)	(ft)	(ft²)
505	80	110	70	37	---	---
(1,657)	(262)	(1,184)	(230)	(398)		
506	91	130	100	64		
(1,660)	(299)	(1,399)	(328)	(689)	0	0
507	100	160	120	100	30	6.5
(1,663)	(328)	(1,722)	(394)	(1,076)	(98)	(70)
508	110	200	130	135	44	18
(1,667)	(361)	(2,153)	(426)	(1,453)	(144)	(194)
509			135	175	52	32
(1,670)			(443)	(1,884)	(171)	(344)
510			142	220	58	50
(1,673)			(466)	(2,368)	(190)	(538)
511			150	260	65	70
(1,677)			(492)	(2,799)	(213)	(753)
512			154	310	70	88
(1,680)			(505)	(3,337)	(230)	(947)
513					76	110
(1,683)					(249)	(1,184)
514					81	135
(1,686)					(266)	(1,453)
515					86	160
(1,690)					(282)	(1,722)
516					91	190
(1,693)					(299)	(2,045)

13. A trapezoidal channel with a base width of 30 ft (9.1 m) and side slopes of 2(horizontal): 1(vertical) has a longitudinal slope of 0.001 ft/ft, a Manning's n of 0.027, and conveys a flow of 2,500 ft³/s (71 m³/s) to a reservoir whose level varies as a function of time. At the point where the channel enters the lake, the invert of the channel is 10 ft (3.0 m) above a datum. Estimate the elevation of the water surface 1 mile upstream of the point where the channel enters the reservoir when
    a. The elevation of the reservoir is 14 ft (4.3 m) above datum.
    b. The elevation of the reservoir is 21 ft (6.4 m) above datum.

14. Rework Problem 13 using HEC-RAS and compare the results with the results from Problem 13.

CHAPTER 7

DESIGN OF CHANNELS

7.1.   INTRODUCTION

A critical topic in the area of open-channel hydraulics is the design of channels capable of transporting water between two points in a safe, cost-effective manner. Although economics, safety, and esthetics must always be considered, in this chapter only the hydraulic aspects of channel design will be examined. In addition, this discussion will be limited to the design of channels for uniform flow, and only three types of channels will be considered: (1) lined or nonerodible; (2) unlined, earthen, or erodible; and (3) grass-lined. In examining the design procedures for these types of channels, there are some basic concepts which are common to all three, and these commonalities will be discussed first.

From the Manning and Chezy equation, it is clear that the conveyance of a channel increases as the hydraulic radius increases or as the wetted perimeter decreases. Thus, from the viewpoint of hydraulics, there is among all channel cross sections of a specified geometric shape and area, an optimum set of dimensions for that shape. Among all possible channel cross sections, the best hydraulic section is a semicircle since, for a given area, it has the minimum wetted perimeter. The proportions of the best hydraulic section of a specified geometric shape can be easily derived (see, for example, Streeter and Wylie, 1975; the geometric elements of these sections are summarized in Table 7.1). It should be noted that from the point of view of applications, the best hydraulic section is not necessarily the most economic section. In practice the following factors must be considered:

1.   The best hydraulic section minimizes the area required to convey a specified flow; however, the area which must be excavated to achieve the flow area required by the best hydraulic section may

be significantly larger if the overburden, which must be removed, is considered.

2. It may not be possible to construct a stable best hydraulic section in the available natural material. If the channel must be lined, the cost of the lining may be comparable with the cost of excavation.

3. The cost of excavation depends not only on the amount of material which must be removed, but also on the ease of access to the site and the cost of disposing of the material removed.

4. The slope of the channel in many cases must also be considered a variable since it is not necessarily completely defined by topographic considerations. For example, while a reduced channel slope may require a larger channel flow area to convey the specified flow, the cost of excavating the overburden may be reduced.

Table 7.1. Geometric elements of best hydraulic sections.

Cross Section	Area A	Wetted perimeter P	Hydraulic radius R	Top width T	Hydraulic depth D
Trapezoid: half of a hexagon	$1.73y^2$	$3.46y$	$0.500y$	$2.31y$	$0.750y$
Rectangle, half of a square	$2y^2$	$4y$	$0.500y$	$2y$	y
Triangle, half of a square	y^2	$2.83y$	$0.354y$	$2y$	$0.500y$
Semicircle	$0.500\pi y^2$	πy	$0.500y$	$2y$	$0.250\pi y$
Parabola: $T = (2\sqrt{2})y$	$1.89y^2$	$3.77y$	$0.500y$	$2.83y$	$0.667y$
Hydrostatic catenary	$1.40y^2$	$2.98y$	$0.468y$	$1.92y$	$0.728y$

The terminology *minimum permissible velocity* refers to the lowest velocity which will prevent both sedimentation and vegetative growth. In general, an average velocity of 2 to 3 ft/s (0.61 to 0.91 m/s) will prevent sedimentation when the silt load of the flow is low. A velocity of 2.5 ft/s (0.76 m/s) is usually sufficient to prevent the growth of vegetation which could significantly affect the conveyance of the channel. It should be recognized that these numbers are, at best only, generalized and, in some cases, very poor estimates of the actual minimum permissible velocity. For example, as early as 1926 Fortier and Scobey (1926) observed that canals carrying turbid waters are seldom bothered by plant growth. While in channels transporting clear water, some plant species flourish at velocities that are significantly in excess of the velocity, which will cause erosion in the channel. The designer who obtains the highest permissible velocity that a channel can withstand from the viewpoint of hydraulic stability will have done all that can be done to prevent sedimentation and plant growth.

In most design problems, the longitudinal slope of the channel is determined by topography, the head required to carry the design flow, and the purpose of the channel. For example, in a hydroelectric power canal, a high head at the point of delivery is desirable, and a minimum longitudinal channel slope should be used.

The side slopes of a channel depend primarily on the engineering properties of the material through which the channel is excavated. From a practical viewpoint, the side slopes should be as steep as possible so that a minimum amount of land is required. In Table 7.2 side slopes for channels excavated through various types of material are suggested. These values are suitable for preliminary design purposes. In deep cuts, side slopes are often steeper above the water surface than they are below the surface. In small drainage ditches, the side slopes are steeper than they would be in an irrigation canal excavated in the same material. In many cases, side slopes are determined by the economics of construction. With regard to this subject, the following general comments are appropriate:

1.   In many unlined earthen canals on federal irrigation projects, side slopes are usually 1.5:1; however, side slopes at steep as 1:1 have been used when the channel runs through cohesive materials.

2.   In lined canals, the side slopes are generally steeper than in an unlined canal. If concrete is the lining material, side slopes greater than 1:1 usually require the use of forms, and with side slopes greater than 0.75:1 the linings must be designed to withstand earth pressures. Some types of lining require side slopes as flat as those used for unlined channels.

Table 7.2. Suitable side slopes for channels built in various types of materials (Chow, 1959).

Material	Side Slope
Rock	Nearly vertical
Muck and peat soils	0.25 : 1
Stiff clay or earth with concrete lining	0.5 : 1 or 1 : 1
Earth with stone lining or earth for large channels	1 :1
Firm clay or earth for small ditches	1.5 : 1
Loose, sand earth	2 : 1
Sandy loam or porous clay	3 : 1

3.   Side slopes through cuts in rock can be vertical, if this is desirable.

The term *freeboard* refers to the vertical distance between either the top of the channel or the top of the channel lining and the water surface, which prevails when the channel is carrying the design flow at normal depth. The purpose of freeboard is to prevent the overtopping of either the lining or the top of the channel by fluctuations in the water surface caused by: (1) wind-driven waves, (2) tidal action, (3) hydraulic jumps, (4) superelevation of the water surface as the flow rounds curves at high velocities, (5) the interception of storm runoff by the channel, (6) the occurrence of greater than design depths of flow caused by canal sedimentation or an increased coefficient of friction, or (7) temporary mis-operation of the canal system.

The freeboard associated with channel linings and the absolute top of the canal above the water surface can be estimated from the empirical curves in Fig. 7.1. In general, these curves apply to a channel lined with either a hard surface, a membrane, or compacted earth with a low coefficient of permeability. For unlined channels, freeboard generally ranges from 1 ft (0.30 m) for small laterals with shallow depths of flow to 4 ft (1.2 m) for channels carrying 3,000 ft³/s (85 m³/s) at relatively large depths of flow (Chow, 1959). A preliminary estimate of freeboard for an unlined channel can be obtained from

$$F = \sqrt{Cy} \qquad (7.1.1)$$

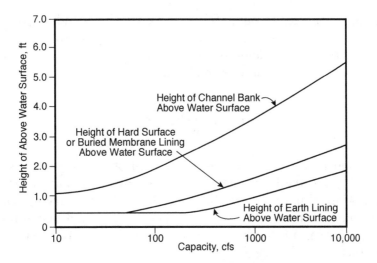

Figure 7.1.   **Freeboard for canal banks and freeboard for hard surface, buried membrane, and earth linings, U.S. Bureau of Reclamation (Anonymous, 1963).**

where F = freeboard, feet
y = design depth of flow, feet
C = coefficient which varies from 1.5 at $Q = 20$ ft³/s (0.57 m³/s) to 2.5 for $Q = 3,000$ ft³/s (85 m³/s)

When a flow moves around a curve, a rise in the water surface occurs at the outer bank with a corresponding lowering of the water surface at the inner bank. In the design of a channel, it is important that this difference in water levels be estimated. If all the flow is assumed to move around the curve at the subcritical average velocity $\bar{u}$, then

$$\Delta h = \frac{\bar{u}^2}{gR} \qquad (7.1.2)$$

where Δh = change in water surface elevation across channel
b = channel width
R = distance from center of curve to centerline of channel.

Equation (7.1.2) always underestimates Δh because of the average velocity assumption. In some cases, this equation may underestimate by as much as 50 percent (Houk, 1956). If Newton's second law of motion is applied to each streamline of the flow as it passes around the curve, then it is possible to demonstrate that the transverse water surface profile is a logarithmic curve of the form

$$h = 2.3 \frac{\overline{u}^2}{g} \log \frac{R_0}{R_i} \tag{7.1.3}$$

where R_0 and R_i are the outer and inner radii of the curve. Woodward (1920, 1941) assumed that the velocity of flow was zero at the banks and had a maximum u_M at the centerline of the curving channel. Between the sides and the center, the velocity varied according to a parabolic curve. Applying Newton's second law with these assumptions,

$$h = \frac{u_M^2}{g} \left[\frac{20}{3} \frac{R}{b} - \frac{16R^3}{b^3} + \left(\frac{4R^2}{b^2} - 1 \right)^2 \ln \left(\frac{2R+b}{2R-b} \right) \right] \tag{7.1.4}$$

Of these three equations, Eq. (7.1.4) provides the best estimate of Δh. The subject of superelevation around curves will be treated in some detail in a subsequent section of this chapter, but the foregoing equations provide the basis for making initial design estimates.

There are no set rules governing the minimum radii of curvature for canals. Shukry (1950), using laboratory data, found that the effects of curves were negligible when the ratio of the radius of curvature to the distance to the center of the canal was greater than 3 times the width of the level portion of the channel bed. In India, the minimum radii of curvature are often longer than those used in the United States. For example, some Indian engineers recommend a minimum radius of 300 ft (91 m) for canals carrying less than 10 ft³/s (0.30 m³/s) to 5,000 ft (1,500 m) for canals carrying more than 3,000 ft³s (85 m³/s) (Houk, 1956).

The width of the banks along a canal are usually governed by a number of considerations which include the size of the canal, the amount of excavation available for bank construction, and the need for maintenance roads. Where roads are needed, the top widths for both lined and unlined canals are 16 ft (5 m) or more. The bank top is usually graded away from the canal so that precipitation will not flow into the canal. Bank widths must also be sufficient to provide for stability against canal water pressure and keep percolating water below the ground level outside the banks.

7.2.   DESIGN OF LINED CHANNELS

Lined channels are built for five primary reasons:

1.   To permit the transmission of water at high velocities through areas of deep or difficult excavation in a cost-effective fashion

2.   To permit the transmission of water at high velocity at a reduced construction cost

3.   To decrease canal seepage, thus conserving water and reducing the waterlogging of lands adjacent to the canal

4.   To reduce the annual costs of operation and maintenance

5.   To ensure the stability of the channel section

The design of lined channels from the viewpoint of hydraulic engineering is a rather elementary process which generally consists of proportioning an assumed channel cross section.   Some typical cross sections of lined channels used on irrigation projects in the United States are summarized in Table 7.3, and a typical lined section of the All-American Canal is shown in Fig. 7.2.   Additional information regarding channel linings can be found in Willison (1958) and Anonymous (1963).   A recommended procedure for proportioning a lined section is summarized in Table 7.4.   In this table, it is assumed that the design flow Q_D, the longitudinal slope of the channel S, the type of channel cross section, e.g., trapezoidal, and the lining material have all been selected prior to the initiation of the channel design process.

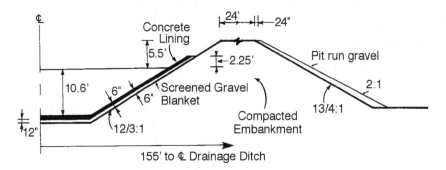

Figure 7.2.   Typical concrete-lined section used on All-American Canal.

Table 7.3. Typical cross sections of lined channels for selected canals in the western United States (Houk, 1956).

Canal	Project	Side slopes	Depth of flow (ft) (m)	Bottom Width (ft) (m)	Ratio of width to depth	Mean Velocity (ft/s) (m/s)	Discharge (ft³/s) (m³/s)	Freeboard (ft)[1] (m)
Lined earth sections								
Contra Costa	Central Valley	1.25:1	1.84 (0.56)	3 (0.91)	1.63	2.33 (0.71)	26 (0.7)	1.16 (0.35)
South Branch	Yakima	1.25:1	3.80 (1.16)	5 (1.52)	1.32	5.94 (1.81)	220 (6.2)	1.00 (0.30)
Ridge	Yakima	1.25:1	7.46 (2.27)	7 (2.13)	0.94	4.93 (1.50)	600 (17.0)	1.04 (0.32)
Heart Mountain	Shoshone	1.25:1	7.51 (2.29)	8 (2.44)	1.06	7.00 (2.13)	914 (25.9)	1.49 (0.45)
Kittitas Main	Yakima	1.25:1	8.99 (2.74)	11 (3.35)	1.22	6.63 (2.02)	1320 (37.4)	1.25 (0.38)
Black Canyon	Boise	1.25:1	9.39 (2.86)	12 (3.66)	1.28	4.86 (1.48)	1089 (30.8)	1.61 (0.49)
Ridge	Yakima	1.25:1	11.20 (3.41)	14 (4.27)	1.25	7.02 (2.14)	2200 (62.3)	1.80 (0.55)
All-American	All-American	1.67:1	10.64 (3.24)	22 (6.71)	2.07	6.62 (2.02)	2800 (79.3)	2.25 (0.69)
Delta-Mendota	Central Valley	1.5:1	16.56[1] (5.05)	48 (14.63)	2.9	---	4600 (130.3)	---
Lined rock sections								
Kittitas Main	Yakima	0.5:1	9.00 (2.74)	14 (4.27)	1.56	7.65 (2.33)	1275 (36.1)	1.25 (0.38)
Main (gravity extension)	Minidoka	0.25:1	13.80 (4.21)	20 (6.10)	1.45	8.20 (2.50)	2695 (76.3)	1.50 (0.46)
Gravity Main	Gila	0.75:1	21.07 (6.42)	32 (9.75)	1.52	5.96 (1.82)	6000 (169.9)	---

[1]To top of lining

Table 7.4. A design procedure for lined channels.

Step	Process
1	Estimate n or C for specified lining material
2	Compute value of section factor $$AR^{2/3} = \frac{nQ}{\phi\sqrt{S}}$$ $\phi = 1.49$ for English units and 1 for SI units
3	Solve section factor equation for y_N given appropriate expressions for A and R (Table 1.1). *Note*: This step *may* require assumptions regarding side slopes, bottom widths, *etc.*
4	If the best hydraulic section is required, compute channel parameters from Table 7.1; otherwise compute channel parameters from Table 1.1 using y_N from Step 3.
5	Check: 1. Minimum permissible velocity if water carries silt and for vegetation 2. Froude number
6	Estimate 1. Required height of lining above water surface, Fig. 7.1 2. Required freeboard, Fig. 7.1
7	Summarize results with a dimensioned sketch.

EXAMPLE 7.1

A lined channel of trapezoidal cross section is to be designed to carry 400 ft^3/s (11 m^3/s). The lining of the canal is to be float-finished concrete, and the longitudinal slope of the canal is 0.0016. Determine appropriate channel proportions.

Solution

Step 1   Estimate Manning's n, (Table 4.8).

$$n = 0.015$$

Step 2   Compute the section factor.

$$Q = \frac{1.49}{n} AR^{2/3} \sqrt{S}$$

$$AR^{2/3} = \frac{nQ}{1.49\sqrt{S}} = \frac{0.015(400)}{1.49\sqrt{0.0016}} = 100.7$$

Step 3   Assume $b = 20$ ft (6.1 m) and $z = 2$. Solve the section factor equation (step 2) for y_N.

$$y_N = 2.5\,\text{ft}\,(0.76\,\text{m})$$

Step 4   Since the best hydraulic section is not required, this step is omitted in Table 7.4.

Step 5   Check minimum permissible velocity and Froude number.

$$A = (b + zy)y = [20 + 2(2.5)] = 62.5\,\text{ft}^2\,(5.8\,\text{m}^2)$$

$$\bar{u} = \frac{Q}{A} = \frac{400}{62.5} = 6.4\,\text{ft}/\text{s}\,(2.0\,\text{m}/\text{s})$$

This velocity should prevent vegetative growth and sedimentation.

$$F = \frac{\bar{u}}{\sqrt{gD}} = 0.78$$

Therefore, this is a subcritical flow.

Step 6   (*a*) Estimate height of the lining above water surface (Fig. 7.1) = 1.2 ft (0.37 m).

(*b*) Estimate height of canal bank above water surface (Fig. 7.1) = 2.9 ft (0.88 m).

The results of this design are summarized in Fig. 7.3.

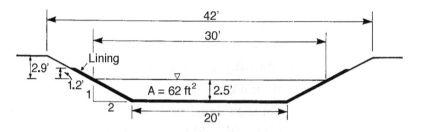

Figure 7.3. Summary of results for Example 7.1.

A primary concern in the design of lined channels is the cost of the lining material and the development of channel dimensions which minimize this cost. The cost of the lining material is a function of the volume of lining material used which in turn is a function of the lining thickness and the magnitude of the wetted perimeter. If a channel is lined with one material of uniform thickness, then the solution which minimizes the lining cost for a channel of arbitrary shape is the best hydraulic section (Table 7.1). However, if the thickness of the lining material changes along the perimeter, then the problem becomes much more complex. Trout (1982) examined the problem of lining material cost minimization for trapezoidal, rectangular, and triangular channels for the special case of the material used to line the base of the channel being different from that used to line the sides. The methodology developed by Trout considers neither placement nor construction costs unless these costs can be evaluated in terms of the channel surface area. The procedure described in the material which follows is particularly useful for long channel sections where construction procedures can be oriented toward minimizing material costs or in situations in which labor costs are low relative to the lining material costs.

As noted previously in this section, the primary design parameter for designing the dimensions of a lined channel is the section factor, Eq. (5.1.5). For trapezoidal, rectangular, and triangular sections, the section factor is

$$AR^{2/3} = \frac{\left(by + zy^2\right)^{5/3}}{\left(b + 2y\sqrt{1+z^2}\right)^{2/3}} = \frac{Qn}{\phi\sqrt{S}} \qquad (7.2.1)$$

where

$$
\begin{aligned}
z &= \text{side slope} \\
b &= \text{bottom width} \\
y &= \text{depth of flow} \\
S &= \text{longitudinal channel slope} \\
n &= \text{Manning's resistance coefficient} \\
Q &= \text{design flow} \\
\phi &= \text{constant equal to 1.48 when English units are} \\
  &\quad \text{used and 1.00 when SI units are used.}
\end{aligned}
$$

Equation (7.2.1) can be rearranged to yield an implicit solution for the depth of flow or

$$
y = \frac{\left(\dfrac{b}{y} + 2\sqrt{1+z^2} \right)^{1/4}}{\left(\dfrac{b}{y} + z \right)^{5/8}} \left(\frac{Qn}{\phi\sqrt{S}} \right)^{3/8}
\tag{7.2.2}
$$

In Equation (7.2.2) if b/y and the side slope z are specified, then the equation can be solved explicitly for y and b.

The cost of the materials used in lining a channel can usually be specified in terms of the volume of material used. Thus, if the thickness of the lining is specified, the cost of the unit length of channel is only a function of the wetted perimeter plus the freeboard and the lining material used in the corners (Fig. 7.4).

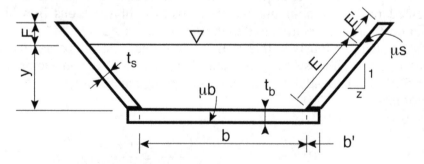

Figure 7.4.   Definition of dimensional terms in cost optimization for a trapezoidal channel (Trout, 1982).

With reference to Fig. 7.4

$$C_b = \mu_b \left(\frac{\text{volume}}{\text{unit length}} \right) = \mu_b t_b \left(b + b' \right) = Bb + k \qquad (7.2.3)$$

$$C_s = \mu_s \left(\frac{\text{volume}}{\text{unit length}} \right) = \mu_s t_s \left(2E + 2E' \right) = 2\Gamma \left(y + F \right) \sqrt{1 + z^2} \qquad (7.2.4)$$

and

$$C = C_b + C_s = Bb + k + 2\Gamma(y + F)\sqrt{1 + z^2} \qquad (7.2.5)$$

where

C	=	total material cost per unit length,
C_b	=	material cost for channel base per unit length,
C_s	=	material cost of sides per unit channel length,
b'	=	bottom corner width,
t_b	=	channel base lining thickness,
t_s	=	channel side lining thickness,
E	=	wetted side length,
E'	=	freeboard side length,
μ_b	=	cost of base lining material per unit volume,
μ_s	=	cost of side lining material per unit volume,
B	=	cost of base lining material for specified thickness per unit area,
k	=	cost of corner materials per unit length,
Γ	=	cost of side lining material for specified thickness per unit area, and
F	=	vertical freeboard requirement.

To determine the dimensions of the minimum-cost trapezoidal, rectangular, or triangular cross section, Eq. (7.2.1) or (7.2.2) must be solved such that the cost function, Eq. (7.2.5), is minimized. As noted by Trout (1982), this open-channel cost optimization problem is analogous to the classic microeconomic problem of the minimizing production costs through input substitution. In this case, the output of the system is the hydraulic capacity; the inputs are the variables which define the channel geometry; and the production function is the equation for the section factor. In the terminology of business, the solution of this optimization problem requires an input mix such that the ratio of the marginal products equals the ratio of the marginal costs. If the channel side slope variable z is assumed constant, then the Lagrange multiplier method can be used to find an explicit algebraic solution to this problem.

As noted above, the dimensional combination in which the ratio of the marginal changes in the section factor are equal to the marginal changes in the costs or

$$\frac{\dfrac{\partial\left(AR^{2/3}\right)}{\partial b}}{\dfrac{\partial\left(AR^{2/3}\right)}{\partial y}} = \frac{\dfrac{\partial C}{\partial b}}{\dfrac{\partial C}{\partial y}} \tag{7.2.6}$$

subject to Eq. (7.2.1), represents the optimal or minimum-cost solution of the problem. If Eqs. (7.2.1) and (7.2.5) are substituted in Eq. (7.2.6) and the result simplified, then the optimal solution is

$$K_1\left(\frac{y}{b}\right)^2 + K_2\left(\frac{y}{b}\right) + K_3 = 0 \tag{7.2.7}$$

where

$$K_1 = 20\left(z^2 + 1\right) - \left(1 + 4\frac{B}{\Gamma}\right)4z\sqrt{z^2 + 1} \tag{7.2.8}$$

$$K_2 = \left(1 - \frac{B}{\Gamma}\right)6\sqrt{1 + z^2} - 10z\frac{B}{\Gamma} \tag{7.2.9}$$

and

$$K_3 = -5\frac{B}{\Gamma} \tag{7.2.10}$$

Then the ratio of b/y for the required solution of Eq. (7.2.2) is

$$\frac{b}{y} = \frac{2K_1}{-K_2 + \left(K_2^2 + 20K_1\dfrac{B}{\Gamma}\right)^{1/2}} \tag{7.2.11}$$

which is a function of z and the ratio of the unit costs of the base to side slope material. The typical solution would proceed in the following steps:

1. Given values of Q, S, n, z, B, and Γ, values of K_1, K_2, and K_3 are determined by Eqs. (7.2.8) to (7.2.10).

2. The minimum-cost value of the ratio b/y is then estimated by Eq. (7.2.11).

3. The minimum-cost depth of flow is estimated by Eq. (7.2.2).

4. The minimum-cost bottom width can then be determined by multiplying y times the ratio b/y.

EXAMPLE 7.2

A lined trapezoidal channel is to be designed to convey 0.08 m³/s (2.8 ft³/s) of water on a slope of 0.001. If $n = 0.014$, $B = \$3.20$ per square meter ($\$0.30$ per square foot), $\Gamma = \$2.00$ per square meter ($\$0.19$ per square foot), $z = 0.5$, the vertical freeboard required is 0.15 m (0.5 ft), and the material cost for the corners is $\$0.35$ per meter ($\$0.11$ per foot), determine the depth of flow and bottom width for which the lining cost is minimized. This example was first presented by Trout (1982).

Solution

This example can be solved by both direct algebraic and graphical techniques. As will be demonstrated, the algebraic technique yields rapid results, but the graphical technique can be used to answer other important questions.

The first step in the direct algebraic technique is to determine values of K_1 and K_2 from the given data. From Eq. (7.2.8)

$$K_1 = 20\left(z^2 + 1\right) - \left(1 + 4\frac{B}{\Gamma}\right)4z\sqrt{z^2 + 1}$$

$$K_1 = 20\left(0.5^2 + 1\right)\left(1 + 4\frac{3.2}{2}\right)(4)(0.5)\left(0.5^2 + 1\right)^{1/2} = 8.45$$

and from Eq. (7.2.9)

$$K_2 = 6\left(1 - \frac{B}{\Gamma}\right)\left(z^2 + 1\right)^{1/2} - 10z\frac{B}{\Gamma}$$

$$K_2 = 6\left(1 - \frac{3.2}{2}\right)\left(0.5^2 + 1\right)^{1/2} - 10\left(0.5\right)\left(\frac{3.2}{2}\right) = -12.0$$

Then, the ratio b/y for the optimal lining costs section is given by Eq. (7.2.11) or

$$\frac{b}{y} = \frac{2K_1}{-K_2 + \left[K_2^2 + 20\left(\dfrac{B}{\Gamma}\right)K_1\right]^{1/2}}$$

$$\frac{b}{y} = \frac{2\left(8.45\right)}{-\left(-12.0\right) + \left[\left(-12.0\right)^2 + 20\left(\dfrac{3.2}{2}\right)\left(8.45\right)\right]^{1/2}} = 0.522$$

With the optimal value of b/y established, Eq. (7.2.2) can be used to find the optimal depth of flow or

$$y = \frac{\left[\dfrac{b}{y} + 2\sqrt{1 + z^2}\right]^{1/4}}{\left(\dfrac{b}{y} + z\right)^{5/8}}\left(\frac{Qn}{\varphi\sqrt{S}}\right)^{3/8}$$

$$y = \frac{\left(0.522 + 2\sqrt{0.5^2 + 1}\right)^{1/4}}{\left(0.522 + 0.5\right)^{5/8}}\left[\frac{0.08\left(0.014\right)}{1.0\sqrt{0.001}}\right]^{3/8} = 0.36\,\text{m}\,\left(1.2\,\text{ft}\right)$$

and therefore

$$b = \frac{b}{y}y = 0.522\left(0.36\right) = 0.19\,\text{m}\,\left(0.62\,\text{ft}\right)$$

The minimum lining cost per unit length of channel can then be computed from Eq. (7.2.5) or

$$C = 0.19(3.20) + 2(2)(0.36 + 0.15)\sqrt{1 + 0.5^2} + 0.35$$
$$= \$3.24 \text{ per meter } (\$0.99 \text{ per foot})$$

Representing Eq. (7.2.2) and (7.2.11) in a graphical form, is a second method of solving this problem. In Fig. 7.5 such a solution is shown for the specific case of $z = 0.5$. In this figure, the curved lines represent values of the section factor while the straight lines radiating from the origin are solutions of Eq. (7.2.11) for various values of B/Γ. With regard to this figure, it is noted, first, that a separate graph is required for each side slope value. Second, any combination of b and y values at the point where the B/Γ line crosses the appropriate section factor line, are the optimum lining cost values.

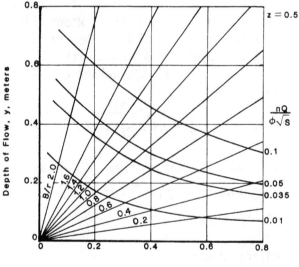

Figure 7.5.   **Lining cost minimization for a trapezoidal channel with $z = 0.5$ (Trout, 1982).**

Using Fig. 7.5 with

$$\frac{nQ}{\varphi\sqrt{S}} = 0.035$$

and

$$\frac{B}{\Gamma} = 1.6$$

it is found that

$$y = 0.36 \text{ m } (1.2 \text{ ft})$$

and

$$b = 0.19 \text{ m } (0.62 \text{ ft})$$

With these values the minimum lining cost in dollars could be calculated as before.

Once a figure such as Fig. 7.5 has been constructed for a specified side slope, other pertinent and important questions can be answered. For example, what additional cost will be incurred for the specified data if the only available construction forms have a bottom width of 0.4 m (1.3 ft)? This question can be answered by following the section factor curve until each intersects b = 0.4 m (1.3 ft). The depth is then determined to be y = 0.25 m (0.82 ft), and the cost per unit length of channel is from Eq. (7.2.5).

$$C = bB + 2\Gamma\left(y + F\right)\sqrt{1 + z^2} + k$$

or

$$C = 0.4(3.20) + 2(2)(0.25 + 0.15)(1 + 0.5^2)^{1/2} + 0.35$$
$$= \$3.42 \text{ per meter } (\$1.04 \text{ per foot})$$

thus, in using the forms available, a

$$\text{Percent increase} = \frac{3.42 - 3.24}{3.24}\left(100\right) = 5.6\%$$

over the minimum lining cost is incurred.

Another question which should be examined is the sensitivity of total lining cost to variations in side slope ratios. In Fig. 7.6, total lining cost is plotted as function of the side slope ratio for the cost and flow data, given in this example. Minimum lining material costs occur at

$z \approx 0.65$. At a side slope ratio of 0.92, the optimum bottom width of the channel has decreased to zero, and the minimum cost channel shape is a triangle. From Figs. 7.5 and 7.6 it can be concluded that lining material costs are rather insensitive to variations in cross-sectional dimensions, if they do not deviate excessively from the optimum dimensions.

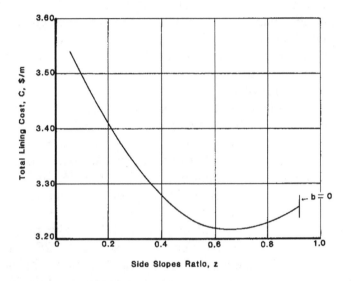

Figure 7.6. **Total lining cost as a function of side slope ratio.**

7.3. DESIGN OF STABLE, UNLINED, EARTHEN CHANNELS: A GENERAL TRACTIVE FORCE DESIGN METHODOLOGY

7.3.1. Introduction

In comparison with the design process typically used for lined channels, the design of stable, unlined or erodible, earthen channels is a complex process involving numerous parameters, most of which cannot be accurately quantified. The complexity of the erodible channel design process results from the fact that in such channels, stability is dependent not only on hydraulic parameters, but also on the properties of the material which composes the bed and sides of the channel. In the context of this discussion, a stable channel section is one in which neither objectionable scour nor deposition occurs. In contrast, there are three types of unstable sections. In the first type, the banks and bed of the channel are scoured but no deposition occurs. This situation can occur when the channel conveys sediment-free water or water which conveys only a very small amount of sediment, but with sufficient energy to erode the channel. The second type of unstable channel is one in which there is deposition but no scour. This situation can result when the water being conveyed carries a large sediment load at a velocity that permits sedimentation. The third type is the case in which both scour and deposition occur. This situation can result when the material through which the channel is excavated is susceptible to erosion and the water being conveyed caries a significant sediment load.

By the mid 1920's, it became obvious that there should exist a relationship between the flow rate or the average velocity, the engineering properties of the material composing the bed and sides of the channel, the amount and type of sediment being transported by the flow, and the stability of the channel section. The American Society of Civil Engineers' Special Committee on Irrigation Hydraulics submitted questionnaires to a number of engineers whose experience qualified them to form authoritative opinions regarding the stability of channels built in various types of material. The hypothesis of this study was that there was a relationship between the average velocity of flow, the channel perimeter material, and channel stability. The results of this survey were published in 1926 (Fortier and Scobey, 1926) and became the theoretical basis for the channel design method known as the *method of maximum permissible velocity*. The primary results of the Fortier and Scobey (1926) report are summarized in columns (1), (3), (5), (7), and (10) of Table 7.5. With regard to these data, the following caveats are noted. First, the data given are for channels with long tangents, and a 25 percent reduction in the maximum permissible velocity is recommended for canals with a sinuous alignment. Second, these data are for depths of flow less than 3.0 ft (0.91 m). For greater depths of flow, the maximum permissible velocity should be increased by 0.5 ft/s (0.15 m/s). Third, the velocity of flow in canals carrying abrasives, such as basalt ravelings, should be reduced by 0.5 ft/s (0.15 m/s). Fourth, channels diverting water from silt-laden rivers such as the Colorado River should be designed for mean design velocities 1 to 2 ft/s (0.30 to 0.61 m/s) greater than would be allowed for the same perimeter material if the water were transporting no sediment.

The pioneering work of Fortier and Scobey (1926) was the basis of channel design for many years; however, it is a design methodology based primarily on experience and observation rather than physical principles. The first step in developing a rational design process for unlined, stable, earthen channels is to examine the forces which cause scour. Scour on the perimeter of a channel occurs when the particles on the perimeter are subjected to forces of sufficient magnitude to cause particle movement.   When a particle rests on the level bottom of a channel, the force acting to cause movement is the result of the flow of water past the particle. A particle resting on the sloping side of a channel is acted on not only by the flow-generated force, but also by a gravitational component which tends to make the particle roll or slide down the slope. If the resultant of these two forces is larger than the forces resisting movement, gravity, and cohesion, then erosion of the channel perimeter occurs.   By definition, the tractive force is the force acting on the particles composing the perimeter of the channel and is the result of the flow of water past these particles. In practice, the tractive force is not the force acting on a single particle, but the force exerted over a certain area of the channel perimeter.   This concept was apparently first stated by duBoys (1879) and restated by Lane (1955).

Recall   that   in   uniform   flow   the   tractive   force   should   be approximated by the effective gravitational force, acting on the water within the control volume (Fig. 4.1), and parallel to the channel bottom, or

$$F_T = \gamma ALS \qquad (7.3.1)$$

where A =   channel cross-sectional area
L =   control volume length
S =   longitudinal slope of channel

The unit tractive force is

$$\tau_0 = \frac{\gamma ALS}{PL} = \gamma RS \qquad (7.3.2)$$

where τ_0 = average value of the tractive force per unit of wetted area of the unit tractive force.   In a wide channel $y_N \simeq R$ and Eq. (7.3.2) becomes $\tau_0 = \gamma y_N S$.

In most channels, the tractive force is not uniformly distributed over the perimeter, and therefore, before an accurate design methodology can be developed, the distribution of the tractive force on the perimeter of the channel must be estimated.

Although many attempts to determine the distribution of the tractive force on a channel perimeter have been made using both field and laboratory data, they have not been successful (Chow, 1959). In Fig. 7.7, the maximum unit tractive force on the sides and bottoms of various

Table 7.5. Maximum permissible velocities recommended by Fortier and Scobey (1926) for straight channels of small slope and after aging.

Material (1)	n (2)	Clear Water				Water transporting colloidal silts				Water transporting noncolloidal silts, sands, gravel or rock fragments	
		$\bar{u}$ ft/s (3)	τ_0 lb/ft² (4)	$\bar{u}$ m/s (5)	τ_0 N/m² (6)	$\bar{u}$ ft/s (7)	τ_0 lb/ft² (8)	$\bar{u}$ m/s (9)	τ_0 N/m² (10)	$\bar{u}$ ft/s (7)	$\bar{u}$ m/s (8)
Fine Sand, non-colloidal	0.020	1.50	0.027	0.457	1.29	2.50	0.075	0.762	3.59	1.50	0.457
Sandy loam, non-colloidal	0.020	1.75	0.037	0.533	1.77	2.50	0.075	0.762	3.59	2.00	0.610
Silt loam, non-colloidal	0.020	2.00	0.048	0.610	2.30	3.00	0.11	0.914	5.27	2.00	0.610
Alluvial silts, non-colloidal	0.020	2.00	0.048	0.610	2.30	3.50	0.15	1.07	7.18	2.00	0.610
Ordinary firm loam	0.020	2.50	0.075	0.762	3.59	3.50	0.15	1.07	7.18	2.25	0.686
Volcanic ash	0.020	2.50	0.075	0.762	3.59	3.50	0.15	1.07	7.18	2.00	0.610
Stiff clay, very colloidal	0.025	3.75	0.26	1.14	12.4	5.00	0.46	1.52	22.0	3.00	0.914
Alluvial silts, colloidal	0.025	3.75	0.26	1.14	12.4	5.00	0.46	1.52	22.0	3.00	0.914
Shales and hardpans	0.025	6.00	0.67	1.83	32.1	6.00	0.67	1.83	32.1	5.00	1.52
Fine gravel	0.020	2.50	0.075	0.762	3.59	4.00	0.32	1.52	15.3	3.75	1.14
Graded loam to cobbles when non-colloidal	0.030	3.75	0.38	1.14	18.2	5.00	0.66	1.52	31.6	5.00	1.52
Graded silts to cobbles when colloidal	0.030	4.00	0.43	1.22	20.6	5.50	0.80	1.68	38.3	5.00	1.52
Coarse gravel, non-colloidal	0.025	4.00	0.30	1.22	14.4	6.00	0.67	1.83	32.1	6.50	1.98
Cobbles and shingles	0.035	5.00	0.91	1.52	43.6	5.50	1.10	1.68	52.7	6.50	1.98

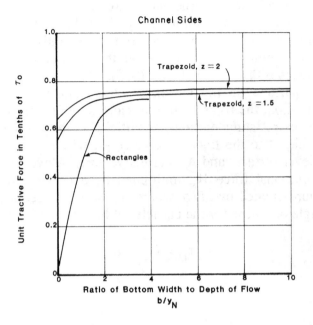

Figure 7.7a.   **Maximum unit tractive force in terms of** $\gamma \, y_N S$ **for channel sides.**

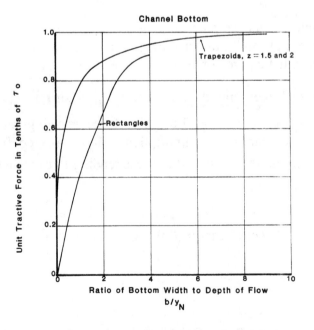

Figure 7.7b.   **Maximum unit tractive force in terms of** $\gamma \, y_N S$ **for channel bottoms.**

channels as determined by mathematical studies are shown as a function of the ratio of the bottom width to the depth of flow. It is noted that for the trapezoidal section, which is the section generally used in unlined canals, the maximum tractive force on the bottom is approximately $\gamma y_N S$ and the sides 0.75 $\gamma y_N S$ (Lane, 1955).

When a particle on the perimeter of a channel is in a state of impending motion, the forces acting to cause motion are in equilibrium with the forces resisting motion. A particle on the level bottom of a channel is subject to the tractive force $A_e \tau_L$, where τ_L = unit tractive force on a level surface, and A_e = effective area. Movement is resisted by the gravitational force W_s, multiplied by a coefficient of friction, which is approximated by $\tan \alpha$ where W_s = submerged article weight and α = angle of repose for the particle. When motion is incipient,

$$A_e \tau_L = W_s \tan \alpha \qquad (7.3.3)$$

or

$$\tau_L = \frac{W_s}{A_e} \tan \alpha \qquad (7.3.4)$$

A particle on the sloping side of a channel is subject to both a tractive force $\tau_s A_e$ and a downslope gravitational component $W_s \sin \Gamma$, where τ_s = unit tractive force on the side slopes, and Γ = side slope angle. These forces and their resultant, $\sqrt{\left(W_s \sin \Gamma\right)^2 + \left(\tau_s A_e\right)^2}$, are shown schematically in Fig. 7.8. The force resisting motion is the gravitational component multiplied by a coefficient of friction $W_s \cos \Gamma \tan \alpha$. Setting the forces tending to cause motion equal to those resisting motion

$$W_s \cos \Gamma \tan \alpha = \sqrt{\left(W_s \tan \Gamma\right)^2 + \left(A_e \tau_s\right)^2} \qquad (7.3.5)$$

or

$$\tau = \frac{W_s}{A_e} \cos \Gamma \tan \alpha \sqrt{1 - \frac{\tan^2 \Gamma}{\tan^2 \alpha}} \qquad (7.3.6)$$

Equations (7.3.4) and (7.3.6) usually combine to form the tractive force ratio

$$K = \frac{\tau_s}{\tau_L}\cos\Gamma \sqrt{1 - \frac{\tan^2\Gamma}{\tan^2\alpha}} = \sqrt{\frac{\sin^2\Gamma}{\sin^2\alpha}}$$

(7.3.7)

where K = tractive force ratio.

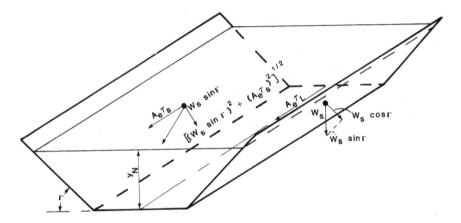

Figure 7.8.   Analysis of forces acting on a particle resisting movement on the perimeter of a channel.

It is noted that the tractive force ratio is a function of both, the side slope angle and the angle of repose of the material composing the channel perimeter. In the case of cohesive materials and fine noncohesive materials, the angle of repose is small and can be assumed to be zero; i.e., for these materials the forces of cohesion are significantly larger than the gravitational component tending to make the particles roll downslope. Lane (1955) found that, in general, the angle of repose is directly proportional to both, the size and angularity of the particle. The laboratory data available to Lane (1955) are summarized in Fig. 7.9. In this figure, the particle size is the diameter of the particle of which 25 percent of all the particles, measured by weight, are larger. With regard to the data summarized in this figure, the following caveats should be noted. First, time was not available when these tests were performed to run a large number of experiments; for this reason there was a large amount of scatter in the raw data. Second, the angles of repose are limited to 41 degrees for the very angular material and 39 degrees for the very rounded material, because of a paucity of data on the larger material.

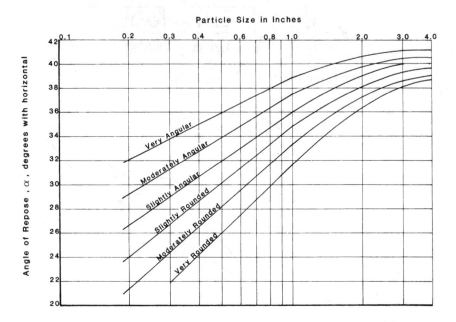

Figure 7.9.   Angles of repose of noncohesive materials (Lane, 1955).

For coarse, noncohesive materials, the laboratory data of Lane (1955) indicated that the maximum permissible tractive force in pounds per square foot was equal to 0.4 times the 25 percent diameter of the particles in inches. In recognition of the fact that actual canals can sustain tractive forces in excess of those predicted by laboratory experiment, Lane (1955) also collected data regarding actual canals. All these field data were in the form of maximum permissible velocities and thus, had to be converted to tractive force data - a process which required numerous assumptions regarding the size of the canal and the depth of flow. For example, in columns (4), (6), (8), and (10) of Table 7.5, the Fortier and Scobey (1926) data have been converted to tractive force data. The results of the Lane field data collection effort are summarized in Fig. 7.10*a*. In this figure, for the *fine* noncohesive, i.e., average diameters less than 5 mm (0.254 in), the size specified is the median size of the diameter of a particle of which 50 percent were larger by weight. Data, regarding the permissible unit tractive forces for canals built in cohesive materials, were presented in Chow (1959) and are summarized in Fig. 7.10*b*. The data in these tables are believed to provide conservative design information, and a factor of safety has already been built into these curves.

Lane (1955) also recognized that sinuous canals scour more easily than canals with straight alignments. To account for this observation in the tractive force design method, Lane developed the following definitions. Straight canals have straight or slightly curved alignments and are typical of canals built in flat plains. Slightly sinuous canals have a degree of curvature which is typical of canals in a slightly undulating topography. Moderately sinuous canals have a degree of curvature which is typical of moderately rolling topography. Very sinuous canals have a degree of curvature which is typical of canals in foothills or mountainous topography. Then, with these definitions, correction factors (see Table 7.6) can be defined.

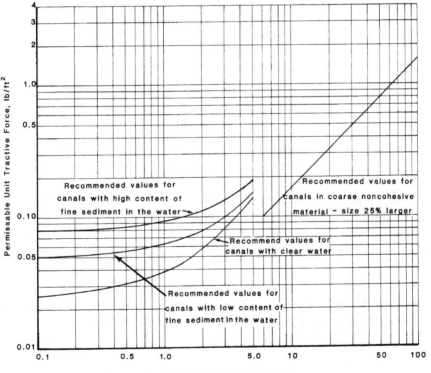

Figure 7.10a. Recommended permissible unit tractive forces for canals built in noncohessive materials (Lane 1955).

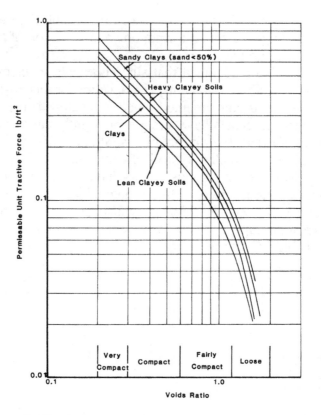

Figure 7.10b. Permissible unit tractive forces for canals in cohesive materials (Chow, 1959).

Even given the limitations of the data available regarding tractive forces, this methodology is superior to the maximum permissible velocity method. In essence, at the level of basic principles, these two methods are analogous. For example, in the method of maximum

Table 7.6. Comparison of maximum tractive forces for canals with varying degrees of sinuosity (Lane, 1955).

Degree of sinuosity	Relative limiting tractive force
Straight canals	1.00
Slightly sinuous canals	0.90
Moderately sinuous canals	0.75
Very sinuous canals	0.60

permissible velocity, it is recognized that if the depth of flow is significantly different from approximately 3 ft (0.91 m), then a correction factor is required (Chow, 1959). Mehrotra (1983) demonstrated that the required corrective factor can be developed from "basic" principles. From Eq. (7.3.1) the critical tractive force on the channel boundary for a depth of flow y_l is

$$\tau_c = \gamma R_1 S_1 \tag{7.3.8}$$

Where R_l and S_l are the hydraulic radius and slope, respectively, corresponding to the depth y_l. Then, for any other uniform depth of flow y_2, in a channel whose bed is composed of the same soil type, the critical tractive force is

$$\tau_c = \gamma R_2 S_2 \tag{7.3.9}$$

Since the same channel perimeter material is involved in both cases, the critical tractive forces must be equal, and hence

$$R_1 S_1 = R_2 S_2$$

or

$$\frac{R_1}{R_2} = \frac{S_2}{S_1} \tag{7.3.10}$$

Considering the Manning uniform flow equation for both situations, it can be demonstrated that

$$\frac{\overline{u_1}}{\overline{u_2}} = \left(\frac{R_2}{R_1}\right)^{2/3}\left(\frac{S_2}{S_1}\right)^{1/2}$$

and substituting Eq. (7.3.10),

$$\frac{\overline{u_2}}{\overline{u_1}} = \left(\frac{R_2}{R_1}\right)^{1/6} = k \tag{7.3.11}$$

The parameter k in Eq. (7.3.11) can be considered a correction factor, which should be applied to the maximum permissible velocity (Table 7.5), if the uniform depth of flow y_2 is different from the depth of flow corresponding to the maximum permissible velocity. If $y_l = 3$ ft (0.91 m) and the channel is wide, Eq. (7.3.11) becomes

$$k = \left(\frac{y_2}{3} \right)^{1/6} \qquad (7.3.12)$$

However, even with these adjustments, it must be recognized that the method of maximum permissible velocity does not consider the tendency of the particles on the side slopes of a channel to roll down the slope under the influence of gravity. Many of the initial investigations of channel stability also noted these same design limitations. For example, Fortier and Scobey (1926) provided channel depth correction factors, and Houk (1926) noted that the maximum permissible velocity seemed to be directly proportional to the depth of flow.

At this point, it is appropriate to note that unlined earthen channels undergo a process known as aging; i.e., the material which composes the perimeter of the canal may be modified with the passage of time. Two examples of this process were noted by Fortier and Scobey (1926). In the Imperial Valley of California, many of the large irrigation canals were initially excavated through sandy soils, which would be eroded at a velocity of 2.5 ft/s (0.76 m/s). The extremely fine silt particles present in the water flowing in these canals have compacted and plastered these noncohesive sands, with a coat of colloidal mud, which can withstand velocities in excess of 5.0 ft/s (1.5 m/s). The coating of the original perimeter materials, in this case, not only reduced the scour potential but also reduced seepage losses by creating a less permeable perimeter. In the mountain valley canals of Colorado and Utah, which were originally constructed in loamy gravel, soils there are now very few traces of loam. The beds now consist of well-graded gravel which are compacted and arranged in such a way that a pavement is formed. It is noted that chemical interactions between the water being transported, the material composing the perimeter, and the sediment can also result in the armoring of a perimeter.

The tractive force method is the procedure recommended for designing unlined earthen canals. Table 7.7 summarizes some typical cross sections of unlined earthen canals used on irrigation projects in the United States, and a typical earthen section of the All-American Canal is shown in Fig. 7.11. A procedure for designing a stable, unlined earthen section is summarized in Table 7.8, where it is assumed that the

Table 7.7. Typical cross sections of unlined, earthen channels for selected canals in the western United States (Houk, 1956)

Canal	Project	Side slopes	Depth of flow (ft) (m)	Bottom Width (ft) (m)	Ratio of width to depth	Mean Velocity (ft/s) (m/s)	Discharge (ft³/s) (m³/s)	Freeboard (ft)[1] (m)
Earth sections									
Lateral	Altus	1.50:1	1.66 (0.51)	4 (1.22)	2.41	1.86 (0.57)	20 (0.6)	1.3 (0.40)
C Line East	Boise	1.50:1	4.00 (1.22)	8 (2.44)	2.00	1.89 (0.58)	106 (3.0)	3.5 (1.07)
C Line East	Boise	1.50:1	7.14 (2.18)	14 (4.27)	1.96	2.25 (0.69)	397 (11.2)	3.4 (1.04)
Altus	Altus	1.50:1	6.20 (1.89)	20 (6.10)	3.23	2.48 (0.76)	450 (12.7)	3.5 (1.07)
Conchas	Tucumcari	1.50:1	8.65 (2.64)	24 (7.32)	2.77	2.19 (0.67)	700 (19.8)	4.3 (1.31)
Kittitas Main	Yakima	1.50:1	11.35 (3.46)	30 (9.14)	2.64	2.47 (0.75)	1,320 (37.4)	5.0 (1.52)
Ridge	Yakima	1.50:1	9.57 (2.92)	40 (12.19)	4.18	2.50 (0.76)	1,300 (36.8)	4.0 (1.22)
Main (gravity extension)	Minidoka	1.50:1	5.60 (1.71)	60 (18.29)	10.71	3.00 (0.91)	1,149 (32.5)	3.0 (0.91)
Coachella	All-American	2:1	10.33 (3.15)	60 (18.29)	5.81	3.00 (0.91)	2,500 (70.8)	6.0 (1.83)
Gravity Main	Gila	2:1	13.54 (4.13)	100 (30.38)	7.38	3.49 (1.06)	6,000 (169.9)	6.0 (1.83)
All-American	All-American	2:1	16.59 (5.06)	130 (39.62)	7.84	3.75 (1.14)	10,155 (287.6)	6.0 (1.83)
All-American	All-American	1.75:1	20.61 (6.28)	160 (48.77)	7.76	3.75 (1.14)	15,155 (429.1)	6.0 (1.83)
Rock sections									
North Unit Main	Deschutes	0.25:1	5.54 (1.69)	20 (6.10)	3.61	4.63 (1.41)	550 (15.6)	2.5 (0.76)
Gravity Main	Gila	0.75:1	21.57 (6.57)	33 (10.06)	1.53	1.79 (0.55)	1,900 (53.8)	Deep cut
Yuma	Yuma	0.50:1	8.46 (2.58)	60 (18.29)	7.10	3.68 (1.12)	2,000 (56.6)	Deep cut
All-American	All-American	0.75:1	20.13 (6.14)	69 (21.03)	3.42	6.00 (1.83)	10,155 (287.6)	Deep cut
All-American	All-American	0.75:1	22.75 (6.93)	94 (28.65)	4.13	6.00 (1.83)	15,155 (429.1)	Deep cut

[1]To top of lining

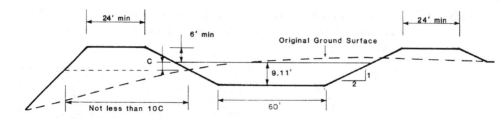

Figure 7.11. Typical earth section used on Coachella Canal, All American Canal project.

Table 7.8. A design procedure for unlined, stable earthen channels.

Step	Process
1	Estimate n or C for specified material composing the perimeter.
2	Estimate the angle of repose for the channel perimeter material (Fig. 7.9)
3	Estimate channel sinuousness from the type of topography through which it will pass and determine the tractive force correction factor (Table 7.6).
4	Assume side slope angle (Table 7.2) and (bottom width)/(normal depth of flow)
5	Assume sides of the channel are the limiting factor in the channel design.
6	Calculate the maximum permissible tractive on the sides, in terms of the unit tractive force. Use the correction factor from Fig. 7.7a and the sinuosity correction factor (Step 3).
7	Estimate tractive force ratio [Eq. (7.3.7)]
8	Estimate permissible tractive force on bottom (Fig. 7.10) and correct for sinuousness (Step 3)
9	Combine the results of Steps 6 to 8 to determine the normal depth of flow y_N.
10	Determine the bottom width with the results from Steps 4 and 9.
11	Compute Q and compare this value with the design flow Q_D, return to Step 4, and repeat the design process with trial b/y ratios until $Q = Q_D$.
12	Compare permissible tractive force on the bottom (Step 8), with actual tractive force given by $\gamma\, y_N S$ and corrected for shape (Fig. 7.7a).
13	Check 1. Minimum permissible velocity if the water carries silt and for vegetation. 2. Froude number
14	Estimate required freeboard [Eq. (7.1.1)] or (Fig. 7.1).
15	Summarize results with a dimensioned sketch.

design flow Q_D, the longitudinal slope of the channel, the relevant engineering properties of the material composing the channel perimeter, and the general shape of the section have been established prior to the initiation of the design process.

EXAMPLE 7.3
A channel, which is to carry 10 m³/s (350 ft³/s) through moderately rolling topography on a slope of 0.0016, is to be excavated in coarse alluvium with 25 percent of the particles being 3 cm (1.2 in) or more in diameter. The material, which will compose the perimeter of this channel, can be described as being moderately rounded. Assuming that the channel is to be unlined and of trapezoidal section, find suitable values of b and z.

Solution
Step 1   Estimate n from Table 4.8

$$n = 0.025$$

Step 2   Estimate the angle of repose.

$$d_{25} = 3 \, cm = \frac{3}{2.54} \, in = 1.18 \, in$$

From Fig. 7.9

$$\alpha = 34^\circ$$

Step 3   Estimate channel sinuousness correction factor (Table 7.6).

$$C_s = 0.75$$

Step 4   Assume side slope 2:1 and $b/y_N = 4$.

Step 5   Assume side slopes are a limiting factor.

Step 6   Find maximum permissible tractive forces on sides (Fig. 7.7a).

$$\tau_s = 0.75 \gamma y_N S$$

Step 7   Estimate tractive force ratio [Eq. (7.3.7)].

$$K = \frac{\tau_s}{\tau_b} = \sqrt{1 - \frac{\sin^2 \Gamma}{\sin^2 \alpha}}$$

$$\Gamma = \tan^{-1}(0.5) = 26.6°$$

$$K = \sqrt{1 - \frac{\sin^2 26.6°}{\sin^2 34°}} = 0.60$$

Step 8   Estimate permissible tractive force on bottom (Fig. 7.10).

$$\tau_b = 0.47\,\text{lb} / \text{ft}^2$$

for $d_{25} = 1.18$ in (30 mm).
    Correct for sinuosity.

$$\tau_b = C_s\tau_b = 0.75(0.47) = 0.35\,\text{lb} / \text{ft}^2\left(17\,\text{N} / \text{m}^2\right)$$

Step 9   Estimate y_N

$$\frac{\tau_s}{\tau_b} = K$$

$$\tau_s = K\tau_b$$

$$0.75\gamma\,y_N S = K\tau_b$$

$$y_N = \frac{0.60(17)}{0.75(9658)(0.0016)} = 0.88\,\text{m}\,(2.9\,\text{ft})$$

Step 10 $b/y_N = 4.$

$$b = 4(0.88) = 3.5\,\text{m}\,(11\,\text{m})$$

Step 11   Determine Q.

$$A = (b + zy)y = \left[3.5 + 2(0.88)\right](0.88) = 4.6\,\text{m}^2\,\left(50\,\text{ft}^2\right)$$

$$P = b + 2y\sqrt{1+z^2} = 3.5 + 2(0.88)\sqrt{5} = 7.4\,\text{m}\,(24\,\text{ft})$$

$$R = \frac{A}{P} = \frac{4.6}{7.4} = 0.60\,\text{m}\,(2.0\,\text{ft})$$

$$Q = \frac{AR^{2/3}}{n}\sqrt{S} = \frac{4.6(0.60)^{2/3}}{0.025}\sqrt{0.0016} = 5.2\,\text{m}^3/\text{s}\,(180\,\text{ft}^3/\text{s})$$

Q is less than Q_D; and therefore, additional computations are required in which b/y_N is variable and C_s, K, permissible τ_b, and z are constant

b/y_N	y_N m	b m	A m²	P m	R m	Q m³/s
5	0.88	4.4	5.4	8.3	0.65	6.5
8.25	0.88	7.3	8.0	11	0.71	10.2
8.15	0.88	7.2	7.9	11	0.71	10

Then, for $z = 2$ and $b/y_N = 8.15$, $y_N = 0.88$ m (2.9 ft), $b = 7.2$ m (24 ft), and $Q = 10$ m³/s (350 ft³/s).

Step 12   Check tractive force on bottom.

Permissible $\tau_b = C_s\tau_b = 17$ N/m²   (see Step 8)

Computed $\tau_b = 0.99\,\gamma\,y_N S$ (Fig. 7.7b)

$$\tau_b = 0.99(9658)(0.88)(0.0016) = 13\ \text{N/m}^2\ (0.27\ \text{lb/ft}^2)$$

Since this is the actual, computed tractive force, the design is acceptable.

Step 13   Check velocity and Froude number.

$$\bar{u} = \frac{Q}{A} = 1.3\,\text{m}/\text{s}\,(4.3\,\text{ft}/\text{s})$$

This velocity should prevent vegetative growth and sedimentation.

$$F = \frac{\bar{u}}{\sqrt{gy}} = 0.48$$

Therefore, this is a subcritical flow.

Step 14   Estimate required freeboard from Eq. (7.1.1).

Design depth of flow = 0.88 m (2.9 ft)

Design flow = 10 m³/s (350 ft³/s)

Estimate *C* in Eq. (7.1.1) as 1.6

Then

$$F = \sqrt{Cy} = \sqrt{1.6(2.9)} = 0.66\,\text{m (2.2 ft)}$$

The results of this design are summarized in Fig. 7.12.

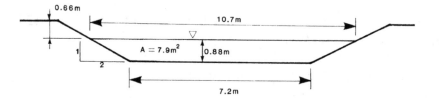

Figure 7.12. Summary of results for Example 7.3.

In the previous paragraphs of this section, the traditional trial and error approach to the design of erodible channels has been presented. Akan (2001) developed a method, using an approximate equation, that facilitates the explicit solution of the problem for the width of the channel. In an erodible channel, the limiting depth of flow, y_{LIM}

$$y_{\text{LIM}} = \frac{C_S K\tau}{ES_0\gamma} \tag{7.3.13}$$

where C_S = correction factor for sinuousness, τ = permissible shear stress on channel material, S_0 = longitudinal channel slope, γ = specific weight of water, and E and K are coefficients. The coefficient E is a function of channel slope and the width to depth ratio. For earthen channels excavated in cohesive soils, the shear stresses on the channel bottom govern the design, and for this case $E = 1.0$ (FHWA, 1988). For earthen channels excavated in cohesion less soils, the shear stresses on the sides govern the design, and for this case $E = 0.76$ (Chaudhry, 1993). In Eq. (7.3.13), the coefficient K accounts for the gravitational forces tending to roll the soil particles down the side slopes. For cohesive soils, $K = 1.0$, and for cohesionless soils

$$K = \sqrt{1 - \frac{1}{\left(1 + z^2\right)\sin^2 \alpha}} \tag{7.3.14}$$

where α = angle of repose of the cohesionless channel material.

Combining Eq. (7.3.14) and the Manning equation, the combined equations can be written in dimensionless form as (Akan, 2001)

$$\frac{nQ\left(ES_0\gamma\right)^{8/3}}{\phi S_0^{0.5}\left(C_S K\tau\right)^{8/3}} = \frac{\left(\dfrac{bES_0\gamma}{C_S K\tau} + z\right)^{5/3}}{\left(\dfrac{bES_0\gamma}{C_S K\tau} + 2\sqrt{1 + z^2}\right)^{2/3}} \tag{7.3.15}$$

and which is plotted in Fig. 7.13. Using a curve fitting technique, Akan (2001) developed the following equation, which approximates Fig. 7.13 that can be solved for the bottom width of a trapezoidal channel, b, explicitly or

$$b = 1.186\left(\frac{C_S K\tau}{ES_0\gamma}\right)\left[\frac{nQ\left(ES_0\gamma\right)^{8/3}}{k S_0^{0.5}\left(C_S K\tau\right)^{8/3}} - \frac{z^{5/3}}{\left(2\sqrt{1 + z^2}\right)^{2/3}}\right]^{0.955} \tag{7.3.16}$$

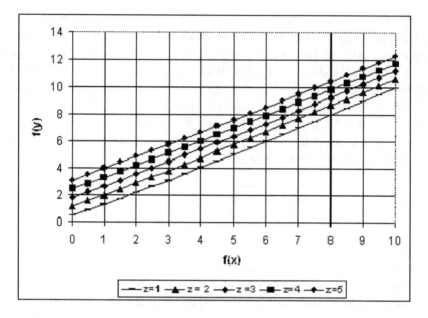

Figure 7.13.   **Design chart for erodible channel width calculation (Akan, 2001) where** $f(x) = (bES_0\gamma)/(C_sK\tau)$ **and** $f(y) = (nQ(ES_0\gamma)^{8/3})/(\phi\sqrt{S_0}(C_sK\tau)^{8/3})$.

EXAMPLE 7.4
Rework Example 7.3, using the approach presented by Akan (2001). From Example 7.3

$$Q = 10 \text{ m}^3/\text{s}$$
$$S = 0.0016 \text{ m/m}$$
$$\alpha = 34°$$
$$z = 2$$
$$C_S = 0.75$$
$$\tau_b = 0.47 \text{ lb/ft}^2 = 22.5 \text{ N/m}^2$$

Solution
Then, for a cohesionless soil [Eq. (7.3.14)]

$$E = 0.76$$

$$K = \sqrt{1 - \frac{1}{(1 + z^2)\sin^2 \alpha}} = \sqrt{1 - \frac{1}{(1 + 4)\sin^2 34°}} = 0.60$$

Then, from Eq. (7.3.16)

$$\frac{C_S K\tau}{ES_0\gamma} = \frac{0.75(0.60)(22.5)}{0.76(0.0016)(9800)} = 0.85$$

$$\frac{nQ(ES_0\gamma)^{8/3}}{k\sqrt{S_0}(C_S K\tau)^{8/3}} = \frac{0.025(10)[0.76(0.0016)(9800)]^{8/3}}{1\sqrt{0.0016}[0.75(0.60)(22.5)]^{8/3}} = 9.88$$

$$\frac{z^{5/3}}{(2\sqrt{1 + z^2})^{2/3}} = \frac{2^{5/3}}{(2\sqrt{5})^{2/3}} = 1.16$$

and

$$b = 1.186(0.85)(9.88 - 1.18)^{0.955} = 7.96\,\text{m}$$

which compares well with the answer obtained by the trial and error procedure in Example 7.3.

It must be emphasized that the approaches discussed above are not the only methodologies which can be applied to the problem of designing stable channel sections. For example, the basic principles and results associated with the concept of threshold of movement can also be applied to this problem. Shields (1936) used an experimental approach to define the threshold of movement, and his results can be stated in terms of two dimensionless parameters:

$$Re_* = \frac{u_* d}{\upsilon} \tag{7.3.17}$$

and

$$F_s = \frac{\tau_c}{\gamma(S_s - 1)d} = \frac{u_*^2}{(S_s - 1)gd} \tag{7.3.18}$$

where Re_* =   a Reynolds number based on the shear velocity and the particle size (this number is also known as the particle Reynolds number)

u_* =   shear velocity

υ =   fluid kinematic viscosity

S_s =   specific gravity of the particles composing the perimeter which is usually taken to be 2.65

d =   diameter of the particles composing the perimeter of the channel

τ_c =   critical shear stress.

For conservative design, d can usually be assumed to be the diameter of the particle of which 25 percent of all the particles, measured by weight, are larger. The results of Shields are usually summarized in a graphical form (Fig. 7.14a), in which the curve delineates the threshold of movement. These results have been confirmed, in a general sense, by the theoretical results of White (1940) and the field results of Lane (1955), which are summarized in Fig., 7.10a.

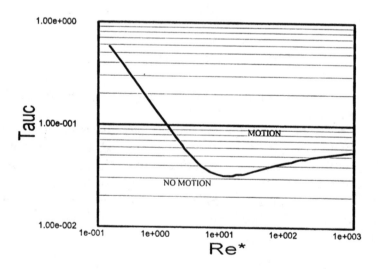

Figure 7.14a   Threshold of movement as a function of particle Reynolds number (Shields, 1936) where *Tauc* is given by Eq. (7.3.18) and *Re** by Eq. (7.3.17).

If it is assumed that $S_s = 2.65$ when R_* exceeds a value of 400, the perimeter particle size must be in excess of 0.25 in (0.006 m). In this case, the channel perimeter material can accurately be classified as coarse alluvium, and Eq. (7.3.18) becomes

$$\frac{\tau_c}{d\gamma\left(S_s - 1\right)} = 0.056 \qquad (7.3.19)$$

where

$$\tau_c = \gamma RS$$

Therefore,

$$\frac{\gamma RS}{\gamma d\left(S_s - 1\right)} = \frac{RS}{d\left(2.65 - 1\right)} = 0.056$$

and

$$d \cong 11RS \qquad (7.3.20)$$

Equation (7.3.20) provides a simple method of estimating the size of the material, which will remain at rest in a channel of specified R and S. Note, *for values of d less than 0.25 in (0.006 m) Eq. (7.3.20) is not valid*, but the curve in Fig. 7.14a can be used to find appropriate values of F_s, and analogous relations can be developed for these sizes.

EXAMPLE 7.5
In Example 7.3, the specified slope was 0.0016 and 25 percent of the channel perimeter particles were 3 cm (1.2 in) or more in diameter. Use the results summarized in Fig. 7.10 to show that the design arrived at in this example represents a conservative design.

Solution
Assume $S_s = 2.65$ and that $R \cong y_N$, for wide channels. Under these assumptions, Eq. (7.3.20) becomes, for level surfaces

$$\tau_c = \frac{\gamma d}{11}$$

From the previous example, the tractive force ratio is

$$\frac{\tau_s}{\tau_b} = K = \sqrt{1 - \frac{\sin^2 \Gamma}{\sin^2 \alpha}} = 0.60$$

Substituting $\tau_b = C_s(\gamma d / 11)$ and $\tau_s = 0.75\gamma y_N S$ in the above equation yields

$$\frac{0.75\gamma y_N S}{C_s\left(\dfrac{\gamma d}{11}\right)} = 0.60$$

or the maximum value of y_N, for a stable channel is

$$y_N = \frac{C_s d(0.60)}{11(0.75)S} = \frac{0.75\left(\dfrac{3}{100}\right)(0.60)}{11(0.75)(0.0016)} = 1.0\,\mathrm{m}\ (3.3\,\mathrm{ft})$$

Thus, the previous computation yields a conservative result for y_N. It should be noted that this result provides a check on the validity of the solution obtained by the recommended design methodology.

The Shields values of τ_c are typically used to identify conditions, under which the bed sediments are stable, but on the verge of moving. It is pertinent to observe that not all practitioners agree with these results. For example, Garcia (1999) gives the critical value of the shear stress as 0.06 for $Re_* \geq 500$, while others have used 0.047. Taylor and Vanoni (1972) reported that small but finite amounts of sediment were transported in flows with values of τ_c given by Fig. 7.14a. The values of τ_c, used for design must take into consideration the particulars of the specific situation. For example, if the particles that are moved can be replaced by others being transported from upstream, then some particle motion is acceptable. However, if the particles cannot be replaced, for example, particles composing a stream bank, then the Shields values of τ_c should be reduced.

For values of $Re_* \leq 400$, Fig. 7.14a is not particularly useful, since, to find τ_c, the critical shear velocity $u_* = \sqrt{(\tau_c / \rho)}$ must be known. Garcia (1999) noted that the relation can be put into an explicit form by plotting τ_c as a function of Re_p where

$$Re_p = \frac{d\sqrt{gd\dfrac{\rho_s - \rho}{\rho}}}{\nu} \qquad (7.3.21)$$

and the internal relationship in the original Shields relationship is

$$\mathrm{Re}_* = \frac{u_* d}{\nu} = \frac{u_*}{\sqrt{gd\frac{\rho_s - \rho}{\rho}}} \cdot \frac{d\sqrt{gd\frac{\rho_s - \rho}{\rho}}}{\nu} = \mathrm{Re}_p \sqrt{\tau} \qquad (7.3.22)$$

A useful fit of the data was provided by Brownlie (1981) or

$$\tau_c = \frac{0.22}{\mathrm{Re}_p^{0.6}} + 0.06\exp\left(\frac{-17.77}{\mathrm{Re}_p^{0.6}}\right) \qquad (7.3.23)$$

With Eq. (7.3.23), values of τ_c can be easily estimated when the properties of the flow and sediment are known. The functional relationship between τ_c and Re_p is shown graphically in Fig. 7.14b.

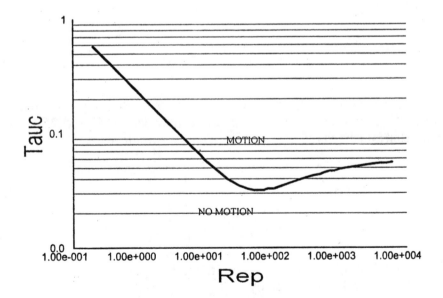

Figure 7.14b. **Threshold of movement as a function of a Reynolds number where *Tauc* is given by Eq. (7.3.18) and *Rep* by Eq. (7.3.21).**

Yang (1996), using the basic principles of fluid mechanics and laboratory data from several researchers (Yang, 1973), developed the following incipient motion criteria

$$\frac{u_{cr}}{\omega} = \frac{2.5}{\log\left(\frac{u_* d}{v}\right) - 0.06} + 0.66 \text{ for } 1.2 < \frac{u_* d}{v} < 70 \qquad (7.3.24)$$

and

$$\frac{u_{cr}}{\omega} = 2.05 \text{ for } 70 \le \frac{u_* d}{v} \qquad (7.3.25)$$

where u_{cr} = average critical velocity at incipient motion, ω = terminal particle fall velocity, and d = particle diameter. There are many approaches to estimating the terminal particle fall velocity, and among these, are the formula introduced by Rubey (1933) for gravel, sand, and silt particles or

$$\omega = F\sqrt{dg\left(\frac{\gamma_s - \gamma}{\gamma}\right)} \qquad (7.3.26)$$

where γ_s = sediment specific weight, and for particles greater than 1 mm settling in water with a temperature between 10°C and 25°C, the parameter F has a value of 0.79. For smaller particle sizes

$$F = \sqrt{\frac{2}{3} + \frac{36v^2}{gd^3\left(\frac{\gamma_s}{\gamma} - 1\right)}} - \sqrt{\frac{36v^2}{gd^3\left(\frac{\gamma_s}{\gamma} - 1\right)}} \qquad (7.3.27)$$

For particle sizes greater than 2 mm, the fall velocity in 16°C water, the fall velocity can be estimated by

$$\omega = 6.01\sqrt{d} \quad \left(\omega \text{ in ft / s, d in ft}\right) \qquad (7.3.28a)$$

or

$$\omega = 3.32\sqrt{d} \quad \left(\omega \text{ in m / s, d in m}\right) \qquad (7.3.28b)$$

EXAMPLE 7.6

In Example 7.3, the specified slope was 0.0016 m/m and 25 percent of the channel perimeter particles were 3 cm (1.2 in) or more in diameter. Use the Yang (1996) critical velocity approach to incipient motion to demonstrate that the results obtained in this example are conservative.

Solution
Given

$$d = 20 \text{ cm} = 200 \text{ mm} = 0.20 \text{ m}$$
$$S = 0.0016 \text{ m/m}$$

From Example 7.3

$$y_N = 0.88 \text{ m}$$
$$R = 0.71 \text{ m}$$

Then, the shear velocity is

$$u_* = \sqrt{gRS} = \sqrt{9.8(0.71)(0.0016)} = 0.11 \text{ m/s}$$

Assume the water temperature is 20°C; and therefore, $\upsilon = 1.003 \times 10^{-6}$ m²/s. Then,

$$\frac{u_* d}{\upsilon} = \frac{0.11(0.2)}{1.003 \times 10^{-6}} = 21,934$$

Therefore, the critical velocity is found using Eq. (7.3.25) or

$$\frac{u_c}{\omega} = 2.05$$

or

$$u_c = 2.05\,\omega$$

For $d = 0.2$ m, Eq. (7.3.28b) can be used to estimate ω

$$\omega = 3.22 d^{0.5} = 3.22(0.2)^{0.5} = 1.44 \text{ m/s}$$

Then

$$u_c = 2.05\,\omega = 2.05(1.44) = 2.95 \text{ m/s}$$

and from Example 7.3

$$u = 1.3 \text{ m/s}$$

Therefore, the design in Example 7.3 is conservative.

7.3.2 The Stable Hydraulic Section

The permissible tractive force design techniques, presented in the previous section, yield a channel cross section in which the tractive force is equal to the permissible value on only a part of the wetted perimeter, usually the sides. It seems logical to attempt to define a channel cross section such that the condition of incipient particle motion prevails at all points of the channel perimeter. The equations defining this channel section were developed by the U.S. Bureau of Reclamation (Glover and Florey, 1951) for erodible channels carrying clear water through noncohesive materials. This methodology yields what is known as stable hydraulic section of maximum hydraulic efficiency.

The assumptions required to develop the equations defining the stable hydraulic section are:

1. The soil particles are held in place in the channel by the component of the submerged weight of the particle acting normal to the bed.

2. At and above the water surface, the channel side slope is the angle of repose of the noncohesive material under the action of gravity.

3. At the centerline of the channel, the side slope is zero, and the tractive force alone is sufficient to produce a state of incipient particle motion.

4. At points between the center and edge of the channel, the particles are kept at a state of incipient motion by the resultant of the tractive force and the gravity component of the submerged weight of the particles.

5. The tractive force acting on an area of the channel is equal to the component of the weight of the water above the area acting in the direction of flow. If this assumption is valid, then there is no lateral transfer of tractive force.

It is noted that with the exception of assumptions 2 and 3, these are the same assumptions which were used in the previous section.

Consider a channel of longitudinal slope S, with the side slope being defined at any point in the section (x, y), by the angle Γ (Fig. 7.15). Then, by assumption 5, the critical tractive stress acting on area AB in a unit length of channel, is

$$\tau_s = \frac{\gamma y S dx}{\sqrt{(dx)^2 + (dy)^2}} = \gamma y S \cos \Gamma \qquad (7.3.29)$$

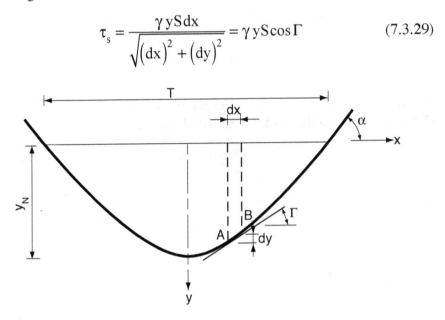

Figure 7.15.   Schematic definition of parameters for stable hydraulic section.

from the previous development, the critical tractive force acting on the side of the channel is given by

$$\tau_s = K\tau_b = \gamma y_N S \cos \Gamma \sqrt{1 - \frac{\tan^2 \Gamma}{\tan^2 \alpha}} \qquad (7.3.30)$$

where $\tau_b = \gamma y_N S$ is the tractive force, at the centerline of the channel, where the depth of flow is y_N. Combining Eqs. (7.3.29 and 7.3.30) and solving for y yields

$$y = \frac{y_N}{\tan \alpha} \sqrt{\tan^2 \alpha - \tan^2 \Gamma} \qquad (7.3.31)$$

Substituting $dy/dx = \tan\Gamma$ into Eq. (7.3.31) yields a differential equation that defines the shape of the cross section

$$\left(\frac{dy}{dx}\right)^2 + \left(\frac{y}{y_N}\right)^2 \tan^2\alpha - \tan^2\alpha = 0 \qquad (7.3.32)$$

Given the condition that $x = 0$, $y = y_N$ the solution of Eq. (7.3.32) is

$$y = y_N \cos\left(\frac{x\tan\alpha}{y_N}\right) \qquad (7.3.33)$$

An alternative form of Eq. (7.3.33) can be obtained by noting that $x = T/2$, $y = 0$. This condition can be satisfied only if

$$\frac{T\tan\alpha}{2y_N} = \frac{\pi}{2} \qquad (7.3.34)$$

or

$$y_N = \frac{T\tan\alpha}{\pi} \qquad (7.3.35)$$

Then, Eq. (7.3.33) becomes

$$y = y_N \cos\left(\frac{\pi x}{T}\right) \qquad (7.3.36)$$

Equations (7.3.33) and (7.3.36) define the channel section which has the smallest width and the largest hydraulic radius for a given area. Thus, this channel has the greatest hydraulic efficiency of all stable, unlined, earthen channels built through noncohesive materials at a longitudinal slope S and carrying clear water.

The flow area of the channel defined by Eqs. (7.3.33) and (7.3.36) is

$$A = 2\int_0^{T/2} y\,dx = 2y_N\int_0^{T/2} \cos\left(\frac{x\tan\alpha}{y_N}\right) dx = \frac{2Ty_N}{\pi} \qquad (7.3.37)$$

and the wetted perimeter

$$P = 2\int_0^{T/2} \sqrt{1 + \left(\frac{dy}{dx}\right)^2}\, dx = \frac{2y_N}{\sin\alpha} E(\sin\alpha) \qquad (7.3.38)$$

where $E(\sin\alpha)$ = a complete elliptic integral of the second kind. $E(\sin\alpha)$ can be evaluated either by standard mathematical tables or by

$$E(\sin\alpha) = \frac{\pi}{2}\left[1 - \left(\frac{1}{2}\right)^2 \sin^2\alpha - \left(\frac{1\bullet 3}{2\bullet 4}\right)^2 \frac{\sin^4\alpha}{3} - \left(\frac{1\bullet 3\bullet 5}{2\bullet 4\bullet 6}\right)^2 \frac{\sin^6\alpha}{5} - \cdots \right]$$ (7.3.39)

The discharge of the channel can then be computed by the Manning equation

$$Q = \frac{2.98 y_N^{8/3} (\cos\alpha)^{2/3} \sqrt{S}}{n(\tan\alpha)\left[E(\sin\alpha)\right]^{2/3}}$$ (7.3.40)

The discharge Q is the flow that would be obtained in a channel designed for the greatest efficiency, in a given noncohesive material, at a specified longitudinal slope. If the design discharge Q_D is larger or smaller than Q, then the channel defined by Eqs. (7.3.33) and (7.3.36) must be modified. If $Q_D > Q$, then additional flow area must be provided; however, the maximum depth of flow can be no greater than the stability depth y_N, since an increase in depth would result in an increased tractive force and instability. Therefore, a rectangular section is added at the center of the theoretical section (Fig. 7.16). The additional width required, T'', is found by a trial-and-error solution of the Manning equation or

$$Q_D = \frac{1.49}{n}\sqrt{S}\left\{ \frac{\left(\dfrac{2y_N^2}{\tan\alpha} + T'y_N\right)^{5/3}}{\left[\dfrac{2y_N E(\sin\theta)}{\sin\alpha} + T'\right]^{2/3}} \right\}$$ (7.3.41)

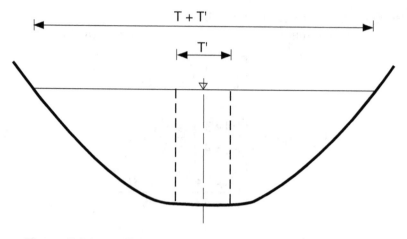

Figure 7.16.   Definition sketch for a stable hydraulic section where $Q < Q_D$.

If $Q_D < Q$, then economy dictates that a portion of the theoretical section must be removed (Fig. 7.17). The width of the channel which must be removed, T''', is also found by a trial-and-error solution of the Manning equation

$$Q = \frac{1.49}{n}\sqrt{S}\,\frac{\left\{\left(\dfrac{2y_N^2}{\tan\alpha}\right)\left[\sin\left(\dfrac{T\tan\alpha}{2y_N}\right) - \sin\left(\dfrac{T''\tan\alpha}{2y_N}\right)\right]\right\}^{5/3}}{\dfrac{2y_N}{\tan\alpha}E\left[\sin\alpha\dfrac{\pi}{2}\left(1 - \dfrac{T''}{T}\right)\right]} \qquad (7.3.42)$$

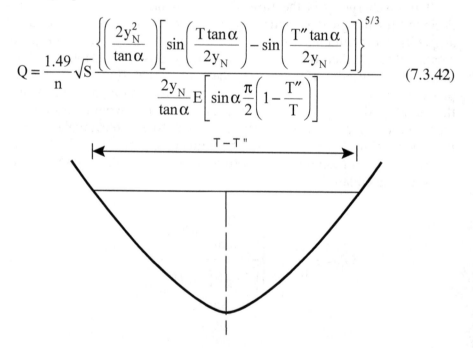

Figure 7.17.   Definition sketch for a stable hydraulic section where $Q > Q_D$.

where $E[sin\alpha, (\pi/2)(1-T''/T)]$ is an incomplete, elliptic integral of the third kind. T'' can be estimated by assuming that the mean velocity in the theoretical section and the adjusted section are equal, thus the discharges are proportional to the flow areas or

$$Q = \frac{2y_N^2}{\tan\alpha}\bar{u} = \frac{2T^2\tan\alpha}{\pi^2}\bar{u} \qquad (7.3.43)$$

and

$$Q_D = \frac{2(T-T'')^2\tan\alpha}{\pi^2}\bar{u} \qquad (7.3.44)$$

Combining Eqs. (7.3.43) and (7.3.44) results in an equation which can be used to estimate T''

$$T'' = T\left(1 - \sqrt{\frac{Q_D}{Q}}\right) \qquad (7.3.45)$$

EXAMPLE 7.7
Design the stable section of greatest hydraulic efficiency for a canal which is to be constructed through a noncohesive material on a slope of 0.0004. Preliminary testing of the natural material, which will compose the channel perimeter, indicates that τ_0 = 0.10 lb/ft^2 (4.8 N/m^2), n = 0.02, and α = 31°. The design flow is 300 ft^3/s (8.5 m^3/s).

Solution
The depth of flow can be determined from Eq. (7.3.2), where it is assumed that $y_N = R$.

$$\tau_0 = \gamma y_N S$$

$$y_N = \frac{\tau_0}{\gamma S} = \frac{0.10}{62.4(0.0004)} = 4.0\,\text{ft}\,(1.2\,\text{m})$$

The shape of the channel is obtained by combining Eqs. (7.3.33) and (7.3.36)

$$y_1 = y_N \cos\left(\frac{x \tan\alpha}{y_N}\right) = 4.0\cos\left(\frac{x \tan 31°}{4.0}\right) = 4.0\cos(0.15x)$$

and

$$T = \frac{\pi y_N}{\tan\alpha} = \frac{4\pi}{\tan 31°} = 21\,\text{ft}\,(6.4\,\text{m})$$

$$A = \frac{2Ty_N}{\pi} = \frac{2(21)(4)}{\pi} = 53\,\text{ft}^2\,(4.9\,\text{m}^2)$$

The discharge of this channel can then be found from Eq. (7.3.40).

$$Q = \frac{2.98y_N^{8/3}(\cos\alpha)^{2/3}\sqrt{S}}{n\tan\alpha\left[E(\sin\alpha)\right]^{2/3}}$$

$$= \frac{2.98(4)^{8/3}(\cos 31°)^{2/3}\sqrt{0.0004}}{0.02(\tan 31°)(1.46)^{2/3}}$$

$$= 140\,\text{ft}^3\,(4.0\,\text{m}^3/\text{s})$$

Since $Q < Q_D$, a trial-and-error solution of Eq. (7.3.41) for T' is required

$$Q_D = \frac{1.49}{n}\sqrt{S}\left\{\frac{\left(\dfrac{2y_N^2}{\tan\alpha} + T'y_N\right)^{5/3}}{\left[\dfrac{2y_N E(\sin\theta)}{\sin\alpha} + T'\right]^{2/3}}\right\}$$

$$\frac{nQ_D}{1.49\sqrt{S}} = \frac{\left(\dfrac{2y_N^2}{\tan\alpha} + T'y_N\right)^{5/3}}{\left[\dfrac{2y_N E(\sin\alpha)}{\sin\alpha} + T'\right]^{2/3}}$$

$$\frac{0.02(300)}{1.49\sqrt{0.0004}} = \frac{\left[\dfrac{4(2)^2}{\tan 31^\circ} + 4T'\right]^{5/3}}{\left[\dfrac{2(4)(1.46)}{\sin 31^\circ} + T'\right]^{2/3}}$$

$$201 = \frac{(53.3 + 4T')^{5/3}}{(22.7 + T')^{2/3}}$$

Trial T′ ft	$\dfrac{(53.3 + 4T')^{5/3}}{(22.7 + T')^{2/3}}$
5	140
10	188
11.5	203

Therefore, the top width required to convey this flow is (21 + 11.5) = 32 ft (9.8 m). The results of this design are summarized in Fig. 7.18.

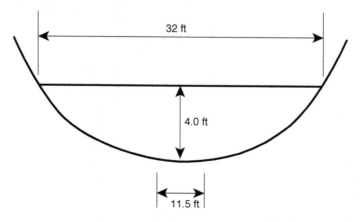

Figure 7.18. Summary of results for Example 7.7.

The foregoing approach is but one of many approaches to the complex problem of designing stable alluvial channels. For example, Kennedy (1895), using data from the irrigation canals in the Upper Bari Doab irrigation system in India, proposed a relationship between the velocity and depth of flow in stable alluvial channels. While the Kennedy (1895) equation is obsolete, it has been replaced by other approaches which are typically known as regime theory. Each regime approach to the problem typically involves three formulas for a stable width, depth, and slope. In the following discussion two regime approaches are discussed.

In a series of papers between 1930 and 1958, Lacey (1930, 1935, and 1958) published three regime equations for alluvial channels. In the first paper (Lacey, 1930), a flow resistance equation was presented in two forms or

$$\bar{u} = 1.15\sqrt{f_L D} \qquad (7.3.46)$$

or

$$\bar{u} = \frac{1.346}{N_R} D^{0.25} \sqrt{RS} \qquad (7.3.47)$$

where f_L = Lacey's silt factor, $\bar{u}$ = mean velocity (ft/s), D = hydraulic depth (ft), and N_R = Lacey's absolute rugosity. f_L and N_R are given by

$$f_L = 1.6\sqrt{d}$$ (7.3.48)

and

$$N_R = 0.0225f^{0.25}$$ (7.3.49)

where d = median bed material size (mm). An examination of Eqs. (7.3.46) through (7.3.49) suggests that Manning's n, in an alluvial channel, is not a constant, but a function of the flow.

The second Lacey regime equation was

$$T = 2.67\sqrt{Q}$$ (7.3.50)

where T is the channel top width and Q = flow rate (ft^3/s). As noted by Chang (1988), the validity of this relationship is contradicted by some rivers where the top width is not only a function of the flow rate but also the longitudinal slope.

The third Lacey regime equation states that the equilibrium longitudinal channel slope is inversely proportional to discharge and directly proportional to the median sediment size or

$$S = \frac{f_L^{5/3}}{1830Q^{1/6}}$$ (7.3.51)

Thus, for a specified discharge and median sediment size, Lacey's three independent equations provide estimates of a stable width, depth, and longitudinal slope. Although Lacey did not provide a specific relationship for the bank slope, it is pertinent to observe that most of the canals used in developing these regime equations had cohesive banks. Chang (1988) noted that Lacey's approach should be limited to bed material sizes between 0.15 and 0.44 mm and flow rates between 5 and 5,000 ft^3/s (0.14 to 142 m^3/s).

Subsequent to the introduction of the Lacey approach more research has been conducted and additional data developed. Blench (1952 and 1970) extended the Lacey approach to include situations where the bank materials ranged from slightly cohesive to highly cohesive. Simons and Albertson (1960) used not only data from India and Pakistan, but additional data collected in Colorado, Wyoming, and Nebraska. These investigators classified the channels into five categories:

1. Sand bed and banks
2. Sand bed and cohesive banks
3. Cohesive bed and banks
4. Coarse, non-cohesive material
5. Same as Category 2 but carrying heavy sediment loads (2,000 - 8,000 ppm).

The Simons and Albertson (1960) regime equations fall into three groups and are English units.

Channel Width

$$P = K_1 \sqrt{Q} \tag{7.3.52}$$

$$\overline{T} = 0.9P = 0.9K_1 \sqrt{Q} \tag{7.3.53}$$

and

$$\overline{T} = 0.92T - 2.0 \tag{7.3.54}$$

where $\overline{T}$ = average width and K_1 = a coefficient, which is a function of the channel type.

Channel Depth

$$R = K_2 Q^{0.36} \tag{7.3.55}$$

$$y = 1.21R \text{ for } R < 7 \text{ ft} \tag{7.3.56}$$

and

$$y = 2 + 0.93R \text{ for } R \geq 7 \text{ ft} \tag{7.3.57}$$

where K_2 = a coefficient, which is a function of the channel type.

Flow Resistance

$$\overline{u} = K_3 \left(R^2 S \right)^m \tag{7.3.58}$$

$$\frac{\overline{u^2}}{gyS} = K_4 \left(\frac{\overline{uT}}{v} \right)^{0.37}$$

(7.3.59)

where m = an exponent and K_3 and K_4 = coefficients. Values for all coefficients and the exponent are tabulated in Table 7.9.

Table 7.9. Summary of exponent and coefficients for the Simons and Albertson (1970) regime equations.

Coefficient	Channel Category				
	1	2	3	4	5
K_1	3.5	2.6	2.2	1.75	1.7
K_2	0.52	0.44	0.37	0.23	0.34
K_3	13.9	16.0	---	17.9	16.0
K_4	0.33	0.54	0.87	---	---
m	0.33	0.33	---	0.29	0.29

EXAMPLE 7.8

Design a stable alluvial canal to convey 2,000 ft³/s (56.6 m³/s) using the regime approaches of Lacey and Simons and Albertson. The channel will have cohesive banks and a sandy bottom with the bed material having a mean size of 0.30 mm. The concentration of sediment in the water entering the canal will be approximately 150 mg/l (150 ppm).

Solution

Lacey: The top width is estimated from Eq. (7.3.50) or

$$T = 2.67\sqrt{Q} = 2.67\sqrt{2000} = 2.67(44.7) = 119\,\text{ft}$$

From Eq. (7.3.48) the Lacey silt factor is

$$f_L = 1.6\sqrt{d} = 1.6\sqrt{0.30} = 1.6(0.55) = 0.88$$

Then from Eq. (7.3.46)

$$\bar{u} = \frac{Q}{A} = 1.15\sqrt{f_L D} = 1.15\sqrt{f_L \frac{A}{T}}$$

and solving for A

$$A = \left(\frac{Q}{1.15\sqrt{\frac{f_L}{T}}}\right)^{2/3} = \left(\frac{2000}{1.15\sqrt{\frac{0.88}{119}}}\right)^{2/3} = 745\,\text{ft}^2$$

To estimate the depth of flow, assume that the channel will have a trapezoidal cross section with $z = 1.5$ and $b = 100$ ft (30.5 m) then

$$A = (b + zy)y$$

$$745 = (100 + 1.5y)y$$

and by trial and error $y = 6.75$ ft (2.1 m). The stable slope is estimated by Eq. (7.3.51)

$$S = \frac{f_L^{5/3}}{1830Q^{1/6}} = \frac{0.88^{5/3}}{1830(2000)^{1/6}} = 0.000124\,\text{ft / ft}$$

Simons and Albertson: From the description of the channel, it concluded that the channel should be considered Category 2; and therefore, from Table 7.9, $K_1 = 2.6$, $K_2 = 0.44$, $K_3 = 16.0$, and $m = 0.33$. With this assumption, the average width is estimated by Eq. (7.3.53) or

$$\bar{T} = 0.9K_1\sqrt{Q} = 0.9(2.6)\sqrt{2000} = 105\,\text{ft}$$

The top width of flow can now be estimated by solving Eq. (7.3.54) for *T* or

$$T = \frac{\bar{T} + 2}{0.92} = \frac{105 + 2}{0.92} = 116\,\text{ft}$$

The hydraulic radius is estimated by Eq. (7.3.55)

$$R = K_2 Q^{0.36} = 0.44(2000)^{0.36} = 6.78 \text{ ft}$$

and the depth of flow is given by Eq. (7.3.56)

$$y = 1.21R = 1.21(6.78) = 8.20 \text{ ft}$$

The average velocity of flow is given by

$$\bar{u} = \frac{Q}{Ty} = \frac{2000}{105(8.20)} = 2.32 \text{ ft}$$

An estimate of the stable slope is found by solving Eq. (7.3.58) for S or

$$S = \left(\frac{\bar{u}}{K_3}\right)^{1/m} \frac{1}{R^2} = \left(\frac{2.32}{16}\right)^{1/0.33} \frac{1}{6.78^2} = 0.000063 \text{ ft / ft}$$

It is observed that the widths, depths, and slopes obtained by both methods compare favorably. As Chang (1988) noted the differences in the stable slopes should not be taken too seriously because they are both very small.

Chang (1980 and 1985) developed an approach to designing stable sand bed channels that are approximately trapezoidal in shape, which is based on physical relationships for sediment transport, and minimum stream power (Chang, 1979 and 1982). The principle of minimum stream power states that for the velocity, width, and depth of a channel to be in dynamic equilibrium, the inflowing water and sediment are transported at the minimum slope that is compatible with the physical relations of flow resistance, sediment transport, and bank stability. The independent variables in this approach are the water flow rate, Q, the sediment inflow rate, Q_s, and the sediment characteristic represented by the median size, d. Chang (1988) asserted that the physical relationship that is required to estimate sediment transport should be a bed load relationship, since bed load transport is responsible for molding the channel shape. Chang (1988) used the DuBoys formula while noting that other models could be used with equal validity. The DuBoys formula states

$$q_{sb} = C_d \tau_0 (\tau_0 - \tau_c)$$ (7.3.60)

where q_{sb} = bed load transport per unit width of the bottom of the channel, C_d = sediment coefficient, and $(\tau_0 - \tau_c)$ = excess shear stress. Straub (1935), using data from small laboratory flumes, provided the following equations to estimate C_d and τ_c

$$C_d = \frac{0.17}{d^{3/4}}\left(m^3 / kg / s\right) = \frac{0.173}{d^{3/4}}\left(ft^6 / lb^2 / s\right)$$ (7.3.61)

and

$$\tau_c = 0.061 + 0.093d\left(kg / m^2\right) = 0.0125 + 0.019d\left(lb / ft^2\right)$$ (7.3.62)

where d is in mm. Note, the DuBoys τ_c is different in value from the Shields τ_c.

Then, for a set of independent variables $(Q, Q_s,$ and $d)$ and an assumed value of z, based on bank stability, the dependent variables $(T, y,$ and $S)$ are estimated in the following fashion:

1. A series of channel top widths, T, are assumed and for each top width the depth of flow, y, is calculated using the flow resistance. Note, Chang (1988) used the Lacey flow resistance relationship, and this approach will be demonstrated in the example that follows.

2. The channel slope is calculated from the bed load model that has been adapted. Using the DuBoys model

$$Q_s = q_{sb} b = \frac{0.173}{d^{3/4}} \gamma RS\left(\gamma RS - \tau_c\right) b$$ (7.3.63)

where b = bottom width of the trapezoidal channel and R = hydraulic radius. Rearranging Eq. (7.3.63) yields quadratic equation for S or

$$c_1 S^2 - c_2 S - c_3 = 0$$

where $c_1 = \gamma^2 R^2$, $c_2 = \gamma R \tau_c$, and $c_3 = (Q_s d^{3/4} / 0.173 b)$. Solution of the quadratic equation for S yields

$$S = \frac{c_2 + \sqrt{c_2^2 + 4 c_1 c_3}}{2 c_1} \tag{7.3.64}$$

3.   The slopes calculated for different widths are compared and the minimum is selected.

   This approach to stable channel design is best illustrated with and example.

EXAMPLE 7.9

Design a stable alluvial canal to convey 1,000 ft^3/s (56.6 m^3/s), using the Chang approach. The channel will have cohesive banks and a sandy bottom, with the bed material having a mean size of 0.30 mm. The concentration of sediment in the water entering the canal will be approximately 50 mg/l (50 ppm). Assume $z = 1.5$ will provide suitable bank stability and that the specific gravity of the bed material is 2.65.

Solution

The first step is to compute the sediment flow rate from the concentration or

$$Q_s = Q C_s 10^{-6} = 1000(50)(10^{-6}) = 0.05 \text{ ft}^3 / s$$

For these values, Lacey's absolute rugosity from Eqs. (7.3.48) and (7.3.49) is

$$f = 1.6 d^{0.5} = 1.6(0.30)^{0.5} = 0.88$$

$$N_R = 0.0225 f^{0.25} = 0.0225(0.88)^{0.25} = 0.022$$

The DuBoys critical shear stress is calculated by Eq. (7.3.62)

$$\tau_c = 0.0125 + 0.019 d = 0.0125 + 0.019(0.3) = 0.018 \text{ lb} / \text{ft}^2$$

Then, with reference to Table

Col. 1   Assumed value of the top width, T, of the flow

Col. 2   Assumed depth of flow, y, in the trapezoidal channel

Col. 3   Bottom width, b, of the trapezoidal channel where (Note, all sample calculations below are based on the first row of Table 7.10)

$$b = T - 2zy = T - 3y = 150 - 3(3) = 141 \text{ ft}$$

Table 7.10. Solution of Example 7.9

T	y	b	A	P	R	c_1	c_2	c_3	S	D	u	Q
(ft)	(ft)	(ft)	(ft²)	(ft)	(ft)	10^{-4}		10^3	10^4	(ft)	(ft/s)	(ft³/s)
(1)	(2)	(3)	(4)	(5)	(6)	(7)	(8)	(9)	(10)	(11)	(12)	(13)
150	3.00	141	437	152	2.88	3.23	3.24	0.831	2.188	2.91	2.02	885
150	3.32	140	481	152	3.16	3.90	3.59	0.837	1.997	3.21	2.08	1000
140	3.48	130	469	143	3.30	4.24	3.75	0.904	1.967	3.35	2.13	1000
130	3.67	119	457	132	3.45	4.64	3.92	0.985	1.938	3.51	2.19	1000
120	3.88	108	443	122	3.62	5.11	4.11	1.08	1.912	3.69	2.26	1000
110	4.13	97.6	429	113	3.81	5.66	4.33	1.20	1.888	3.90	2.33	1000
100	4.43	86.7	413	1103	4.03	6.31	4.57	1.35	1.869	4.13	2.42	1000
90	4.79	74.6	396	92.9	4.27	7.09	4.85	1.55	1.859	4.40	2.52	1000
87	4.91	72.3	391	90.0	4.35	7.35	4.93	1.62	1.858	4.49	2.56	1000
80	5.23	64.3	377	83.2	4.54	8.02	5.15	1.82	1.862	4.72	2.65	1000
70	5.81	52.6	356	73.5	4.84	9.12	5.50	2.23	1.893	5.08	2.81	1000
60	6.59	40.2	330	64.0	5.16	10.4	5.86	2.91	1.982	5.50	3.03	1000
50	7.74	26.8	297	54.7	5.43	11.5	6.17	4.37	2.238	5.94	3.37	1000

Col. 5   Wetted perimeter, P

$$P = b + 2y(1 + z^2)^{0.5} = 141 + 2(3)(1 + 1.5^2)^{0.5} = 152 \text{ ft}$$

Col. 6   Hydraulic radius, R

$$R = A/P = 437/152 = 2.88 \text{ ft}$$

Col. 7   Calculate the first Chang coefficient, c_1 with $\gamma = 62.4$ lb/ft³

$$c_1 = \gamma^2 R^2 = 62.4^2(2.88)^2 = 32,296 = 3.23 \; x10^4$$

Col. 8   Calculate second Chang coefficient, c_2

$$c_2 = \gamma R \tau_c = 62.4(2.88)(0.018) = 3.24$$

Col. 9   Calculate third Chang coefficient, c_3

$$c_3 = \frac{Q_s d^{3/4}}{0.173b} = \frac{0.05(0.3)^{3/4}}{0.173(141)} = 0.831 x 10^{-3}$$

Col. 10   Solve Eq. (7.3.64) for the slope, S

$$S = \frac{c_2 + \sqrt{c_2^2 + 4c_1 c_2}}{2c_1} = \frac{3.24 + \sqrt{3.24^2 + 4(32,300)(0.000831)}}{2(32,300)} = 0.000218$$

Col. 11   Solving for the hydraulic depth, D

$$D = A/T = 437/150 = 2.91 \text{ ft}$$

Col. 12   The velocity is calculated from Eq. (7.3.47)

$$\bar{u} = \frac{1.346}{N_R} D^{0.25} \sqrt{RS} = \frac{1.346}{0.022}(2.91)^{0.25} \sqrt{2.88(0.000218)} = 2.00 \text{ ft / s}$$

Col. 13   Estimated flow rate given the assumptions

$$Q = \bar{u}A = 2.00(437) = 874 \text{ ft}^3 / \text{s}$$

Since the estimated flow rate for the top width and depth of flow assumed does not match that which is required, a new depth of flow for the same top width is assumed, Row 2 of Table 7.10.

In Table 7.10, after the first row only, the correct depths of flow are shown, although for every row a trial and error approach was used to find the correct value of y. In Fig. 7.19, S is plotted as a function of T, and this figure demonstrates that the minimum slope occurs when $T = 87$ ft (26.5 m).

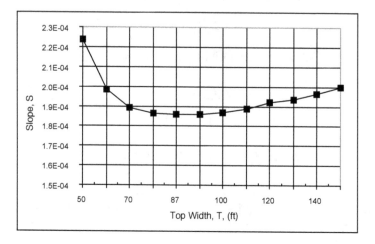

Figure 7.19. S as a function of T for Example 7.9.

7.3.3 Channel Transmission/Seepage Losses

Although a channel may require lining for many reasons (Section 7.2), one of the primary reasons for lining a channel, constructed in materials which would otherwise not require lining, is transmission/seepage losses. The loss of water due to seepage from an unlined channel depends on a variety of factors including, but not limited to, the dimensions of the channel, the gradation of the materials composing the perimeter, and the groundwater conditions. Although a number of attempts to theoretically estimate the seepage from a channel have been made (e.g. Bouwer, 1965), direct measurement of seepage loss is still preferred. There are basically three direct methods of measuring seepage loss:

1. In an existing channel, lined or unlined, selected reaches of the channel may be isolated by dikes to form closed basins of known volume. A mass balance will then suffice to estimate the seepage loss. Since this method usually requires that the channel be removed from service for an extended period of time, these tests are normally performed in the "off season". In such a case, care must be taken to ensure that the losses measured are typical of the season of interest.

2.   If a very careful record of the inflow and outflow to a reach of channel is kept, seepage loss may be estimated from this record. In this method, the canal is not removed from service, but the accuracy of the method is less than that of a previously described ponding method.

3.   In the case of a proposed channel, test reaches of the channel may be constructed and the ponding method used.

A fourth method, which is relatively simple but reliable, is based on historical measurements. In Table 7.11, a set of values originally

Table 7.11.   Seepage losses for canals not affected by the groundwater table (Davis and Sorenson, 1969.)

Perimeter Material	Seepage loss $\dfrac{\text{ft}^2 \text{ of water}}{\text{ft}^2 \text{ wetted perimeter}}$ for a 24-hr period
Impervious clay loam	0.25 - 0.35
Medium clay loam underlain with hardpan at a depth of not over 2-3 ft below bed	0.35 - 0.50
Ordinary clay loam, silt soil, or lava ash loam	0.50 - 0.75
Gravelly clay loam or sandy clay loam, cemented gravel, sand, and clay	0.75 - 1.00
Sandy loam	1.00 - 1.50
Loose sandy soils	1.50 - 1.75
Gravelly sandy soils	2.00 - 2.50
Porous gravelly soils	2.50 - 3.00
Very gravelly soils	3.00 - 6.00

developed by Etcheverry and Harding in 1933 are summarized (Davis and Sorenson, 1969). The values in this table are the result of many field measurements and have been found to be reasonably accurate;

however, it is recommended that these values should be used only as a design guide given a specific site.

In arid and semiarid environments, streams and channels are often ephemeral; that is, flow is occasional and follows precipitation events which are infrequent. When flow does occur in normally dry stream channels, the volume of flow is reduced by infiltration into the bed, banks, and floodplain. These losses, which can be significant, are termed transmission losses and reduce not only the volume of the hydrograph, but also the peak discharge. Lane (1983) developed an approach for estimating transmission losses when data regarding the inflow and outflow to a reach are not available. The assumptions on which this approach is based are as follows:

1. Water is lost in the channel; that is, the channel does not gain water.

2. Infiltration characteristics and other channel properties are uniform in the reach with distance and width.

3. While sediment concentration, water temperature, and antecedent flow affect transmission losses, the approach represents average conditions. For example, if antecedent flow had occurred in the channel, the transmission loss would likely be less than that estimated by this approach.

4. The channel reach is short enough that an average width and duration represent the width and duration of flow for the entire channel reach.

5. Once a threshold volume has been satisfied, outflow volumes are linear with inflow volumes.

6. Once an average loss rate is subtracted and the inflow volume exceeds the threshold volume, peak rates of outflow are linear with peak inflows.

The primary limitations of the approach are

1. Hydrographs are not specifically routed through the reach; estimates are made for volume and peak discharge.

2. Peak flow equations do not consider hydrograph attenuation due to storage or overbank flow and neither is the effect of hydrograph steepening considered.

3. The approach represents average conditions and overall trends.

4. Estimates of effective channel bed hydraulic conductivity are empirically based and represent average rates.

5. Peak rates of outflow are decreased by the average loss rate for the duration of flow.

6. Procedures for floodplain flow are based on a weighted average for the effective hydraulic conductivity of the channel and floodplain.

On the basis of regression analysis of data from the Southwestern United States, Lane (1983) asserted that a prediction equation for the volumetric outflow from a reach is

$$
V_{OUT}(x,w) = \begin{cases} 0 \text{ for } b(x,w) \big] V_{IN} + \dfrac{V_{LAT}}{kw}\big[1 - b(x,w)\big] \le -a(x,w) \\[2mm] a(x,w) + b(x,w) V_{IN} + \dfrac{V_{LAT}}{kw}\big[1 - b(x,w)\big] \end{cases} \tag{7.3.65}
$$

The corresponding equation for peak discharge prediction is

$$
Q_{POUT} = \begin{cases} 0 \text{ for } V_{OUT}(x,w) = 0 \\[2mm] \dfrac{12.1}{D_R}\big\{a(x,w) - \big[1 - b(x,w)\big] V_{IN}\big\} + b(x,w) Q_{PIN} + \\[2mm] \dfrac{5280 Q_{LAT}}{kw}\big[1 - b(x,w)\big] \end{cases} + (7.3.66)
$$

where

$a(x,w)$ and $b(x,w)$	=	parameters estimated for each reach of the channel,
x	=	channel reach length (mi),
w	=	channel width (ft),
$V_{OUT}(x,w)$	=	volumetric outflow at downstream end of the channel reach (acre-ft),
k	=	decay factor (ft-mi)$^{-1}$,
V_{IN}	=	volumetric inflow at the upstream end of the channel reach (acre-ft),

$$V_{LAT} \qquad = \text{uniformly distributed volumetric lateral inflow within the channel reach (acre-ft/mi),}$$

$$Q_{POUT} \qquad = \text{peak flow rate at the downstream end of the channel reach (ft}^3\text{/s),}$$

$$Q_{PIN} \qquad = \text{peak flow rate at the upstream end of the channel reach (ft}^3\text{/s), and}$$

$$Q_{LAT} \qquad = \text{peak rate of lateral inflow (ft}^3\text{/s/ft).}$$

Lane (1983) using data from a number of channels in the Southwest and elsewhere provided the following equations to estimate parameter values in the absence of site specific data

$$b(x,w) = \exp(-kxw) \qquad (7.3.67)$$

$$a(x,w) = \frac{a\left[1 - \exp(-kxw)\right]}{1 - b} = \frac{a\left[1 - b(x,w)\right]}{1 - b} \qquad (7.3.68)$$

where

$$a = -0.00465 K \overline{D_R} \qquad (7.3.69)$$

$$k = -1.09 \ln\left(1 - 0.00545 \frac{K\overline{D_R}}{\overline{V}}\right) \qquad (7.3.70)$$

$$b = \exp(-k) \qquad (7.3.71)$$

K = effective hydraulic conductivity (in/hr), $\overline{D_R}$ = mean flow duration (hr), and $\overline{V}$ = mean volume of flow (acre-ft). Values of K are summarized in Table 7.12 for a number of types of bed material. Of the parameters/variables involved in this approach, $\overline{D_R}$, $\overline{V}$, Q_{PIN}, and Q_{LAT} can be very difficult to estimate if there is not a record of flow in the reach. Some investigators have developed empirical relationships between $\overline{D_R}$ and $\overline{V}$ and the area of the watershed area. For example, Murphey *et al* (1977) using data from watersheds in Arizona developed the following relationships

$$\overline{D_R} = C_1 A_w^{C_2}$$

and

$$\overline{V} = C_3 A_w^{C_4}$$

where A_w = watershed area (mi^2) and the C_i = coefficients and exponents.

TABLE 7.12. Relationships between channel bed material characteristics and parameters for a unit channel, average antecedent conditions (Lane, 1983).

Bed material group	Bed material characteristics	Effective hydraulic conductivity[a] K (in/hr)	Unit channel parameters	
			Intercept a (acre-ft)	Decay factor k (ft-mi)$^{-1}$
1 Very high loss rate	Very clean gravel and large sand $d_{50} > 2$ mm	>5	<- 0.023	>0.030
2 High loss rate	Clean sand and gravel, field conditions $d_{50} > 2$ mm	2.0 - 5.0	-0.0093 to -0.023	0.0120 - 0.030
3 Moderately high loss rate	Sand and gravel mixture with low silt-clay content	1.0 - 3.0	-0.0047 to -0.014	0.0060 - 0.018
4 Moderate loss rate	Sand and gravel mixture with high silt-clay content	0.25 - 1.0	-0.0012 to -0.0047	0.0015 - 0.0060
5 Insignificant to low loss rate	Consolidated bed material; high silt-clay content	0.001 - 0.10	-5 x 10^{-6} to -5 x 10^{-4}	6 x 10^{-6} to 6 x 10^{-4} to

[a]Values of effective hydraulic conductivity reflect the flashy, sediment laden character of episodic flows in ephemeral streams; and thus, do not represent steady flows or clear water infiltration.

EXAMPLE 7.10

Flow occurs over 3.5 hour period in an alluvial channel 150 ft (45.7 m) wide. A stream gage at the upstream end of a 1,000 ft (305 m) reach shows that the total flow during this event was 9.6 acre-ft (11,840 m³). If the effective hydraulic conductivity of the channel bed is 2 in/hr (50.8 mm/hr) and there is no lateral inflow, estimate the channel transmission losses in this channel reach. If the peak flow at the upstream end of the reach is 66.4 ft³/s (1.88 m³/s), estimate the peak flow at the downstream end of the reach.

Solution

The first step is to estimate the threshold volume of flow at the upstream end of the reach; that is, the inflow volume of flow required for there to be flow out of the reach. The threshold volume of flow occurs when (Eq. 7.3.65)

$$V_{OUT}(x,w) = 0 = a(x,w) + b(x,w)V_{IN}$$

or where $V_{IN} = V_T$ or the threshold volume

$$b(x,w)V_T = -a(x,w)$$

Therefore, the threshold volume is given by

$$V_T = -\frac{a(x,w)}{b(x,w)}$$

Then

$$a = -0.00465K\overline{D_R} = -0.00465(2)(3.5) = -0.0326$$

$$k = -1.09\ln\left(1 - \frac{0.00545K\overline{D_R}}{\overline{V}}\right) =$$

$$-1.09\ln\left[1 - \frac{0.00545(2)(3.5)}{9.6}\right] = 0.00434$$

$$b = \exp(-k) = \exp(-0.00434) = 0.996$$

and (where $x = 0.189$ mi)

$$b(x,w) = \exp(-kxw) = \exp[-0.00434(0.189)(150)] = 0.884$$

$$a(x,w) = \frac{a\left[1 - b(x,w)\right]}{1 - b} = \frac{-0.0326\left(1 - 0.884\right)}{1 - 0.996} = -0.945$$

The threshold volume is

$$V_T = -\frac{a(x,w)}{b(x,w)} = -\frac{-0.945}{0.884} = 1.07\,ac - ft\left(1,320\,m^3\right)$$

The second step is to estimate the actual transmission loss in the reach (Eq. 7.3.65)

$$V_{OUT}(x,w) = a(x,w) + b(x,w)V_{IN} =$$
$$- 0.945 + 0.884(9.6) = 7.54ac - ft\left(9,189\,m^3\right)$$

Therefore, the transmission loss, V_L, is

$$V_L = V_{IN} - V_{OUT} = 9.6 - 7.5 = 2.1 \text{ ac-ft } (2,590 \text{ m}^3)$$

Third, the peak flow at the downstream end of the reach is given by

$$Q_{POUT} = \frac{12.1}{D_R}\left\{a(x,w) - \left[1 - b(x,w)\right]V_{IN}\right\}$$

$$+ b(x,w)Q_{PIN} = \frac{12.1}{3.5}\left[-0.945 - \left(1 - 0.884\right)9.8\right] +$$
$$0.884(66.4) = 51.5\,ft^3 / s(1.46m^3 / s)$$

Finally, it is pertinent to observe that transmission losses have been shown to be significant in ephemeral channels (French and Curtis, 1999a and 1999b); and the Lane (1983) approach has been shown to provide reasonable results (French et al., 1999).

7.4.   DESIGN OF CHANNELS LINED WITH GRASS

The grass-lined channel is a common method of transmitting intermittent irrigation flows and controlling erosion in agricultural areas. The grass serves to stabilize the body of the channel, consolidate the soil mass of the channel perimeter, and check the movement of soil particles along the channel bottom. However, grass-lined channels cannot, in general, withstand prolonged inundation and wetness, and their design presents a number of problems which have not been encountered in the preceding sections of this chapter, e.g., the seasonal variation in the resistance coefficient due to the condition of the channel liner.

The Manning roughness coefficient, which is commonly termed the *retardance coefficient* in the literature relating to the grass-lined channel, has been found to be a function of the average velocity, the hydraulic radius, and the vegetal type (Coyle, 1975). n can be represented by a series of empirical curves of n versus $\bar{u}R$, for various degrees of retardance (Fig. 7.20). This figure must be used in conjunction with Table 7.13, which provides an estimate of the degree of retardance which various types of grass provide.

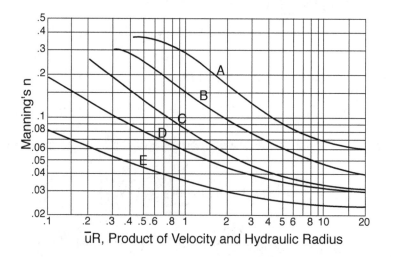

Figure 7.20.   **Manning's** n **as a function of velocity, hydraulic radius, and vegetal retardance (Coyle, 1975).**

Table 7.13. Classification of degree of retardance for various types of grasses (Coyle, 1975).

Retardance	Cover	Condition
A	Reed canary grass	Excellent stand, tall (average height 36 in or 91.4 cm)
	Yellow bluestem *Ischaemum*	Excellent stand, tall (average height 36 in or 91.4 cm)
B	Smooth bromegrass	Good stand, mowed (average height 12 to 15 in or 30.5 to 38.1 cm)
	Bermuda grass	Good stand, tall (average height 12 in or 30.5 cm)
	Native grass mixture (little bluestem, blue gamma, and other long and short midwest grasses)	Good stand, unmowed
	Tall fescue	Good stand, unmowed (average height 18 in or 45.7 cm)
	Lespedeza sericea	Good stand, not woody, tall (average height 19 in or 48.3 cm)
	Grass-legume mixture - timothy, smooth	Good stand, uncut (average height 20 in or 50.8 cm)
	Tall fescue, with bird's foot trefoil or lodino	Good stand, uncut (average height 18 in or 45.7 cm)
	Blue grama	Good stand, uncut (average height 13 in or 33.0 cm)
C	Bahia	Good stand, uncut (6 to 8 in or 15.2 to 20.3 cm)
	Bermuda grass	Good stand, mowed (average height 6 in or 15.2 cm)
	Redtop	Good stand, headed (15 to 20 in or 38.1 to 50.8 cm)
	Grass-legume mixture - summer (orchard grass, redtop, Italian ryegrass, and common *Lespedeza)*	Good stand, uncut (6 to 8 in or 15.2 to 20.3 cm)
	Centipede grass	Very dense cover (average 6 in or 15.2 cm)
	Kentucky bluegrass	Good stand, headed (6 to 12 in or 15.2 to 30.5 cm)
D	Bermuda grass	Good stand, cut to 2.5 in or 6.4 cm height
	Red fescue	Good stand, headed (12 to 18 in or 30.5 to 45.7 cm)
	Buffalo grass	Good stand (3 to 6 in or 7.6 to 15.2 cm)
	Grass-legume mixture - summer (orchard grass, redtop, Italian ryegrass, and common *Lespedeza)*	Good stand, uncut (4 to 5 in or 10.2 to 12.7 cm)
	Lespedeza sericea	After cutting to 2 in (5.1 cm) height; very good stand before cutting
E	Bermuda grass	Good stand, cut to 1.5 in (3.8 cm) height
	Bermuda grass	Burned stubble

The selection of a grass for a specific application depends primarily on the climate and soil conditions which prevail. From the viewpoint of hydraulic engineering, the primary consideration must be channel stability. Table 7.14 summarizes the U.S. Soil Conservation Service recommendations regarding permissible velocities for various types of vegetal covers, channel slopes, and soil types. In addition, the following guidelines should be noted:

1. Where only sparse vegetal cover can be established or maintained, velocities should not exceed 3 ft/s (0.91 m/s).

2. Where the vegetation must be established by seeding, velocities in the range of 3 to 4 ft/s (0.91 to 1.2 m/s) are permitted.

3. Where dense sod can be developed quickly or where the normal flow in the channel can be diverted until a vegetal cover is established, velocities of 4 to 5 ft/s (1.2 to 1.5 m/s) can be allowed.

4. On well-established sod of good quality, velocities in the range of 5 to 6 ft/s (1.5 to 1.8 m/s) are permitted.

5. Under very special conditions, velocities as high as 6 to 7 ft/s (1.8 to 2.1 m/s) are permitted.

The design of grass-lined channels must, in most cases, proceed in two stages. In the first stage, a low degree of retardance, which corresponds to either dormant seasons or the period during which the vegetation is becoming established, is assumed. The second stage provides the pertinent dimensions under the assumption of a high degree of retardance. A recommended design procedure is summarized in Table 7.15, where it is assumed that the longitudinal slope of the channel, the channel shape, grass type, and the design flow have been established, previous to the initiation of the design process. The channel shapes commonly used in this application are trapezoidal, triangular, and parabolic, with the latter two shapes being the most popular. Coyle (1975) has developed extensive design tables for the parabolic shape. The designer must, in designing grass-lined channels, be conscious that a primary consideration may be the ability of farm machinery to easily cross the canal during periods of no flow. This consideration may require that the side slopes of the channel be designed for this purpose rather than hydraulic efficiency or stability.

Table 7.14. Permissible velocities for channels lined with grass (Coyle, 1975).

Cover	Slope Range[2] %	Permissible Velocity[1]	
		Erosion-resistant soils ft/s (m/s)	Easily eroded soils ft/s (m/s)
Bermuda grass	0-5	8 (2.44)	6 (1.83)
	5-10	7 (2.13)	5 (1.52)
	Over 10	6 (1.83)	4 (1.22)
Bahia	----	----	----
Buffalo grass	----	----	----
Kentucky bluegrass	0-5	7 (2.13)	5 (1.52)
Smooth brome	5-10	6 (1.83)	4 (1.22)
Blue gramma	Over 10	5 (1.52)	3 (0.91)
Tall fescue	----		----
Grass mixtures	0-5[2]	5 (1.52)	4 (1.22)
Reed canary grass	5-10	4 (1.22)	3 (0.91)
Lespedeza sericea	----	----	----
Weeping lovegrass	----	----	----
Yellow bluestem	----	----	----
Redtop	0-5[3]	3.5 (1.07)	2.5 (0.76)
Alfalfa	----	----	----
Red fescue	----	----	----
Common *Lespedeza*[4,5]	----	----	----
Sudan grass[4]	0-5	3.5 (1.07)	2.5 (0.76)

[1]Use velocities exceeding 5 ft/s (1.52 m/s) only where good covers and proper maintenance can be obtained.

[2]Do not use on slopes steeper than 10% except for vegetated side slopes in combination with a stone, concrete, or highly resistant vegetative center section.

[3]Do not use on slopes steeper than 10% except for vegetated side slopes in combination with a stone, concrete, or highly resistant vegetative center section.

[4]Annuals - use on mild slopes or as temporary protection until permanent covers are established.

[5]Use of slopes steeper than 5% is not recommended.

Table 7.15. A design procedure for channels lined with grass.

Step	Process
	Stage 1
1	Assume a value of n and determine $\bar{u}R$ which corresponds to this assumption (Fig. 7.20)
2	Select the permissible velocity from Table 7.14, which corresponds to the specified channel slope, lining material, soil then compute a value of R by using the results of Step 1
3	Using the Manning equation and the assumed value of n, compute $$\bar{u}R = \frac{\varphi R^{5/3}\sqrt{S}}{n}$$ where $\phi = 1.49$, if English units are used, and $\phi = 1$ for SI units. The value of R found in Step 2 is used in the right-hand side of this equation.
4	Repeat Steps 1 through 3 until the values $\bar{u}R$, determined in Steps 1 and 3, agree
5	Determine A from the design flow and the permissible velocity (Step 2).
6	Determine the channel proportions for the calculated value of R and A.
	Stage II
1	Assume a depth of flow for the channel determined in Stage 1 and compute A and R.
2	Compute the average velocity $$\bar{u} = Q/A$$ for the A found in Step 1.
3	Compute $\bar{u}R$, using the results of Steps 1 and 2.
4	Using the results of Step 3, determine n from Fig. 7.20.
5	Use n from Step 4, R from Step 1, and the Manning equation to compute $\bar{u}$.
6	Compare the average velocities computed in Steps 2 and 5 and repeat Steps 1 through 5 until they are approximately equal.
7	Add the appropriate freeboard and check the Froude number.
8	Summarize the design in a dimensioned sketch.

EXAMPLE 7.11

Design a triangular channel which will be lined with a mixture of grasses that includes the following types: orchard grass, redtop, Italian ryegrass, and common *Lespedeza*. This channel will be built on a slope of 0.025 through a soil which can be characterized as easily eroded. An intermittent flow of 20 ft³/s (0.57 m³/s) is anticipated.

Solution

For a grass mixture composed of orchard grass, redtop, Italian ryegrass and common *Lespedeza*, minimum retardance occurs during the fall, winter, and spring (curve D, Fig. 7.20). For this type of lining, a specified slope, and the soil, Table 7.14 provides an estimated maximum permissible velocity of 4 ft/s (1.2 m/s). A trial-and-error procedure must then be used to proportion the channel for stage 1 of the design.

Trial No.	Assumed n	$\bar{u}R$ from Fig. 7.20	R	$\bar{u}R = \dfrac{1.49R^{5/3}}{n}\sqrt{S}$
1	0.04	3.0	0.75	3.65
2	0.045	2.0	0.50	1.65
3	0.043	2.5	0.62	2.50

Therefore, $n = 0.43$ and $R = 0.62$ ft (0.19 m). The required flow area is computed by

$$A = \frac{Q}{\bar{u}} = \frac{20}{4} = 5\,\text{ft}^2 \left(0.46\,\text{m}^2\right)$$

Given a triangular channel, these conditions provide for an explicit solution; i.e.

$$A = zy^2 = 5\,\text{ft}^2$$

$$R = \frac{A}{2y\sqrt{1+z^2}} = \frac{5}{2y\sqrt{1+z^2}} = 0.62\,\text{ft}$$

Solving these equations yields

$$z = 2.8$$

and

$$y_N = 1.3 \text{ ft } (0.40 \text{ m})$$

At this point, the stage 1 design must be reviewed from the viewpoint of maximum retardance (curve C, Fig. 7.20).

Trial No.	Assumed y ft	A ft²	R ft	$\bar{u} = \dfrac{Q}{A}$	$\bar{u}R$	n from Fig. 7.20	$\bar{u}R = \dfrac{1.49R^{5/3}}{n}\sqrt{S}$
1	1.4	5.5	0.66	3.6	2.4	0.051	3.5
2	1.5	6.3	0.71	3.2	2.3	0.054	3.5
3	1.45	5.9	0.68	3.4	2.3	0.054	3.4

Thus, the stage II depth of flow is 1.45 ft (0.44 m), and the freeboard may be estimated from Eq. (7.1.1) to be 1.5 ft (0.46 m). If this is an on-farm irrigation ditch which is used intermittently, a smaller freeboard, for example, 1.0 ft (0.30 m), would be acceptable. The results are summarized in Fig. 7.21.

In many cases, the center of a grass-lined channel is lined with either gravel or rock to enhance the stability of the channel and/or to provide adequate drainage. Although Coyle (1975) presents a nomograph for

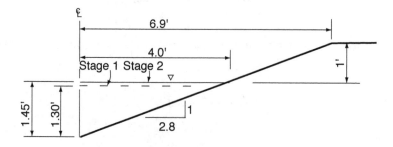

Figure 7.21. Summary of results for Example 7.6.

determining the size of rock, which should be used for a specified flow and channel slope, the threshold of movement concepts discussed in previous sections of this chapter can be used.

EXAMPLE 7.12

A trapezoidal irrigation ditch 8 ft (2.4 m) wide with grass-covered sides must carry prolonged flows of 100 ft^3/s (2.8 m^3/s) on a slope of 0.01. If the side slopes are 2:1 and Manning's n is estimated to be 0.03, determine the size of gravel which should be used to line the center of this channel.

Solution
Determine the normal depth of flow.

$$Q = \frac{1.49}{n} AR^{2/3} \sqrt{S}$$

$$AR^{2/3} = \frac{nQ}{1.49\sqrt{S}} = \frac{0.03(100)}{1.49\sqrt{0.01}} = 20.1$$

$$y_N = 1.6 \text{ ft } (0.49 \text{ m})$$

Then

$$A = (b + zy)y = [8 + 2(1.6)]1.6 = 17.9 \text{ ft}^2 \text{ (1.7 m}^2)$$

$$P = b + 2y(1 + z^2)^{0.5} = 8 + 2(1.6)(5)^{0.5} = 15.2 \text{ ft (4.6 m)}$$

$$R = 1.18 \text{ ft (0.36 m)}$$

From Eq. (7.3.20)

$$d \cong 11 \text{ RS} = 11(1.18)(0.01) = 0.13 \text{ ft} = 1.6 \text{ in (0.04 m)}$$

This is the 25 percent diameter, and size 1.6 in > 0.25 in, the use of Eq. (7.3.20) is valid. Table 4.2 also demonstrates that the assumed value of n is reasonable.

At this point it is relevant to note that there have been a number of studies regarding flow in grass-lined channels whose results should be mentioned. Kouwen and Unny (1973) used flexible plastic strips in a laboratory environment to study the effects of the waving motion and bending of grass in channels on the flow characteristics. The pertinent results of this study were these:

1. Grass-lined channels are apparently characterized by two values of the local friction factor: one for the erect and waving regimes and another for the prone regimes.

2. The friction factor and, hence, Manning's n were found to be functions of the relative roughness for the erect and waving regimes but were a function of the Reynolds number or $\bar{u}R$ for the prone regime.

Phelps (1970) examined the effects of vegetative density and the depth of flow on the characteristics of the flow. The conclusions of this study were as follows:

1. The dissimilarity between the boundaries of flows over vegetation, which is only partially submerged, is significant and must be considered.

2. Recall that for smooth surfaces an actual Reynolds number of approximately 2,000 marks the limit of laminar flow. In the case of flows, over turf and critical Reynolds number, varies as a function of the depth of flow.

Given the complex nature of vegetated surfaces, it is not surprising that the design of such channels is based on empiricism. It is clear that additional research in this topic area is required.

As described in the previous sections, the traditional design procedures for grass-lined open channels has involved a trial and error approach, due to the highly nonlinear functional relationship between the Manning roughness coefficient and the hydraulic radius. Akan and Hager (2001) presented an approach that eliminates the trial and error approach for the quick design (or analysis) of grass-lined channels.

For grass-lined channels, Chen and Cotton (1988) presented an equation relating the Manning resistance coefficient, n, which can be written in a dimensionally homogeneous form as

$$n = \frac{(KR)^{1/6}}{C + 19.97 \log\left(K^{1.4} R^{1.4} S^{0.4}\right)} \tag{7.4.1}$$

where $K = 3.28$ m^{-1} (1.0 ft^{-1}), $R =$ hydraulic radius, $S =$ longitudinal channel slope, and $C =$ dimensionless retardance parameter, depending on the retardance class of the grass (Table 7.13). With this definition of the Manning resistance coefficient, the Manning equation for normal flow becomes

$$Q = \frac{\phi\left[C + 19.97\log\left(K^{1/4}R^{1/4}S^{0.4}\right)\right]}{(KR)^{1/6}}AR^{2/3}S^{1/2} \qquad (7.4.2)$$

where $Q =$ discharge, $\phi = 1.0$ m$^{1/3}$/s (1.49 ft$^{1/3}$/s), and $A =$ flow area.

The Manning equation can be re-written in dimensionless form to facilitate the calculation of the channel width when a depth of flow is specified. For a trapezoidal section with a bottom width b, with side slopes z

$$Q_B = \left(\beta + 19.97\log R_B^{1.4}\right)A_B R_B^{1/2} \qquad (7.4.3)$$

where

$$Q_B = \frac{QK^{1/6}}{\phi S^{1/2}y^{5/2}z} \qquad (7.4.4)$$

$$\beta = C + 19.97\log\left(K^{1.4}S^{0.4}y^{1.4}\right) \qquad (7.4.5)$$

$$A_B = \frac{A}{zy^2} = 1 + B \qquad (7.4.6)$$

$$R_B = \frac{R}{y} = \frac{1+B}{B + 2\sqrt{\dfrac{1+z^2}{z^2}}} \qquad (7.4.7)$$

and

$$B = \frac{b}{zy} \qquad (7.4.8)$$

The dimensionless discharge, Q_B, is a function of β, B, and z. Figure 7.22 is a graphical representation of Eq. (7.4.3) with $[(1 + z^2)/z^2]^{1/2}$ set equal to 1.118, which corresponds to $z = 2$. This results in the overestimation of the channel bottom width by approximately 4% for z = 3.0 and 5% for z = 4.0, which is conservative for design purposes.

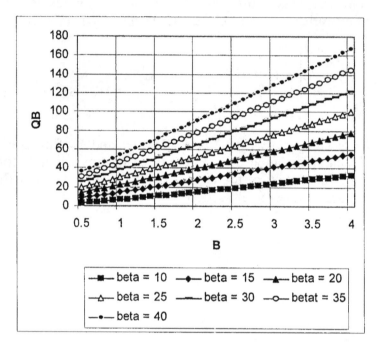

Figure 7.22. Q_B as a function of the channel bottom width parameter, B, and the parameter beta, β.

Figure 7.23 provides a design aid for the calculation of the flow depth for a channel section. The dimensionless parameters used to develop Fig. 7.23 were

$$Q_Y = \left[\alpha + 19.97 \log \left(\frac{zR}{b} \right)^{1/4} \right] \frac{zA}{b^2} \left(\frac{zR}{b} \right)^{1/2} \qquad (7.4.9)$$

where

$$Q_Y = \frac{QK^{1/6} z^{3/2}}{\phi S^{1/2} b^{5/2}} \qquad (7.4.10)$$

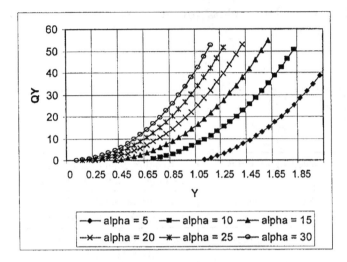

Figure 7.23.   The variable Q_Y as a function of the parameters Y and alpha, α.

$$\alpha = C + 19.97 \log \left(\frac{K^{1.4} b^{1.4} S^{0.4}}{z^{1.4}} \right) \qquad (7.4.11)$$

$$Y = \frac{zy}{b} \qquad (7.4.12)$$

$$\frac{zA}{b^2} = Y\left(1 + Y\right) \qquad (7.4.13)$$

$$\frac{zR}{b^2} = \frac{Y + Y^2}{1 + 2Y\sqrt{\frac{1 + z^2}{z^2}}} \qquad (7.4.14)$$

Figure 7.23 provides a precise solution for $z = 2$, and it overestimates the depth of flow by approximately 2% for $z = 3$ and 3% for $z = 4$. This approximation is conservative and generally acceptable for design purposes.

In Table 7.16, the allowable unit tractive forces for various types of vegetal covers are summarized, and it is pertinent to note that the same

Table 7.16. Retardance class for grass covers (Chen and Cotton, 1988).

Retardance Class (1)	Cover (2)	Condition (3)	C (4)	Allowable τ_{max} N/m² (5)	lb/ft² (6)
A	Weeping lovegrass	Excellent stand, tall [average 76 cm (30 in)]	15.8	177.2	3.70
A	Yellow bluestem ischaemum	Excellent stand, tall [average 91 cm (36 in)]	15.8	177.2	3.70
B	Bermuda grass	Good stand, tall [average 30 cm (12 in)]	23.0	100.5	2.20
B	*Lespedeza sericea*	Good stand, not woody, tall [average 48 cm (19 in)]	23.0	100.5	2.20
B	Weeping lovegrass	Good stand, unmowed [average 33 cm (13 in)]	23.0	100.5	2.20
C	Crabgrass	Fair stand, uncut [25-120 cm (10-48 in)]	30.2	47.9	1.00
C	Bermuda grass	Good stand, mowed [average 15 cm (6 in)]	30.2	47.9	1.00
C	Common *Lespedeza*	Good stand, uncut [average 28 cm (11 in)]	30.2	47.9	1.00
D	Bermuda grass	Good stand, cut to 6 cm (2.5 in) height	34.6	28.7	0.60
D	Common *Lespedeza*	Excellent stand, uncut [average 11 cm (4.5 in)]	34.6	28.7	0.60
D	*Lespedeza sericea*	After cutting to 5 cm (2 in) height, very good stand before cutting	34.6	28.7	0.60
E	Bermuda grass	Good stand, cut to 4 cm (1.5 in) height	37.7	16.7	0.35

type of grass can belong to various retardance classes, depending on the maturity of the grass. In Table 7.16, Classes A and E represent the highest and lowest retardances, respectively. In a grassed channel, the channel bottom is the critical part of the section and the maximum shear stress on the bottom is estimated as

$$\tau_{max} = \gamma \, yS \tag{7.4.15}$$

As was the case for the trial and error approaches described previously, the design of a channel lined with grass is completed in two stages. In the first stage, the channel is sized for stability using the condition of the lowest retardance of the lining, and in the second stage the design is modified for conveyance under the highest retardance condition of the grass.

EXAMPLE 7.13 (Akan and Hager, 2001)
A trapezoidal channel lined with Bermuda grass must be designed to convey 0.70 m³/s. The channel will have a slope of 0.004 m/m and side slopes of $z = 3$.

Solution
   Stage 1:
   The lowest retardance class for Bermuda grass is E (Table 7.16), for which $C = 37.7$ and $\tau_{max} = 16.7$ N/m². The permissible depth of flow can be estimated from Eq. (7.4.15) or

$$y = \frac{\tau_{max}}{\gamma S} = \frac{16.7}{9810(0.004)} = 0.43 \, m$$

Note, this depth represents the maximum allowable depth of flow under low retardance conditions, and any smaller depth of flow could be safely used. For simplicity, let us use $y = 0.40$ m and then from Eqs. (7.4.4) and (7.4.5)

$$Q_B = \frac{QK^{1/6}}{\phi S^{1/2} y^{5/2} z} = \frac{0.70(3.28)^{1/6}}{1(0.004)^{1/2}(0.4)^{5/2}(3)} = 44.4$$

and

$$\beta = C + 19.97 \log\left(K^{1.4} S^{0.4} y^{1.4}\right) =$$
$$37.7 + 19.97 \log\left[3.28^{1.4}(0.004)^{0.4}(0.4)^{1.4}\right]$$

$$\beta = 21.8$$

Then from Figure 7.22 for $Q_B = 44.4$ and $\beta = 21.8$, $B = 2$. Then, from Eq. (7.4.8)

$$b = Bzy = 2(3)(0.4) = 2.4\,m$$

Note, and width ≥ 2.4 m can be used and $b = 2.5$ will be used in the computations that follow.

Stage 2:
The highest retardance class for Bermuda grass is E (Table 7.13), for which $C = 23$. From Eqs. (7.4.10) and (7.4.11)

$$Q_Y = \frac{QK^{1/6}z^{3/2}}{\phi S^{1/2}b^{5/2}} = \frac{0.7(3.28)^{1/6}(3)^{3/2}}{1(0.004)^{1/2}(2.50)^{5/2}} = 7.1$$

$$\alpha = C + 19.97\log\left(\frac{K^{1.4}b^{1.4}S^{0.4}}{z^{1.4}}\right) = 23.0 + 19.97\log\left[\frac{3.28^{1.4}(2.50)^{1.4}(0.004)^{0.4}}{3^{1.4}}\right]$$

$$\alpha = 16.1$$

Then from Figure 7.23 for $Q_Y = 7.1$ and $\alpha = 1.61$, $Y = 0.77$ and from Eq. (7.4.12)

$$y = \frac{bY}{z} = \frac{2.5(0.77)}{3.0} = 0.64\,m$$

This depth of flow determines the depth of the channel when freeboard is added.

7.5.   BIBLIOGRAPHY

Akan, A.O., "Tractive Force Design Aid," *Canadian Journal of Civil Engineering*, Vol. 28, 2001, pp. 865-867.

Akan, A.O. and Hager, W.H., "Design Aid for Grass-Lined Channels," ASCE, *Journal of Hydraulic Engineering*, vol. 127, no. 3, March 2001, pp. 236-237.

Anonymous, "Linings for Irrigation Canals," U.S. Bureau of Reclamation, Washington, 1963.

Blench, T., "Regime Theory Design of Canals with Sand Beds," ASCE, *Journal of the Irrigation and Drainage Division*, Vol. 96, No. IR2, 1970, pp. 205-213.

Blench, T., "Regime Theory for Self-Formed Sediment-Bearing Channels," *Transactions of the American Society of Civil Engineers*, Vol. 117, 1952, pp. 383-408.

Bouwer, H., "Theoretical Aspects of Seepage from Open Channels," *Proceedings of the American Society of Civil Engineers, Journal of the Hydraulics Division*, vol. 91, no. HY3, May 1965, pp. 37-59.

Brownlie, W.R., "Prediction of Flow Depth and Sediment Discharge in Open Channels," Report No. KH-R-43A, Keck Laboratory of Hydraulics and Water Resources, California Institute of Technology, Pasadena, CA, 1981.

Chang, H.H., *Fluvial Processes in River Engineering*, John Wiley & Sons, New York, 1988.

Chang, H.H., "Design of Stable Alluvial Canals in a System," ASCE, *Journal of Irrigation and Drainage Engineering*, Vol. 111, No. 1, 1985, pp. 36-43.

Chang, H.H., "Mathematical Model for Erodible Channels," ASCE, *Journal of the Hydraulics Division*, Vol. 108, No. HY5, 1982, pp. 678-689.

Chang, H.H., "Stable Alluvial Canal Design," ASCE, *Journal of the Hydraulics Division*, Vol. 106, No. HY5, 1980, pp. 873-891.

Chang, H.H., "Minimum Stream Power and River Channel Patterns," *Journal of Hydrology*, Vol. 41, 1979, pp. 303-327.

Chaudhry, M.H., *Open Channel Flow*, Prentice-Hall, Englewood Cliffs, N.J., 1993.

Chen, Y.H. and Cotton, G.K, "Design of Roadsite Channels with Flexible Linings," *Hydraulic Engineering Circular No. 15*, Federal Highway Administration, Washington, D.C., 1988.

Chow, V.T., *Open Channel Hydraulics*, McGraw-Hill Book Company, New York, 1959.

Coyle, J.J., "Grassed Waterways and Outlets," *Engineering Field Manual*, U.S. Soil Conservation Service, Washington, April 1975, pp. 7-1-7-43.

Davis, C., and Sorenson, K.E., "Canals and Conduits," *Handbook of Applied Hydraulics*, 3d Ed., McGraw-Hill Book Company, New York, 1969, pp. 7-1-7-34.

duBoys, P., "The Rhone and Streams with Moveable Beds," *Annales des ponts et chaussees*, section 5, vol. 18, 1879, pp. 141-195.

FHWA, "Design of Roadside Channels with Flexible Linings," Hydraulic Engineering Circular No. 15, Federal Highway Administration, McLean, VA., 1988.

Fortier, S., and Scobey, F.C., "Permissible Canal Velocities," *Transactions of the American Society of Civil Engineers*, vol. 89, 1926, pp. 940-984.

French, R.H. and Curtis, S., "The Precipitation Event of 23-24 February 1998," Report No. 45170, Division of Hydrologic Sciences, Desert Research Institute, Reno, NV, 1999a.

French, R.H. and Curtis, S., "Serendipity: Capturing a Design Level Precipitation Event," *Proceedings of the ASCE Water Resources Engineering Division Conference*, Seattle, WA, 1999b (CD-ROM).

French, R.H., Buchanan, T.L., Hokett, S. and Curtis, S., "Calibration of a Hybrid Rainfall-Runoff Model in an Arid Environment," *Proceedings of the XXVIII IAHR Congress*, Graz, Austria, 1999 (CD-ROM).

Garcia, M., "Sedimentation and Erosion Hydraulics," in *Hydraulic Design Handbook*, L.W. Mays, ed., McGraw-Hill, New York, NY, 1999, pp. 6.1-6.113.

Glover, R.E., and Florey, Q.L., "Stable channel Profiles," *Hydraulic Laboratory Report No. Hyd-325*, U.S. Bureau of Reclamation, Washington, September 1951.

Houk, I.E., "Discussion of Permissible Canal Velocities," *Transactions of the American Society of Civil Engineers*, vol. 89, 1926, pp. 971-977.

Houk, I.E., *Irrigation Engineering*, vol. II, John Wiley & Sons, New York, 1956.

Kennedy, R.G., "The Prevention of Silting in Irrigation Canals," *Proceedings, Institution of Civil Engineers, London*, Vol. 119, 1895, pp. 281-290.

Kouwen, H., and Unny, T.E., "Flexible Roughness in Open Channels," *Proceedings of the American Society of Civil Engineers, Journal of the Hydraulics Division*, vol. 99, no. HY5, May 1973, pp. 713-728.

Lacey, G., "Flow in Alluvial Channels with Sand Mobile Beds," *Proceedings, Institution of Civil Engineers, London*, Vol. 9, and discussion, Vol. 11, 1958.

Lacey, G., "Uniform Flow in Alluvial Rivers and Canals," *Proceedings, Institution of Civil Engineers, London*, Vol. 237, 1935.

Lacey, G., "Stable Channels in Alluvium," *Proceedings, Institution of Civil Engineers, London*, Vol. 229, 1930.

Lane, E.W., "Design of Stable Channels," *Transactions of the American Society of Civil Engineers*, vol. 120, 1955, pp. 1234-1279.

Lane, L.J., "Transmission Loss," *National Engineering Handbook, Chapter 19*, U.S. Department of Agriculture, Soil Conservation Service, Washington, 1983.

Mehrotra, S.C., "Permissible Velocity Correction Factors," *Proceedings of the American Society of Civil Engineers, Journal of the Hydraulic Engineering*, Vol. 109, No. 2, February 1983, pp. 305-308.

Murphey, J.B., Wallace, D.E., and Lane, L.J., "Geomorphic Parameters Predict Hydrograph Characteristics in the Southwest," AWRA, *Water Resources Bulletin*, Vol. 13, No. 1, 1977, pp. 25-38.

Phelps, H.O., "The Friction Coefficient for Shallow Flows over a Simulated Turf Surface," *Water Resources Research*, vol. 6, no. 4, August 1970, pp. 1220-1226.

Rubey, W.W., "Settling Velocities of Gravel, Sand, and Silt Particles," *American Journal of Science*, Vol. 25, No. 148, 1933, pp. 325-338.

Shields, A., "Anwendung der Aehnlichkeitsmechanik und der Turbulenzforschung auf Geschiebebegung" (Application of Similarity Principles and Turbulence Research to Bed-Load Movement), *Mitteilungen der Preuss Versuchsanst fur Wasserbau und Schiffbau*, No. 26, Berlin, 1936. (Available in translation by W.P. Ott and J.C. vanUchelen, Soil Conservation Service Cooperative Laboratory, California Institute of Technology, Pasadena.)

Shukry, A., "Flow around Bends in an Open Flume," *Transactions of the American Society of Civil Engineers*, vol. 115, 1950, pp. 751-788.

Simons, D.B. and Albertson, M.L., "Uniform Water Conveyance Channels in Alluvial Material," *Proceedings of the American Society of Civil Engineers, Journal of the Hydraulic Engineering*, Vol. 86, No. HY5, May, 1960, pp. 33-71.

Straub, L.G., "Missouri River Report," House Document 238, Appendix XV, U.S. Army Corps of Engineers, 73rd United States Congress, 2nd Session, 1935, p. 1156.

Streeter, V.L., and Wylie, E.B., *Fluid Mechanics*, McGraw-Hill Book Company, New York, 1975, pp. 592-595.

Taylor, B.D. and Vanoni, V.A., "Temperature Effects in Low-Transport, Flat-Bed Flows," *Proceedings of the American Society of Civil Engineers, Journal of the Hydraulic Engineering*, Vol. 98, No. HY12, December, 1972, pp. 2191-2206.

Trout, T.J., "Channel Design to Minimize Lining Material Costs," *Proceedings of the American Society of Civil Engineers, Journal of the Irrigation and Drainage Division*, vol. 108, no. IR4, December 1982, pp. 242-249.

White, C.M. "Equilibrium of Grains on the Bed of a Stream," *Proceedings of the Royal Society of London*, Series A, Vol. 174, 1940, pp. 322.

Willison, R.J., "USBR's Lower-Cost Canal Lining Program," *Proceedings of the American Society of Civil Engineers, Journal of the Irrigation and Drainage Division*, vol. 84, no. IR2, April 1958, pp. 1589-1-1589-29.

Woodward, S.M., "Hydraulics of the Miami Flood Control Project," *Miami Conservancy District Technical Report*, part VII, Dayton, Ohio, 1920.

Woodward, S.M., and Posey, C.J., *Hydraulics of Steady Flow in Open Channels*, John Wiley & Sons, New York, 1941.

Yang, C.T., *Sediment Transport: Theory and Practice*, The McGraw-Hill Companies, Inc., New York, NY, 1996.

Yang, C.T., "Incipient Motion and Sediment Transport,' *Proceedings of the American Society of Civil Engineers, Journal of the Hydraulics Division*, vol. 99, no. HY10, 1973, pp. 1679-1704.

7.6.   PROBLEMS

1.   Demonstrate that the best hydraulic section for

   a.   A rectangular channel is half a square
   b.   A trapezoidal channel is half a hexagon

2.   Calculate the dimensions of the most economical trowel-finished, concrete-lined trapezoidal channel to convey 250 m³/s (8800 ft³/s) with a slope of 0.0003 m/m.

3.   A trapezoidal channel with $b = 20$ ft (6.1 m), $z = 2$ on a slope of 0.001 ft/ft conveys a uniform flow at a normal depth of 6 ft (1.8 m). Given the above, calculate

   1.   The *average* unit tractive force and shear velocity on the channel bottom.
   2.   The maximum tractive force and shear velocity on the channel side slopes.
   3.   The maximum tractive force and shear velocity on the channel bottom.

4.   A trapezoidal channel has been designed to convey 8.5 m³/s (300 ft³/s) at a slope of 0.0015 m/m in noncolloidal coarse gravels with, $z = 2$ and $b = 6.1$ m (20 ft). As the senior design engineer, you are to examine the appropriateness of this design. What is you conclusion?

5.   Use the method of maximum permissible velocity to design an unlined trapezoidal channel in fine gravel that will convey a discharge of 600 ft³/s (17 m³/s). The channel will have a longitudinal slope of 0.0016 ft/ft. Compare the designs developed

assuming clear water and water transporting colloidal silts. What are the freeboard requirements for each design?

6. Design a nonerodible (lined) channel to carry 200 ft³/s (5.7 m³/s) with n = 0.020 and S = 0.0020 ft/ft. Use engineering judgement and list all assumptions. Do not forget to estimate the lining height and the freeboard required. Show the final design in a sketch.

7. The All-American canal is designed to divert 15,155 ft³/s (430 m³/s) of desilted water from the Colorado River to irrigate the Imperial Valley in Southern California. The canal is 80 miles (130 km) long. The typical maximum section has a bottom width of 160 ft (49 m), a top width at the water surface of 232 ft (71 m), a water depth of 20.6 ft (6.3 m), a minimum freeboard of 6 ft (1.8 m), and a bank width of 27 to 30 ft (8.2 to 9.1 m). The terminal capacity of this canal is 2600 ft³/s (74 m³/s). The canal is excavated primarily in alluvial materials, ranging from light sandy or silty loams to adobe and having an average particle size of 0.0025 in (0.063 mm). Review the hydraulic design of this section.

8. Determine the cross section and discharge of the stable hydraulic section of a channel excavated in noncohesive material with τ_0 = 0.10 lb/ft², S = 0.0004 ft/ft, θ = 31°, and n = 0.020.

9. For the data given in Prob. 8, determine the modified channel section if the channel is to carry

a. 75 ft³/s (2.1 m³/s)
b. 300 ft³/s (8.5 m³/s)

10. Design a waterway lined with Bermuda grass on erosion resistant soil to carry 200 ft³/s (5.7 m³/s). The average slope of the channel is 3 percent. Use the curve for moderate vegetal retardance.

11. The channel section shown below is to convey 2,000 ft³/s (57 m³/s) of heavily silted water. It is excavated on a slope of 0.001 ft/ft through country composed of moderately rounded alluvium. The canal will be armored with selected rock averaging 2 in (51 mm) in diameter. Do you foresee any problems with this design?

12. A trapezoidal channel with a bottom width of 4 ft (1.2 m), side slopes of 4:1, and a longitudinal slope of one percent is lined with Bermuda grass in easily erodible soil. If the grass is kept mowed to a height of 2 to 4 in (51 to 102 mm) and the design flow is 100 ft³/s (2.8 m³/s), evaluate the adequacy of the design.

13. A trapezoidal channel with a bottom width of 4 ft (1.2 m), side slopes of 3:1, and a longitudinal slope of 0.005 ft/ft must convey a continuous flow of 30 ft³/s (0.85 m³/s) and a flood flow of 170 ft³/s (4.8 m³/s). If the channel is to be built in sandy loam and the flood water conveys fine silts, evaluate the following lining options if supercritical flow is not acceptable:

   a. No lining

   b. Complete concrete lining

   c. Combination of mowed Bermuda grass and concrete lining as shown below

   If none of the above designs are acceptable, recommend and justify a new design using a combination of concrete and mowed Bermuda grass linings.

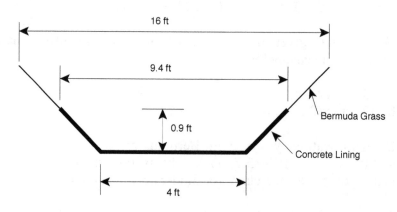

CHAPTER 8

TURBULENT DIFFUSION AND DISPERSION IN OPEN CHANNEL FLOW

8.1.   INTRODUCTION

The topic of riverine water quality modeling is almost inseparable from the topic of open channel hydraulics, with the linkage being established by a common and essential interest in the mechanics of mass transport. In this chapter, turbulent diffusion and dispersion are discussed from the viewpoint of quantifying these processes rather than from the viewpoint of quantifying the environmental effects of pollutant transport. Thus, no actual pollutants will be mentioned; environmentally neutral tracers will be discussed. This chapter will also be in the form of a summary. A complete treatment of turbulent diffusion and dispersion transport processes from the viewpoint of hydraulic engineering is available in Fischer et al. (1979); from an environmental engineering point of view in Krenkel and Novotny (1980); and from an interdisciplinary viewpoint in French et al.(1999).

One of the basic problems in discussing transport processes has been a semantics problem which originated many years ago with imprecise and non-consensus definitions of the processes involved.   Following Fischer et al. (1979), under the broad definition of mixing processes, there are two categories of interest in open channel flow - turbulent diffusion and dispersion. The terminology *turbulent diffusion* refers to the random scattering of particles in a flow by turbulent motions. *Dispersion* is the scattering of particles by the combined effects of shear and transverse turbulent diffusion.   In this chapter, we will discuss advective transport processes; i.e., transport by imposed velocity systems, rather than convective transport, which connotes transport due to the hydrodynamic instabilities that are usually associated with temperature gradients. *Shear* refers to the advection of a fluid at different velocities at different positions within the flow. For example, the vertical velocity profile defined by Eq. (1.4.14) results in shear since the fluid particles near the surface are moving faster than those near the bottom.

8.2.   GOVERNING EQUATIONS

To begin, consider the control volume defined in Fig. 8.1 and assume that the molecular motion is causing mass transfer of a tracer through the control surfaces in the x direction. If it is assumed that the transport of mass through the control surface is proportional to the concentration gradient, then an application of the law of conservation of mass yields

$$\frac{\partial c}{\partial t} = D \frac{\partial^2 c}{\partial x^2} \qquad (8.2.1)$$

where c = tracer concentration and D = molecular diffusion coefficient with the tacit assumption that $D \neq f(x)$.

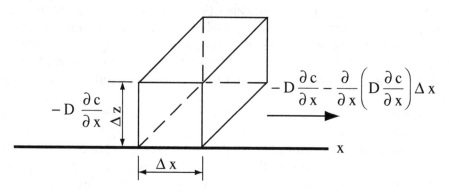

Figure 8.1. Molecular diffusion.

Equation (8.2.1) is a mathematical statement of Fick's first law of diffusion. The fundamental solution of Fick's first law is obtained under the following conditions:

$$c(x,0) = \delta(x)$$

and

$$c(\pm\infty, t) = 0$$

where $\delta(x)$ designates the Dirac delta function. This function has the properties of being zero everywhere except at $x = 0$ and that

$$\int_{-\infty}^{\infty} \delta(x)dx = 1$$

With these initial and boundary conditions, the solution of Eq. (8.2.1) is

$$c(x,t) = \frac{1}{\sqrt{4\pi Dt}} \exp\left(-\frac{x^2}{4Dt}\right) \qquad (8.2.2)$$

Equation (8.2.2) is a Gaussian distribution, and if the mass under the distribution curve is unity, it is known as the normal distribution. In the material which follows, we will often be concerned with the various moments of Eq. (8.2.2), where these moments are defined by

$$M_n = \int_{-\infty}^{\infty} x^n c(x,t)dx \qquad (8.2.3)$$

Thus, the 0 moment of the fundamental solution is given by

$$M_0 = \int_{-\infty}^{\infty} c(x,t)dx$$

The mean, μ, and the variance, σ^2, of the normal or Gaussian distribution can be defined in terms of the moments or

$$\mu = \frac{M_1}{M_0} \qquad (8.2.4)$$

and

$$\sigma^2 = \frac{M_2}{M_0} - \mu^2 \qquad (8.2.5)$$

For a normal distribution,

$$\mu = 0$$

and

$$\sigma^2 = 2Dt$$

At this point, it is noted that

1.   The variance of the distribution is a measure of the spread of the distribution about the origin. Recall from statistics (see, for example, Benjamin and Cornell, 1970) that for a Gaussian distribution a spread of 4σ includes approximately 95 percent of the total area under the concentration curve.

2.   In applied studies, the diffusion coefficient is often determined from

$$\frac{d\sigma^2}{dt} = 2D \qquad (8.2.6)$$

3.   Fick's first law and its normal boundary conditions belong to a set of differential equations in which the principle of superposition can be applied; i.e., the equation and the usual boundary conditions are linear. This is an important observation, since in most practical applications of Eq. (8.2.1), there exist boundaries across which diffusion cannot take place. Suppose, for example, there exist boundaries at $x = \pm L$ across which there is no diffusion (Fig. 8.1). According to Eq. (8.2.1) the appropriate boundary conditions are

$$\frac{\partial c}{\partial x} = 0 \quad \text{at} \quad x = \pm L$$

To satisfy the boundary condition of $x = -L$, an image source of tracer is added at $x = -2L$; in an analogous fashion, to satisfy the boundary condition at $x = L$, an image source is added at $x = 2L$.

The image source at $x = -2L$ makes $\partial c / \partial x = 0$ at $x = -L$ but causes a positive gradient at $x = +L$; therefore, another image source must be added at $x = +4L$. Again, analogous reasoning requires additional image sources at $x = -6L, +8L$, *ad infinitum*. The solution is then

$$c(x,t) = \frac{1}{\sqrt{4\pi Dt}} \sum_{n=-\infty}^{+\infty} \exp\left(-\frac{x + 2nL}{4Dt}\right) \qquad (8.2.7)$$

Now consider a two dimensional channel flow (Fig. 8.2) defined such that no parameter varies in the direction perpendicular to the plane of paper. At time t, an inert tracer is injected into the flow which is moving from left to right with the velocity distribution shown. As the tracer moves downstream, it spreads out, and the maximum concentration is reduced by dilution. To derive an equation which represents the transport in this situation, define an elemental control volume with dimensions dx by dz and area dA (Fig. 8.2). The principle of conservation of mass can then be applied, and an equation which describes the situation derived.

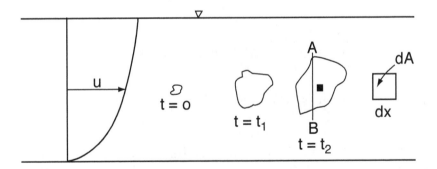

Figure 8.2. Transport in two-dimensional flow (Holley, 1969).

For this initial development, assume that the spreading of the tracer cloud is due only to molecular diffusion. Two methods of accounting for the movement of the tracer through the control volume suggest themselves. First, the movement of each molecule in the tracer cloud could be described. Although such an approach has theoretical merit, it becomes cumbersome as the movement of each molecule is described. The second method is a continuum mechanics approach similar to that used for the derivation of Fick's first law; i.e., it is assumed that the fluid carries the tracer with it at a rate which depends on both the velocity in the longitudinal direction u, and the concentration c. However, the continuum mechanics approach is only approximate since many of the tracer molecules have a velocity different from u. Therefore, it is necessary to introduce the molecular diffusion coefficient D to correct for the approximate nature of this solution. The application of the continuum mechanics approach to formulating the required mass balance yields

$$\frac{\partial c}{\partial t} + u \frac{\partial c}{\partial x} = D \frac{\partial^2 c}{\partial x^2} + D \frac{\partial^2 c}{\partial z^2} \qquad (8.2.8)$$

where

$$\frac{\partial c}{\partial t} = \text{Time rate of change of concentration at a point}$$

$$u \frac{\partial c}{\partial x} = \text{Advection of tracer with fluid}$$

$$D \frac{\partial^2 c}{\partial x^2} + D \frac{\partial^2 c}{\partial z^2} = \text{Molecular diffusion}$$

which is also known as Fick's second law (Holley, 1969).

Now, assume that the fluid is turbulent. In Fig. 8.3a and b, the velocity and tracer concentration in the incremental control volume defined in Fig. 8.2 are shown. In these figures, the straight solid lines represent the time-averaged values of u and c. In a manner similar to that used in the case of advective molecular diffusion, the mass balance for the situation is

$$\frac{\partial c}{\partial t} + u \frac{\partial c}{\partial x} + v \frac{\partial c}{\partial z} = D \frac{\partial^2 c}{\partial x^2} + D \frac{\partial^2 c}{\partial z^2} \qquad (8.2.9)$$

where u and v are the instantaneous values of the turbulent velocities in the x and z directions, respectively, and c is an instantaneous value of the tracer concentration. It is noted that a term for advection in the z direction is included since, even though the primary direction of flow is in the x direction, there is a turbulent motion in the z direction. Equation (8.2.9) takes the effect of transport by turbulent diffusion into account directly, because u and v are instantaneous turbulent velocity components. Since instantaneous values of u and v are not usually available, this direct incorporation of the effects of turbulence into the transport equation results in an equation which cannot be solved.

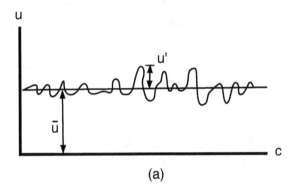

(a)

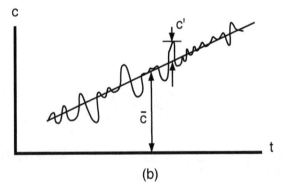

(b)

Figure 8.3.   Turbulent fluctuation of velocity and concentration in an elemental control volume.

From an applied viewpoint, the effect of the turbulent transport of trace through the incremental control volume dA must be expressed in terms of the time-averaged value of u, v, and c or

$$u = \bar{u} + u'$$

$$v = v'$$

and

$$c = \bar{c} + c'$$

where the overbar indicates a time-averaged value of the variable, and variables with primes are turbulent fluctuations of the variable. Also, recall that there is no average velocity in the z direction by definition. Substitution of these equations for u, v, and c in Eq. (8.2.9) and averaging the resulting equation with respect to time yields

$$\frac{\partial \bar{c}}{\partial t} + \bar{u}\frac{\partial \bar{c}}{\partial t} = D\frac{\partial^2 \bar{c}}{\partial x^2} + D\frac{\partial^2 \bar{c}}{\partial z^2} + \frac{\partial\left(-\overline{u'c'}\right)}{\partial x} + \frac{\partial\left(-\overline{v'c'}\right)}{\partial z} \qquad (8.2.10)$$

The turbulent diffusion coefficients ϵ_x and ϵ_z are then defined in analogy with molecular diffusion or

$$\overline{u'c'} = -\epsilon_x \frac{\partial c}{\partial x} \qquad (8.2.11)$$

and

$$\overline{v'c'} = -\epsilon_z \frac{\partial c}{\partial z} \qquad (8.2.12)$$

If it is assumed that $\epsilon_x \neq f(x)$ and $\epsilon_z \neq f(z)$, the substitution of these variables in Eq. (8.2.10) yields

$$\frac{\partial \bar{c}}{\partial t} + \bar{u}\frac{\partial \bar{c}}{\partial t} = \left(D + \epsilon_x\right)\frac{\partial^2 \bar{c}}{\partial x^2} + \left(D + \epsilon_z\right)\frac{\partial^2 \bar{c}}{\partial z^2} \qquad (8.2.13)$$

Equation (8.2.13) represents the transport of tracer due to both molecular and turbulent diffusion. The terminology *turbulent diffusion* arises, as was the case with molecular diffusion, because the advective transport term written in terms of $\bar{u}$ does not represent the complete advective transport process; therefore, corrective factors are required.

In Fig. 8.4, the tracer cloud is shown progressing through time until time t_4. By this time, turbulent transport has mixed the tracer completely in the vertical direction. After complete mixing in the vertical direction has occurred, the primary variation of the tracer concentration in the two-dimensional problem defined here is in the longitudinal direction. Although Eq. (8.2.13) remains valid, it provides more information than is required, since it describes the variation of c in both the x and z directions. For efficiency, a simplification of this equation is sought. Define

$$\overline{u} = U + u' \tag{8.2.14}$$

and

$$\overline{c} = C + c' \tag{8.2.15}$$

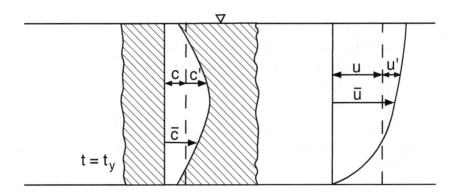

Figure 8.4. One-dimensional transport (Holley, 1969).

where U and C are the average values of the velocity and tracer concentration, respectively, *in the cross section*. Substitution of Eqs. (8.2.14) and (8.2.15) in Eq. (8.2.13) yields, after cross sectional averaging,

$$\frac{\partial C}{\partial t} + U\frac{\partial C}{\partial x} = \left(D + \varepsilon_x\right)\frac{\partial^2 C}{\partial x^2} + \frac{\partial\left(\overline{-u'c'}\right)}{\partial x} \tag{8.2.16}$$

where the double overbar indicates a cross sectional average. Taylor (1953, 1954) demonstrated that under certain conditions u' is proportional to the longitudinal gradient of C or

$$E\frac{\partial C}{\partial x} = -\overline{u'c'} + \left(D + \varepsilon_x\right)\frac{\partial C}{\partial x} \tag{8.2.17}$$

The transport defined by Eq. (8.2.17) is termed *longitudinal dispersion*, and the coefficient E is the coefficient of longitudinal dispersion. Substitution of Eq. (8.2.17) in Eq. (8.2.16) yields the one-dimensional equation of longitudinal dispersion or

$$\frac{\partial C}{\partial t} + U\frac{\partial C}{\partial x} = E\frac{\partial^2 C}{\partial x^2} \tag{8.2.18}$$

With regard to Eqs. (8.2.17) and (8.2.18), the following should be noted:

1. This equation applies whenever both the vertical and transverse mixing are complete; i.e., when the tracer completely fills the cross section and the primary variation in concentration is in the longitudinal direction.

2. Equation (8.2.17) includes both the effect of $\overline{u'c'}$ and molecular and turbulent diffusion. After complete mixing is achieved, diffusion acts to smooth the distribution; i.e., after the point of complete mixing is reached, the effect of diffusion is small relative to the effect of the velocity gradient.

3. The role of the velocity gradient in the dispersion process is paramount and is reflected in the term $\overline{u'c'}$.

4. Equations (8.2.17) and (8.2.18) are limited because Taylor's assumption was based on the fact that after an initial period of time the rate of advective transport u', either upstream or downstream from a vertical plane which moves at a velocity U, depends only on the tracer concentration gradient and E, which is constant for a uniform flow. In general, Eq. (8.2.18) is valid only after an initial period of time, since before the elapse of this presently unspecified time, Taylor's assumption is not valid.

5.   Equation (8.2.18) governs a special case in which the flow is steady and uniform, the channel prismatic, and E constant. A much more general case is represented by

$$\frac{\partial}{\partial t}(AC) + \frac{\partial}{\partial x}(UAC) = \frac{\partial}{\partial x}\left(EA\frac{\partial C}{\partial x}\right) \qquad (8.2.19)$$

In this situation, U, A, and E can vary with distance and/or time.

At this point it is convenient to note some of the solution of the partial differential equation

$$\frac{\partial C}{\partial t} + U\frac{\partial C}{\partial x} = \theta\frac{\partial^2 C}{\partial x^2} \qquad (8.2.20)$$

where both U and θ are constants. The solution of this equation is dependent on the initial or boundary conditions which are specified. For example, if

$$C(x,0) = \begin{array}{l} 0 \text{ for } x \succ 0 \\ C_0 \text{ for } x \prec 0 \end{array}$$

then the solution of Eq. (8.2.20) is

$$C(x,t) = \frac{C_0}{2}\left[1 - \text{erf}\left(\frac{x - Ut}{\sqrt{4\theta t}}\right)\right] \qquad (8.2.21)$$

where *erf* designates the error function. The above initial condition could, for example, represent the case where one fluid in an open channel is being displaced by a second fluid which has a tracer in it with a concentration C_0 and the fluid moves at a cross sectional average velocity U.

A much more complex and important case arises when the boundary and initial conditions are

$$C(0,t) = C_0 \text{ for } 0 \prec t \prec \infty$$

$$C(x,0) = 0 \text{ for } 0 \prec x \prec \infty$$

In this case, the solution of Eq. (8.2.18) is

$$C(x,t) = \frac{C_0}{2}\left[\text{erfc}\left(\frac{x - Ut}{\sqrt{4\theta t}} \right) + \text{erfc}\left(\frac{x + Ut}{\sqrt{4\theta t}} \right) \exp\left(\frac{Ux}{\theta} \right) \right] \qquad (8.2.22)$$

where *erfc* designates the complementary error function. This set of boundary and initial conditions might represent the situation in which a tracer of concentration C_0 is introduced at the origin of a coordinate system at time zero and continued. It should be obvious that in this case as $t \to \infty$, $C(x,t) \to C_0$.

Another problem of interest to the hydraulic engineer is that the spread of a tracer in three-dimensional flow. The governing equation for this situation is

$$\frac{\partial C}{\partial t} + U\frac{\partial C}{\partial x} = \theta_x \frac{\partial^2 C}{\partial x^2} + \theta_y \frac{\partial^2 C}{\partial y^2} + \theta_z \frac{\partial^2 C}{\partial z^2}$$

where θ_x, θ_y, and θ_z are the coefficients in the x, y, and z directions, respectively. This governing equation can be significantly simplified if it is assumed that $\theta_x = \theta_y = \theta_z = \theta$. The governing equation is then

$$\frac{\partial C}{\partial t} + U\frac{\partial C}{\partial x} = \theta\left(\frac{\partial^2 C}{\partial x^2} + \frac{\partial^2 C}{\partial y^2} + \frac{\partial^2 C}{\partial z^2} \right) \qquad (8.2.23)$$

Although a general solution of Eq. (8.2.23) can be obtained in some situations, it is usually possible to reduce the three-dimensional problem to a two-dimensional problem. For example, in open channel flow diffusion in the longitudinal direction can usually be neglected (e.g., Fischer *et al.*, 1979). If this is the case, then for an instantaneous injection of tracer of mass M at the origin of the coordinate system, the resulting downstream concentration is

$$C(x,y,z) = \frac{M}{4\pi\theta x}\exp\left[-\frac{\left(y^2 + z^2 \right)U}{4x\theta} \right] \qquad (8.2.24)$$

In the case where the tracer injection is not instantaneous but is added continuously at a mass flow rate of M^*, the approximate solution of the problem is given by

$$C(x,y) = \frac{M^*}{U\sqrt{\dfrac{4\pi\theta x}{U}}} \exp\left(-\frac{y^2 U}{4x\theta}\right) \qquad (8.2.25)$$

It is noted that the solutions provided by Eqs. (8.2.24) and (8.2.25) are valid only when the following condition is satisfied:

$$t \succ\succ\succ \frac{2\theta}{U^2} \qquad (8.2.26)$$

In most cases, this condition is satisfied at very small values of t.

8.3. VERTICAL AND TRANSVERSE TURBULENT DIFFUSION AND LONGITUDINAL DISPERSION

When a tracer is injected into a homogeneous channel flow, the complete advective transport process can be conveniently viewed as being composed of three stages. In the first stage, the tracer is diluted by the flow in the channel because of its initial momentum. In the second stage, the tracer is mixed throughout the cross section by turbulent transport processes. And, in the third stage, longitudinal dispersion tends to erase any longitudinal variations in the tracer concentration. In some cases, the second stage is eliminated because the tracer discharge has a significant amount of initial momentum; however, in many cases, the tracer flow is small and has an insignificant amount of momentum associated with it. In the latter case, the first transport stage is eliminated. In this section, only the second and third transport stages will be treated, with the implied assumption that if there is a first stage, it can be treated separately.

To develop a quantitative expression for the vertical, turbulent diffusion coefficient, consider a relatively shallow flow in a wide, rectangular channel (Fig. 8.5). From Chapter 1, it is known that the vertical transport of momentum in such a flow is given by

$$\tau = \varepsilon_z \rho \frac{du}{dz} \qquad (8.3.1)$$

where τ = shear stress a distance z above the bottom boundary, ρ = fluid density, ε_z = vertical turbulent diffusion coefficient, and u = longitudinal velocity.

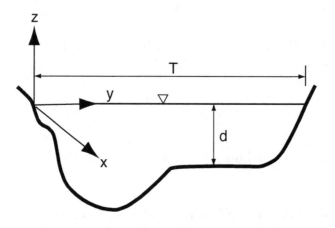

Figure 8.5. Coordinate system definition.

Further, in a two-dimensional flow, such as that defined in Fig. 8.4, the vertical velocity profile can be assumed to be given by

$$u = \frac{u_*}{k}\left(1 + \ln\frac{z}{d}\right) \qquad (8.3.2)$$

where u_* = shear velocity, k = von Karman's turbulence coefficient ($k \simeq 0.40$), and d = depth of flow. A shear stress distribution similar to that which prevails in a circular pipe is assumed or

$$\tau = \tau_0\left(1 - \frac{z}{d}\right) \qquad (8.3.3)$$

where $\tau_0 = \rho\, gdS$ = bottom boundary shear stress. Then, as demonstrated by Elder (1959), combining Eqs. (8.3.1) to (8.3.3) yields

$$\varepsilon_z = ku_*dz'(1 - z') \tag{8.3.4}$$

where $z' = z/d$. The depth average value of $\varepsilon_{z\,z}$ is

$$\overline{\varepsilon_z} = \frac{1}{d}\int_0^d \varepsilon_z dz = 0.067du_* \tag{8.3.5}$$

In the infinitely wide channel hypothesized for the derivation of Eq. (8.3.4), there is not a transverse velocity profile. Thus, a quantitative expression for ε_y, the transverse, turbulent diffusion coefficient, cannot be derived in a manner analogous to that used for ε_z. Instead, expressions for ε_y must be derived from experimental results. A large of number of such experiments have been performed and the results of this work are summarized in Fischer *et al.* (1979) and Lau and Krishnappen (1977). In straight, rectangular channels, an approximate average of the results available is

$$\varepsilon_y \cong 0.15du_* \pm 50\% \tag{8.3.6}$$

where ±50 percent indicates the magnitude of the error incurred in estimating ε_y. In natural channels, the value of ε_y is significantly higher than the value estimated by Eq. (8.3.6). For channels which can be classified as "slowly meandering" with only moderate boundary irregularities

$$\varepsilon_y \cong 0.60du_* \pm 50\% \tag{8.3.7}$$

If the channel has curves of small radii, rapid changes in channel geometry, or severe bank irregularities, e.g., groins, then the value of ε_y will be larger than that which is estimated by Eq. (8.3.7). For example, Fischer (1969) estimated that

$$\varepsilon_y = 25\frac{\overline{u^2}d^3}{R_c^2 u_*} \tag{8.3.8}$$

where R_c = curve radius. At present, whether a channel should be considered slowly meandering and Eq. (8.3.7) used to estimate ε_y of whether a larger value of ε_y should be used is a matter of judgement.

Fischer *et al.* (1979 suggested that a slowly meandering channel is one in which

$$\frac{T\overline{u}}{R_c u_*} \prec 2 \tag{8.3.9}$$

where T = channel width, $\overline{u}$ = average velocity of flow, and R_c = curve radius. At best, Eq. (8.3.9) provides only a crude estimate of channel sinuosity relative to ε_y.

As noted in the first paragraph of this section, the complete advective transport process in a two-dimensional channel flow can be conveniently viewed as being composed of three stages. In the second stage, the primary transport mechanism is turbulent diffusion, with the diffusion coefficients being defined by Eqs. (8.3.5) and (8.3.7), or in some cases Eq. (8.3.8). A comparison of Eqs. (8.3.5) and (8.3.7) demonstrates that

$$\varepsilon_y \cong 10\varepsilon_z$$

or that the transverse mixing coefficient is roughly 10 times larger than the vertical mixing coefficient. Thus, the rate at which a plume of tracer spreads laterally is an order of magnitude larger than the rate of spread in the vertical direction. However, in almost all cases the channel will be much wider than it is deep; e.g., typical natural channel dimensions might be width 100 ft (30 m) and depth 3.0 ft (1.0 m). An examination of the equation governing the spread of a plume, Eq. (8.2.25) for example, demonstrates that a mixing time can be defined as proportional to the square of the distance divided by the mixing coefficient. Thus, in a typical channel it will take approximately 90 time as long for a plume to spread completely across the channel as it will to mix in the vertical dimension.   In most applied problems, it is both convenient and appropriate to begin by assuming that the tracer is uniformly distributed over the vertical.

Recall from Section 8.2 that, in a diffusional process where the tracer is added at a constant mass flow rate M^* to an unbounded channel, the downstream concentration of tracer is given approximately by

$$C = \frac{M^*}{\overline{ud}\sqrt{\dfrac{4\pi\varepsilon_y x}{\overline{u}}}}\exp\left[-\frac{y^2\overline{u}}{4x\varepsilon_y}\right] \qquad (8.3.10)$$

If the channel has width T, then the boundary conditions are

$$\frac{\partial C}{\partial y} = 0 \quad \text{at} \quad y = 0$$

and

$$\frac{\partial C}{\partial y} = 0 \quad \text{at} \quad y = T$$

must be satisfied by the solution. At this point, it is convenient to define the following dimensionless variables:

$$C' = \frac{M^*}{\overline{ud}T}$$

$$x' = \frac{x\varepsilon_y}{\overline{u}T^2}$$

and

$$y' = \frac{y}{T}$$

If the source of the tracer is located at $y = y_0$, where $y' = y_0'$, then the downstream concentration is

$$\frac{C}{C_0} =$$

$$\frac{1}{\sqrt{4\pi x'}}\sum_{n=-\infty}^{\infty}\left\{\exp\left[-\frac{\left(y' - 2n - y_0'\right)^2}{4x'}\right] + \exp\left[\frac{\left(y - 2n + y_0'\right)^2}{4x'}\right]\right\}$$

$$\qquad (8.3.11)$$

where the principle of superposition (Section 8.2) has been used to meet the boundary conditions. In Fig. 8.6, Eq. (8.3.11) is plotted for a centerline injection of tracer, i.e. $y_0' = 0.5$, and the relative concentration of tracer along the centerline of the channel and the side of the channel are shown as functions of x'. In this figure, it is noted that for $x' \geq 0.1$ the concentration of C is within 5 percent of C_0 throughout the channel cross section. This situation is often referred to as *complete mixing*, and for a centerline injection, the distance to the point where complete mixing is achieved is given by

$$L = \frac{0.1\overline{u}T^2}{\varepsilon_y} \tag{8.3.12}$$

If the tracer is injected at the side of the channel, the width over which the plume must mix is double that for a centerline injection, but otherwise the boundary conditions are identical and the equations available can be used to obtain a solution.

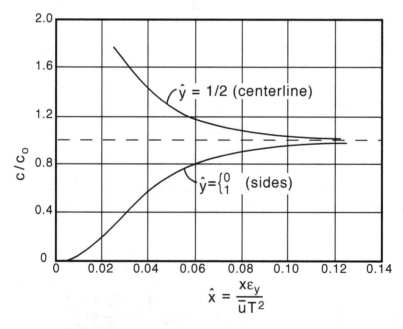

Figure 8.6. Downstream tracer concentrations resulting from continuous centerline injection.

EXAMPLE 8.1

A tracer is injected at the center of a slowly meandering channel at a flow rate of 0.15 m³/s (5.3 ft³/s). If the concentration of the tracer at the point of injection is 300 mg/L (300 parts per million), the width of the channel is 300 m (984 ft), the depth of flow 10 m (33 ft), the slope of the channel is 0.0001, and the total flow rate in the channel is 1600 m³/s (57,000 ft³/s), determine the maximum concentration of the tracer 500 m (1600 ft) downstream of the point of injection and estimate the width of the tracer plume at this point.

Solution

It is assumed that the tracer is completely mixed over the vertical dimension at the point of injection. The shear velocity of the flow is then estimated from

$$u_* = \sqrt{gRS} = \sqrt{9.8(10)(0.0001)} = 0.099\,\mathrm{m/s}\left(0.32\,\mathrm{ft/s}\right)$$

where R, the hydraulic radius, is taken as the depth of flow. From Eq. (10.3.7) the turbulent, transverse diffusion coefficient is

$$\varepsilon_y = 0.60 du_* = 0.60(10)(0.099) = 0.59\,\mathrm{m^2/s}\left(6.4\,\mathrm{ft^2/s}\right)$$

The mass flow rate of tracer is

$$M^* = Q_T C = 0.15(300) = 45\frac{\mathrm{m^3 - mg}}{\mathrm{s - L}}\left(1600\frac{\mathrm{ft^3 - ppm}}{\mathrm{s}}\right)$$

where Q_T = the flow rate associated with the tracer. The average velocity of the flow is estimated by assuming that the channel cross section is approximately rectangular in shape; hence

$$\bar{u} = \frac{Q}{A} = \frac{1600}{300(10)} = 0.53\,\mathrm{m/s}\left(1.7\,\mathrm{ft/s}\right)$$

The width of the plume at x = 500 m (1640 ft) is estimated first to determine if the lateral boundaries of the channel can be ignored in the calculation of the maximum tracer concentration at this distance. The width of the trace plume is given approximately by

$$b = 4\sigma = 4\sqrt{2\varepsilon_y t}$$

where t can be estimated from $t = \dfrac{x}{u}$. Then

$$b = 4\sqrt{2(0.59)\left(\frac{500}{0.53}\right)} = 130\,\text{m}\,(430\,\text{ft})$$

Therefore, the width of the trace plume is less than the width of the channel at $x = 500$ m (1640 ft), and the effect of the lateral channel boundaries can be ignored in computing the maximum tracer concentration at this distance. The maximum tracer concentration in this situation is estimated from Eq. (8.2.25)

$$C_{max} = \frac{45}{0.53(10)\sqrt{\dfrac{4\pi(0.59)(500)}{0.53}}} \exp\left(-\frac{0(0.53)}{4(500)(0.59)}\right)$$
$$= 0.10\,\text{mg}\,/\,\text{L}\,(1\,\text{ppm})$$

Note, y is taken as zero above since an estimate of the maximum concentration is required.

$$C_{max} = \frac{M^*}{\bar{ud}\sqrt{\dfrac{4\pi\varepsilon_y x}{\bar{u}}}} \exp\left[-\frac{y^2\bar{u}}{4x\varepsilon_y}\right]$$

EXAMPLE 8.2
For the data specified in Example 8.1, determine the longitudinal length required to achieve complete mixing in the traverse direction; that is, the location where the concentration varies no more than 5 percent over the cross section, for (a) tracer injection at the channel centerline, and (b) tracer injection at the side of the channel. How would this distance to the point of complete mixing be affected if the mixing took place as the channel was rounding a bend with a radius of 50 m (160 ft)?

Solution

Again, it is assumed that the tracer is completely mixed over the vertical dimension at the point of injection. Then, the distance to the point of complete mixing can be computed from Eq. (8.3.12) with $\bar{u} = 0.53$ m/s (1.7 ft/s) and $\varepsilon_y = 0.59$ m²/s (6.4 ft²/s).

1.   When the trace injection is at the centerline of the channel, the appropriate value of T is 300 m (980 ft), and

$$L = \frac{0.1\bar{u}T^2}{\varepsilon_y} = \frac{0.1(0.53)(300)^2}{0.59} = 8100\,\text{m}\,(27,000\,\text{ft})$$

2.   When the tracer is injected at the side of the channel, $2T$ is used in Eq. (8.3.12) for the width or

$$L = \frac{0.1\bar{u}(2T)^2}{\varepsilon_y} = \frac{0.1(0.53)\left[2(300)\right]^2}{0.59} = 32,000\,\text{m}\,(105,000\,\text{ft})$$

When the mixing takes place on a curve, the coefficient of turbulent transverse mixing is increased, and the distance to the point of complete mixing is decreased. For the data given, ε_y is, by Eq. (8.3.8)

$$\varepsilon_y = 25\frac{\bar{u}^2 d^3}{R_c^2 u_*} = 25\frac{0.53^2(10)^3}{50^2(0.099)} = 25\,\text{m}^2/\text{s}\left(300\,\text{ft}^2/\text{s}\right)$$

Then, for centerline injection

$$L = \frac{0.1\bar{u}(2T^2)}{\varepsilon_y} = \frac{0.1(0.53)\left[\left[2(300)\right]\right]^2}{0.59} = 8100\,\text{m}\,(27,000\,\text{ft})$$

and for injection at the side of the channel

$$L = \frac{0.1\bar{u}(2T)^2}{\varepsilon_y} = \frac{0.1(0.53)\left[2(300)\right]^2}{28} = 680\,\text{m}\,(2200\,\text{ft})$$

Thus, the distance in both cases required for complete mixing is significantly reduced.

Before a discussion of longitudinal dispersion is initiated, it is appropriate to note that most investigations of turbulent, transverse diffusion have addressed the problem from the viewpoint of channels with a free surface. Although the transverse diffusion process in ice-covered channels has received some attention (see, for example, Engmann, 1977), the results available at the present time must be considered indicative rather than conclusive. Apparently, the effect of an ice cover is to cause approximately a 50 percent reduction in the absolute value of ε_y when similar hydraulic conditions - constant depth, width, and discharge - are compared. Thus, in an ice-covered channel, a tracer plume will satisfy the complete mix criterion much further downstream that a tracer plume in a comparable channel with a free surface. Engmann's data also indicate that the dimensionless diffusion coefficient; i.e., ε_y/Ru_*, for ice-covered channels does not differ significantly from the corresponding coefficient for open water conditions.

In Section 8.2 the equation of longitudinal dispersion [Eq. (8.2.18)] was developed following the approach used by Elder (1959). At that time, it was noted that this equation is subject to a number of caveats. The most crucial of these is that Eq. (8.2.18) cannot be applied until an initial period of time has elapsed and the rate of advective transport either upstream or downstream of a vertical plane, which moves at the cross-sectional average velocity, depends only on the tracer concentration and E. This initial period during which Eq. (8.2.18) is not valid in terms of a dimensionless distance is given by

$$0 < x' < 0.4$$

where

$$x' = \frac{x\varepsilon_y}{uT^2} \tag{8.3.13}$$

with x being the longitudinal distance.

Elder (1959) in his development assumed that the velocity varied only in the vertical dimension and according Eq. (8.3.2). With this assumption, a theoretical expression for the longitudinal dispersion coefficient can be derived

$$E = 5.93du_* \tag{8.3.14}$$

In Table 8.1, a number of field- and laboratory-determined values of E are summarized along with values of E estimated by Eq. (8.3.14). From this table, it is evident that the Elder equation provides poor estimates of E in both natural and artificial channels. This is a result of Elder's neglecting to include the effect of the transverse velocity profile in his analysis. Fischer (1967) presented the following equation to estimate E in a natural channel when a cross section has been defined and gaged:

$$E = \frac{1}{A\frac{\partial c}{\partial x}} \iint_A u'c'\,dA = -\frac{1}{A}\int_0^T u'd\int_0^y \frac{1}{\varepsilon_y d}\int_0^y u'\,dy\,dy\,dy \qquad (8.3.15)$$

In practice, the integrals of Eq. (8.3.15) are replaced with summations or

$$E = -\frac{1}{A}\sum_{k=2}^N q'_k \Delta y \left[\sum_{j=2}^k \frac{\Delta y}{\varepsilon_y d_j}\left(\sum_{i=1}^{j-1} q'_i \Delta y\right)\right] \qquad (8.3.16)$$

where

$$q'_i = u'_i \frac{d_i + d_{i+1}}{2}$$

u_i = mean velocity of the ith element of the cross section

u'_i = $u_i - \bar{u}$

$\bar{u}$ = cross-sectional average velocity of flow

d_i = depth at beginning of the ith cross-sectional element

Δy = width of element (constant)

ε_y = $0.6du_*$ = transverse, turbulent diffusion coefficient between $(i-1)$ and ith cross-sectional elements

N = number of cross-sectional elements

Table 8.1. Summary of dispersion experiments (Fischer *et al.*, 1979).

Source	Channel	Depth (ft)	Width (ft)	Average Velocity (ft/s)	Shear Velocity (ft/s)	E observed (ft²/s)	Eq. (8.3.14)	Eq. (8.3.15)	Eq. (8.3.17)
Thomas (1958)	Chicago Ship Canal	26.5	160	0.88	0.0627	30	10	---	130
State of CA (Anon, 1962)	Sacramento River	13.1	---	1.7	0.17	160	13	---	---
Owens *et al.* (1964)	River Derwent	0.82	---	1.2	0.46	50	2.2	---	---
Glover (1964)	South Platte River	1.5	---	2.2	0.23	170	2.0	---	---
Schuster (1965)	Yuma Mesa A Canal	11.3	---	2.2	1.1	8.2	74	---	---
Fischer (1967)	Laboratory Channel	1.1	1.3	0.82	0.066	1.3	0.43	1.4	0.17
		1.5	1.4	1.5	0.12	2.7	1.1	2.7	0.36
		1.1	1.3	1.5	0.12	4.5	0.78	4.0	0.32
		1.1	1.1	1.4	0.11	2.7	0.72	2.7	0.22
		0.69	1.1	1.5	0.11	4.3	0.45	4.8	0.39
		0.69	0.62	1.5	0.13	2.4	0.53	1.8	0.11
Fischer (1968a)	Green-Duwamish River, WA.	3.6	66	---	0.16	70-91	3.4	84	---
Yotsukura *et al.* (1970)	Missouri River	8.9	660	5.1	0.24	16×10^3	14	---	58×10^3
Godfrey and Frederick (1970) [Predicted values of E from Fischer (1968b)]	Copper Creek, VA (below gage)	1.6	52	0.89	0.26	220	2.5	64	57
		2.8	59	2.0	0.33	230	5.5	300	170
		1.6	52	0.85	0.26	100	2.5	120	52
		2.8	59	2.0	0.33	150	5.5	160	170
	Clinch River, TN	6.9	200	3.1	0.34	580	14	920	1800
		6.9	170	2.7	0.35	510	14	590	960
		1.3	62	0.52	0.38	32	2.9	30	23
	Copper Creek, VA (above gage)	2.8	110	0.49	0.18	100	3.0	98	63
	Powell River, TN	1.9	120	0.69	0.16	87	1.8	320	250
	Clinch River, VA	5.1	79	2.3	0.14	100	4.2	42	510
	Coachella Canal, CA								
McQuivey and Keefer (1974)	Bayou Anacoco, CO	3.1	85	1.1	0.22	360	4.0	—	141
		3.0	120	1.3	0.22	420	4.0	---	400
	Nooksack River	2.5	210	2.2	0.89	380	13	—	1100

Source	Channel	Depth (ft)	Width (ft)	Average Velocity (ft/s)	Shear Velocity (ft/s)	E observed (ft²/s)	Eq. (8.3.14)	Eq. (8.3.15)	Eq. (8.3.17)
	Wind-Bighorn Rivers	3.6	190	2.9	0.39	450	8.3	–	2400
		7.1	230	5.1	0.56	1700	23	–	3800
	John Day River	8.1	110	2.7	0.59	700	58	–	200
		1.9	82	3.3	0.46	150	5.2	–	920
	Comite River	1.4	52	1.2	0.16	150	1.3	–	190
	Sabine River	6.7	340	1.9	0.16	3400	6.3	–	4300
		16	420	2.1	0.26	7200	25	–	2100
		7.7	230	1.4	0.33	1200	15	–	450
	Yadkin River	13	240	2.5	0.43	2800	33	---	710

EXAMPLE 8.3

For the cross section defined in Fig. 8.7 and Table 8.2, estimate E by Eq. (8.3.16) if $u_* = 0.100$ m/s (0.33 ft/s).

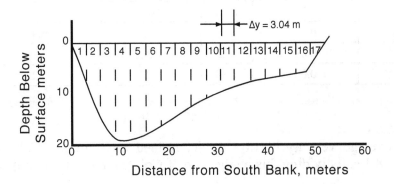

Figure 8.7. Cross section for Example 8.3.

Table 8.2   Channel definition for Example 8.3.

Element	Average velocity	Area	Element flow	Depth left side of element
	(m/s)	(m²)	(m³/s)	(m)
(1)	(2)	(3)	(4)	(5)
1	0.24	12.5	3.0	0
2	0.30	34.4	10	8.2
3	0.37	49.2	18	14
4	0.40	56.7	23	18
5	0.49	57.1	28	19
6	0.43	54.8	24	18
7	0.43	48.8	21	18
8	0.40	40.9	16	14
9	0.37	35.3	13	12
10	0.34	30.2	10	11
11	0.37	26.5	9.8	9.1
12	0.30	23.7	7.1	8.2
13	0.27	21.4	5.8	7.3
14	0.27	20.0	5.4	6.7
15	0.24	19.0	4.6	6.4
16	0.09	17.2	1.5	6.1
17	0.03	7.90	0.23	5.2

Solution

The average velocity of flow is determined by summing Col. (4) in Table 8.1 and dividing this by the sum of Col. (3) in the same table or

$$\bar{u} = \frac{\sum\limits_{i=1}^{N} q_i}{\sum\limits_{i=1}^{N} A_i} = \frac{200}{555} = 0.36\,\text{m/s}\left(1.2\,\text{ft/s}\right)$$

Then, values for Cols. (3) to (5) of Table 8.3 can be calculated. The values of transverse, turbulent diffusion coefficient in Col. (6) of Table

8.3, are estimated by using the depth of the left-hand side of the element; or for element 2

$$\varepsilon_y = 0.6du_* = 0.6(8.2)(0.100) = 0.49 \text{ m}^2/\text{s} \ (5.2 \text{ ft}^2/\text{s})$$

Table 8.3. Tabular solution of Example 8.3.

Element	Depth left side of element	u'_i	$u'_i A_i = q'_i$	$\sum q'_i$	ε_y	*
	(m)	(m/s)	(m³/s)	(m³/s)	(m²/s)	(m)
(1)	(2)	(3)	(4)	(5)	(6)	(7)
1	0	-0.12	-1.5	-1.5	0	0
2	8.2	-0.06	-2.1	-3.6	0.49	-1.1
3	14	0.01	0.49	-3.1	0.84	-2.1
4	18	0.04	2.3	-0.8	1.1	-2.5
5	19	0.13	7.4	6.6	1.1	-2.7
6	18	0.07	3.8	10.4	1.1	-1.6
7	18	0.07	3.4	13.8	1.1	-0.05
8	14	0.04	1.6	15.4	0.84	3.5
9	12	0.01	0.35	15.7	0.72	8.9
10	11	-0.02	-0.60	15.1	0.66	16
11	9.1	0.01	0.26	15.4	0.55	25
12	8.2	-0.06	-1.4	14.0	0.49	36
13	7.3	-0.09	-1.9	12.1	0.44	50
14	6.7	-0.09	-1.8	10.3	0.40	63
15	6.4	-0.12	-2.3	8.0	0.39	76
16	6.1	-0.27	-4.6	3.4	0.37	87
17	5.2	-0.33	-2.6	0.8	0.31	93

$$* \sum_{j=2}^{n} \frac{\Delta y}{\varepsilon_y d_j} \sum_{i}^{j-1} q'_i$$

The values of Col. (7) in Table 8.3 are computed as a running total. The value of this sum is by definition zero for element 1. For element 2, the value of Col. (7) is given by

$$\frac{\Delta y}{\varepsilon_y d_2} q_1 = \frac{3.0}{0.49(8.2)}(-1.5) = -1.1$$

and for element 3

$$\frac{\Delta y}{\varepsilon_y d_3} q_2 + \frac{\Delta y}{\varepsilon_y d_4} q_1 = -1.1 + \frac{3.0}{0.84(14)}(-3.6) = -2.0$$

The longitudinal dispersion coefficient is then estimated as

$$E = -\frac{\sum (\text{col } 4)(\text{col } 7)}{\text{flow area}} = -\frac{1100}{555} = 2.0\, \text{m}^2 / \text{s} \left(22\, \text{ft}^2 / \text{s} \right)$$

While estimates of E yielded by Eq. (8.3.15) are superior to those obtained with Eq. (8.3.14), they are not, in all cases, accurate. The primary reason for the discrepancies, which can be noted in Table 8.1, appear to result from the fact that no natural channel completely fulfills the assumptions inherent in the development of Eq. (8.3.15). For example, all natural channels and many constructed channels have bends, changes in cross sectional shape, pools, and many other irregularities, all of which contribute significantly to the dispersion process. Fischer *et al.* (1979) presented the following equation which can be used to estimate an approximate value of E:

$$E = \frac{0.011 \overline{u}^2 T^2}{d u_*} \tag{8.3.17}$$

This equation has the distinct advantage of estimating E from parameters which are usually known. In Table 8.1, value of E estimated by this method can be compared with observed values.

At this point, it is appropriate to note the presence and presumed effect of what has in the literature been termed *dead zones*. This terminology has been used to refer to areas along the boundaries of the channel, in which tracer is trapped, as the main cloud of tracer passes. The trapped tracer is released from the dead zone slowly and can cause measurable tracer concentrations to be observed downstream long after the main cloud of tracer has passed. The result is downstream concentration versus time plots which are not Gaussian and are characterized by a long tail (e.g., Fig. 8.8). At the present time, there is no effective method of quantifying the effect of dead zones either on the dispersion coefficient or the movement of a tracer plume downstream. One method, which has been used by a number of investigators, is to treat the tail of the concentration versus time plot as if it did not exist with the justification being that the mass of tracer in this tail is small (see, for example, Fischer, 1968a).

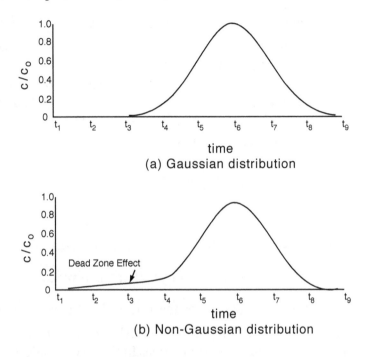

(a) Gaussian distribution

(b) Non-Gaussian distribution

Figure 8.8. Effect of dead zones on concentration versus time plots.

In an examination of the results summarized in Table 8.1, none of the methods which have been suggested for estimating E yields completely satisfactory results. To some extent, the difference between the predicted and observed values of the longitudinal dispersion coefficient are due to the inaccuracy of the method used to determine the observed values. For example, Fischer *et al.* (1979) have suggested that the observed value is usually accurate only within a factor of 2. Thus, the estimated values of E in Table 8.1 may be better than only a cursory examination would indicate.

There are essentially three methods of calculating a value of E from field data. First, from Section 8.2, by definition, the longitudinal dispersion coefficient measures the rate of change of the variance of the tracer cloud or

$$E = 0.5 \frac{d\sigma_x^2}{dt} \qquad (8.3.18)$$

where σ_x^2 = variance of the concentration distribution, with respect to distance along the channel. The application of Eq. (8.3.18) to field data is somewhat complicated by the fact that normal field data consist of tabulated tracer concentrations, at a specified distance downstream from the point of injection as a function of time. A typical data base is contained in Table 8.4. Equation (8.3.18) can be transformed to a form which is useful with the typical field data set or

$$E = \frac{\overline{u^2}}{2} \left(\frac{\sigma_{t2}^2 - \sigma_{t1}^2}{\overline{t_2} - \overline{t_1}} \right) \qquad (8.3.19)$$

where σ_{t1}^2 and σ_{t2}^2 = variance of the concentration versus time curves at upstream (station 1) and downstream (station 2), respectively, and $\overline{t_1}$ and $\overline{t_2}$ = mean times of passage of the tracer cloud past the upstream and downstream stations, respectively. Although Eq. (8.3.19) is theoretically correct, a number of practical limitations should be noted. First, a large initial dose of tracer is required to yield measurable downstream concentrations. Second, a long reach of prismatic channel is required. Third, the presence of dead zones can have a significant effect on the value of E, which is estimated by this method, since the long tail characteristic of dead zones will affect the value of σ^2.

EXAMPLE 8.4

A tracer is injected into a flow occurring in a natural channel, and as the resulting tracer cloud passes two downstream stations, it is sampled. The results of the sampling program are summarized in Table 8.4 and Fig. 8.9. If in the reach between the two sampling stations the average depth, width, and shear velocity are 2.77 ft (0.84 m), 60 ft (18 m), and 0.33 ft/s (0.10 m/s), respectively, estimate the longitudinal dispersion coefficient for this reach of channel using Eq. (8.3.19). *Note*: The data for this example were first presented by Fischer (1968b).

Table 8.4. Tracer concentration profiles at two stations for dispersion coefficient estimation (Fischer, 1968b).

Time relative to 1st observation (min)	Tracer concentration (mg/L)
Station 1: 7870 ft (2400 m) from point of injection	
0	0
3	0.26
6	0.67
9	0.95
11	1.09
13	1.13
15	1.10
17	1.04
19	0.95
24	0.72
29	0.50
34	0.31
39	0.21
49	0.08
59	0.02
Station 2: 13,550 ft (4100 m) from point of injection	
32	0
37	0
42	0.07
47	0.22
52	0.40
56	0.50
60	0.58
62	0.59
64	0.59
68	0.54
75	0.44
84	0.27
94	0.14
104	0.06
114	0.03
124	0.025
134	0.02
144	0

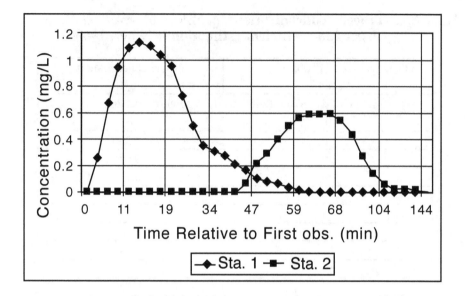

Figure 8.9.   **Tracer concentration at two stations as a function of time, from data in Table 8.4.**

Solution
The only caveat associated with the use of Eq. (8.3.19) is that both sampling stations must be sufficiently downstream for Eq. (8.2.18) to apply or

$$x' = \frac{x\varepsilon_y}{uT^3} > 0.4$$

With reference to Fig. 8.9, the average velocity of flow can be estimated as

$$\overline{u} = \frac{\Delta x}{\Delta t}$$

where Δx = distance between the sampling stations and Δt = travel time between the sampling stations. Δt can be approximated as the time difference between the tracer peaks or Δt = 49 min. Then,

$$\overline{u} = \frac{13,550 - 7870}{49} = 116\,\text{ft / min} = 1.94\,\text{ft / s}\left(0.59\text{m / s}\right)$$

and

$$\varepsilon_y = 0.6du_* = 0.6(2.77)(0.33) = 0.55 \text{ ft}^2/\text{s} \ (0.051 \text{ m}^2/\text{s})$$

x' is then computed as

$$x' = \frac{7870(0.55)}{1.94(60)^2} = 0.62$$

and therefore, Eq. (8.3.19) can be used to estimate E. The parameters $(\sigma_{ti})^2$ can be quickly estimated by

$$\sigma_{ti}^2 = \frac{\displaystyle\sum_{i=1}^{N} C_i t_i^2}{\displaystyle\sum_{i=1}^{N} C_i} - \left(\frac{\displaystyle\sum_{i=1}^{N} C_i t_i}{\displaystyle\sum_{i=1}^{N} C_i} \right)^2$$

where the time base used for the computations is not a factor, but the indicated summations must be performed over uniformly space readings. From Fig. 8.9, Table 8.5 is constructed, and we have the following results

	Station 1	Station 2
ΣC_i	9.08	4.26
$\Sigma C_i t_i$	181	298
$\Sigma C_i t_i^2$	4776	21,988
σ_{ti}^2	129	268

Although Fischer (1968b) was not precise in his definition of $\bar{t}_1$ and $\bar{t}_2$,

Levenspiel and Smith (1957) used the following definition

$$\bar{t}_i = \frac{\displaystyle\int_0^\infty Ct\,dt}{\displaystyle\int_0^\infty C\,dt}$$

Table 8.5. Tracer concentration profiles at two stations estimated from the data in Fig. 8.9.

Station 1		Station 2	
t (min)	C (mg/L)	t (min)	C (mg/L)
0	0	37	0
3	0.26	42	0.07
6	0.67	47	0.22
9	0.95	52	0.40
12	1.1	57	0.54
15	1.1	62	0.59
18	1.0	67	0.55
21	0.87	72	0.48
24	0.72	77	0.39
27	0.59	82	0.30
30	0.45	87	0.22
33	0.33	92	0.16
36	0.26	97	0.11
39	0.21	102	0.07
42	0.16	107	0.05
45	0.14	112	0.03
48	0.10	117	0.02
51	0.07	122	0.02
54	0.05	127	0.02
57	0.03	132	0.01
60	0.02	137	0.01
63	0	142	0

Application of this definition yields

$$\overline{t_1} = 17.8 \, \text{min}$$

and

$$\overline{t_2} = 68.9 \, \text{min}$$

Substituting these data in Eq. (8.3.19) yields

$$E = \frac{\overline{u^2}}{2}\left(\frac{\sigma_{t2}^2 - \sigma_{t1}^2}{\overline{t_2} - \overline{t_1}}\right) = \frac{116^2}{2}\left(\frac{268 - 129}{68.9 - 17.8}\right) = 18{,}300\,\text{ft}^2 \,/\,\text{min}\left(1700\,\text{m}^2 \,/\,\text{min}\right)$$

$$E = 300\,\text{ft}^2 \,/\,\text{s}\left(28\,\text{m}^2 \,/\,\text{s}\right)$$

The second method of estimating a value of E for a reach of channel uses what is termed in the literature a *routing procedure*. This methodology involves matching an observed concentration distribution at a distance x_2, below the point of tracer injection, with a distribution predicted by a model which uses a concentration distribution observed at a distance x_1, below the point of injection, where $x_1 < x_2$. From Fischer (1968b)

$$C\left(x_2,t\right) = \int_{-\infty}^{\infty} \overline{u}\,C\left(x,\tau\right)\frac{\exp\left\{\dfrac{-\left[\overline{u}\left(\overline{t_2} - \overline{t_1} - t + \tau\right)\right]^2}{4E\left(\overline{t_2} - \overline{t_1}\right)}\right\}}{\sqrt{4\pi E\left(\overline{t_2} - \overline{t_1}\right)}}\,d\tau \qquad (8.3.20)$$

where $C(x_1,\tau)$ =   observed tracer concentration at a distance x_1, below point of tracer injection at time τ

$\quad C(x_1,\tau)$ =   estimated tracer concentration at a distance x_2, below point of tracer injection at time t

$\quad \tau$ =   time variable of integration.

Although Eq. (8.3.20) indicates that the limits of integration are $-\infty < \tau < \infty$, in practice, the integration needs only to be performed over the interval $t(1) \le \tau \le t(2)$, where $t(1)$ and $t(2)$ are the beginning and ending points of the concentration distribution at x_1 or

For $\tau \le t(1)$ $C(x_1,\tau) = 0$

For $\tau \ge t(2)$ $C(x_1,\tau) = 0$

The use of Eq. (8.3.20) to estimate E for a reach of channel involves the following steps:

1.   A trial value of E is estimated using Eq. (8.3.19).

2.   The trial value of E, determined in the previous step, is used with Eq. (8.3.20) to estimate a concentration distribution at x_2.

3.   The predicted and observed concentration distributions at the downstream station are compared by computing the mean squared concentration difference between the two curves.

4.   A search is then made, by trial and error, to determine if there is a value of E, which minimizes the mean squared differences between the observed and predicted concentration distribution curves at the downstream station.

EXAMPLE 8.5

For the data given for Example 8.4, use Eq. (8.3.20) to estimate a value of E.

Solution

The most effective method of solving this problem is to develop a simple computer program which will perform the integration indicated in Eq. (8.3.20) to determine $C(x_2,t)$, when t, $\bar{u}$, $\bar{t}_1$, $\bar{t}_2$, E, and a matrix of $C(x_1,t)$ values are specified. From Example 8.4, the following variable values are available

$$\bar{u} = 116 \, \text{ft} \, / \, \text{min} \left(35 \text{m} \, / \, \text{min}\right)$$

$$\bar{t}_1 = 17.8 \, \text{min}$$

$$\bar{t}_2 = 68.9 \, \text{min}$$

The analysis of the data in Table 8.4 also provides an estimate of the correct value of E or

$$E = 18,300 \, \text{ft2} \, / \, \text{min} \, (1700 \, \text{m2} \, / \, \text{min})$$

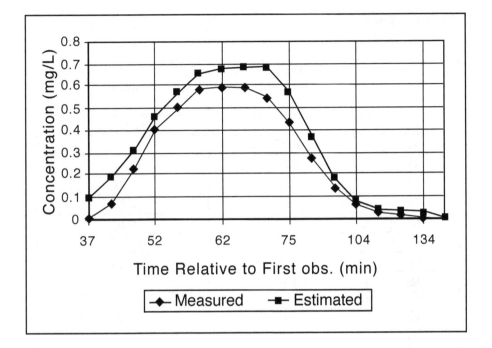

Figure 8.10.   Measured and estimated values of $C(x_2,t)$ for E = 19,500 ft^2/min.

By use of these data, value of $C(x_2,t)$ can be estimated (Table 8.6 and Fig. 8.10). With regard to the data summarized in this table and figure, the following are noted:

1.   The estimated peak value arrives at Station 2 after the measured peak value; and therefore, it might be concluded that the estimated value of $\bar{u}$ is low. No adjustment to this value can be made without further information.

2.   A dispersion coefficient, which minimizes some goodness of fit parameter, must be searched for; e.g.,

3. $$D = \frac{\left[C^*\left(x_2,t\right) - C\left(x_2,t\right)\right]^2}{N}$$

where $C^*(x_2,t)$ =   estimated concentration at Station 2
$C(x_2,t)$ =   measured concentration at Station 2
N   =   number of points considered

Table 8.6.   Values of $C(x_2,t)$ estimated by Eq. (8.3.20) for $E = 18,300$ ft^2/min.

Time (min)	Measured value of $C(x_2,t)$ (mg/L)	Estimated value of $C(x_2,t)$ (mg/L)
37	0	0.097
42	0.07	0.19
47	0.22	0.31
52	0.40	0.46
56	0.50	0.57
60	0.58	0.65
62	0.59	0.67
64	0.59	0.68
68	0.54	0.68
75	0.44	0.57
84	0.27	0.37
94	0.14	0.18
104	0.06	0.072
114	0.03	0.023
124	0.025	0.0055
134	0.02	0.00088
144	0	0

For $E = 18,300$ ft^2/min (1700 m^2/min)

$$D = \frac{0.1088}{17} = 0.0064$$

Additional trial value of E, both larger and smaller than 18,300 ft^2/min (1700 m^2/min) must be considered.

The third method of estimating E from field data is based on the solution of Eq. (8.2.18) for a pulse input of tracer (Krenkel, 1960 and Krenkel and Novotny, 1980) or

$$C = \frac{M}{A\sqrt{4\pi Et}} \exp\left[-\frac{\left(x - \overline{u}t\right)^2}{4Et}\right] \qquad (8.3.21)$$

where M = weight of tracer and A = flow area. Rearranging and taking the logarithm of both sides of this equation yields

$$\log\left(C\sqrt{t}\right) = \log\left(\frac{M}{A\sqrt{4\pi E}}\right) - \frac{\left(x - \bar{u}t\right)^2}{4Et}\log\left(e\right)$$

Then, a plot of $\log\left(C\sqrt{t}\right)$ versus $\left(x - \bar{u}t\right)^2 / t$ theoretically results in a straight line whose slope is $[log(e)/4E]$ from which E can be estimated. Although this approach to estimating E requires only a concentration versus time curve from a single downstream station, t is measure from the time of tracer injection. For this reason, the field data for either station in Table 8.4 cannot be used to estimate E, since the time of tracer injection is not known.

At this point, some observations regarding dye studies designed to determine a value of the dispersion coefficient are appropriate. First, it is tacitly assumed that the dye injected in the flow behaves in exactly the same manner as the water. In selecting a dye for a dispersion study, the following characteristics should be considered: (1) detectability, (2) toxicity, (3) solubility, (4) stability, and (5) cost (Kilpatrick *et al.*, 1970). Fluorescent dyes have generally been preferred in dispersion studies (e.g., Kilpatrick *et al.*, 1970 and Wilson, 1968). *Fluorescence* refers to the response of a substance to a light source of a given wavelength. Both rhodamine BA and rhodamine WT have been successfully used in such studies. In general rhodamine WT dye is preferred even though it costs more than rhodamine BA because rhodamine BA is absorbed readily by almost everything it comes in contact with; e.g., aquatic plants, suspended clays, and the channel boundaries. It is also noted that the U.S. Geological Survey recommends that (1) since most dyes have an affinity for plastics, particularly rhodamine WT, glass bottles should be used for sample collection, and (2) maximum dye concentration at a water supply intake should be limited to 10 μ/L (0.010 ppm).

Second, several empirical equations can be used to estimate the quantity of dye required to produce a specified peak concentration at a sampling site. The general form of these equations is

$$V = \phi\left(\frac{Q_{max}L}{\bar{u}}\right)^{\Gamma} C_p \qquad (8.3.22)$$

where V =   dye volume (L)

              Q_{max} =   maximum discharge in reach (ft³/s)

              L =   distance from point of injection to sampling point (mi)

              $\bar{u}$ =   average reach velocity (ft/s)

              C_p =   desired peak concentration at sampling point (μ /L)

              ϕ and Γ =   empirical coefficients

For example, if rhodamine WT 20 percent dye is used, it has been found

$$\phi = 3.4 \times 10^{-4} \tag{8.3.23}$$

and

$$\Gamma = 0.93 \tag{8.3.24}$$

Third, a minimum of two sampling points is required. The first point must be far enough downstream from the point of dye injection that complete lateral and vertical mixing have taken place. The second point of sampling is then at a convenient but sufficient distance downstream of the first sampling point.

Fourth, as noted by Kilpatrick *et al.* (1970), the schedule for collecting samples at a sampling point is the most uncertain aspect of any study plan. At a specified distance downstream from the point of injection, the time from injection until the peak concentration arrives at the sampling location can be estimated by

$$T_p = 1.47 \frac{L}{u} \tag{8.3.25}$$

Then, from Fig. 8.11, the duration, in hours, of the dye cloud passage at that point can be estimated. This figure also recommends the sampling interval which should be used at the downstream station. If the dye cloud is assume to be symmetrical in shape, then the leading edge of the dye cloud will arrive at the sampling station at

$$T_e = T_P - 0.5T_D \tag{8.3.26}$$

where T_e = expected time of arrival of the leading edge of the dye cloud, in hours, after the time of injection.

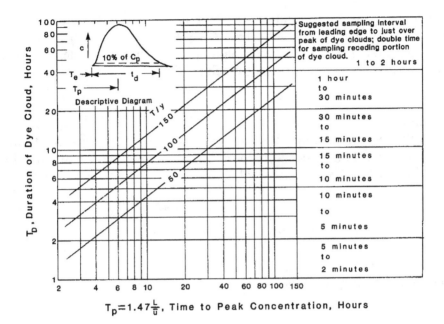

Figure 8.11.   **Duration of dye cloud as a function of travel time to peak concentration and the average channel width to depth ratio (Kilpatrick *et al.*, 1970).**

EXAMPLE 8.6

For the situation defined in Fig. 8.12, estimate the following: (1) the volume of rhodamine WT 20% dye which must be injected at mile 0 to

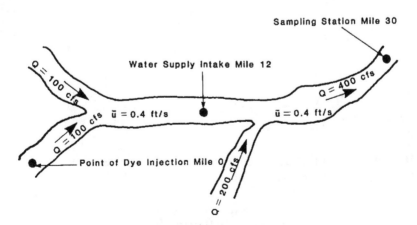

Figure 8.12. Example 8.6.

produce a peak concentration of $2\mu g/L$ at mile 30; (2) the peak concentration that can be expected at the water supply inlet located at mile 12; and (3) the time when sampling should begin at mile 30, if the mean width to depth ratio for this reach is 100.

Solution

The volume of rhodamine WT 20% dye, which must be injected at mile 0 to produce a peak concentration of $2\mu g/L$ at mile 30, is estimated by combining Eqs. (8.3.22) to (8.3.24) or

$$V = 3.4 \times 10^{-4} \left(\frac{Q_{max} L}{u} \right)^{0.93} C_p$$

The peak discharge for this reach is 400 ft³/s (11 m³/s); and therefore,

$$V = 3.4 \times 10^{-4} \left[\frac{400(30)}{0.4} \right]^{0.93} (2) = 9.9 \, L$$

The peak dye concentration, at the water supply inlet located 12 miles downstream from the point of dye injection, can also be estimated from Eqs. (8.3.22) to (8.3.24), where the average velocity is 0.4 ft/s (0.12 m/s), but Q_{max} = 200 ft³/s (5.7 m³/s). Rearranging the equation developed above yields

$$\frac{V}{C_p} = 3.4 \times 10^{-4} \left(\frac{Q_{max} L}{u} \right)^{0.93}$$

and

$$\frac{V}{C_p} = 3.4 \times 10^{-4} \left[\frac{200(12)}{0.4} \right]^{0.93} = 1.1$$

If 9.9 L of dye is injected at mile 0, then the expected concentration at the water supply diversion would be

$$C_p = \frac{9.9}{1.1} = 9.0 \mu g / L$$

The time at which sampling should commence 30 miles below the point of injection is estimated by Eq. (8.3.25) and (8.3.26) and Fig. 8.11. From Eq. (8.3.25)

$$T_p = 1.47 \frac{L}{u} = 1.47 \frac{30}{0.4} = 110 \, h$$

From Fig. 8.11, the length of time required for the dye cloud to pass the sampling station is estimated to be, for a width to depth ratio (T/y) of 100,

$$T_D = 45 \, h$$

Therefore, it is estimated by Eq. (8.3.26) that the leading edge of the dye cloud will arrive at sampling station

$$T_e = T_p - 0.5 T_D = 110 - 0.5(45) = 87 \, h$$

after the dye is injected upstream. Note, at best, this is a crude estimate of the time of arrival. From Fig 8.11, it is estimated samples should be taken at least every hour and perhaps every 30 min.

As noted previously, in the reach of channel defined by $0 < x' < 0.4$, where x' is defined by Eq. (8.3.13), the one-dimensional equation for longitudinal dispersion, Eq. (8.2.18), is not valid. In many respects, this limitation is crucial since this may be the critical channel reach from the viewpoint of satisfying environmental regulations. Although there are a number of techniques available for modeling the movement of a trace plume in this reach, space considerations dictate that only two of the many methods be discussed.

Fischer (1968a) developed a simple model of the spread of a tracer plume in this reach of channel. This model has the advantage of requiring that velocities be specified at only one cross section of the reach under consideration, but it also has the disadvantage of requiring that the channel be prismatic and the flow uniform. The steps involved in implementing this model are:

1.   The cross section is divided into N subsections of area A_1, A_2, ..., A_N (Fig. 8.13).

2.   The relative or advective velocity, u_j, of each channel subsection is given by

$$u'_j = u_j - \bar{u} \tag{8.3.27}$$

where u_j = measured velocity in the jth subsection and $\bar{u}$ = cross sectional average velocity. In performing this calculation, care must be used to ensure that the following condition is satisfied

$$\sum_{j=1}^{N} u'_j A_j = 0$$

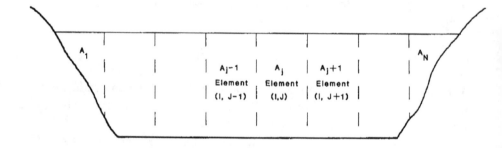

Figure 8.13. Definition of notation.

3.   The longitudinal distance step is defined by

$$\Delta x = \left| u'_j (\text{max}) \right| \Delta t$$

where Δt = selected time step and $u_j'(\text{max})$ = the largest advective velocity without regard to sign, Eq. (8.3.27). Thus, the average flow in the subsection where $u_j'(\text{max})$ occurs, moves at the rate of one longitudinal grid point per time step.

4.   A two-dimensional grid of tracer concentration, $C^k(i,j)$ is defined where the index i refers to the longitudinal distance in a coordinate system, which moves at the average velocity of the cross section, and the index j refers to the transverse location of

the subsection in the cross section (Fig. 8.13). The superscript k references the current time step.

5.   The computation advances in time by performing two sets of computations. First, the tracer is advected and then mixed in the transverse direction.

The advective movement of the tracer, in each subsection either upstream or downstream, is computed according to the value of u_j'. In performing the advective computation, it is usually advantageous to convert the advective velocities to units of mesh points per time step or

$$U_j = u_j' \frac{\Delta t}{\Delta x}$$

This operation results in a two-dimensional scratch matrix $D(i,j)$, which contains temporary computation values of the tracer concentration. The $D(i,j)$ matrix is defined by

If $U_j > 0$, then

$$D(i,j) = U_j \left[C^k(i-1,j) - C^k(i,j) \right] \tag{8.3.28}$$

If $U_j < 0$, then

$$D(i,j) = U_j \left[C^k(i,j) - C^k(i+1,j) \right] \tag{8.3.28}$$

Tracer concentrations, which are advected a fraction of a grid space, are divided between the grid point involved inversely as the distance from each. It should be noted that the use of Eq. (8.3.28) introduces what is termed *numerical dispersion*.
Transverse mixing between the channel subsections is accomplished by a second set of computations. The change in concentration at a grid point (i,j) due to transverse mixing is given by

$$\Delta C(i,j) = \frac{1}{A_j \Delta x} \left(\Delta M_{j,j+1} - \Delta M_{j-1,j} \right)$$

where the variable M designates the mass transport between the subsections. M is calculated under the assumption that the mass transport for the duration of the time step is given by

$$\Delta M_{j,j+1} = h_j \varepsilon_j \left(\frac{\Delta C_j}{s_j} \right) \Delta x \Delta t$$

where h_j = the area of the plane dividing subsections j and $j + 1$ per unit of downstream length, s_j = distance separating the centroids of the subsections j and $j + 1$, ΔC_j = difference in concentration between subsections j and $j + 1$ or

$$\Delta C_j = C(i,j + 1) - C(i,j)$$

and ε_j = turbulent transverse diffusion coefficient given by

$$\varepsilon_j = 0.6 d_j u_*$$

6.   When the above operations are combined, the tracer concentration at the grid point (i,j) and at time (k + 1) is

$$C^{k+1}\big((i,j)\big) = D(i,j) +$$

$$\left\{ \frac{h_j \varepsilon_j}{s_j} \Big[D(i,j+1) - D(i,j) \Big] - \frac{h_{j-1} \varepsilon_{j-1}}{s_{j-1}} \Big[D(i,j) - D(i,j-1) \Big] \right\} \frac{\Delta t}{A_j} \quad (8.3.29)$$

Fischer (1968a) noted that the numerical scheme defined by the above operations is stable for all values of j if

$$\frac{\varepsilon_j \Delta t}{s_j^2} < 0.5$$

The obvious limitations of the method described above are that the channel must be prismatic and the flow uniform. There are other approaches to solving this problem. For example, Harden and Shen (1979) developed a method that takes into account channel cross sectional irregularities but which involves a degree of mathematical sophistication which may not be warranted in many practical situations.

8.4.   NUMERICAL DISPERSION

In the foregoing sections of this chapter, the concepts of turbulent diffusion and dispersion have been summarized from the viewpoint of rather idealized situations.   In most of the cases treated, analytic solutions were available; however, in practice, situations where analytic solutions are applicable are rare; in general, a numerical solution of the governing equations is required. The use of finite difference or finite element methods to solve mass transport problems results in a fictitious transport of mass which is termed numerical dispersion, as was noted above. Although a detailed treatment of numerical methods is beyond the scope of this discussion, some consideration is required if the problem of numerical dispersion is to be understood. Additional information regarding finite difference methods is available in Forsyth and Wasow (1960) and Street (1973).

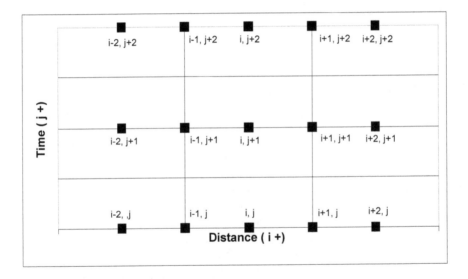

Figure 8.14.   One-dimensional   finite   difference   grid network.

In finite difference schemes, the partial differential equation to be solved is written in terms of a finite difference representation of each derivative. For example, consider the time-averaged, one dimensional dispersion equation for a nonprismatic channel or

$$A\frac{\partial C}{\partial t} + Q\frac{\partial C}{\partial x} = \frac{\partial}{\partial x}\left(EA\frac{\partial C}{\partial x}\right)$$

A grid system for the solution of this equation is shown in Fig. 8.14. In this equation, A, Q, and E are constant in time but variable in space. For the development of a finite difference scheme, it is convenient to represent the value of the tracer concentration at the ith longitudinal node and at the jth node in time as $C_{i,j}$. At time $t = 0$ or $j = 1$, values of C are specified for all values of i, and it is required that $C_{i,j}$ for $j \neq 1$ be determined for all i. In performing this task by finite differences, the terminology of explicit and implicit difference schemes arises. The phrase *explicit difference scheme* refers to a computational scheme in which values of $C_{i,j}$, $j \neq 1$ are determined from the solution of explicit equations; i.e., all derivatives are expressed in terms of known values except in the derivative describing the variation of C with time. In explicit difference schemes, the following expressions can be used:

1.   *Backward Difference Operator*

$$\frac{\partial C}{\partial x} \approx \frac{C_{i,j} - C_{i-1,j}}{\Delta x}$$

2.   *Forward Difference Operator*

$$\frac{\partial C}{\partial x} \approx \frac{C_{i+1,j} - C_{i,j}}{\Delta x}$$

3.   *Central Difference Operator*

$$\frac{\partial C}{\partial x} = \frac{C_{i+1,j} - C_{i-1,j}}{\Delta x}$$

The phrase *implicit difference scheme* refers to a computational scheme in which the values of $C_{i,j}$ are determined by solving a system of simultaneous equations; i.e., the expressions for the derivatives are written in terms of unknown values. For example,

$$\frac{\partial C}{\partial x} = \frac{1}{2}\left(\frac{C_{i+1,j+1} - C_{i-1,j+1}}{2\Delta x} + \frac{C_{i+1,j} - C_{i-1,j}}{2\Delta x} \right)$$

Explicit difference schemes are preferred because of their computational simplicity, but they are usually unstable from a numerical viewpoint unless a relatively small time step is used. Implicit schemes, while requiring the solution of a set of simultaneous equations, are stable from a numerical viewpoint, and a much larger time step can be used, which reduces the computational time required to obtain a solution. The characteristic of both schemes which gives rise to numerical dispersion is that mass concentrations can exist only at nodes within the system (Fig. 8.14). Thus, if a mass is advected and dispersed only part of the distance between two nodes, some mass is returned numerically to the point of origin, and some of the mass is numerically advected and dispersed to the next node.

As an example of the significance numerical dispersion, consider a case of pure advection or

$$\frac{\partial C}{\partial t} = -\overline{u}\frac{\partial C}{\partial x}$$

Let the concentration at node (i,j) be 1, and zero elsewhere. If $\overline{u}\dfrac{\Delta t}{\Delta x} = 0.5$ and a backward difference operator is used, then

$$C_{i,j+1} = C_{i,j} - \overline{u}\frac{\Delta t}{\Delta x}\left(C_{i,j} - C_{i-1,j}\right) = 1 - 0.5\left(1 - 0\right) = 0.5$$

and

$$C_{i+1,j+1} = C_{i+1,j} - \overline{u}\frac{\Delta t}{\Delta x}\left(C_{i+1,j} - C_{i,j}\right) = 0 - 0.5\left(0 - 1\right) = 0.5$$

Thus, in this example after one time step, half of the original concentration of the tracer remains at node i and half has moved to node $i + 1$. In fact, all of the tracer should be located between these nodes. At the beginning of the computation, the variance of the concentration distribution was 0, while at the end of the computation, the variance of the concentration distribution is

$$\sigma^2 = \int_{-\infty}^{\infty}\left(x - \overline{x}\right)^2 C dx = \left(-0.5\Delta x\right)^2 0.5 + \left(0.5\Delta x\right)^2 0.5 = 0.25\left(\Delta x\right)^2$$

If the effect of the numerical procedure on the distribution of tracer is assumed equivalent to a diffusive process, then by Eq. (8.2.6)

$$\frac{d\sigma^2}{dt} = 2E'$$

where E' = numerical dispersion coefficient and

$$E' = \frac{0.25(\Delta x)^2}{2\Delta t} = \frac{0.125(\Delta x)^2}{\Delta t}$$

This, then, is what is meant by the term *numerical dispersion*.

A variety of methodologies for dealing with the problem of numerical dispersion have been developed. First, if the numerical dispersion is small compared with the actual dispersion, then it can be ignored. Second, if E' can be accurately estimated, the value of E can be adjusted downward so that the physically correct result is obtained. Third, numerical schemes which minimize the amount of numerical dispersion in a particular situation can be adapted (e.g., Thatcher and Harleman, 1972).

8.5.   VERTICAL, TURBULENT DIFFUSION IN A CONTINUOUSLY STRATIFIED ENVIRONMENT

In Section 8.3 a theoretical expression which has been verified with both laboratory and field data for the vertical, turbulent diffusion coefficient was derived, Eq. (8.3.4). In a continuously, stably stratified flow, the vertical diffusion of both momentum and mass is inhibited by the stratification; and Eq. (8.3.4) is not valid. In such a situation, the eddy viscosity (i.e., the turbulent momentum diffusion coefficient) is larger than the eddy diffusivity (i.e., the turbulent diffusion coefficient) of heat and mass. In this section, the problem of estimating values of the eddy viscosity and diffusivity, in a continuously stratified fluid, is briefly treated; however, it must be emphasized that at the present time there does not exist an expression for either the eddy viscosity or diffusivity which is universally considered valid.

Rossby and Montgomery (1935) proposed an equation relating the stratified eddy viscosity for stratified flow ε_z^s to the corresponding value for homogenous conditions ε_z of the form

$$\frac{\varepsilon_z^s}{\varepsilon_z} = \frac{1}{1 + \beta_{RM} Ri} \qquad (8.5.1)$$

where β_{RM} = a coefficient and Ri = gradient Richardson number, Eq. (1.2.2). In the development of Eq. (8.5.1), it was assumed that the change in kinetic energy per unit mass in going from a neutral or unstratified condition to a stably stratified condition was equal to the potential energy due to displacement over the mixing length from the equilibrium position with a different density. Kent and Pritchard (1957) also used a conservation of energy argument to develop an equation or

$$\frac{\varepsilon_z^s}{\varepsilon_z} = \frac{1}{\left(1 + \beta_{KP} Ri\right)^2} \qquad (8.5.2)$$

where β_{KP} = a coefficient. Holzman (1943) hypothesized a critical value of the gradient Richardson number above which turbulence to accomplish mixing existed, and this hypothesis resulted in

$$\frac{\varepsilon_z^s}{\varepsilon_z} = \left(1 - \beta_{HRi} Ri\right) \qquad (8.5.3)$$

where β_H = a coefficient. Mamayev, in Anonymous (1974), argued that the ratio of ε_z^s to ε_z should decrease exponentially with increasing values of Ri or

$$\frac{\varepsilon_z^s}{\varepsilon_z} = e^{\beta_M Ri} \qquad (8.5.4)$$

where β_M = a coefficient. Munk and Anderson (1948) proposed a generalized form of the Rossby and Montgomery (1935) and Holzman (1943) equations or

$$\frac{\varepsilon_z^s}{\varepsilon_z} = \left(1 + \beta_{MA} Ri\right)^{\alpha_{MA}} \qquad (8.5.5)$$

where β_{MA} and α_{MA} = coefficients.   Equation (8.5.5) must be considered an empirical equation with two free coefficients.   French (1979) proposed a second empirical relationship of the form

$$\varepsilon_z^s = \beta_F \left(\frac{\varepsilon_z}{1 + R_0} \right)^{\alpha_F}$$

(8.5.6)

where β_F and α_F = coefficients and

$$R_0 = \frac{gD \dfrac{\Delta\rho}{\rho}}{u_*^2}$$

(8.5.6)

Finally, Odd and Rodger (1978) developed a hybrid model based on the original hypothesis of Rossby and Montgomery (1935) or

1.   If there is a significant peak in the vertical profile of Ri, at a distance $z = z_0$ from the bottom boundary where θ peak gradient Richardson number, then

$$\frac{\varepsilon_z^s}{\varepsilon_z} = \frac{1}{1 + \beta_{OR}\theta} \quad \text{for } \theta \le 1$$

and

$$\frac{\varepsilon_z^s}{\varepsilon_z} = \frac{1}{1 + \beta_{OR}} \quad \text{for } \theta > 1$$

(8.5.7)

where β_{OR} = a coefficient.

2.   If there is no significant peak in the vertical profile of Ri, then

$$\frac{\varepsilon_z^s}{\varepsilon_z} = \frac{1}{1 + \beta_{OR}Ri} \quad \text{for } Ri \le 1$$

and

$$\frac{\varepsilon_z^s}{\varepsilon_z} = \frac{1}{1+\beta_{OR}} \quad \text{for Ri>1} \tag{8.5.8}$$

Equations (8.5.7) and (8.5.8) are applied throughout the vertical dimension, but near the boundaries if $\varepsilon_z^s > \varepsilon_z$ then ε_z is used.

The problem with all of the above methodologies is that they cannot be shown to be universally valid. Values for some of the coefficients in the above equations are summarized in Table 8.7. With regard to the data summarized in this table, the following should be noted:

Table 8.7. Summary of coefficient values for turbulent vertical diffusion of momentum in continuously stratified channel flow.

Equation	β	α	Source
8.5.1	2.5 5.0 30.3	– – --	Nelson (1972) Anonymous (1974) French and McCutcheon (1983)
8.5.2	9.8	--	French and McCutcheon (1983)
8.5.3	3.3	--	Nelson (1972)
8.5.4	0.4	--	Anonymous (1974)
8.5.5	10 30	-0.5 -0.5	Munk and Anderson (1948) Anonymous (1974)
8.5.6	0.31 0.062	0.747 0.379	French (1979) French and McCutcheon (1983)
8.5.7	180	--	Odd and Rodger (1978)

1.   Nelson (1972) used published oceanographic, atmospheric, and pipe flow data for his analysis, and the same was true of the analysis by the Delft Hydraulics Laboratory (Anonymous,

1974). Thus, these investigators had no control over the quality of their data.

2.   The data used by French (1979) were taken under laboratory conditions, but the flume used for these experiments had small width-to-depth ratio, and the results may be unduly affected by this fact.

3.   Odd and Rodger (1978) used field data from a reach of channel affected by the tide. The Odd and Rodger (1978) data set is perhaps the best data regarding the turbulent vertical diffusion of momentum under stratified conditions presently available.

4.   French and McCutcheon (1983) used the Odd and Rodger (1978) data set for their analysis. The coefficient value determined in this work for Eq. (8.5.7) and (8.5.8) differs from that of Odd and Rodger (1978), because of a difference in the definition of goodness of fit.

5.   In the past, Eq. (8.5.1) has been the most commonly used method of estimating ε_z^s (Nelson, 1972); however, the methods of Odd and Rodger (1978) and French and McCutcheon (1983) or French (1979) may be superior.

6.   The Delft Hydraulics Laboratory (Anonymous, 1974) concluded that when $Ri < 0.7$, the scatter of the data available is so great that no best-fit equation can be selected.

   A number of models for the eddy diffusivity in stratified flow have also been proposed. One of the most frequently used is

$$\frac{\left(\varepsilon_z^s\right)^M}{\varepsilon_z} = c\left(1 + \beta' Ri\right)^\alpha \tag{8.5.9}$$

where $\left(\varepsilon_z^s\right)^M$ = stratified flow, vertical eddy diffusivity, and c, β', and α = coefficients. Munk and Anderson (1948) estimated that $c = 1$, $\beta' = 3.33$, and $\alpha = 1.5$.

At this time, it is appropriate to observe that stratification apparently also acts to reduce the value of the turbulent transverse diffusion coefficient; however, the results presently available in this area (Sumer, 1976), are not yet extensive enough to warrant inclusion.

8.6.   BIBLIOGRAPHY

Anonymous, "Sacramento River Water Pollution Survey," Bulletin No. 111, State of California, Department of Water Resources, Sacramento, 1962.

Anonymous, "Momentum and Mass Transfer in Stratified Flows," Report No. R880, Delft Hydraulics Laboratory, Delft, The Netherlands, December 1974.

Benjamin, J.R., and Cornell, C.A., *Probability, Statistics, and Decision for Civil Engineers*, McGraw-Hill Book Company, New York, 1970.

Elder, J.W., "The Dispersion of Marked Fluid in Turbulent Shear Flow," *Journal of Fluid Mechanics*, vol. 5, 1959, pp. 544-560.

Engmann, E.O., "Turbulent Diffusion in Channels with a Surface Cover," *Journal of Hydraulic Research*, International Association for Hydraulic Research, vol. 15, no. 4, 1977, pp. 327-335.

Fischer, H.B., "The Mechanics of Dispersion in Natural Streams," *Proceedings of the American Society of Civil Engineers, Journal of the Hydraulics Division*, vol. 93, no. HY6, November 1967, pp. 187-216.

Fischer, H.B., "Methods for Predicting Dispersion Coefficients in Natural Streams with Application to Lower Reaches of the Green and Duwamish Rivers," Geological Survey Professional Paper 582-A, U.S. Geological Survey, Washington, 1968a.

Fischer, H.B. "Dispersion Predictions in Natural Streams," *Proceedings of the American Society of Civil Engineers, Journal of the Sanitary Engineering Division*, vol. 94, no. SA6, October 1968b, pp. 927-943.

Fischer, H.B., "The Effect of Bends on Dispersion in Streams," *Water Resources Research*, vol. 5, no. 2, 1969, pp. 496-506.

Fischer, H.B., H.B., List, E.J., Koh, R.C.Y., Imberger, J., and Brooks, N.H., *Mixing in Inland and Coastal Waters*, Academic Press, New York, 1979.

Forsythe, G.E., and Wasow, W.R., *Finite Difference Methods for Partial Differential Equations*, Wiley, New York, 1960.

French, R.H., "Vertical Mixing in Stratified Flows," *Proceedings of the American Society of Civil Engineers, Journal of the Hydraulics Division*, vol. 105, no. HY9, September 1979, pp. 1087-1101.

French, R.H., McCutcheon, S.C. and Martin, J.L., "Environmental Hydraulics," in *Hydraulic Design Handbook*, L.W. Mays, ed., McGraw-Hill, New York, NY, 1999.

French, R.H., and McCutcheon, S.C., "Vertical Momentum Transfer in Continuously Stratified Channel Flow," Water Resources Center, Desert Research Institute, Las Vegas, NV, 1983.

Glover, R.E., "Dispersion of Dissolved or Suspended Materials in Flowing Streams," Geological Survey Professional Paper 433-B, U.S. Geological Survey, Washington, 1964.

Godfrey, R.G., and Frederick, B.J., "Dispersion in Natural Streams," Geological Survey Professional Paper 433-K, U.S. Geological Survey, Washington, 1970.

Harden, T.O., and Shen, H.R., "Numerical Simulation of Mixing in Natural Rivers," *Proceedings of the American Society of Civil Engineers, Journal of the Hydraulics Division*, vol. 105, no. HY4, April 1979, pp. 393-408.

Holley, E.R., "Unified View of Diffusion and Dispersion," *Proceedings of the American Society of Civil Engineers, Journal of the Hydraulics Division*, vol. 95, no. HY2, March 1969, pp. 621-631.

Holzman, B., "The Influence of Stability on Evaporation," *Boundary Layer Problems in the Atmosphere and Ocean*, W.G. Valentine (ed.), vol. XLIV, article 1, 1943, pp. 13-18.

Kent, R.E., and Pritchard, D.W., "A Test of Mixing Length Theories in a Coastal Plain Estuary," *Journal of Marine Research*, vol. 1, 1957, pp. 456-466.

Kilpatrick, F.A., Martens, L.A., and Wilson J.F., Jr., "Measurement of Time of Travel and Dispersion by Dye Tracing," *Techniques of Water-Resources Investigations of the United States Geological Survey*, chapter A9, book 3, U.S. Geological Survey, Washington, 1970.

Krenkel, P.A., "Turbulent Diffusion and the Kinetics of Oxygen Absorption," Ph.D. thesis, University of California, Berkeley, 1960.

Krenkel, P.A., and Novotny, V., *Water Quality Management*, Academic Press, New York, 1980.

Lau, Y.L., and Krishnappen, B.G., "Transverse Dispersion in Rectangular Channels," *Proceedings of the American Society of Civil Engineers, Journal of the Hydraulics Division*, vol. 103, no. HY10, October, 1977, pp. 1173-1189.

Levenspiel, O., and Smith, W.K., "Notes on the Diffusion Type Model for the Longitudinal Mixing of Fluid in Flow," *Chemical Engineering Science*, vol. 6, 1957, pp. 227-233.

McQuivey, R.S. and Keefer, T., "Simple Method for Predicting Dispersion in Streams," *Proceedings of the American Society of Civil Engineers, Journal of the Environmental Engineering Division*, vol. 100, no. EE4, August, 1974, pp. 997-1011.

Munk, W.H., and Anderson, E.R., "Notes on the Theory of Thermocline," *Journal of Marine Research*, vol. 1, 1948, p. 276.

Nelson, J.E., "Vertical Turbulent Mixing in Stratified Flow - A Comparison of Previous Experiments," Report No. WHM-3, Hydraulic Engineering Laboratory, University of California, Berkeley, December, 1972.

Odd, N.V.M., and Rodger, J.G., "Vertical Mixing in Stratified Tidal Flows," *Proceedings of the American Society of Civil Engineers, Journal of the Hydraulics Division*, vol. 104, no. HY3, March 1978, pp. 337-351.

Owens, M., Edwards, R.W., and Gibbs, J.W., "Some Reaeration Studies in Streams," *Air-Water Pollution Institute Journal*, vol. 8, 1964, pp. 469-486.

Rossby, C.G., and Montgomery, R.B., "The Layer of Friction Influence in Wind and Ocean Currents," *Papers in Physical Oceanography and Meteorology*. Vol. 3, no. 3, 1935.

Schuster, J.C., "Canal Discharge Measurements with Radioisotopes," *Proceedings of the American Society of Civil Engineers, Journal of the Hydraulics Division*, vol. 91, no. HY2, March 1965, pp. 101-124.

Street, R.L., *Analysis and Solution of Partial Differential Equations*, Brooks/Cole, Monterey, CA, 1973.

Sumer, S.M., "Transverse Dispersion in Partially Stratified Tidal Flow," Report No. WHM-20, Hydraulic Engineering Laboratory, University of California, Berkeley, May 1976.

Taylor, G.I., "Dispersion of Soluble Matter in Solvent Flowing Slowly through a Tube," *Proceedings of the Royal Society of London*, series A, vol. 219, August 25, 1953, pp. 186-203.

Taylor, G.I., "The Dispersion of Matter in a Turbulent Flow through a Pipe," *Proceedings of the Royal Society of London*, series A, vol. 223, May 20, 1954, pp. 446-468.

Thatcher, M., and Harleman, D.R.F., "A Mathematical Model for the Prediction of Unsteady Salinity Intrusion in Estuaries," R.M. Parsons Laboratory Report No. 144, Massachusetts Institute of Technology, Cambridge, Mass., 1972.

Thomas, I.E., "Dispersion in Open Channel Flow," Ph.D. thesis, Northwestern University, Evanston, IL, 1958.

Wilson, J.F., "Fluormetric Procedures for Dye Tracing," *Techniques of Water-Resources Investigations of the U.S. Geological Survey*, chapter A12, book 3, U.S. Geological Survey, Washington, 1968.

Yotsukura, N., Fischer, H.B., and Sayre, W.W., "Measurements of Mixing Characteristics of the Missouri River between Sioux City, Iowa and Portsmouth, Nebraska," Water Supply Paper 1899-G, U.S. Geological Survey, Washington, 1970.

8.7.   PROBLEMS

1.   A facility discharges 5 ft³/s (0.14 m³/s) of wastewater containing 100 ppm (100 mg/L) of a conservative substance into the center of a very wide meandering channel. The stream is 25 ft (7.6 m) deep, the mean velocity of flow is 1.75 ft/s (0.53 m/s), and the shear velocity is 0.15 ft/s (0.05 m/s). Assuming this discharge is well mixed in the vertical, estimate the maximum concentration of the substance 500 ft (152 m), 1000 ft (305 m), and 1500 ft (457 m) downstream.

2.   A facility discharges a conservative substance at the side of a straight rectangular channel. If the channel is 610 m (2000 ft) wide, $n = 0.02$, $S = 0.0002$ m/m, and conveys a flow of 500 m³/s (19,800 ft³/s), estimate the length of channel required for complete mixing.

3.   An industry discharges 4 million gallons per day (0.174 m³/s) of effluent containing 400 ppm (400 mg/L) of a conservative substance at the center of a slowly meandering channel 200 ft (61 m) wide and 20 ft (6.1 m) deep. If the shear velocity is 0.2 ft/s (0.06 m/s) and the average velocity is 1 ft/s (0.30 m/s), then
   a.   What is the width of the plume and maximum concentration 1000 ft (305 m) downstream?
   b.   What is the distance to the pint of complete mixing?

4.   A conservative substance is discharged at the bottom and in the center of a straight rectangular channel in which
        Average velocity of flow = 0.61 m/s (2 ft/s)
           Depth of flow = 1.5 m (4.9 ft)
           Slope = 0.0002 m/m
           Width = 60 m (197 ft)
   a.   If the substance is completely mixed in the transverse direction, what is the length of channel required for complete vertical mixing?
   b.   If the substance is completely mixed in the vertical direction, what is the length of channel required for complete transverse mixing?

5.   For the cross section and field data summarized below, for a
      meandering channel estimate the value of the dispersion
      coefficient if $\Delta y = 2.5$ ft (0.76 m) and $u_* = 0.33$ ft/s (0.10 m/s).

Element	Dist. from. left bank to start of Element	Element mean velocity	Depth, left side of element
	(ft)	(ft/s)	(ft)
(1)	(2)	(3)	(4)
1	0	0.15	0
2	2.5	0.25	1.00
3	5.0	0.35	2.15
4	7.5	0.42	2..61
5	10.0	0.65	3.30
6	12.5	1.35	3.41
7	15.0	1.80	3.32
8	17.5	2.30	3.14
9	20.0	2.35	3.00
10	22.5	2.40	3.01
11	25.0	2.50	3.10
12	27.5	2.55	2.99
13	30.0	2.70	2.78
14	32.5	2.75	2.64
15	35.0	2.65	2.59
16	37.5	2.45	2.66
17	40.0	2.30	2.90
18	42.5	2.15	2.98
19	45.0	1.85	2.84
20	47.5	1.50	2.78
21	50.0	1.10	2.72
22	52.5	0.70	2.39
23	55.0	0.40	1.82
24	57.5	0.20	0.84
25	60.0	0.10	0.34

6.   For the cross section and field data summarized below, estimate
      the value of the dispersion coefficient if $S = 0.001$ ft/ft and the

channel meanders. Is there any problem apparent with the data provided?

Element	Element Avg. Velocity. (ft/s)	Element area (ft²)	Element flow (ft³/s)	Depth, left side of element (ft)
(1)	(2)	(3)	(4)	(5)
1	0.08	400	32	0
2	0.19	630	120	8
3	0.32	1800	580	19
4	0.48	3200	1500	32
5	0.43	880	380	40
6	0.43	3300	1400	40
7	0.42	1200	500	38
8	0.42	1100	460	35
9	0.52	3200	1700	33
10	0.38	1300	490	26
11	0.38	1200	460	23
12	0.52	1100	570	20
13	0.44	1300	570	18
14	0.30	1100	330	16
15	0.32	800	260	13
16	0.35	960	340	11
17	0.18	880	160	11

7.   For the cross section defined in the following table, estimate the dispersion coefficient if $\Delta y = 15$ ft (4.6 m) and $u_* = 0.4$ ft/s (0.12 m/s). Compare the estimated values obtained by using Eq. (8.3.17).

Element	Element Avg. Vel. (ft/s)	Element area (ft²)	Element flow (ft³/s)	Depth, left side of element (ft)
(1)	(2)	(3)	(4)	(5)
1	0.80	202	162	0
2	1.0	548	548	27
3	1.2	788	946	46
4	1.3	907	1180	59
5	1.6	907	1450	62
6	1.4	885	1240	59
7	1.4	788	829	59
8	1.3	638	674	46
9	1.2	562	544	39
10	1.1	495	514	36
11	1.2	428	382	30
12	1.0	382	310	27
13	0.90	345	290	24
14	0.90	322	243	22
15	0.79	308	83	21
16	0.30	278	12.8	20
17	0.10	128	160	17

8.   A facility accidently discharges a slug of pollutant into a river with a natural flow section that is approximated as a rectangle 200 ft (61 m) wide and 5 ft (1.5 m) deep. If the velocity of flow is 2 ft/s (0.61 m/s) and the longitudinal slope is 0.002 ft/ft, estimate the distance downstream before a dispersion analysis can be used. Estimate an appropriate value of the dispersion coefficient for this problem.

9.   For the stream reach and flow conditions shown in the following figure, estimate the volume of Rhodamine WT 20% dye that should be injected at Mile 0 to produce a peak concentration of 5 μ/L at Mile 30.

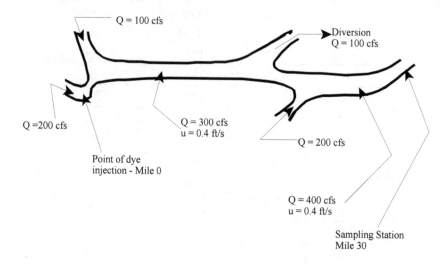

Q = 100 cfs

Diversion
Q = 100 cfs

Q = 200 cfs

Q = 300 cfs
u = 0.4 ft/s

Q = 200 cfs

Point of dye
injection - Mile 0

Q = 400 cfs
u = 0.4 ft/s

Sampling Station
Mile 30

CHAPTER 9

UNSTEADY FLOW: HYDROLOGIC AND HYDRAULIC APPROACHES

9.1. INTRODUCTION

Many open-channel flow phenomena of great importance to the hydraulic engineer involve flows that are unsteady; i.e., either the depth of flow and/or the velocity of flow varies with time. Although a limited number of gradually varied, unsteady flow problems can be solved analytically, most problems in this category require a numerical solution of the governing equations and boundary and initial conditions. In this chapter, only gradually varied, unsteady flow phenomena are discussed. This terminology refers to flows in which the curvature of the wave profile is mild; the change of depth with time is gradual; and the vertical acceleration of the water particles is negligible in comparison to the total acceleration; and the effect of boundary friction must be taken into account. Examples of gradually varied, unsteady flows include flood waves, tidal flows, and waves generated by the slow operation of control structures such as sluice gates and navigation locks.

The mathematical models available to treat gradually varied, unsteady flow problems can generally be divided into two categories: (1) hydraulic models which solve the St. Venant equations, and (2) hydrologic models which solve various approximations of the St. Venant equations. In this chapter, both types of models are considered.

The St. Venant equations, which were derived in Chapter 1, consist of the equation of continuity

$$\frac{\partial y}{\partial t} + y \frac{\partial u}{\partial x} + u \frac{\partial y}{\partial x} = 0 \qquad (9.1.1)$$

and the one-dimensional conservation of momentum equation

$$\frac{\partial u}{\partial t} + u\frac{\partial u}{\partial x} + g\frac{\partial y}{\partial x} - g\left(S_x - S_f\right) = 0 \tag{9.1.2}$$

Alternate but equally valid forms of these equations are

$$T\frac{\partial y}{\partial t} + \frac{\partial\left(Au\right)}{\partial x} = 0 \tag{9.1.3}$$

and

$$\frac{1}{g}\frac{\partial u}{\partial t} + \frac{u}{g}\frac{\partial u}{\partial x} + \frac{\partial y}{\partial x} + S_f - S_x = 0 \tag{9.1.4}$$

where
u	=	velocity in the longitudinal direction
x	=	longitudinal coordinate
T	=	top width
A	=	flow area
S_x	=	slope of the channel bed in the longitudinal direction
S_f	=	friction slope
g	=	acceleration of gravity.

Equations (9.1.1) and (9.1.2) or (9.1.3) and (9.1.4) compose a group of gradually varied, unsteady models which will be termed *complete, one-dimensional dynamic models*. Being complete, this group of models can provide accurate results regarding unsteady flow; but, at the same time, they can be very demanding of computer resources. The models in this group are also limited by the assumptions required in the development of the St. Venant equations and the assumptions required to apply them to a specific problem; e.g., assumptions regarding channel irregularities are usually required. From the group of models termed here as complete, dynamic models, two groups of simplified models can be derived by making various assumptions regarding the relative importance of various terms in the conservation of momentum equation [Eq. (9.1.2) or (9.1.4)].

The development and understanding of approximate models can, to some degree, be facilitated by rearranging Eq. (9.1.3) into the form of a rating equation which relates the discharge directly to the depth of flow (Weinmann and Laurenson, 1979). In general, the flow rate is given by

$$Q = \Gamma A R^m \sqrt{S_f} \tag{9.1.5}$$

where Γ =   empirical resistance coefficient
    R =   hydraulic radius
    m $\doteq$ empirical exponent.

In unsteady flow, S_f varies with both the slope of the wave and the depth of flow. In the case of a steady, uniform flow, the normal discharge is given by

$$Q = Q_N = \Gamma A R^m \sqrt{S_x}$$

or

$$\Gamma A R^m = \frac{Q_N}{\sqrt{S_x}} \tag{9.1.6}$$

Substituting Eq. (9.1.6) in Eq. (9.1.5) yields

$$Q = Q_N \sqrt{\frac{S_f}{S_x}} \tag{9.1.7}$$

Solving Eq. (9.1.4) for S_f

$$S_f = S_x - \frac{1}{g}\frac{\partial u}{\partial t} - \frac{u}{g}\frac{\partial u}{\partial x} - \frac{\partial y}{\partial x}$$

and substituting this expression in Eq. (9.1.7) yields

$$Q = Q_N \left(1 - \frac{1}{S_x}\frac{\partial y}{\partial x} - \frac{u}{S_x g}\frac{\partial u}{\partial x} - \frac{1}{S_x g}\frac{\partial u}{\partial t} \right)^{1/2} \tag{9.1.8}$$

Equation (9.18) is termed a *looped rating curve* (Fig. 9.1). In this figure, the points *A* and *B* indicate the points of maximum flow and maximum depth, respectively. The width of the loop and, therefore, the order of accuracy achieved by the approximate methods depend on the magnitude of the secondary terms in Eq. (9.1.8).

$$Q = Q_N \left(1 - \frac{1}{S_x} \frac{\partial y}{\partial x} - \frac{u}{S_x g} - \frac{1}{S_x g} \frac{\partial u}{\partial t} \right)^{1/2} \tag{9.1.8}$$

Kinematic wave

Diffusion analogy

Complete dynamic

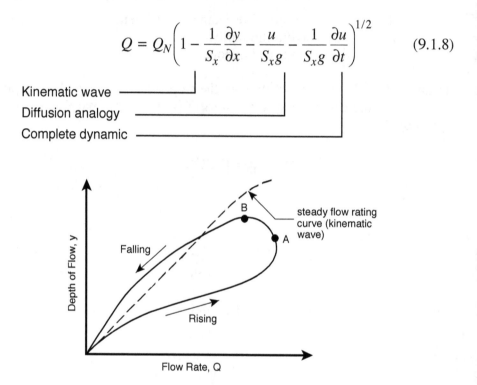

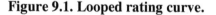

Figure 9.1. Looped rating curve.

The group of models known as the diffusion analogy models is based on the continuity equation

$$\frac{\partial A}{\partial t} + \frac{\partial Q}{\partial x} = 0 \tag{9.1.9}$$

which is a variant of Eq. (9.1.3) and the indicated terms of Eq. (9.1.8). These two equations can then be combined to yield a single equation

$$\frac{\partial Q}{\partial t} + c \frac{\partial Q}{\partial x} = D \frac{\partial^2 Q}{\partial x^2} \tag{9.1.10}$$

where c = a coefficient responsible for the translation characteristics of the wave, and D = the diffusion coefficient, which simulates the attenuation of the wave as it passes down the channel. For regular channels

$$c = \frac{1}{T}\frac{dQ}{dy} \qquad (9.1.11)$$

and

$$D = \frac{Q}{2TS_x} \qquad (9.1.12)$$

The similarity of Eqs. (9.1.10) and (8.2.20) is noted. If c and D are evaluated by fitting them to observed hydrographs, they can account for the effects of channel irregularities and flood plain storage.

In essence, the diffusion model assumes that, in the momentum equation, the inertia terms are negligible relative to the pressure, friction, and gravity terms. Ponce *et al.* (1978), in evaluating the range of applicability of the diffusion type of model, claimed that the diffusion model yields reasonable results in comparison with the complete dynamic model when

$$T_p S_x \left(\frac{g}{y_N} \right)^{1/2} \geq 30 \qquad (9.1.13)$$

where T_P = wave period of sinusoidal perturbation of steady, uniform flow and y_N = steady, uniform depth of flow. Rearranging Eq. (9.1.13) yields

$$T_p \geq \frac{30}{S_x} \left(\frac{y_N}{g} \right)^{1/2} \qquad (9.1.14)$$

If Eq. (9.1.14) is satisfied, a diffusion model will accurately approximate the unsteady flow, and such a model can be used in place of the complete dynamic model.

The kinematic wave group of models assumes that the discharge is always equal to the normal discharge; therefore, the discharge is always a single-valued function of the depth of flow (Fig. 9.1). The St. Venant equations are thus reduced to Eq. (9.1.9) and the indicated terms in Eq. (9.1.8). Combining these equations yields

$$\frac{1}{c}\frac{\partial Q}{\partial t} + \frac{\partial Q}{\partial x} = 0 \tag{9.1.15}$$

where the coefficient c = kinematic wave speed and may, at a particular cross section and for a specified flow rate, be estimated from

$$c = \frac{dQ}{dA} = \frac{1}{T}\frac{dQ}{dy} \tag{9.1.16}$$

The kinematic wave models assume that the inertia and pressure terms are negligible compared with the gravity and friction terms. Thus, the kinematic wave travels without attenuation but with a change of shape at the kinematic wave speed. It is should also be noted that there are two subgroups of kinematic models. If c is assumed constant, then the model is linear; however, if c is a variable, then the model is nonlinear.

Ponce *et al.* (1978) evaluated the range of applicability of kinematic wave models compared with diffusion models. If the kinematic wave model is to be 95 percent accurate in predicting wave amplitude after one wave propagation period, then

$$T_p \geq \frac{171 y_N}{S_x u_N} \tag{9.1.17}$$

where u_N = normal flow velocity. An examination of the limit defined by Eq. (9.1.17) demonstrates that the wave period must be very long for the kinematic wave models to be applicable in a channel of mild slope. In general the steeper the slope of the channel, the shorter the wave period needs to be to satisfy the kinematic wave assumption. Kinematic wave models have been used extensively in investigating overland flow and in simulating slow-rising flood waves, which travel unchanged in form.

Before the advent of the high-speed digital computer, necessity required that unsteady flow problems be addressed with approximate models. Even with the advent of the high-speed digital computer, it was still necessary that approximate models be used in many cases because of the limited computer resources available. Today, with the availability of what, in essence, is almost unlimited computer resources, the need for and the usefulness approximate models are limited; but still critical.

In this chapter, although a number of the approximate models are considered, the emphasis is on basic numerical methods to accurately solve the equations describing the complete dynamic model of gradually varied, unsteady flow. In addition, HEC-RAS (USACE, 2001) is used extensively in the example problems and in the homework.

9.2.   HYDROLOGIC APPROACHES

9.2.1.   Level Pool Routing

A flood wave passing through a lake or reservoir is attenuated as it enters and spreads over the surface. Water stored in the lake or reservoir is gradually released as pipe flow through turbines or outlet works, normally called principal spillways, or under extreme flow conditions over an emergency spillway. The primary element in addressing this problem is the balancing of inflow, outflow, and volume of water stored in the reservoir; that is satisfying the equation of continuity or

$$I_i + I_{i+1} + \left(\frac{2V_i}{\Delta t} - O_i \right) = \frac{2V_{i+1}}{\Delta t} + O_{i+1} \tag{9.2.1}$$

where I_i and I_{i+1} = the inflows at times i and $i+1$, respectively, O_i and O_{i+1} = outflows, and V_i and V_{i+1} = reservoir volumes. For any time increment, the only unknown is the term on the right hand side of Eq. (9.2.1). One approach to solving this equation is to pick pairs of trial value of S_{i+1} and O_{i+1} that satisfy the equation and confirm these trial values with the storage-outflow curve. Rather than use a trial and error solution, a value of Δt is selected; and then, the values on the storage-outflow curve are replotted as a storage indication curve. The storage indication graph allows a direct determination of the outflow O_{i+1} once a value of the ordinate

$$\frac{2S_{i+1}}{\Delta t} + O_{i+1}$$

has been calculated from Eq. (9.2.1). This approach is best illustrated with an example.

EXAMPLE 9.1

Route the inflow hydrograph given in Table 9.1 through a reservoir with the physical characteristics summarized in Table 9.2, assuming that the initial water level is at the emergency spillway level 350 m (1,148 ft), and that the principal spillway is plugged with debris. The emergency spillway has a width of 150 m (492 ft) and a discharge coefficient of 1.65 and a discharge exponent of 1.5.

Table 9.1. Inflow to the reservoir as a function of time.

Time (hr)	I (m³/s)
0.0	0
0.5	100
1.0	310
1.5	300
2.0	140
2.5	45
3.0	13
3.5	3
4.0	0.3
4.5	0

Table 9.2. Reservoir physical characteristics

Elevation (m)	Pool Area (m²) x 10^4	Storage Volume (m³) x 10^4
339	0	0
342	7.9	11.85
348	34.8	139.95
353	91.0	454.45
354	100.3	550.10
355	109.6	655.05
356	118.9	769.30
357	128.2	892.85
358	138.0	1,025.95
360	232.2	1,396.15

Solution

The first step is to construct a table of flow over the emergency spillway as a function of time where the flow over the emergency spillway is given by

$$O = CLH^x = 1.65(150)H^{1.5} = 247.5H^{1.5}$$

Taking Δt as 0.5 hr (1,800 seconds), Table 9.3 is constructed as a function of reservoir pool elevation.

Table 9.3.   Outflow and the parameter 2V/Δt + O as a function of reservoir elevation.

Elevation (m)	Outflow (m³/s)	Storage Volume (m³) x 10⁴	$\frac{2V}{\Delta t} + O$ (m³/s)
350	0	0[1]	0[2]
353	1286	188.70	3383
354	1980	284.35	5139
355	2767	389.30	7093
356	3637	503.55	9232
357	4584	627.10	11552
358	5600	760.20	14047
360	7827	1130.40	20387

[1] The interpolated storage volume at this elevation is 265.75 x 10⁴ m³
[2] Everything below this elevation is dead storage and therefore is not part of active storage; and therefore, values of S are estimated by subtracting 265.75 x 10⁴ m³ from the storage volumes given in Table 9.2.

The data in the above table are plotted in Figure 9.2, and the solution is summarized in Table 9.4 and Figure 9.3.

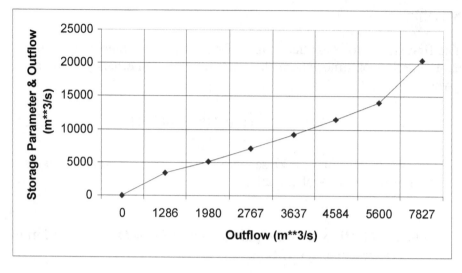

Figure 9.2. 2V/Δt + O versus O.

Table 9.4.   Reservoir outflow as a function of time.

Time (hr) (1)	i (2)	I_i (m³/s) (3)	$I_i + I_{i+1}$ (m³/s) (4)	$\dfrac{2V_i}{\Delta t} - O_i$ (m³/s) (5)	$\dfrac{2V_{i+1}}{\Delta t} + O_{i+1}$ (m³/s) (6)	O_{i+1} (m³/s) (7)
0	1	0	100	0	----	0
0.5	2	100	410	24	100	38
1.0	3	310	610	104	434	165
1.5	4	300	440	172	714	271
2.0	5	140	185	146	612	233
2.5	6	45	58	79	331	126
3.0	7	13	16	33	137	52
3.5	8	3	3.3	11	49	19
4.0	9	0.3	0.3	3.5	14.3	5.4
4.5	10	0	0	1.0	3.8	1.4
5.0	11	0	0	0.2	1.0	0.4
5.5	12	0	0	0	0.2	0.1
6.0	13	0	0		0	0

With regard to the solution summarized above, the following notes are pertinent:

1.   The entries in Columns (1) and (3) are known from the specified inflow hydrograph.
2.   The entries in Column (4) are the additions of $I_i + I_{i+1}$ from Column (3).
3.   The $2V/\Delta t + O$ versus O table or plot are entered with known values of $2V/\Delta t + O$ to estimate values of O.
4.   From the values in Column (6), twice the number in Column (7) is subtracted to enter a value of $2V/\Delta t + O$ in Column (5).
5.   Add the value in Column (5) to the advanced sum in Column (4) and enter the result in Column (6) for the new time period.
6.   The new outflow is again estimated as in Note 3.

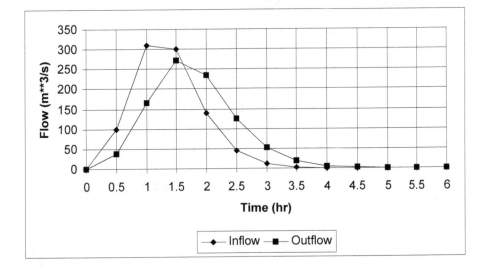

Figure 9.3. Inflow and outflow hydrographs for Example 9.1.

9.2.2.   Muskingum Routing

The unsteady flow approach known as the Muskingum routing method was developed in the 1930's (McCarthy, 1938) in connection with the design of flood protection for the Muskingum River Basin in Ohio. This method is likely the most widely used hydrologic stream channel routing method.

The Muskingum method is based on the differential equation of storage or

$$\frac{dV}{dt} = I - O \tag{9.2.2}$$

where I = inflow, O = outflow, and V = storage. In an ideal channel, storage is a function of the inflow and outflow. In the Muskingum method, storage is assumed to be a linear function of inflow and outflow or

$$V = K\left[XI + (1-X)O\right] \tag{9.2.3}$$

where K = time constant or storage coefficient and X = dimensionless weighting factor. It is pertinent to observe that with $X = 0$, Eq. (9.2.2) is reduced to a form of the level pool routing approach.

Discretizing Eq. (9.2.2) yields

$$\frac{I_i + I_{i+1}}{2} - \frac{O_i + O_{i+1}}{2} = \frac{S_{i+1} - S_i}{\Delta t} \tag{9.2.4}$$

where

$$S_i = K\left[XI_i + (1-X)O_i\right] \tag{9.2.5}$$

$$S_{i+1} = K\left[XI_{i+1} + (1-X)O_{i+1}\right] \tag{9.2.6}$$

where the subscripts i and $i+1$ indicate times. Substituting Eqs. (9.2.5) and (9.2.6) in Eq. (9.2.4) yields

$$\frac{I_i + I_{i+1}}{2} - \frac{O_i + O_{i+1}}{2} =$$
$$\frac{K\left[XI_{i+1} + (1-X)O_{i+1}\right] - K\left[XI_i + (1-X)O_i\right]}{\Delta t} \tag{9.2.7}$$

With algebraic manipulation, it can be shown that Eq. (9.2.7) can be represented as follows

$$O_{i+1} = C_0 I_{i+1} + C_1 I_i + C_2 O_i \tag{9.2.8}$$

where

$$C_0 = \frac{\dfrac{\Delta t}{K} - 2X}{2(1-X) + \dfrac{\Delta t}{K}} \quad (9.2.9)$$

$$C_1 = \frac{\dfrac{\Delta t}{K} + 2X}{2(1-X) + \dfrac{\Delta t}{K}} \quad (9.2.10)$$

and

$$C_2 = \frac{2(1-X) - \dfrac{\Delta t}{K}}{2(1-X) + \dfrac{\Delta t}{K}} \quad (9.2.11)$$

It is pertinent to note that $C_0 + C_1 + C_2 = 1$, and these routing coefficients can be interpreted as weighting factors. K accounts for the peak flow translation along a channel reach; and therefore, is related to the reach length and the mean channel flow velocity. One approach to estimating K is to use the Manning equation to estimate the velocity of flow assuming a wide rectangular channel. Then,

$$K = \frac{L}{\psi u_{wave}} \quad (9.2.12)$$

where L = reach length, Ψ = time conversion factor, and u_{wave} = velocity of the flood wave through the reach. It is pertinent to observe that the uniform velocity of flow is not the translation velocity. Table 9.5 provides an approximate relationship between the normal velocity of flow and the velocity of the wave of translation. The weighting factor X generally varies from 0 to 0.5, with 0 a reservoir type peak reduction and 0.5 representing no peak flow reduction.

Table 9.5. **Relationship between the uniform velocity of flow (u) and the velocity of the translation wave (CCRFCD, 1999).**

Channel Shape	Wave Velocity u_{wave}
Wide rectangular	0.625u
Triangular	1.33u
Wide parabolic	1.22u

EXAMPLE 9.2

An inflow hydrograph to a channel reach is given in the table below. Assuming the base flow is 352 m^3/s, $K = 2$ days, $X = 0.1$, and $\Delta t = 1$ day, predict the outflow hydrograph using the Muskingum approach. This example is based on one provided in Ponce (1989).

Table 9.6. **Inflow as a function of time for Example 9.2.**

Time (days)	Inflow (m^3/s)	Time (days)	Inflow (m^3/s)
0	352	13	3007
1	587	14	2357
2	1353	15	1779
3	2725	16	1405
4	4408	17	1123
5	5987	18	952
6	6704	19	730
7	6951	20	605
8	6839	21	514
9	6207	22	422
10	5346	23	352
11	4560	24	352
12	3861	25	352

Solution
From the given information

$$C_0 = \frac{\frac{\Delta t}{K} - 2X}{2(1-X) + \frac{\Delta t}{K}} = \frac{\frac{1}{2} - 2(0.1)}{2(1-0.1) + \frac{1}{2}} = 0.13$$

$$C_1 = \frac{\frac{\Delta t}{K} + 2X}{2(1-X) + \frac{\Delta t}{K}} = \frac{\frac{1}{2} + 2(0.1)}{2(1-0.1) + \frac{1}{2}} = 0.304$$

and

$$C_2 = \frac{2(1-X) - \frac{\Delta t}{K}}{2(1-X) + \frac{\Delta t}{K}} = \frac{2(1-0.1) - \frac{1}{2}}{2(1-0.1) + \frac{1}{2}} = 0.565$$

Then, the solution is given in Table 9.7.

The inflow and outflow results are shown graphically in Fig. 9.4, and the following observations regarding the solution are pertinent. First, the solution was extended to the point where the inflow and outflow were nearly equal; that is steady state. Second, a quick check of the numerical solution can be accomplished by comparing the volume that flowed into the system to the volume that flowed out of system; that is, continuity is satisfied. When Δt is constant, the inflow volume is

$$V_{IN} = \frac{I_1}{2} + \sum_{i=2}^{n-1} I_i + \frac{I_n}{2}$$

Table 9.7. Solution of Example 9.2.

n (days)	I_i	I_{i+1} (m³/s)	C_0I_{i+1} (m³/s)	C_1I_i (m³/s)	C_2O_i (m³/s)	O_{i+1} (m³/s)
0	352	587	77	107	199	383
1	587	1353	176	179	216	571
2	1353	2725	355	412	323	1090
3	2725	4408	575	829	616	2020
4	4408	5987	781	1342	1142	3264
5	5987	6704	874	1822	1845	4542
6	6704	6951	907	2040	2567	5514
7	6951	6839	892	2116	3117	6124
8	6839	6207	810	2081	3462	6353
9	6207	5346	697	1889	3591	6177
10	5346	4560	595	1627	3491	5713
11	4560	3861	504	1388	3229	5121
12	3861	3007	392	1175	2894	4462
13	3007	2357	307	915	2522	3744
14	2357	1779	232	717	2116	3066
15	1779	1405	183	541	1733	2458
16	1405	1123	146	428	1389	1963
17	1123	952	124	342	1110	1576
18	952	730	95	290	891	1275
19	730	605	79	222	721	1022
20	605	514	67	184	578	829
21	514	422	55	156	468	680
22	422	352	46	128	384	559
23	352	352	46	107	316	469
24	352	352	46	107	265	418
25	352	352	46	107	236	389
26	352	352	46	107	220	373
27	352	352	46	107	211	364
28	352	352	46	107	206	359
29	352	352	46	107	203	356
30	352	352	46	107	201	354
31	352	352	46	107	200	353

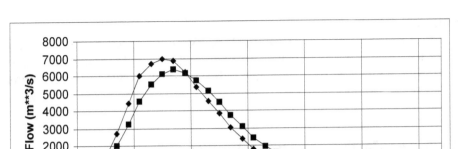

Figure 9.4. Inflow and outflow versus time for Example 9.2.

where n = number of values. From the above table, V_{IN} = 71,590 (m³-day)/s and V_{OUT} = 71,573 (m³-day)/s; and therefore, the solution satisfies continuity.

Traditionally, the Muskingum parameters K and X are calibrated using stream flow records; that is, simultaneous inflow-outflow discharge measurements for a channel reach are coupled with a trial and error procedure to estimate values of K and X. The basis of the procedure is that a linear relationship is assumed to exist between V/K and $[XI + (1 - X)O]$. Therefore, trial values of X are assumed until the difference between V/K and $[XI + (1 - X)O]$ is minimized. This procedure is best illustrated with an example in which the inflow and outflow data from Example 9.2 will be used.

EXAMPLE 9.3

Storage is defined by discretizing the continuity equation or

$$\frac{V_{i+1} - V_i}{\Delta t} = \frac{I_i + I_{i+1}}{2} - \frac{O_i + O_{i+1}}{2} \qquad (9.2.12)$$

and solving for V_{i+1}

$$V_{i+1} = V_i + \frac{\Delta t}{2}\left(I_i + I_{i+1} - O_i - O_{i+1}\right) \qquad (9.2.13)$$

Solution

At the first time increment, channel storage is assumed to be zero; and from Example 9.2, $\Delta t = 1$ day. In Table 9.8, channel storage values are computed using Eq. (9.2.13) and the inflow-outflow values in Columns 2 and 3, for $i = 0$ to $i = 3$.

Table 9.8. Channel storage as a function of time.

Time i (days) (1)	Inflow I (m³/s) (2)	Outflow O (m³/s) (3)	Storage V_{i+1} (m³-days/s) (4)
0	352	352	$V_1 = 0$
1	587	383	$V_2 = 0 + 0.5(352+587-352-571)=102$
2	1353	571	$V_3 = 102+0.5(587+1353-383-571)=595$
3	2725	1090	$V_4 = 595+0.5(1353+2725-571-1090)=1804$

Then, from Eq. (9.2.5)

$$\frac{V}{K} = XI + (1 - X)O$$

For the first trial, assume $X = 0.3$. Given this assumption, the values of V/K and $[XI + (1 - X)O]$ for $i = 0$ to $i = 3$ are summarized in Table 9.9.

The complete results for $X = 0.3$ are plotted in Fig. 9.5, and it is observed there is not a linear relationship between S and $[XI + (1 - X)O]$. Therefore, the next trial is $X = 0.1$, and the results for $i = 0$ to $i = 3$ are summarized in the following table.

Table 9.9. Values of $[XI + (1 -X)O]$ as a function of time for $X =0.3$.

Time i (days) (1)	Inflow I (m^3/s) (2)	Outflow O (m^3/s) (3)	Storage V_{i+1} (m^3-days/s) (4)	$XI + (1 - X)O$ $0.3I + 0.7O$ (m^3/s) (5)
0	352	352	0	
1	587	383	102	0.3(587)+0.7(383)=444
2	1353	571	595	0.3(1353)+0.7(571)=806
3	2725	1090	1804	0.3(2725)+0.7(1090)=1580

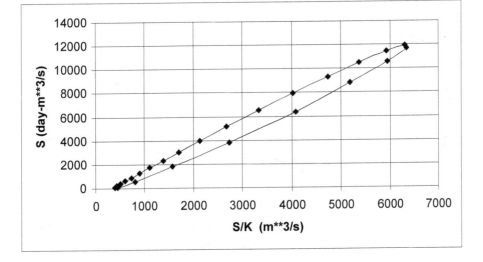

Figure 9.5. Storage versus S/K for $X =0.3$.

Table 9.10. Values of [*XI* + *(1 -X)O*] as a function of time for *X* =0.1.

Time i (days) (1)	Inflow I (m³/s) (2)	Outflow O (m³/s) (3)	Storage V_{i+1} (m³-days/s) (4)	$XI + (1 - X)O$ $0.1I + 0.9O$ (m³/s) (5)
0	352	352	0	
1	587	383	102	0.1(587)+0.9(383)=403
2	1353	571	595	0.1(1353)+0.9(571) = 649
3	2725	1090	1804	0.1(2725)+0.9(1090)=1254

The complete results for $X = 0.1$ are plotted in Fig. 9.6, and it is observed there is a linear relationship between S/K and [$XI + (1 - X)O$].

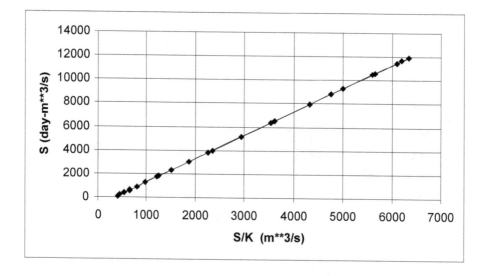

Figure 9.6. Storage versus *S/K* for *X* =0.1.

Although graphical visualization of the linearity between S/K and [$XI + (1 - X)O$] is one method for finding the correct value of X, other numerical measures are available. For example, one measure of goodness of fit that could be used is root mean square of the differences or

$$RMS = \sqrt{\frac{\sum_{i=1}^{n}\left\{V_i - \left[XI + \left(1 - X\right)O_i\right]\right\}^2}{n - 1}} \qquad (9.1.31)$$

with the least value of the *RMS* indicating the best fit. With the best fit value of *X* established, a corresponding value of *K* can be estimated by

$$K = \frac{V_P - V_1}{\left[XI + \left(1 - X\right)O_P\right] - \left[XI + \left(1 - X\right)O_1\right]} = \frac{1804 - 102}{1254 - 403} = 2\,\text{days}$$

where the values are from the table above.

9.3.   HYDRAULIC APPROACHES

9.3.1.   Governing Equations and Basic Numerical Techniques

From Chapter 1, the equations governing gradually varied, unsteady flow are the equation of continuity and the conservation of momentum equation. In this section, the application and solution of these equations in a rectangular channel for the case of one dimensional, subcritical, unsteady flow, are considered.

For a rectangular channel, the equation of continuity is

$$\frac{\partial y}{\partial t} + y\frac{\partial u}{\partial x} + u\frac{\partial y}{\partial x} = 0 \qquad (9.3.1)$$

where y =   depth of flow
          t =   time
          u =   average velocity of flow
          x =   longitudinal distance.

The conservation of momentum equation is

$$\frac{\partial u}{\partial t} + u\frac{\partial u}{\partial x} + g\frac{\partial y}{\partial x} - g\left(S_x - S_f\right) = 0 \qquad (9.3.2)$$

where S_x = channel slope in the longitudinal direction and S_f = friction slope. Equations (9.3.1) and (9.3.2) are a set of simultaneous equations which can be solved for the two unknowns u and y, given appropriate initial and boundary conditions. The solution of this set of equations is virtually impossible without the aid of a high-speed digital computer.

The most direct method of simultaneously solving Eqs. (9.3.1) and (9.3.2) is an explicit finite difference scheme with a fixed time step. Given the schematic definition of a rectangular finite difference scheme in Fig. 9.7, the derivatives appearing in the governing equations can be approximated by

$$\left(\frac{\partial u}{\partial x}\right)_M = \frac{u_R - u_L}{2\Delta x} \tag{9.3.3}$$

$$\left(\frac{\partial u}{\partial t}\right)_P = \frac{u_P - u_M}{\Delta t} \tag{9.3.4}$$

$$\left(\frac{\partial y}{\partial x}\right)_M = \frac{y_R - y_L}{2\Delta x} \tag{9.3.5}$$

$$\left(\frac{\partial y}{\partial t}\right)_P = \frac{y_P - y_M}{\Delta t} \tag{9.3.6}$$

where Δx = longitudinal distance between nodes, Δt = distance in time between nodes, and the subscript designates the node at which the variable is being evaluated.

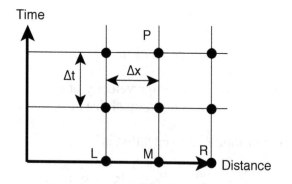

Figure 9.7. Definition of finite difference network.

Substitution of the above expressions for the derivatives in Eq. (9.3.1) yields

$$\frac{y_P - y_M}{\Delta t} + y_M\left(\frac{u_R - u_L}{2\Delta x}\right) + u_M\left(\frac{y_R - y_L}{2\Delta x}\right) = 0$$

and solving for y_P

$$y_P = y_M + \frac{\Delta t}{2\Delta x}\left[u_M\left(y_L - y_R\right) + y_M\left(u_L - u_R\right)\right] \qquad (9.3.7)$$

Substituting the finite difference approximations for the derivatives in Eq. (9.3.2) yields

$$\frac{u_P - u_M}{\Delta t} + u_M\left(\frac{u_R - u_L}{2\Delta x}\right) + g\left(\frac{y_R - y_L}{2\Delta x}\right) = g\left(S_x - S_f\right) \qquad (9.3.8)$$

In the computation of unsteady flow, it is usually assumed that the friction slope S_f can be estimated from the Manning or Chezy resistance equations. Use of the Manning equation yields

$$S_f = \frac{u|u|n^2}{\phi^2 R^{4/3}} \qquad (9.3.9)$$

where R = hydraulic radius
         n = Manning resistance coefficient
         ϕ = coefficient dependent on system of units used (1.49 for English system and 1 for SI system).

It is noted that the absolute value of the velocity of flow is used in combination with a signed value of the velocity to ensure that frictional resistance always opposes the motion. If the rectangular channel is wide, then $R \sim y$, and the friction slope can be represented in finite difference form as

$$S_f = \frac{u_P|u_P|n^2}{\phi^2 y_P^{4/3}} \qquad (9.3.10)$$

Substitution of Eq. (9.3.10) in Eq. (9.3.8) yields

$$\frac{u_P - u_M}{\Delta t} + u_M\left(\frac{u_R - u_L}{2\Delta x}\right) + g\left(\frac{y_R - y_L}{2\Delta x}\right) = g\left(S_x - \frac{u_P|u_P|n^2}{\phi^2 y_P^{4/3}}\right) \quad (9.3.11)$$

For notational convenience, define

$$\Gamma = \frac{\phi^2 y_P^{4/3}}{g\Delta t n^2}$$

Simplifying and rearranging Eq. (9.3.11) yields

$$u_P|u_P| + \Gamma u_P + \Gamma\left[\frac{u_M \Delta t}{2\Delta x}(u_R - u_L) + \frac{g\Delta t}{2\Delta x}(y_R - y_L) - g\Delta t S_x\right] = 0 \quad (9.3.12)$$

Equation (9.3.12) is a quadratic equation in u_P and can be solved by the quadratic formula or

$$u_P = \frac{-\Gamma + \sqrt{\Gamma^2 - 4\beta}}{2} \quad (9.3.13)$$

where

$$\beta = \Gamma\left[\frac{u_M \Delta t}{2\Delta x}(u_R - u_L) + \frac{g\Delta t}{2\Delta x}(y_R - y_L) - g\Delta t S_x\right]$$

with y_P given by Eq. (9.3.7). Thus, the solution of an unsteady flow problem, by an explicit finite difference technique with a fixed time step in a wide, rectangular channel, proceeds by determining y_P at and advanced time step by Eq. (9.3.7) and using this value in Eq. (9.3.13), determine u_P.

A second numerical technique, which has often been used to solve unsteady flow problems, involves the solution of the governing differential equations by the use of characteristics. Again, assuming a wide, rectangular channel and rearranging Eqs. (9.3.1) and (9.3.2) yields

$$H_1 = \frac{\partial y}{\partial t} + y\frac{\partial u}{\partial x} + u\frac{\partial y}{\partial x} = 0 \qquad (9.3.14)$$

and

$$H_2 = \frac{\partial u}{\partial t} + u\frac{\partial u}{\partial x} + g\frac{\partial y}{\partial x} - g(S_x - S_f) = 0 \qquad (9.3.15)$$

where H_1 and H_2 with an unknown multiplier λ can be combined in a linear combination to form a new function H

$$H = \lambda H_1 + H_2 \qquad (9.3.16)$$

which, for any two real, distinct values of λ, will produce two equations in u and y that will retain all the attributes of Eqs. (9.3.14) and (9.3.15). Then, combining Eqs. (9.3.14) and (9.3.15) according to Eq. (9.3.16) yields

$$H = \frac{\partial u}{\partial t} + u\frac{\partial u}{\partial x} + g\frac{\partial y}{\partial x} - g(S_x - S_f) + \lambda\left(\frac{\partial y}{\partial t} + y\frac{\partial u}{\partial x} + u\frac{\partial y}{\partial x}\right)$$

or

$$H = \left[(u + \lambda y)\frac{\partial u}{\partial x} + \frac{\partial u}{\partial t}\right] + \lambda\left[\left(u + \frac{g}{\lambda}\right)\frac{\partial y}{\partial x} + \frac{\partial y}{\partial t}\right] - g(S_x - S_f) \qquad (9.3.17)$$

In Eq. (9.3.17), the first and second terms are the total derivatives of the velocity and depth of flow or

$$\frac{du}{dt} = \frac{\partial u}{\partial x}\frac{dx}{dt} + \frac{\partial u}{\partial t} \quad \text{if} \quad \frac{dx}{dt} = u + \lambda y \qquad (9.3.18)$$

and

$$\frac{dy}{dt} = \frac{\partial y}{\partial x}\frac{dx}{dt} + \frac{\partial y}{\partial t} \quad \text{if} \quad \frac{dx}{dt} = u + \frac{g}{\lambda} \qquad (9.3.19)$$

Equation (9.3.17) can then be rewritten as

$$H = \frac{du}{dt} + \lambda \frac{dy}{dt} - g\left(S_x - S_f\right)u + \lambda y = u + \frac{g}{\lambda} \qquad (9.3.20)$$

Equating the expressions for *dx/dt* in Eqs. (9.3.18) and (9.3.19)

$$u + \lambda y = u + \frac{g}{\lambda}$$

and after solving yields

$$\lambda = \pm \sqrt{\frac{g}{y}} \qquad (9.3.21)$$

The two real, distinct roots for, λ specified by Eq. (9.3.21), can be used to transform Eqs. (9.3.14) and (9.3.15) into a pair of ordinary differential equations, subject to the restrictions regarding *dx/dt* specified in Eqs. (9.3.18) and (9.3.19) or

$$du + dy\sqrt{\frac{g}{y}} + g\left(S_f - S_x\right)dt = 0 \qquad (9.3.22)$$

$$dx = \left(u + \sqrt{gy}\right)dt \qquad (9.3.23)$$

$$du - dy\sqrt{\frac{g}{y}} + g\left(S_f - S_x\right)dt = 0 \qquad (9.3.24)$$

$$dx = \left(u - \sqrt{gy}\right)dt \qquad (9.3.25)$$

At this point, it is noted that the curve defined by Eq. (9.3.23) is termed the *positive characteristic (C+)*(curve *LP* in Fig. 9.8). The curve defined by Eq. (9.3.25) is termed the *negative characteristic (C-)*(curve *RP* in Fig. 9.8).

The solution of Eqs. (9.3.22) to (9.3.25) must be accomplished by numerical methods. Using a first order, explicit finite difference scheme, Eqs. (9.3.22) to (9.3.25) become

$$u_P - u_L + \sqrt{\frac{g}{y_L}}\left(y_P - y_L\right) + g\left(t_P - t_L\right)\left(S_{fL} - S_x\right) = 0 \quad (9.3.26)$$

$$x_P - x_L = \left(u_L + \sqrt{gy_L}\right)\left(t_P - t_L\right) \quad (9.3.27)$$

$$u_P - u_R - \sqrt{\frac{g}{y_R}}\left(y_P - y_R\right) + g\left(t_P - t_R\right)\left(S_{fR} - S_x\right) = 0 \quad (9.3.28)$$

$$x_P - x_R = \left(u_R - \sqrt{gy_R}\right)\left(t_P - t_R\right) \quad (9.3.29)$$

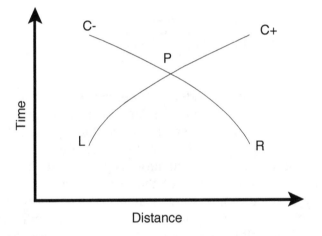

Figure 9.8. Definition of characteristic curves.

Unlike the explicit finite difference scheme with a fixed time step, which was previously discussed, in this scheme the length of the time step is determined by subtracting Eq. (9.3.29) from Eq. (9.3.27) or

$$t_P = \frac{x_L - x_R + t_R\left(u_R - \sqrt{gy_R}\right) - t_L\left(u_L - \sqrt{gy_L}\right)}{u_R - u_L - \sqrt{gy_L} - \sqrt{gy_R}} \quad (9.3.30)$$

The distance step in the longitudinal direction is determined from Eq. (12.2.27)

$$x_P = x_L + \left(u_L + \sqrt{gy_L}\right)\left(t_P - t_L\right) \qquad (9.3.31)$$

The depth at the node P is determined by subtracting Eq. (9.3.28) from Eq. (9.3.26) and solving for y_P

$$y_P = \frac{y_R\sqrt{\dfrac{g}{y_R}} + \left(t_P - t_R\right)\left(S_{fR} - S_x\right) - \left(t_P - t_L\right)\left(S_{fL} - S_x\right)}{\sqrt{\dfrac{g}{y_L}} + \sqrt{\dfrac{g}{y_R}}} \qquad (9.3.32)$$

From Eq. (12.2.26), the velocity of flow is given by

$$u_P = u_L - \left(y_P - y_L\right)\sqrt{\dfrac{g}{y_L}} - \left(t_P - t_L\right)\left(S_{fL} - S_x\right) \qquad (9.3.33)$$

Before either of the methodologies described in this section can be applied to a problem, initial and boundary conditions must be specified. The term *initial condition* refers to the initial state of flow in the channel; i.e., the depth, velocity, or discharge along the reach under consideration at the time computation begins. The term *boundary condition* refers to the definition of the depth, velocity, or discharge at the upper and lower ends of the reach, for all times, after the computation begins. For the general case, the initial condition is usually specified as uniform flow or a gradually varied flow profile (Chapter 6). A typical upstream boundary condition is the specification of a hydrograph, while a downstream condition could be critical depth.

The primary difficulty with explicit finite difference techniques is the problem of numerically unstable solutions. Unstable conditions usually result if Δt is large relative to Δx. The Courant stability condition requires

$$\Delta t \leq \frac{\Delta x}{u + c} \qquad (9.3.34)$$

where c = wave celerity. However, it has been found that for the type of explicit difference schemes discussed in this section, Δt should be approximately 20 percent of the value given by Eq. (9.2.34). Viessman *et al.* (1972) noted that more stable solutions can be obtained if a diffusing difference approximation is used; i.e., in the foregoing equations, substitute

$$\left(\frac{\partial u}{\partial t}\right)_M = \frac{u_P - 0.5\left(u_L + u_R\right)}{\Delta t} \qquad (9.3.35)$$

$$\left(\frac{\partial y}{\partial t}\right)_M = \frac{y_P - 0.5\left(y_L + y_R\right)}{\Delta t} \qquad (9.3.36)$$

$$S_f = \frac{S_{fL} + S_{fR}}{2} \qquad (9.3.37)$$

This formulation of the derivatives allows the size of the time increment to be significantly increased; but the Courant stability condition [Eq. (9.3.34)] must still be met. In addition, the diffusing scheme also imposes a friction criteria or

$$\Delta t \leq \frac{\phi^2 R^{4/3}}{gn^2 |u|} \qquad (9.3.38)$$

For stability, the lesser value of Δt, between Eqs. (9.3.34) and (9.3.38), is used.

9.3.2   Kinematic Wave
The kinematic wave approach to gradually varied unsteady flow is based on the conservation of mass and this implies

1. Local and convective inertia are ignored,
2. Pressure gradients are ignored, and
3. Momentum source terms are ignored.

Beginning with the Manning equation

$$Q = \frac{\phi}{n} AR^{2/3} \sqrt{S} \qquad (9.3.39)$$

and then rearranging Eq. (9.3.39) into a discharge-area rating form

$$Q = \alpha A^{\beta} \qquad (9.3.40)$$

where

$$\alpha = \frac{\phi}{n} \frac{\sqrt{S}}{P^{2/3}}$$

and

$$\beta = \frac{5}{3}$$

Then, differentiating Eq. (9.3.40) with respect to A, where it is tacitly assumed that A and P are independent variables (a wide, rectangular channel)

$$\frac{dQ}{dA} = \beta \alpha A^{\beta-1} = \beta \left(\frac{\phi}{n} \frac{\sqrt{S}}{P^{2/3}} \right) A^{2/3} = \left(\frac{\frac{\phi}{n} \frac{\sqrt{S}}{P^{2/3}} A^{5/3}}{A} \right) = \beta \frac{Q}{A} \quad (9.3.41)$$

Therefore,

$$\frac{dQ}{dA} = \beta \frac{Q}{A} = \beta u \qquad (9.3.42)$$

where u = mean velocity of flow.

One form of the continuity equation

$$\frac{\partial Q}{\partial A} + \frac{\partial A}{\partial t} = 0$$

and multiplying both sides of the equation by dQ/dA

$$\frac{dQ}{dA} \frac{\partial Q}{\partial x} + \frac{dQ}{dA} \frac{\partial A}{\partial t} = \frac{dQ}{dA} \frac{\partial Q}{\partial x} + \frac{\partial Q}{\partial t} = \frac{\partial Q}{\partial t} + \beta u \frac{\partial Q}{\partial x} = 0 \quad (9.3.43)$$

Equation (9.3.43) is a first order, partial differential equation describing the movement of waves that are kinematic; that is, these waves do not attenuate with distance. Further, Eq. (9.3.43), in the general case, requires a numerical solution.

With reference to Fig. 9.9, where j is a distance and i is a time step, a first order numerical approximation (backward differences) to Eq. (9.3.43) is given by

$$\frac{Q^{i+1}_{j+1} - Q^i_{j+1}}{\Delta t} + \beta u \frac{Q^{i+1}_{j+1} - Q^{i+1}_j}{\Delta x} = 0 \qquad (9.3.44)$$

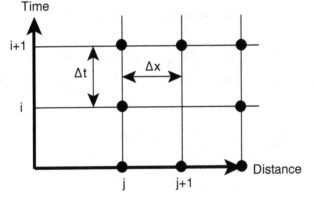

Figure 9.9.   Definition of notation for kinematic wave development.

Rearranging and solving for Q^{i+1}_{j+1}

$$Q^{i+1}_{j+1}\left(1 + \beta u \frac{\Delta t}{\Delta x}\right) = Q^i_{j+1} + \beta u \frac{\Delta t}{\Delta x} Q^{i+1}_j$$

$$Q^{i+1}_{j+1} = \frac{\beta u \dfrac{\Delta t}{\Delta x}}{1 + \beta u \dfrac{\Delta t}{\Delta x}} Q^{i+1}_j + \frac{1}{1 + \beta u \dfrac{\Delta t}{\Delta x}} Q^i_{j+1} \qquad (9.3.45)$$

Notice that the parameter $\beta u (\Delta u / \Delta t)$ is similar to the Courant stability condition, Eq. (9.3.34); and in general there will be minimal numerical diffusion if this parameter has a value of approximately one.

Again, with reference to Fig. 9.9, a second order approach to solving Eq. (9.3.43) can be developed using central differences or

$$\frac{\dfrac{Q_{j+1}^{i+1} + Q_{j}^{i+1}}{2} - \dfrac{Q_{j+1}^{i} + Q_{j}^{i}}{2}}{\Delta t} + \beta u \left(\frac{\dfrac{Q_{j+1}^{i} + Q_{j+1}^{i+1}}{2} - \dfrac{Q_{j}^{i} + Q_{j}^{i+1}}{2}}{\Delta x} \right) = 0 \quad (9.3.46)$$

Rearranging, simplifying, and solving for Q_{j+1}^{i+1}

$$\frac{Q_{j+1}^{i+1}}{2\Delta t} + \frac{Q_{j}^{i+1}}{2\Delta t} - \frac{Q_{j+1}^{i}}{2\Delta t} - \frac{Q_{j}^{i}}{2\Delta t} + \frac{\beta u Q_{j+1}^{i}}{2\Delta x} + \frac{\beta u Q_{j+1}^{i+1}}{2\Delta x} - \frac{\beta u Q_{j+1}^{i+1}}{2\Delta x} - \frac{\beta u Q_{j}^{i+1}}{2\Delta x} = 0$$

$$Q_{j+1}^{i+1} \left(1 + \beta u \frac{\Delta t}{\Delta x} \right) = Q_{j}^{i+1} \left(\beta u \frac{\Delta t}{\Delta x} - 1 \right) +$$

$$Q_{j}^{i} \left(1 + \beta u \frac{\Delta t}{\Delta x} \right) + Q_{j+1}^{i} \left(1 - \beta u \frac{\Delta t}{\Delta x} \right)$$

and

$$Q_{j+1}^{i+1} = \frac{\beta u \dfrac{\Delta t}{\Delta x} - 1}{1 + \beta u \dfrac{\Delta t}{\Delta x}} Q_{j}^{i+1} + Q_{j}^{i} + \frac{1 - \beta u \dfrac{\Delta t}{\Delta x}}{1 + \beta u \dfrac{\Delta t}{\Delta x}} Q_{j+1}^{i} \quad (9.3.47)$$

This method is stable when $\beta u (\Delta t / \Delta x)$ is one or less.

Yet a third numerical approach to solving Eq. (9.3.43) can be developed using a forward-in-time and backward-in-space finite difference scheme (the Convex method) or

$$\frac{Q_{j+1}^{i+1} - Q_{j+1}^{i}}{\Delta t} + \beta u \frac{Q_{j+1}^{i} - Q_{j}^{i}}{\Delta x} = 0 \quad (9.3.48)$$

Rearranging and solving for Q_{j+1}^{i+1}

$$Q_{j+1}^{i+1} = Q_{j+1}^{i}\left(1 - \beta u \frac{\Delta t}{\Delta x}\right) + \beta u \frac{\Delta t}{\Delta x} Q_j^i \qquad (9.3.49)$$

and this method is stable when $\beta u (\Delta t / \Delta x)$ is one or less.

The kinematic wave approach to unsteady flow can be easily modified to take into account lateral inflows whether they be concentrated (tributaries) or distributed (groundwater). With lateral inflows, the equation of continuity is

$$\frac{\partial Q}{\partial A} + \frac{\partial A}{\partial t} = q_L \qquad (9.3.50)$$

where q_L = lateral inflow per unit length of the channel. The formulation is the same regardless of whether the inflow is concentrated or distributed.

Then, as before

$$\frac{dQ}{dA} = \beta u$$

and

$$\frac{dQ}{dA}\frac{\partial Q}{\partial x} + \frac{dQ}{dA}\frac{\partial A}{\partial t} = \frac{dQ}{dA} q_L$$

$$\beta u \frac{\partial Q}{\partial x} + \frac{\partial Q}{\partial t} = \beta u q_L \qquad (9.3.51)$$

If q_L is positive, the flow is into the system; and if q_L is negative, the flow is out of the system.

9.3.3   Diffusion Wave

The kinematic wave approach can be enhanced by allowing a small amount of physical (as opposed to numerical) diffusion in the model formulation. Diffusion is present in most natural, unsteady flows; and therefore, this is a reasonable approach.

In deriving the diffusion wave model, it is assumed that the friction slope is equal to the water surface slope (equivalent to steady, nonuniform flow) or

$$Q = \frac{\phi}{n} AR^{2/3} \sqrt{S_0 - \frac{dy}{dx}} \qquad (9.3.52)$$

where $(S_0 - dy/dx)$ is the water surface slope. The difference between the kinematic and diffusion wave approaches is in the term dy/dx. From a physical standpoint, the term dy/dx accounts for the natural diffusion processes present in unsteady flow.

Rearranging Eq. (9.3.52),

$$S_0 - \frac{dy}{dx} = \left(\frac{nQ}{\phi AR^{2/3}} \right) \qquad (9.3.52)$$

Define

$$m = \left(\frac{n}{\phi AR^{2/3}} \right)^2$$

and substituting in Eq. (9.3.52) yields

$$mQ^2 = S_0 - \frac{dy}{dx} \qquad (9.3.53)$$

Recall for a channel of arbitrary shape

$$dA = T \, dy$$

or

$$dy = \frac{dA}{T}$$

where T = top width of the channel; and therefore,

$$\frac{dy}{dx} = \frac{1}{T} \frac{dA}{dx}$$

Substituting this relationship in Eq. (9.3.53) yields

$$\frac{1}{T}\frac{dA}{dx} + mQ^2 - S_0 = 0 \qquad (9.3.54)$$

Equation (9.3.54) together with the equation of continuity

$$\frac{\partial Q}{\partial x} + \frac{\partial A}{\partial t} = 0 \qquad (9.3.55)$$

define the diffusion wave model.

Ponce (1989) demonstrated that Eqs. (9.3.54) and (9.3.55) can be linearized and the result is

$$\frac{\partial Q}{\partial t} + \frac{\partial Q}{\partial A}\frac{\partial Q}{\partial x} = \frac{Q}{2TS_0}\frac{\partial^2 Q}{\partial x^2} \qquad (9.3.56)$$

The left hand side of Eq. (9.3.56) is the kinematic wave equation and the right hand side accounts for physical diffusion of the flood wave. It is pertinent to observe that the term

$$\frac{Q}{2TS_0}$$

has units of $[L^2T^{-1}]$ and is referred to as the hydraulic or channel diffusivity. The hydraulic diffusivity is small for steep slopes (mountain streams) and large for mild slopes (tidal rivers).

9.3.2. Full Dynamic Equation Set

With the advent and general availability of high speed, digital computers, there was great interest in developing general numerical solutions of the St. Venant unsteady flow equations. For example, Amein (1966) and Fletcher and Hamilton (1967) developed numerical approaches using the method of characteristics, while Amein et al. (1968, 1970, 1975) developed a rapidly convergent and accurate implicit technique using a finite difference scheme. Fread (1976 and 1982) developed equations that took into account the interaction that takes place between the channel and the floodplain.

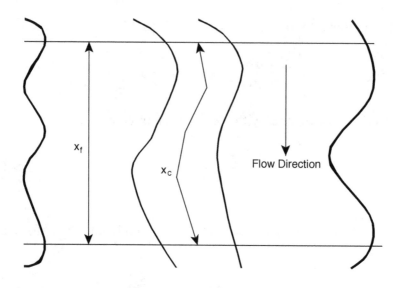

Figure 9.10. A river reach after Barkau(1997).

Figure 9.10 illustrates the interaction that takes place between the channel and the floodplain that makes the analysis of the movement of a flood through a river valley a two dimensional problem. As the river rises, water flows laterally from the channel and inundates the floodplain and storage areas. As the depth increases, the water generally flows along a shorter flow path than if it remained completely in the main channel; i.e., $\Delta x_f < \Delta x_c$. When the river stage is falling, flow occurs from the floodplain and storage areas to the main channel. A common approach to this problem in the past was to ignore the overbank conveyance and assume that only storage takes place there. While this assumption may be suitable for channels where the flow is confined by levees, it is not a suitable assumption in the general case. Because in the general case the flow is still downstream, the two-dimensional problem can, in many cases, be approximated as a one-dimensional problem. Fread (1976 and 1982) and Smith (1978) divided the system into two separate channels: writing continuity and momentum equations, for each channel. Three assumptions were also used: 1) a horizontal water surface at each cross section normal to the direction of flow; 2) exchange of momentum between the channel and the floodplain was negligible; and 3) the flow in the channel and floodplain was distributed according to the conveyance, or

$$Q_c = \phi Q \tag{9.3.57}$$

where Q_c = flow in the channel, Q = total flow, $\phi = \dfrac{K_c}{K_c + K_f}$, K_c = channel conveyance and K_f = floodplain conveyance, where the subscripts c and f refer to the channel and floodplain, respectively. With these assumptions, the one-dimensional equations of motion are

$$\frac{\partial A}{\partial t} + \frac{\partial (\phi Q)}{\partial x_c} + \frac{\partial \left[(1-\phi)Q \right]}{\partial x_f} = 0 \qquad (9.3.58)$$

and

$$\frac{\partial Q}{\partial t} + \frac{\partial \left(\dfrac{\phi^2 Q^2}{A_c} \right)}{\partial x_c} + \frac{\partial \left[\dfrac{(1-\phi)^2 Q^2}{A_f} \right]}{\partial x_f} + \qquad (9.3.59)$$

$$gA_c \left(\frac{\partial y}{\partial x_c} + S_{fc} \right) + gA_f \left(\frac{\partial y}{\partial x_f} + S_{ff} \right) = 0$$

These non-linear equations were approximate using implicit finite differences and solved using a Newton-Raphson iteration technique.

Barkau (1997) expanded on the work of Fread and Smith and defined a new set of equations that were computationally more convenient. The numerical solution of these equations became the public domain model known as HEC-UNET (Barkau, 1997). The HEC-UNET model was subsequently incorporated into the HEC-RAS modeling system (USACE, 2001). The use of the unsteady flow module of HEC-RAS is best illustrated with an example.

EXAMPLE 9.4

A 20 ft (6.1 m) wide rectangular channel, 2 miles (3,220 m) long, with n = 0.02, and a longitudinal slope of 0.0015 conveys a uniform steady flow at a depth of 6 ft (1.8 m). At the upstream end the flow increases over a period of 20 minutes to 2000 ft³/s (56.6 m³/s), and the flow then decreases to the initial flow depth over a period of 40 minutes. Plot the inflow and outflow hydrographs as a function of time. At the upstream end, the elevation of the channel invert is 100 ft above an arbitrary datum.

Solution
Step 1: Compute the uniform flow rate from the Manning equation or

$$Q = \frac{1.49}{n} AR^{2/3}\sqrt{S} = \frac{1.49}{0.02}(120)\left(\frac{120}{32}\right)^{2/3}\sqrt{0.0015}$$

$$= 836\,\text{ft}^3/\text{s}\left(23.7\,\text{m}^3/\text{s}\right)$$

Step 2: Compute the inflow hydrograph assuming that the time increment is 5 minutes, Table 9.11.

Table 9.11. Inflow as a function of time.

Time (min)	Inflow (ft³/s)
0	836
5	1127
10	1418
15	1709
20	2000
25	1854.5
30	1709
35	1563.5
40	1418
45	1272.5
50	1127
55	981.5
60	836

Step 3: Build the channel cross sections for HEC-RAS.   For the specified slope, the fall over a 2 mile reach is

$$\text{Fall} = 2(5280)(0.0015) = 15.84 \text{ ft } (4.83 \text{ m})$$

Since the channel is prismatic and the slope uniform, only the upstream and downstream cross sections need to be specified and intermediate cross sections can be interpolated within HEC-RAS.   Assume that upstream channel invert has an elevation of 100 ft and therefore the downstream channel invert has an elevation of (100 - 15.84 =) 84.16 ft.

Also assume that the channel is 20 ft deep. Then, the upstream and downstream channel geometry is summarized in Table 9.12.

Table 9.12. Summary of HEC-RAS upstream and downstream sections channel geometry.

Distance from Right Bank (ft)	Elevation (ft)
Upstream Station (Station 10,560 ft)	
0	120
0	100
20	100
20	120
Downstream Station (Station 0 ft)	
0	104.16
0	84.16
20	84.16
20	104.16

Step 4: Boundary Conditions. The upstream boundary condition is a hydrograph and the downstream boundary condition is uniform flow.

The HEC-RAS solution is shown in Table 9.13 and Figure 9.11; and with regard to the model and the output, the following notes are pertinent. First, the time increment for the upstream hydrograph was five minutes; but the computational time in the model was one minute. Second, extended periods of steady flow were added at the beginning and end of the hydrograph. Third, the computational time for the output is five minutes; therefore, Profile 2 is at zero minutes, Profile 3 is at five minutes; *etc.*

Table 9.13. HEC-RAS for Example 9.4.

Profile Output Table - rhf-1
HEC-RAS Plan: Plan 01 River: rio ras Reach: 1

# Rivers	= 1		# Plans	= 1
# Hydraulic Reaches	= 1		# Profiles	= 61
# River Stations	= 2			

Reach	River Sta	Q Total (cfs)	Profile	Min Ch El (ft)	W.S. Elev Max (ft)	Chl Dpth (ft)	E.G. Elev (ft)	Vel chnl (ft/s)	Flow Area (sq ft0	Top Width (ft)	Chl Froude #
1.	10560	836.00	2	100.00	106.01	6.01	106.76	6.96	120.20	20.00	0.50
1.	10560	836.00	3	100.00	106.01	6.01	106.76	6.96	120.18	20.00	0.50
1.	10560	836.00	4	100.00	106.01	6.01	106.76	6.96	120.18	20.00	0.50
1.	10560	836.00	5	100.00	106.01	6.01	106.76	6.96	120.17	20.00	0.50
1.	10560	836.00	6	100.00	106.01	6.01	106.76	6.96	120.17	20.00	0.50
1.	10560	836.00	7	100.00	106.01	6.01	106.76	6.96	120.17	20.00	0.50
1.	10560	836.00	8	100.00	106.01	6.01	106.76	6.96	120.17	20.00	0.50
1.	10560	836.00	9	100.00	106.01	6.01	106.76	6.96	120.17	20.00	0.50
1.	10560	836.00	10	100.00	106.01	6.01	106.76	6.96	120.17	20.00	0.50
1.	10560	836.00	11	100.00	106.01	6.01	106.76	6.96	120.17	20.00	0.50
1.	10560	836.00	12	100.00	106.01	6.01	106.76	6.96	120.17	20.00	0.50
1.	10560	836.00	13	100.00	106.01	6.01	106.76	6.96	120.17	20.00	0.50
1.	10560	836.00	14	100.00	106.01	6.01	106.76	6.96	120.17	20.00	0.50
1.	10560	836.00	15	100.00	106.01	6.01	106.76	6.96	120.17	20.00	0.50
1.	10560	836.00	16	100.00	106.01	6.01	106.76	6.96	120.17	20.00	0.50
1.	10560	836.00	17	100.00	106.01	6.01	106.76	6.96	120.17	20.00	0.50
1.	10560	836.00	18	100.00	106.01	6.01	106.76	6.96	120.17	20.00	0.50
1.	10560	836.00	19	100.00	106.01	6.01	106.76	6.96	120.17	20.00	0.50
1.	10560	836.00	20	100.00	106.01	6.01	106.76	6.96	120.17	20.00	0.50
1.	10560	1137.78	21	100.00	107.06	7.06	108.07	8.05	141.29	20.00	0.53
1.	10560	1407.22	22	100.00	108.19	8.19	109.34	8.59	163.80	20.00	0.53
1.	10560	1709.00	23	100.00	109.41	9.41	110.69	9.08	188.21	20.00	0.52
1.	10560	1994.61	24	100.00	110.60	10.60	111.97	9.41	211.97	20.00	0.51
1.	10560	1859.89	25	100.00	110.60	10.60	111.79	8.77	211.95	20.00	0.48
1.	10560	1709.00	26	100.00	110.23	10.23	111.31	8.35	204.64	20.00	0.46
1.	10560	1558.11	27	100.00	109.71	9.71	110.71	8.02	194.23	20.00	0.45
1.	10560	1423.39	28	100.00	109.16	9.16	110.10	7.77	183.14	20.00	0.45
1.	10560	1272.50	29	100.00	108.51	8.51	109.38	7.47	170.30	20.00	0.45
1.	10560	1121.61	30	100.00	107.82	7.82	108.62	7.17	156.43	20.00	0.45
1.	10560	986.89	31	100.00	107.15	7.15	107.89	6.90	142.93	20.00	0.46
1.	10560	836.00	32	100.00	106.40	6.40	107.06	6.53	127.97	20.00	0.46

Reach	River Sta	Q Total (cfs)	Profile	Min Ch El (ft)	W.S. Elev Max (ft)	Chl Dpth (ft)	E.G. Elev (ft)	Vel chnl (ft/s)	Flow Area (sq ft0	Top Width (ft)	Chl Froude #
1.	10560	836.00	33	100.00	106.17	6.17	106.88	6.78	123.37	20.00	0.48
1.	10560	836.00	34	100.00	106.08	6.08	106.81	6.88	121.59	20.00	0.49
1.	10560	836.00	35	100.00	106.04	6.04	106.79	6.92	120.83	20.00	0.50
1.	10560	836.00	36	100.00	106.02	6.02	106.77	6.94	120.49	20.00	0.50
1.	10560	836.00	37	100.00	106.02	6.02	106.77	6.95	120.33	20.00	0.50
1.	10560	836.00	38	100.00	106.01	6.01	106.76	6.95	120.25	20.00	0.50
1.	10560	836.00	39	100.00	106.01	6.01	106.76	6.95	120.21	20.00	0.50
1.	10560	836.00	40	100.00	106.01	6.01	106.76	6.96	120.19	20.00	0.50
1.	10560	836.00	41	100.00	106.01	6.01	106.76	6.96	120.18	20.00	0.50
1.	10560	836.00	42	100.00	106.01	6.01	106.76	6.96	120.18	20.00	0.50
1.	10560	836.00	43	100.00	106.01	6.01	106.76	6.96	120.17	20.00	0.50
1.	10560	836.00	44	100.00	106.01	6.01	106.76	6.96	120.17	20.00	0.50
1.	10560	836.00	45	100.00	106.01	6.01	106.76	6.96	120.17	20.00	0.50
1.	10560	836.00	46	100.00	106.01	6.01	106.76	6.96	120.17	20.00	0.50
1.	10560	836.00	47	100.00	106.01	6.01	106.76	6.96	120.17	20.00	0.50
1.	10560	836.00	48	100.00	106.01	6.01	106.76	6.96	120.17	20.00	0.50
1.	10560	836.00	49	100.00	106.01	6.01	106.76	6.96	120.17	20.00	0.50
1.	10560	836.00	50	100.00	106.01	6.01	106.76	6.96	120.17	20.00	0.50
1.	10560	836.00	51	100.00	106.01	6.01	106.76	6.96	120.17	20.00	0.50
1.	10560	836.00	52	100.00	106.01	6.01	106.76	6.96	120.17	20.00	0.50
1.	10560	836.00	53	100.00	106.01	6.01	106.76	6.96	120.17	20.00	0.50
1.	10560	836.00	54	100.00	106.01	6.01	106.76	6.96	120.17	20.00	0.50
1.	10560	836.00	55	100.00	106.01	6.01	106.76	6.96	120.17	20.00	0.50
1.	10560	836.00	56	100.00	106.01	6.01	106.76	6.96	120.17	20.00	0.50
1.	10560	836.00	57	100.00	106.01	6.01	106.76	6.96	120.17	20.00	0.50
1.	10560	836.00	58	100.00	106.01	6.01	106.76	6.96	120.17	20.00	0.50
1.	10560	836.00	59	100.00	106.01	6.01	106.76	6.96	120.17	20.00	0.50
1.	10560	836.00	60	100.00	106.01	6.01	106.76	6.96	120.17	20.00	0.50
1.	10560	836.00	61	100.00	106.01	6.01	106.76	6.96	120.17	20.00	0.50
1.	10560	836.00	62	100.00	106.01	6.01	106.76	6.96	120.17	20.00	0.50
1	0	841.31	2	84.16	90.20	6.04	90.95	6.97	120.70	20.00	0.50
1.	0	839.15	3	84.16	90.18	6.02	90.94	6.96	120.49	20.00	0.50
1.	0	837.80	4	84.16	90.18	6.02	90.93	6.96	120.35	20.00	0.50
1.	0	837.00	5	84.16	90.17	6.01	90.93	6.96	120.27	20.00	0.50
1.	0	836.55	6	84.16	90.17	6.01	90.92	6.96	120.23	20.00	0.50
1.	0	836.30	7	84.16	90.17	6.01	90.92	6.96	120.20	20.00	0.50
1.	0	836.16	8	84.16	90.17	6.01	90.92	6.96	120.19	20.00	0.50

Reach	River Sta	Q Total (cfs)	Profile	Min Ch El (ft)	W.S. Elev Max (ft)	Chl Dpth (ft)	E.G. Elev (ft)	Vel chnl (ft/s)	Flow Area (sq ft0	Top Width (ft)	Chl Froude #
1.	0	836.09	9	84.16	90.17	6.01	90.92	6.96	120.18	20.00	0.50
1.	0	836.05	10	84.16	90.17	6.01	90.92	6.96	120.18	20.00	0.50
1.	0	836.03	11	84.16	90.17	6.01	90.92	6.96	120.17	20.00	0.50
1.	0	836.01	12	84.16	90.17	6.01	90.92	6.96	120.17	20.00	0.50
1.	0	836.01	13	84.16	90.17	6.01	90.92	6.96	120.17	20.00	0.50
1.	0	836.01	14	84.16	90.17	6.01	90.92	6.96	120.17	20.00	0.50
1.	0	836.00	15	84.16	90.17	6.01	90.92	6.96	120.17	20.00	0.50
1.	0	836.00	16	84.16	90.17	6.01	90.92	6.96	120.17	20.00	0.50
1.	0	836.00	17	84.16	90.17	6.01	90.92	6.96	120.17	20.00	0.50
1.	0	836.00	18	84.16	90.17	6.01	90.92	6.96	120.17	20.00	0.50
1.	0	836.00	19	84.16	90.17	6.01	90.92	6.96	120.17	20.00	0.50
1.	0	836.00	20	84.16	90.17	6.01	90.92	6.96	120.17	20.00	0.50
1.	0	837.05	21	84.16	90.17	6.01	90.93	6.96	120.28	20.00	0.50
1.	0	866.34	22	84.16	90.32	6.16	91.09	7.03	123.22	20.00	0.50
1.	0	980.41	23	84.16	90.89	6.73	91.72	7.28	134.66	20.00	0.49
1.	0	1184.00	24	84.16	91.87	7.71	92.79	7.67	154.28	20.00	0.49
1.	0	1439.44	25	84.16	93.06	8.90	94.08	8.09	177.99	20.00	0.48
1.	0	1630.47	26	84.16	93.92	9.76	95.01	8.35	195.30	20.00	0.47
1.	0	1696.67	27	84.16	94.22	10.06	95.32	8.43	201.19	20.00	0.47
1.	0	1678.07	28	84.16	94.14	9.98	95.24	8.41	199.54	20.00	0.47
1.	0	1612.40	29	84.16	93.84	9.68	94.92	8.32	193.68	20.00	0.47
1.	0	1520.30	30	84.16	93.43	9.27	94.47	8.20	185.36	20.00	0.47
1.	0	1409.59	31	84.16	92.92	8.76	93.93	8.04	175.28	20.00	0.48
1.	0	1290.64	32	84.16	92.37	8.21	93.33	7.86	164.25	20.00	0.48
1.	0	1167.45	33	84.16	91.80	7.64	92.70	7.64	152.72	20.00	0.49
1.	0	1055.58	34	84.16	91.26	7.10	92.12	7.44	141.92	20.00	0.49
1.	0	972.05	35	84.16	90.85	6.69	91.67	7.26	133.85	20.00	0.49
1.	0	918.18	36	84.16	90.58	6.42	91.38	7.15	128.43	20.00	0.50
1.	0	883.89	37	84.16	90.41	6.25	91.19	7.07	124.99	20.00	0.50
1.	0	863.04	38	84.16	90.30	6.14	91.07	7.02	122.89	20.00	0.50
1.	0	851.04	39	84.16	90.24	6.08	91.00	6.99	121.68	20.00	0.50
1.	0	844.28	40	84.16	90.21	6.05	90.97	6.98	121.00	20.00	0.50
1.	0	840.51	41	84.16	90.19	6.03	90.95	6.97	120.62	20.00	0.50
1.	0	838.44	42	84.16	90.18	6.02	90.93	6.96	120.42	20.00	0.50
1.	0	837.32	43	84.16	90.18	6.02	90.93	6.96	120.30	20.00	0.50
1.	0	836.71	44	84.16	90.17	6.01	90.92	6.96	120.24	20.00	0.50
1.	0	836.38	45	84.16	90.17	6.01	90.92	6.96	120.21	20.00	0.50
1.	0	836.20	46	84.16	90.17	6.01	90.92	6.96	120.19	20.00	0.50

Reach	River Sta	Q Total (cfs)	Profile	Min Ch El (ft)	W.S. Elev Max (ft)	Chl Dpth (ft)	E.G. Elev (ft)	Vel chnl (ft/s)	Flow Area (sq ft0	Top Width (ft)	Chl Froude #
1.	0	836.11	47	84.16	90.17	6.01	90.92	6.96	120.18	20.00	0.50
1.	0	836.06	48	84.16	90.17	6.01	90.92	6.96	120.18	20.00	0.50
1.	0	836.03	49	84.16	90.17	6.01	90.92	6.96	120.17	20.00	0.50
1.	0	836.02	50	84.16	90.17	6.01	90.92	6.96	120.17	20.00	0.50
1.	0	836.01	51	84.16	90.17	6.01	90.92	6.96	120.17	20.00	0.50
1.	0	836.01	52	84.16	90.17	6.01	90.92	6.96	120.17	20.00	0.50
1.	0	836.00	53	84.16	90.17	6.01	90.92	6.96	120.17	20.00	0.50
1.	0	836.00	54	84.16	90.17	6.01	90.92	6.96	120.17	20.00	0.50
1.	0	836.00	55	84.16	90.17	6.01	90.92	6.96	120.17	20.00	0.50
1.	0	836.00	56	84.16	90.17	6.01	90.92	6.96	120.17	20.00	0.50
1.	0	836.00	57	84.16	90.17	6.01	90.92	6.96	120.17	20.00	0.50
1.	0	836.00	58	84.16	90.17	6.01	90.92	6.96	120.17	20.00	0.50
1.	0	836.00	59	84.16	90.17	6.01	90.92	6.96	120.17	20.00	0.50
1.	0	836.00	60	84.16	90.17	6.01	90.92	6.96	120.17	20.00	0.50
1.	0	836.00	61	84.16	90.17	6.01	90.92	6.96	120.17	20.00	0.50
1.	0	836.00	62	84.16	90.17	6.01	90.92	6.96	120.17	20.00	0.50

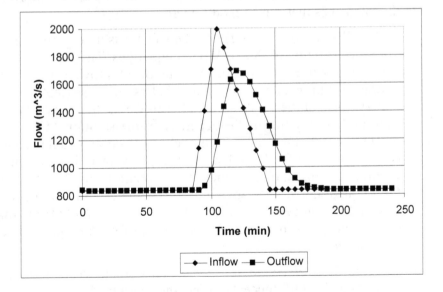

Figure 9.11.   Inflow and outflow hydrographs for Example 9.4.

The use of HEC-RAS to solve unsteady flow problems requires that the user exercise both judgment and care. In the following paragraphs some of the subjects that must be considered are discussed. First, the user must specify the computation interval, and this interval is how the computation proceeds. This interval must be small enough to accurately describe the rise and fall of the hydrograph. In general, a computation interval less than the time of rise of the hydrograph divided by 24 will provide satisfactory results (USACE, 2001). For example, if the hydrograph rises from a base flow rate to a peak value in 24 hours, then a computation interval of one hour or less is satisfactory. Additional consideration must be given to the computation interval when hydraulic structures (bridges, culverts, weirs, and gated spillways) are part of the system being modeled. In the vicinity of these structures, the water surface can change rapidly; and these rapid changes can cause the numeric solution of the unsteady flow equations to become unstable. One solution to this problem is to use a very small computation interval, one to 5 minutes. In these situations, the time step should be adjusted to find the largest computation interval that will provide an accurate solution of the equations.

Second, the user can also override the default settings for calculation tolerances. These tolerances are used in the solution of the momentum equation; and increasing the values of these tolerances can result in computational errors in the calculation of the water surface profile. Among the tolerances that can be modified are:

Implicit Weighting Factor Theta: This factor is used in the finite difference solution of the St. Venant equations. The value of this factor ranges from 0.6 to 1.0. A value of 0.6 provides the most accurate solution of the equations, but is more susceptible to numeric instabilities. A value of 1.0 yields the most stable solution, but may impair the accuracy of the solution. USACE (2001) suggests that once the model is set up and running the user should modify the value of *theta*, toward 0.6; and if the solution remains stable, a value of 0.6 should be used. USACE (2001) also noted that in some cases there will not be an appreciable change in the results when changing *theta* from 1.0 to 0.6, but experimentation is required to find the appropriate value of *theta*.

Water surface elevation tolerance: This tolerance is used in comparing the difference between the computed and assumed water surface elevations at a cross section. If the difference is greater than the tolerance, then additional iterations are performed. Alternatively, if the difference is less than this tolerance, then the program assumes that a valid numerical solution has been obtained. The default value is 0.02 feet.

Maximum number of iterations: This variable sets the maximum number of iterations performed in attempting to solve the unsteady flow equations given the specified tolerances. The default value is 20, and the allowable range is from 0 to 40.

Maximum number of warm-up time steps: During the warm-up period, the program performs a series of calculations with a constant inflow. This is done to smooth the water surface profile before the unsteady flow simulation begins. The default number of warm up steps is 20, and the allowable range is from 20 to 40.

Time step during the warm up period: During the warm-up period, it may be necessary to use a computation interval that is smaller than that used during the unsteady flow computations. The default is to leave this field blank which means that the computation interval is the same as for the unsteady flow modeling.

Minimum time step for interpolation: This program has the ability to interpolate between time steps when a steep rise in the inflow hydrograph is found or when there is a rapid change in the water surface elevation at any cross section. This option provides the user the opportunity to set the minimum computational interval during the interpolation process and prevents the program from using too small a time step in this process.

Maximum number of interpolated time steps: This option defines the maximum number of interpolated time steps allowed.

9.4. BOUNDARY AND INITIAL CONDITIONS

As noted previously in this chapter, the solution of the equations governing unsteady flow requires that boundary conditions be specified throughout the simulation period. One set of boundary conditions be specified throughout the simulation period. One set of boundary conditions must be specified at the physical extremities of the system, and additional conditions are required; e.g., at nodes where tributaries join the main channel.

Various combinations of boundary conditions can be specified at the physical extremities of a network. External boundary conditions, which are commonly specified, are; zero discharge, discharge as a function of time, stage as a function of time, or a known, unique stage-discharge relationship. Fread (1975) noted that when there is no significant flow disturbance downstream of the routing reach which can propagate into the reach and influence the flow, a rating equation is always available - the Manning or Chezy uniform flow equation.

The solution of the equations governing unsteady and the specified boundary conditions also requires that initial values of the unknown variables be specified. Initial values may be obtained from measurements or estimated by computation; e.g., uniform flow or gradually varied flow approximations or previous unsteady flow simulations. While reasonable initial conditions are desirable, estimates can be used if a sufficient period of time is provided for the unsteady flow model to dissipate the errors and converge to the true solution. Models of flow systems having high rates of dissipation will converge more rapidly than models of systems having low rates of energy dissipation (Lai, 1965a, 1965b).

9.5.   CALIBRATION AND VERIFICATION

As is the case with all models, unsteady flow models must be calibrated and verified. These are two processes which require accurate prototype data. As noted by French and Krenkel (1980), the calibration and verification processes require two independent and statistically reliable sets of prototype data. One data set is used to establish the optimum values of the "free" coefficients, and the second set is used to verify that the calibration process established valid coefficient values.

In the calibration process, prototype data are used to refine the values of the least quantifiable variables; e.g., the flow resistance coefficient. The objective of the process is to adjust these variables so that the prototype system for a range of flow conditions is accurately replicated. Although the reproduction of the water surface elevations by the model is of value, the replication of the prototype discharge hydrograph in both time and amplitude is much more important. If a high level of calibration can be achieved and verified, then it may be possible to extend the application of the model beyond the limits of the data used in the calibration and verification processes. In general, the data for the calibration and verification of an unsteady flow model consists of two time series of measured discharges with simultaneously measured water surface elevations. The goal of the calibration and verification processes is to satisfy a goodness of fit criterion; e.g., minimizing the sum of the squares of the differences between the measured and simulated values, Eq. (9.1.31).

While in theory all aspects and variables in a model may be subject to adjustment and refinement, in practice, there are variables which cannot be determined directly. For example, in the case of unsteady flow models, channel geometry data are not usually adjusted since such data can be measured with reasonable accuracy. In such models, attention is usually centered on the value of the resistance coefficient, a variable which is not normally measured directly.

The difficulty in determining an accurate value of the resistance coefficient results primarily from the fact that in applied hydraulics the energy dissipation relation used is the result of empiricism. In addition, the dissipation formulation used in unsteady flow models is an approximation borrowed from the semi-empirical formula developed for steady, uniform flow. There is little factual information regarding the effect of perimeter resistance under unsteady flow conditions available.

Schaffranek *et al.* (1981) have suggested that the calibration of an unsteady flow model begins with a flow data set gathered during steady or nearly steady flow conditions. If such a data set is available, then the resistance coefficient can be either determined from the Manning or Chezy uniform flow equation or simply estimated. At this point, the following observations regarding the calibration process are pertinent:

1. The use of a resistance coefficient that is too small reduces the flow resistance and increases the momentum or inertia. The resulting simulations will have peaks which are too high and minimum points which are too low. In general, there will also be a phase shift which may or may not be desirable. The use of a resistance coefficient which is too large has the opposite effect on the simulation results.

2. The recorded water surface elevations may be in error due to datum errors, surveying inaccuracies, or vertical displacement of the gaging structure. The result of an error of this type is an increase or decrease in the water surface slope throughout the reach in which the error occurs. An increased water surface slope yields increased flow in the downstream direction while a decreased water surface slope has the opposite effect.

3. The use of too large a cross-sectional area results in magnified peaks and troughs while using a cross sectional area which is too small has the opposite effect.

Schaffranek *et al.* (1981) noted that the calibration of unsteady flow models may range from simple to difficult depending on the complexity of the flow regime being simulated and the interconnection of the channels within the prototype. Channel networks in which the flow has only one path to travel between any two locations are generally easier to calibrate than channel networks where it is possible for the flow to use more than one route between locations. In this latter situation, erroneous circulations may appear internally and render the model useless.

As can be discerned from the foregoing discussion, the determination of an optimum value of the flow resistance coefficient is the primary task in the calibration of a one-dimensional unsteady flow model. An obvious and widely used method of accomplishing this is a trial-and-error technique in which the governing equations are repeatedly solved for different values of the flow resistance coefficient. The value of the coefficient which produces the best fit, as measured by some goodness-of-fit criterion, between the measured and simulated values is the optimum value. This trial-and-error search for optimum values can be tedious, difficult, and expensive. In addition, in the case of a model for a river system, Fread and Smith (1978) have noted that modification of the resistance coefficient value in a tributary can affect the flow both upstream and downstream of the tributary in the main stem of the river as well as the flow in the other tributaries. This is particularly true in large rivers with major tributaries and/or when the bottom slope of the river is mild; e.g., 2 ft/mile (0.38 m/km).

The determination of parameter values in a set of partial differential equations for a specified set of boundary and initial conditions is termed the *inverse problem*. Fread and Smith (1978) have developed an optimization methodology for determining the value of the flow resistance coefficient in one dimensional unsteady flow models. In this technique, the resistance coefficient is assumed to vary with the stage and discharge in the channel, but is not a function of longitudinal distance. The goodness-of-fit criterion used by Fread and Smith (1978) is the minimization of the absolute value of the sum of the differences between the observed and modeled values of either the stage or discharge.

9.6. BIBLIOGRAPHY

Amein, M., Streamflow Routing on Computer by Characteristics, Water Resources Research, vol. 2, no. 1, 1966, pp. 123-130.

Amein, M., An Implicit Method for Numerical Flood Routing, Water Resources Research, vol. 4, no. 4, 1968, pp. 719-726.

Amein, M. and Fang, C.S., Implicit Flood Routing in Natural Channels, Proceedings of the American Society of Civil Engineers, Journal of the Hydraulics Division, vol. 96, no. HY12, December 1970, pp. 2481-2500.

Amein, M. and Chu, H.L., Implicit Numerical Modeling of Unsteady Flows, Proceedings of the American Society of Civil Engineers, Journal of the Hydraulics Division, vol. 101, no. HY6, June 1975, pp. 717-731.

Barkau, R.L., "UNET One-Dimensional Unsteady Flow Through a Network of Open Channels, Users Manual," Report No. CPD-66, U.S. Army Corps of Engineers, Hydrologic Engineering Center, Davis, CA., 1997.

Fread, D.L., Discussion of "Comparison of Four Numerical Methods for Flood Routing," by R.K. Price, *Proceedings of the American Society of Civil Engineers, Journal of the Hydraulics Division*, vol. 101, no. HY3, 1975, pp. 565-567.

Fread, D.L., "Flood Routing in Meandering Rivers with Flood Plains," *Proceedings, Rivers '76*, Third Annual Symposium of Waterways, Harbors and Coastal Engineering Division, American Society of Civil Engineers, vol. 1, 1976, pp. 16-35.

Fread, D.L., "DAMBRK: The NWS Dam Break Flood Forecasting Model," Office of Hydrology, National Weather Service, Silver Spring, MD., 1978.

Fread, D.L. and Smith, G.F., "Calibration Technique for 1-D Unsteady Flow Models," *Proceedings of the American Society of Civil Engineers, Journal of the Hydraulics Division*, vol. 104, no. HY7, July 1978, pp. 1027-1044.

French, R.H. and Krenkel, P.A., "Effectiveness of River Models," *Progress in Water Technology*, vol. 13, 1980, pp. 99-113.

Lai, C., "Flows of Homogeneous Density in Tidal Reaches: Solution by Method of Characteristics," U.S. Geological Survey, Open File Report, Washington, 1965a.

Lai, C., "Flows of Homogeneous Density in Tidal Reaches: Solution by Implicit Method," U.S. Geological Survey, Open File Report, Washington, 1965b.

McCarthy, G.T., "The Unit Hydrograph and Flood Routing," *Conference of the North Atlantic Division*, U.S. Army Corps of Engineers, New London, CN, 1938.

Ponce, V.M., *Engineering Hydrology: Principles and Practices*, Prentice Hall, Englewood Cliffs, NJ, 1989.

Ponce, V.M., Li, R.M., and Simons, D.B., "Applicability of Kinematic and Diffusion Models," *Proceedings of the American Society of Civil Engineers, Journal of the Hydraulics Division*, vol. 104, no. HY3, March, 1978, pp. 353-360.

Proceedings of the American Society of Civil Engineers, Journal of the Hydraulics Division, vol. 96, no. HY12, December 1970, pp. 2481-2500.

Proceedings of the American Society of Civil Engineers, Journal of the Hydraulics Division, vol. 96, no. HY12, December 1970, pp. 2481-2500.

Schaffranek, R.W., Baltzer, R.A., and Goldberg, D.E., "A Model for Simulation of Flow in Singular and Interconnected Channels," chapter C3, *Techniques of Water Resources Investigations of the United States Geological Survey*, Washington, 1981.

Smith, R.H., "Development of a Flood Routing Model for Small Meandering Rivers," Ph.D. Dissertation, Department of Civil Engineering, University of Missouri at Rolla, MO.

USACE, "HEC-RAS River Analysis System, Hydraulic Reference Manual, Version 3.0," U.S. Army Corps of Engineers, Hydrologic Engineering Center, Davis, CA, 2001.

Viessman, W., Jr., Knapp, J.W., Lewis, G.L., and Harbaugh, T.E., *Introduction to Hydrology*, 2nd edition, Harper& Row, New York, 1972.

Weinmann, D.E. and Laurenson, E.M., "Approximate Flood Routing Methods: A Review," *Proceedings of the American Society of Civil Engineers, Journal of the Hydraulics Division*, vol. 105, no. HY12, December, 1979, pp. 1521-1536.

9.7.   PROBLEMS

1.   Given a vertical wall reservoir with a surface area of 500 acres (2,023,428 m²). The reservoir has an emergency spillway with width of 90 ft (27.4 m) and a discharge coefficient of 3.0. *H* is the elevation of the water surface above the spillway crest, and the initial inflow and outflow are both 75 ft³/s (2.1 m³/s), and this steady discharge continues throughout the flood event (Part d). Initially the reservoir is at the elevation of the emergency spillway.

    a. In acre-feet and ft³, determine the values of reservoir storage *V* corresponding to values of *H:* 0, 0.5, 1, 1.5, 2, 3, 4 ft.

    b. Determine the values of the emergency spillway *Q* corresponding to the elevations in Part a of this problem.

    c. Plot and label (Cartesian co-ordinate system) a storage indication versus discharge graph.

    d. Determine the outflow rates over the spillway at the ends of successive days corresponding to the following instantaneous inflow rates, at the end of successive days, of 100, 400, 1200, 1500, 2000,1500, 1100, 700, 400, 300, 200, 100, 100, 100 ft³/s. Use a routing table to accomplish this, and continue the procedure until the outflow over the spillway is approximately 0 ft³/s.

e.   Graph the inflow and outflow hydrographs.

2.   It is asserted in the text that Eq. (9.1.24) can be transformed to Eq. (9.1.25). Demonstrate that this is a true statement

3.   It is asserted in the text with regard to the Muskingum method that $C_0 + C_1 + C_2 = 1$, prove this assertion.

4.   Given the following inflow hydrograph to a channel reach, calculate the outflow by the Muskingum method with $K = 1$ hr, $X = 0.2$, $\Delta t = 1$ hr and assuming the base flow is 350 ft^3/s.

Time (hr)	Flow (ft^3/s)
0	350
1	710
2	1400
3	2800
4	4200
5	5300
6	4200
7	2100
8	1800
9	1400
10	1100
11	710
12	350
13	359
14	350
15	350
16	350
17	350
18	350
19	350
20	350

5.   Given the inflow and outflow hydrographs for a channel reach summarized below, estimate values of the Muskingum parameters K and X.

Time (hrs)	Inflow (ft³/s)	Outflow (ft³/s)
0	2520	2520
1	3870	2643
2	4560	3598
3	6795	4500
4	8975	6367
5	9320	8295
6	7780	8900
7	6520	7971
8	5340	6808
9	4105	5628
10	3210	4439
11	2520	3482
12	2520	2782
13	2520	2592
14	2520	2540
15	2520	2525
16	2520	2521
17	2520	2520

6.   For the triangular channel defined below, calculate the kinematic wave parameter β using the Manning equation.

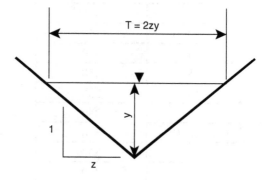

7.   Develop a second order kinematic wave numerical model that includes lateral inflow.

8.   The hydrograph summarized in the table below was measured at Section 1, and it is desired to estimate the hydrograph that will be experienced at a downstream section using 1st and 2nd order kinematic wave routing models for the variables and parameters specified below. Note, the hydrograph is triangular in shape and $\beta = 5/3$, $u = 1.2$ m/s and $\Delta t = 1$ hr.

Flow at Section 1 as a function of time.

Time (hr)	Flow (m³/s)
0	0
1	30
2	60
3	90
4	120
5	150
6	120
7	90
8	60
9	30
10	0

a.   First order model with $\Delta x = 7{,}200$ m.
b.   Second order model with $\Delta x = 7{,}200$ m.
c   First order model with $\Delta x = 4{,}800$ m.
d.   Second order model with $\Delta x = 4{,}800$ m.
e.   Compare and contrast the estimates of the downstream hydrograph and discuss the results relative to the value of Courant number $\beta u(\Delta t/\Delta x)$.

9.   The Department of Transportation (DoT) in building a new road must cross an irrigation/flood control channel that is 10 m (32.8 ft) wide and 6 m (19.7 ft) deep and rectangular in cross section. This channel conveys a base flow of 3 m³/s (106 ft³/s) and has a longitudinal slope of 0.001 m/m. The Manning resistance coefficient is 0.02. The DoT has proposed a free span bridge with abutments protruding into the channel as shown the figure below. The lower chord of the bridge must be at an elevation such that it does not interfere with the flood hydrograph tabulated below for a station 2000 m (6560 ft) upstream of the

bridge, and the elevation of the channel invert at this location 100 m above a datum. Therefore, what is the minimum elevation of the lower chord of the bridge above the bottom of the channel at the location of the bridge? Also, plot the stage as a function of time at the bridge.

Time (hr)	Flow (m³/s)
0	3
1	3
2	3
3	5
4	10
5	15
6	20
7	25
8	30
9	25
10	20
11	20
12	15
13	10
14	5
15	3
16	3
17	3
18	3

A rectangular concrete channel ($n = 0.02$) with a bottom width of 20 ft (6.1 m) on a slope of 0.001 ft/ft connects a wastewater treatment plant to a tidal lagoon 10 mi (16 km) downstream. The channel is 30 ft (9.1 m) deep). Under normal operating conditions, the flow from the wastewater treatment plant is 500 ft³/s (14.2 m³/s). The elevation of the channel invert at the downstream end is 100 ft (30.5 m), and the water surface elevation (in feet) at this point as a function of time is given by

$$\text{W.S. Elevation} = 106 + 5\sin\left(\frac{2\pi t}{24}\right)$$

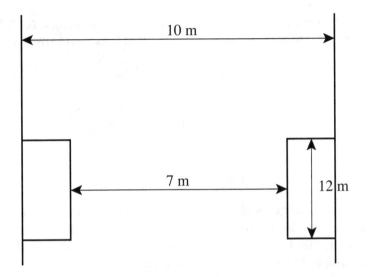

where t is time in hours.

a.   Are the stage and discharge at the waste treatment plant influenced by the tide in the lagoon? A suggested simulation time is 48 hours? Plot the stage and discharge at the waste treatment plant as a function of time.

b.   Plot the stage and discharge at the point where the channel meets the tidal lagoon as a function of time.

c.   Does the existence of the tidal lagoon adversely affect discharges from the wastewater treatment plant?

11.   A rectangular concrete channel (n = 0.02) with a bottom width of 20 ft (6.1 m) on a slope of 0.001 ft/ft connects a wastewater treatment plant to a tidal lagoon 10 mi (16 km) downstream. The channel is 30 ft (9.1 m) deep). Under normal operating conditions, the flow from the wastewater treatment plant is 500 ft³/s (14.2 m³/s). The elevation of the channel invert at the downstream end is 90 ft (27.4 m), and the water surface

$$\text{W.S. Elevation} = 106 + 5\sin\left(\frac{2\pi t}{24}\right)$$

where t is time in hours.

The wastewater treatment plant serves a city with a combined sewer system; and therefore, during flood events it discharges not only wastewater but also storm water. During a recent flood event, the discharge from the treatment was as follows

$$0 \le t \le 6 \quad Q = 500 \, \text{ft}^3 / s$$

$$6 < t \le 18 \quad Q = 500 + 1000 \sin\left[\frac{2\pi(t-6)}{24}\right]$$

$$18 < t \le 48 \quad Q = 500 \, \text{ft}^3 / s$$

a. Are the stage and discharge at the waste treatment plant influenced by the tide in the lagoon? A suggested simulation time is 48 hours? Plot the stage and discharge at the waste treatment plant as a function of time.

b. Plot the stage and discharge at the point where the channel meets the tidal lagoon as a function of time.

c. Does the existence of the tidal lagoon adversely affect discharges from the wastewater treatment plant?

d. Plot the stage and discharge curves at a distance 2,000 ft (610 m) upstream of where the channel meets the tidal lagoon.

CHAPTER 10
HYDRAULIC MODELS

10.1. INTRODUCTION

Unlike many other fields of scientific endeavor, modern hydraulic engineering has been and remains to a large extent based on experiment. A study of the history of hydraulics (see, for example, Rouse and Ince, 1963) indicates that periods of experimental investigation have alternated with periods of analysis. For example, as noted by the American Society of Civil Engineers' (ASCE) Hydraulics Division Committee (Anonymous, 1942), it was not until the eighteenth century that sufficient experimental data had been accumulated to permit the foretelling of future progress, and another 100 years passed before the results marking the achievement of this progress were available.

During the twentieth century, hydraulic engineering greatly benefitted from the developments of boundary layer and turbulence theory in the field of fluid mechanics. Before these developments and the advent of the high-speed digital computer, many problems in the field of hydraulics could be resolved only with model studies. Today, the dual approach of theoretical and model studies is available to the engineer.

The use of models in the laboratory environment to solve hydraulic engineering problems requires a clear and accurate understanding of the principles of similitude. With regard to similarity, previously discussed in Chapter 1, there are three distinct viewpoints.

1. *Geometric Similarity:* Two objects are said to be geometrically similar if the ratios of all corresponding dimensions are equal. Thus, the *geometric similarity* refers only to similarity in form.

2. *Kinematic Similarity:* Two motions are said to be kinematically similar if (a) the patterns or paths of motion are geometrically similar and (b) the ratios of the velocities of the particles involved in the two motions are equal.

3. *Dynamic Similarity:* Two motions are said to be dynamically similar if (a) the ratio of the masses of the objects involved are equal and (b) the ratios of the forces which affect the motion are equal.

At this point, it is noted that while geometric and kinematic similarity can be achieved in most modeling situations, complete dynamic similarity is an ideal which can seldom, if ever, be achieved in practice (Anonymous, 1942).

As noted in Section 1.5, one approach to developing appropriate parameters to ensure dynamic similarity is the scaling of the governing equations (Section 1.4). A second method of achieving this objective is the use of the concept of dimensionless analysis and the Buckingham II theorem (see, for example, Streeter and Wylie, 1975).

Experience dictates that in most models it is necessary to confine similarity to a single force. Since in most open-channel flow problems the force of gravity is the primary force, this chapter focuses on models designed for similitude with respect to the force of gravity. Further, the modeler is almost invariably constrained to use water as the model fluid. It should be mentioned that there are, of course, exceptions to these statements when special circumstances require exceptions (see, for example, Walski, 1979). In addition, in most models of interest to hydraulic engineers, the flow should be turbulent rather than laminar. Recall that turbulent flow generally exists when the Reynolds number, based on the hydraulic radius, exceeds 2000. Flow is laminar with the Reynolds number is less than 500. Between these limits lies the transition zone which should be avoided in designing models.

In addition to theoretical considerations, only a few of which have been discussed in the foregoing paragraphs, the modeler must be concerned with the following practical limitations: (1) time, (2) money, (3) laboratory space, and (4) the available water supply. These four practical considerations may have a crucial and significant impact on any modeling effort.

The foregoing theoretical and practical considerations combine to produce model scale ratios in the following ranges:

1. Models of spillways, conduits, and other appurtenances, the prototypes of which have relatively smooth surfaces and have scale ratios that range from 1:50 (model distance prototype distance) to 1:15. Such models should never be distorted.

2. Models of rivers, harbors, estuaries, and reservoirs, the prototypes of which have relatively rough surfaces and have horizontal-scale ratios that range from 1:2000 to 1:200 and vertical-scale ratios that range from 1:150 to 1:50. Models in this category are almost always distorted.

Models with distorted-scale ratios depart from strict dynamic similarity, and such a departure is acceptable only when the requirement of strict similitude can be waived.

The need for distorted-scale models, or what is more accurately termed geometrically distorted models, is usually found only in the study of open-channel phenomena and arises from two sets of circumstances.

1. The area required for an undistorted model would be so large that space and economic considerations dictate that the horizontal scale of the model must be made small. In such a situation, if the vertical scale were equal to the horizontal scale, the depth of flow would be so small that it could not be measured satisfactorily. Given these circumstances, it is usually considered desirable to increase the vertical scale of the model relative to the horizontal scale.

2. In the case of a movable-bed model whose purpose is the simulation of the movement of bed material, the slopes and fluid velocities encountered in an undistorted model would usually be too small to move any of the materials typically available for use as bed materials. As before, it is usually considered desirable to use a vertical scale which is larger than the horizontal scale.

Although in the subsequent sections of this chapter the primary principle used will be the Froude law, it is appropriate to consider, at this point, not only the implication of the Froude law but also a number of other specialized modeling laws.

Froude law models assert that the primary force causing fluid motion is gravity and that all other forces such as fluid friction and surface tension can be neglected. As noted in Section 1.5 if only Froude number similarity is required, then

$$F_M = F_P \qquad (10.1.1)$$

where F = Froude number and the subscripts M and P designate the model and prototype Froude numbers, respectively. Equation (10.1.1) can be solved to yield

$$U_R = \frac{U_M}{U_P} = \left(\frac{g_M}{g_P} \frac{L_M}{L_P} \right)^{1/2} = \sqrt{g_R L_R} \qquad (10.1.2)$$

where
R = subscript indicating ratio of model to prototype variables
U_R = velocity ratio
L_R = length scale ratio
g_R = gravity ratio

Since from a practical viewpoint the acceleration of gravity cannot be altered

$$g_R = 1$$

and Eq. (10.1.2) becomes

$$U_R = \sqrt{L_R} \qquad (10.1.3)$$

Since the velocity of flow can be expressed in terms of distance and time, the time-scale ratio can be derived from Eq. (10.1.3)

$$\frac{U_M}{U_P} = \frac{\dfrac{L_M}{T_M}}{\dfrac{L_P}{T_P}} = \frac{T_P L_M}{T_M L_P} = \frac{L_R}{T_R}$$

Substitution of Eq. (10.1.3) in this result yields

$$T_R = \sqrt{L_R} \qquad (10.1.4)$$

where T_R = time scale ratio. Previously in this section it was noted that in some cases Froude models with distorted scales must be used. In such cases, the distorted model should be designed so that

$$U_R = \sqrt{Y_R} \qquad (10.1.5)$$

where Y_R = vertical-scale ratio. The time scale is

$$T_R = \frac{L_R}{\sqrt{Y_R}} \tag{10.1.6}$$

where L_R = horizontal-scale ratio. Although Eqs. (10.1.5) and (10.1.6) are recommended formulations for distorted models (see, for example, Anonymous 1942), many distorted -scale models depart from these recommendations.

Reynolds law models assert that viscosity is the primary force that must be considered and, hence, that gravitational and surface tension forces can be safely neglected. In models based on the Reynolds law

$$\text{Re}_R = 1 \tag{10.1.7}$$

where Re = Reynolds number, therefore

$$U_R = \frac{\mu_R}{L_R \rho_R} \tag{10.1.8}$$

and

$$T_R = \frac{L_R \rho_R}{\mu_R} \tag{10.1.9}$$

where ρ_R = density ratio and μ_R = dynamic viscosity ratio.

Weber law models ensure similitude between the model and prototype with regard to surface tension effects. Although this modeling law is of paramount importance in the study of droplet formation, it can also be a crucial consideration in some open-channel models. For example, surface tension effects are noticeable in flow over weirs when the head on the weir is small. In addition some wave phenomena noticeable in models are due to surface tension. The Weber number is defined by

$$W = \frac{\rho U^2 L}{\sigma} \tag{10.1.10}$$

where U = characteristic velocity
L = characteristic length
ρ = fluid density
σ = surface tension force

Equating the Weber numbers for a model and prototype yields the following relationships for the velocity and time ratios:

$$U_R = \left(\frac{\sigma_R}{L_R \rho_R} \right)^{1/2}$$

(10.1.11)

and

$$T_R = \left(\frac{L_R^3 \rho_R}{\sigma_R} \right)^{1/2}$$

(10.1.12)

Table 10.1 summarizes the relations that must exist between the several properties of the fluids used in the model and prototype if perfect similitude is to be obtained under the various modeling laws discussed. Each expression in Table 10.1 is a combination of ratios, model to prototype, between an arbitrary length and force characteristic of each law. Time is generally considered the dependent ratio and becomes fixed once the length ratio and fluids for the model and prototype are selected.

In open-channel flow, the presence of the free surface ensures that gravitational forces are always dominant; therefore, the Froude law is the primary principle on which almost all open-channel models are built. Surface tension effects are important only when the radius of curvature of the free surface or the distances from solid boundaries are very small. Thus, surface tension effects are negligible in most prototypes, and care must be taken to ensure that they are also negligible in the models. From a practical viewpoint, if model depths of flow are greater than 2 in (0.05 m), surface tension effects can usually be ignored. Viscosity is much more important and can exert its influence in many different situations. The only method of keeping viscous effects exactly the same in an open-channel model and prototype is to keep both the Reynolds and Froude numbers equal in both the model and prototype. From a practical viewpoint, this requirement cannot be satisfied. However, if the flow in the model is turbulent, then the form drag will be accurately modeled, but surface drag will not be accurately modeled. Thus, the modeler must at all times be concerned with what are termed *scale effects*, i.e., distortions introduced by forces other than the dominant force on which the model is based.

Table 10.1. Fluid property scales (Anonymous, 1942)

Characteristic	Dimension	Scale Ratio for the laws of			
		Froude	Reynolds	Weber	Cauchy
Force Characteristics		Gravity	Viscosity	Surface Tension	Elasticity
Length	L	L_R	L_R	L_R	L_R
Area	L^2	$(L_R)^2$	$(L_R)^2$	$(L_R)^2$	$(L_R)^2$
Volume	L^3	$(L_R)^3$	$(L_R)^3$	$(L_R)^3$	$(L_R)^3$
Kinematic Properties					
Time	T	$[(L\rho/\gamma_R]^{1/2}$	$(L^2\rho/\mu)_R$	$[L^3\rho/\sigma)_R]^{1/2}$	$[L(\rho/K)^{1/2}]_R$
Velocity	L/T	$(L\gamma\rho)_R$	$(\mu/L\rho)_R$	$[(\sigma/L\rho)_R]^{1/2}$	$[(K/\rho)^{1/2}]_R$
Acceleration	L/T^2	$(\gamma\rho)_R$	$(\mu^2/\rho^2L^3)_R$	$(\sigma/L^2\rho)_R$	$(K/L\rho)_R$
Discharge	L^3/T	$[L^{5/2}\gamma/\rho)^{1/2}]_R$	$(L\mu/\rho)_R$	$[L^{3/2}(\sigma/\rho)^{1/2}]_R$	$[L^2(K/\rho)^{1/2}]_R$
Kinematic viscosity	L^2/T	$[L^{3/2}\gamma/\rho)^{1/2}]_R$	$(\mu/\rho)_R$	$[(L\sigma/\rho)_R]^{1/2}$	$[L(K/\rho)^{1/2}]_R$
Dynamic Properties					
Mass	M	$(L^3\rho)_R$	$(L^3\rho)_R$	$(L^3\rho)_R$	$(L^3\rho)_R$
Force	ML/T^2	$(L^3\gamma)_R$	$(\mu^2/\rho)_R$	$(L\sigma)_R$	$(L^2K)_R$
Density	M/L^3	ρ_R	ρ_R	ρ_R	ρ_R
Specific weight	M/L^2T^2	γ_R	$(\mu^2/L^3\rho)_R$	$(\sigma/L^2)_R$	$(K/L)_R$
Dynamic viscosity	M/LT	$[L^{3/2}(\rho\gamma^{1/2}]_R$	μ_R	$[(L\rho\sigma)_R]^{1/2}$	$[L(K\rho)^{1/2}]_R$
Surface tension	M/T^2	$(L^2\gamma_R$	$(\mu^2/L$	σ_R	$(LK)_R$
Volume elasticity	M/LT^2	$(L\gamma_R$	$(\mu^2/L^2\rho)_R$	$(\sigma/L)_R$	K_R
Pressure intensity	M/LT^2	$(L\gamma_R$	$(\mu^2/L^2\rho)_R$	$(\sigma/L)_R$	K_R
Momentum impulse	ML/T	$[L^{7/2}(\rho\gamma^{1/2}]_R$	$(L^2\mu)_R$	$[L^{5/2}(\rho\sigma)^{1/2}]_R$	$[L^3(K\rho)^{1/2}]_R$
Energy and work	ML^2/T^2	$(L^4\gamma_R$	$(L\mu^2/\rho)_R$	$(L^2\sigma)_R$	$(L^3K)_R$
Power	ML^2/T^3	$(L^{7/2}\gamma^{3/2}/\rho^{11/2})_R$	$(\mu^3/L\rho^2)_R$	$[\sigma^{3/2}(L/\rho)^{1/2}]_R$	$(L^2K^{3/2}/\rho^{1/2})_R$

10.2.   FIXED-BED RIVER OR CHANNEL MODELS

For river and channel studies in which the motion of the bed is unimportant, either an undistorted or a distorted fixed-bed model can be used, depending on the characteristics of the flow which must be represented. An undistorted model is recommended if the study involves the reproduction of supercritical flow, transitions, wave patterns, or water surface profiles. If the model is to reproduce channel capacity or channel storage capacity, then a distorted-scale model is satisfactory.

The use of an undistorted model presents the modeler with a minimum of design and analytical problems. In a distorted-scale model, the primary difficulty is ensuring that the model is sufficiently rough so that conversions of kinetic to potential energy and vice versa are not distorted. To demonstrate this difficulty, consider a distorted-scale model in which the vertical-scale ratio is Y_R and the horizontal-scale ratio is L_R. The average velocities of flow in the prototype and model are given by the Manning equation or

$$\overline{u_P} = \frac{\varphi R_P^{2/3} \sqrt{S_P}}{n_P} \tag{10.2.1}$$

and

$$\overline{u_M} = \frac{\varphi R_M^{2/3} \sqrt{S_M}}{n_M} \tag{10.2.2}$$

where R = hydraulic radius

S = channel bottom slope

n = Manning resistance coefficient

φ = coefficient whose value depends on system of units used (1 for SI system, 1.49 for English system)

Then, combing these equations to form the velocity scale ratio yields

$$U_R = \frac{\overline{u_M}}{\overline{u_P}} = \frac{R_R^{2/3} \sqrt{S_R}}{n_R}$$

and since $S_R = Y_R/L_R$

$$U_R = \frac{R_R^{2/3} \sqrt{Y_R}}{n_R \sqrt{L_R}} \tag{10.2.3}$$

The volumetric discharge ratio can be obtained by noting that

$$Q_R = L_R Y_R U_R$$

or

$$Q_R = \frac{R_E^{2/3} \sqrt{L_R} Y_R^{3/2}}{n_R} \qquad (10.2.4)$$

A time scale can then be defined as follows

$$T_R = \frac{\text{volume scale}}{\text{discharge scale}} = \frac{L_R^2 Y_R}{Q_R} \qquad (10.2.5)$$

Note: In a distorted hydraulic model, the hydraulic radius cannot be estimated analytically but must be computed on a section-by-section basis.

EXAMPLE 10.1

If

$$
\begin{array}{lcl}
L_R & = & 1/200 \\
Y_R & = & 1/80 \\
Q_R & = & 1/128{,}000 \\
n_P & = & 0.024
\end{array}
$$

and corresponding channel sections of the prototype and model are as shown in Fig. 10.1, determine n_M and T_R. This example was first presented in Anonymous (1942).

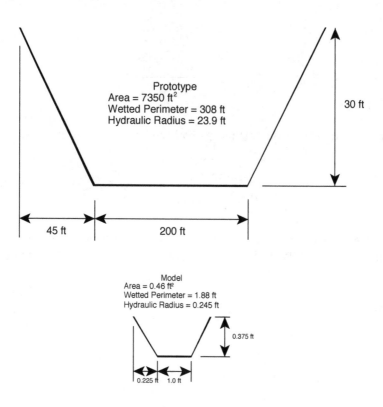

Figure 10.1. Schematic for Example 10.1.

Solution
The slope ratio is given by

$$S_R = \frac{Y_R}{L_R} = \frac{1/80}{1/200} = \frac{200}{80} = 2.5$$

From the information in Fig. 10.1

$$R_R = \frac{R_M}{R_P} = \frac{0.245}{23.9} = \frac{1}{97.5} = 2.5$$

From Eq. (10.2.4)

$$Q_R = \frac{1}{n_R} R_R^{2/3} Y_R^{3/2} \sqrt{L_R} = \frac{1}{n_R}\left(\frac{1}{97.5}\right)^{2/3}\left(\frac{1}{80}\right)^{3/2}\sqrt{\frac{1}{200}}$$

$$= \frac{1}{128,000} = \frac{1}{n_R}4.66 \times 10^{-6}$$

and

$$n_R = 4.66 \times 10^{-6}\left(128,000\right) = 0.596$$

The required value of n for the model is

$$n_M = 0.596 n_P = 0.596\left(0.024\right) = 0.014$$

The time-scale ratio for this set of circumstances is estimated by Eq. (10.2.5) or

$$T_R = \frac{L_R^2 Y_R}{Q_R} = \frac{\left(\frac{1}{200}\right)^2\left(\frac{1}{80}\right)}{\left(\frac{1}{128,000}\right)} = 0.04 = \frac{1}{25}$$

Thus, 1 hr of prototype time corresponds to 2.4 min in the model.

In the preceding example, a close examination of the computations demonstrates that under some circumstances it is conceivable that the model boundaries would be rougher than the prototype boundaries. This would be the case if $Q_R < 1/214,590$. Such situations are often encountered in practice and are resolved in some cases by setting pebbles in the model boundaries or by placing folded metal strips in the boundary to increase the roughness.

At this point, it is constructive to consider the model of the Mississippi River which has been built by the U.S. Army Corps of Engineers at Vicksburg, Mississippi. The Mississippi River basin extends approximately 2000 miles (3200 km) in each direction. Therefore, a horizontal scale of 1:2000 was chosen, and the resulting model covers an area of approximately 1 mi² (2.6 km²). Even at this rather large scale, the model river is only 2 to 3 ft wide (0.61 to 0.91 m) where the prototype river is 1 mile wide (1.6 km). In the vertical dimension, a scale of 1:2000 would result in extremely shallow depths of flow with the attendant problems of surface tension effects and laminar motion; therefore, in the model a vertical-length scale ratio of 1:100 was chosen. This length scale results in flow depths of at least 1 in (0.025 m), eliminates surface tension effects, and makes the flow turbulent. The scale distortion does not seriously distort the gross flow pattern and in most cases yields satisfactory results.

There are a number of hydraulic processes, which cannot be modeled with distorted models, for example, the study of waste dispersion in a river or estuary. Recall from Section 8.2 that dispersion in a one-dimensional open channel is governed by

$$\frac{\partial}{\partial t}(AC) + \frac{\partial}{\partial x}(\bar{u}AC) = \frac{\partial}{\partial x}\left(EA\frac{\partial c}{\partial x}\right) \qquad (10.2.6)$$

where c = concentration of conservative dissolved substance
t = time
$\bar{u}$ = average velocity in longitudinal direction
A = cross-sectional flow area
E = longitudinal dispersion coefficient

Equation (10.2.6) uses only quantities which are averaged over the cross section and assumes that all mass transport other than that caused by advection is proportional to the concentration gradient and the dispersion coefficient. In Chapter 8, it was also asserted that E is a function of the velocity gradients in both the vertical and transverse directions. Given these observations, it might be expected that a distorted hydraulic model, especially if exaggerated roughness elements are used, will not correctly model dispersive processes.

By definition, the correct modeling of pollutant dispersion is a simulation in which the relative concentration at each point in the model is equal to the relative concentration at each geometrically corresponding point in the prototype. The relative concentration is the actual concentration divided by the mass input per unit volume of the system. For the correct modeling of a pollutant cloud, a necessary, though not a sufficient condition, is

$$\left(\sigma_x\right)_R = L_R \qquad (10.2.7)$$

where σ_x = longitudinal variance of the dispersing cloud, and, as before L_R = horizontal-length scale ratio. Then, by Eq. (10.2.6)

$$E = \frac{1}{2}\frac{d}{dt}\sigma_x^2$$

Combining Eq. (10.2.6) with Eqs. (10.1.6) and (10.2.7) yields a necessary condition for correct modeling or

$$E_R = \frac{L_R^2}{T_R} = \frac{L_R}{\sqrt{Y_R}} \qquad (10.2.8)$$

If Eq. (10.2.8) is satisfied, then the concentration profiles are simply transposed between the prototype and the model.

Fischer and Holley (1971) examined the problem of correctly reproducing a dispersing cloud in a distorted-scale hydraulic model by noting that the overall dispersion coefficient E can be viewed as having two components, namely, (1) E_v , which is the dispersion coefficient component associated with the vertical velocity distribution, and (2) E_t which is the dispersion coefficient component associated with the transverse velocity distribution. In a steady flow dominated by vertical velocity gradients, the dispersion coefficient can be estimated by

$$E_v = 5.9yu_* \qquad (10.2.9)$$

where u_* = $(gRS)^{1/2}$
       y = depth of flow
      R = hydraulic radius
      S = longitudinal channel slope

Fischer and Holley (1971) assumed that

$$\left(u_*\right)_R = \sqrt{g_R Y_R S_R} = \frac{Y_R}{\sqrt{L_R}} \qquad (10.2.10)$$

where $g_R = 1$. Combining Eqs. (10.2.9) and (10.2.10) yields

$$\left(E_v\right)_R = \frac{Y_R^2}{\sqrt{L_R}} \qquad (10.2.11)$$

A comparison of Eqs. (10.2.8) and (10.2.11) demonstrates that even in this elementary case a distorted hydraulic model will not correctly model longitudinal dispersion.   For flows dominated by transverse velocity gradients, Fischer (1967) demonstrated that

$$E_t = \frac{0.3\langle u'' \rangle^2 b^2}{Ru_*}$$ (10.2.12)

where b = distance on the water surface from the thread of maximum velocity to the most distant bank, u'' = difference between the point velocity and the cross-sectional mean velocity, and $\langle \ \rangle$ indicates a cross-sectional average of the variable enclosed. If it is assumed that

$$\frac{\langle u''_R \rangle}{u^2_R} = 1$$

and Eq. (10.2.12) combined with this assumption (Fischer and Holley, 1971), then

$$\left(E_t \right)_R = \frac{L_R^{5/2}}{Y_R}$$ (10.2.13)

Equation (10.2.13) indicates that a distorted-scale hydraulic model will not model the dispersion of a pollutant cloud correctly when the flow is dominated by transverse velocity gradients. The addition of roughness strips to most distorted-scale models increases the discrepancy that will be noted between model and prototype. Based on the results presented above and additional work, Fischer and Holley (1971) concluded that:

1.   In steady flow in a distorted hydraulic model, the dispersive effects of the vertical velocity distribution are magnified while the dispersive effects of the transverse gradients of velocity are diminished. Although it is conceivable that in specific cases these effects may cancel and yield a correct simulation, the chances of this occurring must be considered minimal.

2.   In oscillatory flows such as those found in tidal estuaries, the dispersive effects of the vertical velocity distribution are magnified

while the effects of the transverse velocity distribution may be magnified, diminished, or correctly modeled, depending on the value of the dimensionless time scale for the prototype.

3. In very wide tidal estuaries, the dispersive effects of both the transverse and vertical velocity gradients are magnified; therefore, dispersion occurs much more quickly in the model than in the prototype.

Other investigators (see, for example, Moretti and McLaughlin, 1977, and Roberts and Street, 1980) have also reached these conclusions. Moretti and McLaughlin (1977) concluded that similarity criteria could be used to develop models which modeled either transverse or vertical mixing correctly, but that a model which simulated both processes simultaneously was not feasible with a distorted-scale model. Roberts *et al.* (1979) described an initial series of experiments in what they termed a variable distortion facility (see also Roberts et al, 1979) and reached the conclusion that at present the details of the mixing processes in a turbulent flow and how they vary with geometric-scale distortion are unknown. Thus, the general conclusion is that distorted hydraulic models are not appropriate to the study of phenomena in which the dispersion of a pollutant is a critical issue.

10.3.   MOVABLE-BED MODELS

When the movement of the materials which compose the sides and bed of a channel is of paramount importance, a movable-bed model is used. Such models can be used to address the following problems:

1. General river morphology: Changes in river grades, cross-sectional shape, meanders, erosion, sedimentation, and changes in discharge and sediment yield associated with upstream hydraulic structures or land use changes

2. River training: Elimination of bends or meanders, relocation of the main channel, and the optimum location and design of groins

3. Flood plain development

4. Bridge pier location and design

5.   Scour below dams

6.   Pipeline crossings

   In comparison with the fixed-bed models described in the foregoing section of this chapter, the design and operation of a movable-bed model are much more complex. Two major difficulties are:

1.   The boundary roughness of the model is not controlled by design but is a dynamic variable controlled by the motion of the sediment and the bed forms which are developed.

2.   The model must correctly simulate not only the prototype water movement but also the prototype sediment movement.

There are three methods of designing a movable-bed model which range from the purely empirical trial and error to the essentially theoretical method developed by Einstein and Chien (1956b).

10.3.1.   Trial-and-Error Design

The earliest known movable-bed hydraulic model study was conducted in 1875 by Louis Jerome Fargue for improvements in the Garonne River at Bordeaux (Zwamborn, 1967). In 1885, Osborne Reynolds constructed a physical model with a movable bed to study the estuary of the River Mersey in England. These early models were constructed and effectively used many decades before the development of modern theoretical modeling principles. The basis for design in these early models was the hypothesis that if a model can be adjusted to reproduce events that *have* occurred in the prototype, then the model should also reproduce events which *will* occur in the prototype. The accuracy of the results obtained by Fargue and Reynolds proves the validity of this hypothesis, which basically requires the development of a model by trial and error coupled with verification.

   The trial-and-error approach to the design of movable-bed hydraulic models involves the selection of a bed material. The velocity of flow required to move the selected material and the resistance coefficient which characterizes the material are established. Then, the vertical scale of the model required to scour the bed material is estimated. The model is then constructed, and the model variables are adjusted as required to achieve agreement between the results in the model and prototype for a verification event or a series of verification events.

The verification event or events should satisfy the following criteria:

1. The event or events from the prototype which are used to verify the model should involve phenomena pertinent to the goals and objects of the proposed model study. For example, a model which is to be used to study bed load movement should not be verified with flood flow data.

2. The event or events on which the verification is to be based should be continuous and of reasonable duration.

3. The event or events on which the verification is based should not be extreme unless the study of such events is the purpose of the modeling program.

4. The greater the departure of the conditions which are to be modeled from the conditions under which the model was verified, the greater the uncertainty in the modeling results.

As indicated above, the development of a movable-bed hydraulic model by this design process, i.e., comparison of model and prototype data, is a trial-and-error process which is repeated until satisfactory results are achieved. Among the model variables which can be adjusted and manipulated are:

1. Discharge scale

2. Water surface slope

3. Type of bed material

4. Magnitude of the time scale

5. Roughness of the fixed boundary

At this point a number of miscellaneous comments regarding movable-bed models can be made:

1. In practice, it has been found that the low velocities of flow typically found in a model are not sufficient to scour many bed materials. Therefore, a model bed material which is significantly less dense than natural sand (specific gravity $\approx$ 2.65) is required.

Bituminous materials, e.g., coal dust (specific gravity 1.3 or less), or plastics (specific gravity $\approx$ 1.2 or less), are commonly used.

2. While in fixed-bed models geometric distortion can be used in many cases without creating unsolvable problems, in a movable-bed model geometric-scale distortion many result in channel side slopes that exceed the angle of repose of the model bed material. If the side slopes are stable in the prototype, then the side slopes in the model can be constructed of rigid materials.

3. Geometric-scale distortion also results in an increase in the longitudinal slope of the model and requires an increase in the boundary roughness. However, the boundary roughness is not only a function of the material selected but also a function of the movement of the bed itself. Thus, in movable-bed models, geometric-scale distortion should be either avoided or minimized.

10.3.2.   Theoretical Methods

Although movable-bed hydraulic models can be constructed by trial and error, there are explicit design techniques available which are based on reasonably well-established theoretical and semitheoretical concepts. For example, consider Fig. 7.13, which delineates the threshold of particle movement as a function of two dimensionless parameters Re_* and F_s where

$$\mathrm{Re}_* = \frac{u_* d}{\nu} \qquad (10.3.1)$$

$$F_s = \frac{\tau_0}{(S_s - 1)d} = \frac{u_*^2}{(S_s - 1)gd} \qquad (10.3.2)$$

where Re_* = Reynolds number based on shear velocity and particle size
u_* = shear velocity
ν = fluid kinematic viscosity
S_s = specific gravity of particles composing perimeter
d = diameter of particles composing perimeter of channel

As suggested earlier in this chapter, the state of the channel bed is governed by Fig. 7.13 and the location of the Re_* - F_s line in this figure. Therefore, it is asserted that if F_s and Re_* are the same in the model and prototype, then the state of the bed will be the same in both cases. In addition, it can be assumed that the equivalent roughness-to-particle-size ratio will be the same in both the prototype and the model.

In a flow which transports sediment, the sediment suspended in the flow reduces the flow resistance, but this effect is small in comparison with the effect of the formation of dunes in the bed of the channel. Einstein and Barbarossa (1956a) assumed that the shear stress on the bed of the channel was composed of two components: $\tau_0' =$ a shear stress due to the intrinsic roughness of the sediment composing the bed and τ_0'' = a shear stress due to the bed or form roughness. Einstein and Barbarossa (1956a) further assumed that the total flow area A and the total wetted perimeter P could also be divided into components; and therefore,

$$\tau_0' = \gamma R'S \tag{10.3.3}$$

$$\tau_0'' = \gamma R''S \tag{10.3.4}$$

and

$$\tau = \tau_0' + \tau_0'' = \gamma (R' + R'')S \tag{10.3.5}$$

where R' and R'' are the component hydraulic radii associated with τ_0' and τ_0'' , respectively. In 1923, Strickler asserted (see, for example, Chow, 1959, or Henderson, 1966) or Section 4.3 that

$$n = 0.034d^{1/6} \tag{10.3.6}$$

where n = Manning's n and d = diameter of the sediment. Then, if as stated in the previous paragraph, the equivalent roughness-to-particle-size ratio is the same in both the model and prototype

$$n_R = d_R^{1/6} \tag{10.3.7}$$

and

$$\left(\frac{R'}{R''}\right)_R = 1 \tag{10.3.8}$$

from Eq. (10.2.3)

$$U_R = \frac{R_R^{2/3}\sqrt{Y_R}}{n_R\sqrt{L_R}} \tag{10.3.9}$$

$$n_R = \frac{R_R^{2/3}}{\sqrt{L_R}}$$

where $U_R = Y_R^{1/2}$. Combining Eqs. (10.3.7) and (10.3.9) yields

$$d_R^{1/6} = \frac{R_R^{2/3}}{\sqrt{L_R}} \tag{10.3.10}$$

The implication of Eq. (10.3.8) is

$$\tau_0 = \gamma_R R_R S_R = \frac{\gamma_R R_R Y_R}{L_R} \tag{10.3.11}$$

It then follows from Eqs. (10.3.1) and (10.3.2) that

$$\frac{R_R Y_R}{\alpha_R L_R d_R} = 1 \tag{10.3.12}$$

and

$$\frac{R_R Y_R d_R^2}{L_R v_R^2} = 1 \tag{10.3.13}$$

where $\alpha = S_s - 1$. R_R is a function of Y_R and L_R, and if $v_R = 1$, then Eqs. (10.3.10) to (10.3.13) are a set of three equations in four unknowns. Thus, in designing a movable-bed model by this method, one model ratio is selected and the other three model ratios are determined from Eqs. (10.3.10), (10.3.12), and (10.3.13).

With regard to Eqs. (10.3.10), (10.3.12), and (10.3.13) the following implications are noted:

1. If $Y_R = L_R$, that is, the model is undistorted, then a scale model cannot be constructed, and any model will have to have the same scale as the prototype.

2. If the specific gravity ratio is selected in advance, then from Eqs. (10.3.12) and (10.3.13)

$$d_R = \alpha_R^{-1/3} \qquad (10.3.14)$$

If $v_R = 1$. Equation (10.3.14) implies that if the sediment material used in the model has a specific gravity less than the material present in the prototype, then the size of the material used in the prototype should be larger.

3. If R_R cannot be assumed equal to Y_R, then the scale ratios L_R and R_R must be found by trial and error. However, if $R_R = Y_R$, then

$$L_R = \alpha_R^{5/3} \qquad (10.3.15)$$

and

$$Y_R = \alpha_R^{7/6} \qquad (10.3.16)$$

4. It can be shown that the ratio of the rate of sediment transport per unit width to the flow per unit width must be larger in the model than in the prototype.

5. In most cases of the design of movable-bed models, the Froude number model law is not considered an absolute rule. For example, if the Froude number is low, then gravitational effects are

not pronounced; therefore, slight changes in the Froude number between the model and prototype can be tolerated.

6.   The particle size ratio referred to in Eq. (10.3.10) characterizes the roughness of the bed of the channel, while the particle size ratio used in Eqs. (10.3.12) and (10.3.13) is characteristic of the sediment transport properties of the particles. It should be recognized that these ratios which characterize the particle size distributions in the model and prototype are not necessarily the same.

EXAMPLE 10.2

A channel reach that is approximately parabolic in section has a surface width of 60m (197 ft) when the depth of flow is 1.5 m (4.9 ft). In the prototype, the quartzite sand which composes the bed has a specific gravity of 2.65.   If a movable-bed model of this reach is to be constructed of material which has a specific gravity of 1.2, determine the appropriate particle size in the model and the length-scale ratios L_R and Y_R.

Solution
If the Einstein and Chien (1956b) method of analysis is used and $U_R = 1$, then the particle size ratio (model to prototype) is given by Eq. (10.3.14) or

$$d_R = \alpha_R^{-1/3} = \left(\frac{1.2 - 1}{2.65 - 1}\right)^{-1/3} = 2.0$$

Therefore, the size of the particles used in the model must be approximately twice the size of the particles in the prototype.

The scale ratios L_R and Y_R are then determined by the simultaneous solution of Eqs. (10.3.12) and (10.3.13).   Since $R_R = f(L_R, Y_R)$, the solution of these equations must be by trial and error.   Initial trial values for L_R and Y_R are determined by assuming $R_R = Y_R$; hence, by Eqs. (10.3.15) and (10.3.16)

$$L_R = \alpha_R^{5/3} = \left(\frac{1.2 - 1}{2.65 - 1}\right)^{5/3} = 0.030$$

and

$$Y_R = \alpha_R^{7/6} = \left(\frac{1.2-1}{2.65-1}\right)^{7/6} = 0.085$$

For the parabolic channel section defined, the hydraulic radius in the prototype is (see Table 1.1)

$$R_P = \frac{2T^2 y}{3T^2 + 8y^2} = \frac{2(60)^2(1.5)}{3(60)^2 + 8(1.5)^2} = 1.00\,\text{m}\,(3.28\,\text{ft})$$

The hydraulic radius in the model for the assumed geometric scale is

$$R_M = \frac{(2TL_R)^2 yY_R}{(3TL_R)^2 + 8(yY_R)^2} = \frac{2[60(0.03)]^2(1.5)(0.085)}{3[60(0.03)]^2 + 8[1.5(0.085)]^2} = 0.084\,\text{m}\ (0.28\,\text{ft})$$

Therefore,

$$R_R = \frac{R_M}{R_P} = \frac{0.084}{1.0} = 0.084$$

At this point, it must be demonstrated that the assumed geometric-scale ratios satisfy Eqs. (10.3.12) and (10.3.13). Equation (10.3.12) requires

$$\frac{R_R Y_R}{\alpha_R L_R d_R} = 1$$

and

$$= \frac{0.084(0.085)}{\left[\dfrac{1.2-1}{2.65-1}\right](0.030)(2.0)} = 0.98$$

Equation (10.3.13) requires

$$\frac{R_R Y_R d_R^2}{L_R v_R^2} = 1$$

and

$$= \frac{0.084\big(0.085\big)\big(2.0\big)^{2}}{0.03\big(1.0\big)^{2}} = 0.95$$

Since Eqs. (10.3.12) and (10.3.13) are satisfied almost exactly, it is concluded that

$$L_R = 0.03$$

and

$$Y_R = 0.085$$

A few investigators [see for example, Bogardi 1959, or Bogardi (in Henderson, 1966), and Raudkivi, 1976] have asserted that the requirement that F_s and Re_* have exactly the same value in the model and prototype can and should be relaxed somewhat. When $Re_* \geq 100$, the flow around the sediment grains is fully turbulent, and, as is the case with other models, it is not necessary for $(Re_*)_M = (Re_*)_P$ As long as $(Re_*)_M \geq 100$. It is also asserted that the bed formation is a function only of the parameter β which is defined by

$$\frac{1}{\beta} = d^{\,0.88}\left(\frac{u_*^2}{gd}\right)$$

(10.3.17)

where the unit of length used is the centimeter. Henderson (1966) noted that all the evidence quoted in support of Eq. (10.3.17) results from grain water systems in which $S_s = 2.65$ and $\nu = 0.011$ cm^2/s (1.18 x 10^{-5} ft^2/s). Therefore, Eq. (10.3.2) can be rewritten as

$$F_s = \frac{u_*^2}{\big(S_s - 1\big)gd} = \frac{456}{\beta^{1.41}}\left(\frac{u_*d}{\nu}\right)^{-5/6}$$

(10.3.18)

The values of β, in centimeters, given by Bogardi for the various bed formations, are summarized in Table 10.2. These values, when used in Eq. (10.3.18), define a series of bands in the F_s - Re_* plane (Fig. 10.2) which graphically define the bed forms. In Fig. 10.2 the lines defining the bed form bands all terminate at $F_s = 0.05$, which is the threshold of motion. Figure 10.2 also demonstrates a contradiction in the theory of Bogardi. Specifically, it was claimed earlier that if $Re_* \geq 100$, then the bed forms in the model and prototype would be similar, however, Fig. 10.2 demonstrates that even if $R_* \geq 100$, the bed forms in the model and prototype could be different. The conclusion is that, to achieve similarity between the model and prototype, the values of F_s and Re_* do not have to be exactly equal, but the points representing the model and prototype should be within the same bed form band of Fig. 10.2.

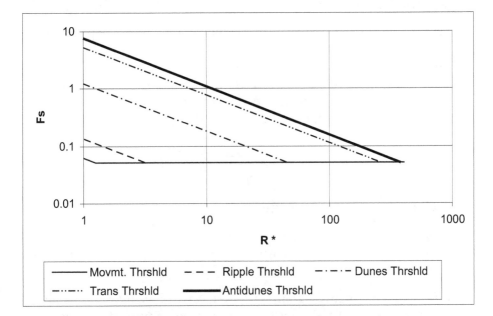

Figure 10.2.   Relation of bed formation to position on the F_s - R_* (R* - Fs) plane.

Table 10.2. Bed form type as a function of β (Henderson, 1966).

Bed form	β (cm)
Threshold of motion	550
Ripples form	322
Dunes develop	66
Transition occurs	24
Antidunes develop	18.5

Zwamborn (1966, 1967, 1969) has developed a set of similarity criteria for movable-bed models which have been validated by model-prototype correlations. The similarity criteria are:

1. Flow in natural rivers is always fully turbulent. To ensure that the flow in the model is also fully turbulent, the Reynolds number for the model should exceed 600.

2. Dynamic similarity between the model and prototype is achieved when the model-to-prototype ratios of the inertial, gravitational, and frictional forces are all equal. Similarity is achieved when:

    a. the Froude numbers for the model and prototype are equal,

    b. the friction criteria are satisfied when three conditions are met. First, the product of the grain roughness coefficient and the inverse square root of the slope are equal in the model and prototype or

$$C_R^{'} = \frac{1}{\sqrt{S}_R} \qquad (10.3.19)$$

where C' = grain roughness coefficient

$$C' = 5.75\sqrt{g} \log\left(\frac{12R}{d_{90}}\right) \qquad (10.3.20)$$

d_{90} = 90 percent smaller grain diameter, and R = hydraulic radius. Second, the ratio of the shear to the settling velocity in the model and prototype are equal or

$$\left(\frac{u_*}{u_s}\right) = 1 \qquad (10.3.21)$$

where $u_* = (gRS)^{1/2}$ = shear velocity and u_s = settling velocity. The settling velocities of selected model sediment materials are summarized in Fig. 10.3 and Table 10.3 as a function of grain size and water temperature. Third, the Reynolds number based on the sediment grain diameter in the model should be about one-tenth of the grain Reynolds number for the prototype or

$$\left(Re_*\right)_R < \frac{1}{10} \qquad (10.3.22)$$

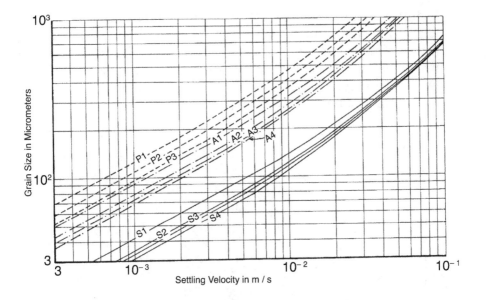

Figure 10.3. Settling velocity curves (Zwamborn, 1969).

Table 10.3. Identification of curves in Figure 10.3.

Perspex (sg gr 1.18)	Anthracite (sp gr 1.35)	Sand (sp gr 2.65)
P1 = 4.5° C (40° F)	A1 = 4.5° C (40° F)	S1 = 4.5° C (40° F)
P2 = 15.5° C (60° F)	A2 = 15.5° C (60° F)	S2 = 15.5° C (60° F)
P3 = 26.7° C (80° F)	A3 = 21.1° C (70° F)	S3 = 21.1° C (70° F)
	A4 = 26.7° C (80° F)	S4 = 26.7° C (80° F)

where $Re_* = ud/v =$ grain Reynolds number
$u =$ velocity
$v =$ kinematic viscosity

c. The geometric-scale distortion should not be large; i.e.

$$\frac{Y_R}{L_R} < \frac{1}{4}$$

(10.3.23)

3.   Sediment motion is closely related to the bed form type which is, in turn, a function of u_*/u_s, Re_* and F. For specified values of u_*/u_s and F, similar bed forms and consequently similar models of bed movement are ensured when the value of Re_* for the model falls within the range as determined by u_*/u_s and the appropriate range of F defining a type of bed form (Fig. 10.4).

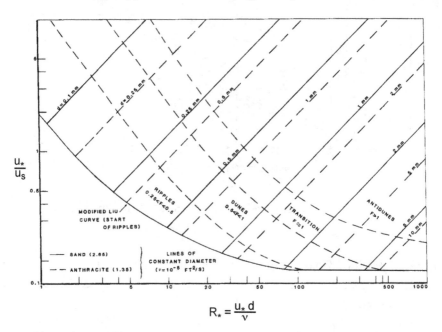

$$R_* = \frac{u_* d}{\nu}$$

Figure 10.4. Bed shape criteria (Zwamborn, 1969).

4.   Given the similarity criteria defined above, the following scale factors can be derived.

Flow Rate

$$Q_R = L_R Y_R^{1.5} \qquad\qquad (10.3.24)$$

Hydraulic Time

$$T_R = \frac{L_R}{U_R} \qquad\qquad (10.3.25)$$

Sedimentological Time

$$\left(T_s\right)_R = \frac{L_R Y_R \Delta_R}{\left(q_s\right)_R} \qquad (10.3.26)$$

where T_s =   sediment time scale
        Δ =   relative submerged sediment density
        q_s =   sediment bed load per unit width

From Eq. (10.3.26) it is evident that the determination of the time scale for the sediment requires that the bed load in both the model and the prototype be known. Although the bed load of the prototype can be estimated (see for example, Rouse, 1958), Zwamborn (1966, 1967 and 1969) gave the following approximate relationship:

$$\left(T_s\right)_R \approx 10 T_R \qquad (10.3.27)$$

The validity of the similarity criteria stated above has been confirmed by comparison of the results achieved in models with those observed in prototypes (see for example, Zwamborn, 1966, 1967, and 1969).

EXAMPLE 10.3

In the prototype $u_*/u_s = 5.5\text{m } \mathbf{F} = 0.5$, and the bed shape is dunes. Using the criteria of Zwamborn (1966, 1967, and 1969) which are summarized in Fig. 10.4, determine the appropriate range of grain Reynolds numbers for the model.

Solution
The Froude numbers for the model and prototype are equal, and Eq. (10.3.21) requires

$$\left(\frac{u_*}{u_s}\right)_R = 1$$

Therefore, from Fig. 10.4, if the bed form is dunes

$$2 < \left(\text{Re}_*\right)_M < 5$$

10.3.3.   Regime Method

A channel is said to be in regime when it has adjusted its slope and shape to an equilibrium condition. The implication of a channel being in regime is that although the bed of the channel is in motion it is a stable movement since the rate of transport equals the rate of sediment supply. The regime hypothesis is a completely empirical approach which incorporates no theoretical considerations. This hypothesis was developed in India from field measurements taken in canals of human origin which were rather uniform in character but had widely varying sizes. Three equations were developed which can be rearranged into a form that implies that any one of three variables - average depth, average velocity, and cannel width - is a unique function of discharge. These equations were developed by plotting the available data on log-log graphs and determining the best-fit straight lines. As noted by Einstein and Chien (1956b) in reply to a discussion by Blench (1956), because the field data were obtained from channels which were either empty or running at their design discharge, the data were regular.

In particular, the regime hypothesis asserts that

$$T \propto \sqrt{Q} \qquad (10.3.28)$$

$$Y \propto Q^{1/3} \qquad (10.3.29)$$

$$S \propto Q^{-1/6} \qquad (10.3.30)$$

From Eqs. (10.3.28) to (10.3.29), appropriate model scale ratios can be derived by induction or

$$L_R = \sqrt{Q_R} \qquad (10.3.31)$$

and

$$Y_R = Q_R^{1/3} \qquad (10.3.32)$$

As a result of Eqs. (10.3.30) and (10.3.31), the following modeling ratios can be derived:

$$S_R = \frac{Y_R}{L_R} = \frac{Q_R^{1/3}}{\sqrt{Q_R}} = Q_R^{-1/6} \qquad (10.3.33)$$

$$Y_R = L_R^{2/3} \qquad (10.3.34)$$

and

$$Q_R = X_R Y_R^{3/2} \qquad (10.3.35)$$

With regard to this method of movable-bed model design, the following observations can be made:

1. From Eq. (10.3.34), it is obvious that the model has distorted geometric scales.

2. Equation (10.3.35) agrees with the Froude modeling law.

3. If this design method is used, the bed material in the model is tacitly assumed to be the same as that found in the prototype.

Thus, it is concluded that the regime hypothesis yields results which are surprisingly in accordance with the requirements of similarity. Two objections to this design process must be noted. First, as asserted by Einstein and Chien (1956b), theories or hypotheses which satisfactorily represent flow in designed channels may be unsatisfactory when applied to natural rivers. Einstein and Chien (1956b), using data from U.S. rivers, clearly demonstrated that the application of Eqs. (10.3.28) to (10.3.30) to natural rivers can lead to significant errors. Second, if the bed material present in this prototype is used in the model, then Eqs. (10.3.15) and (10.3.16) predict that X_R - Y_R, and hence the model is the prototype.

10.4.   MODEL MATERIALS AND CONSTRUCTION

In constructing a physical, hydraulic model, the modeler must be aware that the model must:

1. Be an accurate, scaled geometric replica of the prototype

2. Retain its geometric consistency and accuracy during operation

3.   Contain the appurtenances necessary to control and measure the flow

4.   Be amenable to quick and easy changes in detail

5.   Be consistent with the purpose of the study and its cost allocations

Models are generally constructed from materials which are readily available, e.g., wood, concrete, metal, wax, paraffin, plastic, sand, and coal. The shop facilities required to prepare and shape these materials must also be available. In the following paragraphs of this section, a select number of topics will be discussed from the viewpoint of materials and methods of model construction. In many cases, most of the time required to conduct a model study is used in the design and construction processes. It is noted that additional information regarding materials and construction and construction techniques can be found in Anonymous (1942) and Sharp (1981).

10.4.1.   Materials

Wood: The best-suited types of wood for the construction of hydraulic models are sugar pine, cypress, redwood, mahogany, and marine grade plywood. In general, since wood of all types is to a greater or lesser degree affected by water, the wood should be waterproofed in some manner; e.g., an initial coat of linseed oil followed by one or more coats of varnish is usually satisfactory. Wood has the advantages of being relatively inexpensive and easily shaped; in the case of plywood, it can be bent or warped slightly. It has the disadvantages of being affected dimensionally by water content and warping and being opaque.

Sheet Metal: Sheet metal is also relatively inexpensive and can be easily shaped. It may be cut to any shape and can be attached to wood with countersunk screws. The disadvantages of sheet metal are that it tends to buckle under unequal stress and under thermal stress; it cannot be shaped in warped surfaces; its surfaces must be protected from corrosion by either painting or galvanizing; and it is opaque.

Plastics: Plastics such as lucite and plexiglass are transparent and can be molded with heat, machined, or cut and glued into many shapes. They are particularly useful when one of the goals of the modeling program is flow visualization. Plastics have the disadvantage of being affected by humidity, temperature, and sunlight.

Wax: Wax had the advantages of being very workable and unaffected by water, and this material may also be formed and shaped into a wide variety of geometric configurations. Wax is particularly well suited for constructing curved surfaces which are required to be very accurate. The primary disadvantages of wax are that it is fragile, devoid of structural strength, opaque, and easily damaged by temperature changes.

Cement: Cement has the advantages of being workable when wet, unaffected by water, and unaffected by temperature changes. With templates, it can be molded into many shapes and can be made very smooth by working with a trowel until the initial set occurs. Experience has shown that the best mix is one part cement to one to three parts sand (Anonymous, 1942). This material is well suited to models involving warped surfaces or large areas. In general, even after the final set, cement can be easily cut and patched. The primary disadvantage of cement is that it is opaque. The designer should also be aware that in a newly constructed cement model, salts will be dissolved from the cement, and instrumentation systems which measure the conductivity of water may be affected by this dissolution process.

Fiberglass and Resins: When complex shapes must be constructed, they may be constructed from fiberglass or cast from resins. It should be noted that the continuous exposure of technicians to epoxy resin fumes may result in significant health problems (Sharp, 1981).

Polyurethane Foam: In recent years, high-density polyurethane foam (6 to 12 lb/ft^3 or 940 to 1900 N/m^3) has been used to model small hydraulic structures. Although this is a rather expensive construction material, the labor costs are reduced by the ease with which this material can be cut and formed to the correct shape. Building models with polyurethane foam usually involves gluing the sheets of foam together with epoxy resin, forming them into the proper shape, and finally sealing the model. Although various sealers may be used, metal paint primer appears to yield satisfactory results (Sharp, 1981). The primary disadvantage of this material is that when glues are used, separation at the joints may occur. This problem is apparently overcome when a resin, for example, fiberglass, is used. Some separation may occur even if fiberglass resin is used when the foam sheets are less than 1 in (2.5 cm) in thickness.

10.4.2.   Construction

As the above discussion indicates, there are many materials which can be used in the construction of a model. Each material has certain advantages and also a number of disadvantages which preclude its use as a universal model material.

In the actual construction of a hydraulic model, reliable and accurate horizontal and vertical controls must be established. The method of horizontal control is usually strongly dependent on the size and shape of the model. For example, if the model is long and narrow, then horizontal control may be established with a traverse which approximates the centerline of the model. In the case of a model which is both long and wide, horizontal control may be established with a traverse around the perimeter of the model. If the model is very large, then a perimeter traverse supplemented with a Cartesian grid system is used.

The establishment of vertical control is usually based on the location of a benchmark located in the vicinity of the model. If the model is large, then supplementary benchmarks may be required.

Once horizontal and vertical control has been established, the topography must be established. There are three commonly used methods of accomplishing this:

1. *Female Templates:* Female templates represent prototype cross sections of the channel which have been reduced to an appropriate size by the geometric-scale ratios during the design process (Fig. 10.5). These templates are plotted and cut with precision and then are securely anchored in models using the horizontal and vertical control nets to establish their location. The surface of the model is then molded in accordance with the surface established by the templates.

    The materials used to make the templates include sheet metal, masonite, and wood. It is recommended that the templates be spaced no farther apart than 2 ft (0.6 m) and no closer than 6 in (0.15 m).

2. *Male Templates:* The channel topography may also be established with the aid of male templates (Fig. 10.5). In this case the templates sit on graded rails and are removed when the channel surfaces have been shaped.

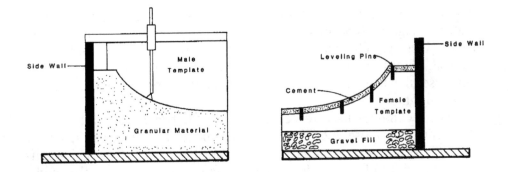

Figure 10.5.   Schematic illustrating use of male and female templates and leveling pegs in model construction.

3. *Pegs:* The topography of an area may also be established by pegs which are placed in the model such that their tops reflect the appropriate elevation. The surface is then molded in accordance with these peg elevations.

4. *Horizontal Templates:* Although the use of male or female templates, which compose the group termed *vertical templates*, can be used in most situations, in some cases horizontal templates cut to the shape of the contour lines are appropriate. If the thickness of the template is identical to the scaled vertical distance between the contours, then the templates can be laid one on top of another and a solid model formed. Although plywood and polyurethane have been used successfully in this manner, it is more common to use thin templates separated by vertical spacers to ensure correct elevations. If this latter methodology is used, then the templates must be fairly rigid plywood or sheet metal and covered with a material such as fiberglass.

In considering the model construction techniques discussed above, it must be realized that none of the techniques will result in a model which is geometrically exact. Although in all the methods there is an emphasis on horizontal and vertical control between the points, lines, and planes of control, the topography must be interpreted by the human hand and eye. At this point, the following three points are noted. First, the female and male template construction techniques are usually considered optimum when large areas with considerable topographic relief must be covered. Second, when the male or female template method is used, the volume between the templates is usually filled with an inexpensive filler material; e.g., a 50/50 mixture of clay and fine sand has been found to be satisfactory. This material may then be covered with a thin layer of cement. Third, the horizontal template technique has been found to be very useful when the horizontal extent of the model is large and the relief slight.

The techniques used to construct movable-bed models are slightly different from those used for fixed-bed models. The construction of movable-bed models generally requires positioned guides or guide rails along the sides of the model. These rails or guides may be the model tank walls if they have been accurately constructed. The male template method of construction may then be used; i.e. the templates are suspended from the guides or guide rails and the proper topography formed. Depending on the size of the model basin and the presence and distribution of bedrock in the prototype, the bottom of the basin can be filled with inexpensive filler materials as was suggested in the case of fixed-bed models. In the top layers of a movable-bed model, the model sediment is placed. In the placement of this material, care must be taken to ensure that the depth of the mobile model sediment is greater than the greatest feasible depth of scour.

10.5. PHYSICAL MODEL CALIBRATION AND VERIFICATION

As is the case with numerical models, a model has no value unless it can be used to predict prototype behavior. Thus, after a physical hydraulic model has been constructed, it must be calibrated and verified; i.e., a determination must be made whether prototype events are accurately reproduced in the model. In numerical models, prototype model agreement is usually achieved by adjusting the coefficients used in the model until adequate agreement between the prototype and model is achieved. In a physical model, prototype model agreement is attained by adjusting physical characteristics such as the bed roughness, the discharge, and/or the water levels.

The successful calibration of either a numerical or a physical model requires accurate prototype data regarding channel geometry, water surface elevations, discharge, sediment transport, and velocities. It is noted that the collection of a database suitable for calibrating a model can be an expensive and time-consuming undertaking. With regard to the calibration process, the following should be noted:

1.   In the case of models in which the conditions for exact similarity are closely met, very little calibration will be required.   For example, the gross flow characteristics at a hydraulic structure can be modeled at a relatively large scale and with very little, if any, geometric distortion.   Thus, the model will reproduce prototype behavior with very little calibration.

2.   When frictional resistance is important, e.g., in channel flow, the problem of calibration becomes much more complex.   At a minimum, water surface elevations between the model and prototype must be compared for discharges close to the operating discharge for the model. If the model is to simulate discharges in the prototype over a wide range, then the model should be calibrated for a wide range of discharges.

3.   Velocity distribution measurements are also valuable even though in models with geometric-scale distortions it may be unrealistic to expect the model to reproduce these distributions accurately. However, the comparison of velocity distributions provide valuable information regarding the overall accuracy of the model.

4.   If the model is to simulate point velocities in a reach, then accurate field data regarding the velocity of flow at these points is required.

5   If the model is to simulate unsteady flow events, then the movement of the unsteady event through the reach should be reproduced, at the appropriate spatial and temporal scales, in the model.

6.   If the model is to simulate sediment transport, then the calibration process requires accurate data regarding sediment transport and bed forms in the prototype.   Sharp (1981) noted that although steady-state flow information regarding sediment transport is useful, measurements of sediment transport under unsteady conditions for model calibration are particularly valuable.

After reasonable agreement is obtained for the calibration event or events, the verification process begins.   Verification requires data independent of the calibration data and seeks to confirm that the model has been correctly calibrated.   In many cases, a rigorous verification process is not performed because of the high cost involved in obtaining adequate and reliable field data. The calibration and verification processes are essential if the modeling program is to yield useful and reliable information.

10.6.   SPECIAL-PURPOSE MODELS

Ice

As has been discussed in Section 5.4, ice can be an important and critical problem in the design and analysis of open channels in northern climates. Given the importance of ice problems in many areas of the world, it is crucial that physical models of ice problems be considered. Two different types of models have been developed. When the internal strength of an ice sheet comes in contact with bridge piers, not only must the hydrodynamics of the flow be correctly modeled, but there must also be hydroelastic similarity between the prototype and the model. On the other hand, ice jams which result from the accumulation of drifting ice can in many cases be adequately addressed without considering hydroelastic similarity.

In the case where the internal strength of the ice must be considered, the manner in which the ice load is applied to the structure (see, for example, Neill, 1976) is important. For example, when ice pushes against a vertical pier, it fails by crushing (Fig. 10.6). If it is assumed that the flow is turbulent and that gravity does not enter into the problem because the ice sheet neither bends nor rides up on the pier, then

$$F = g(h,\ d,\ u,\ \sigma_c,\ \varepsilon,.\ E) \qquad (10.6.1)$$

where
F = force on pier
h = ice sheet thickness
d = characteristic size of pier
u = ice sheet velocity
σ_c = compressive strength of ice at strain rate ε
E = Young's modulus of elasticity

Application of the Buckingham II theorem to the functional formulation assumed in Eq. (10.6.1) yields

$$\frac{F}{\sigma_c dh} = g\left(\frac{h}{d}, \frac{u}{\varepsilon h}, \frac{E}{\sigma_c}\right) \qquad (10.6.2)$$

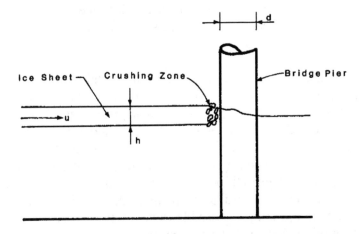

Figure 10.6. Ice failure at a vertical bridge pier.

Sharp (1981) noted that this formulation also assumes that the ice crystal size is small relative to the width of the pier; the ice is broken into small fragments; and ε must be included because of the known variation of the compressive strength of ice with the rate of deformation.

In the case of ice pushing against an inclined pier (Fig. 10.7), the ice can be assumed to fail by a combination of crushing, shear, and bending. In this case, gravitational forces must be considered because the ice rides up on the pier face. Neill (1976) has suggested that in this case

$$\frac{F}{\sigma_h h^2} = g\left(\frac{d}{h}, \Gamma, \frac{u^2}{gh}, \frac{E}{\sigma_f}\right) \qquad (10.6.3)$$

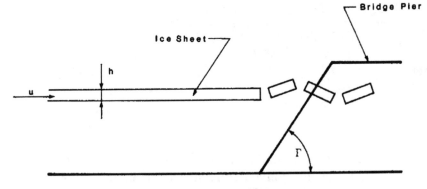

Figure 10.7. Ice failure at inclined bridge pier.

where $\quad\quad\quad\quad \sigma_f =$ bending strength of ice
$\quad\quad\quad\quad\quad\quad\quad \Gamma =$ pier force slope angle
$\quad\quad\quad\quad\quad\quad\quad g =$ acceleration of gravity

    The primary difficulty associated with modeling ice problems in which the internal strength of ice must be considered is the development of a model ice in which all the mechanical properties of prototype ice are properly scaled. A number of techniques have been developed to form model ice, and it has been demonstrated that the flexural strength of the ice sheet depends on: (1) the salinity of the water in which the ice forms; (2) the rate of growth of the ice sheet; and (3) the temperature at which the ice sheet is held after formation. If these variables are suitably controlled, then it is possible to develop ice sheets with bending strengths ranging between prototype values and 0.01 of prototype values. However, the primary problem is in developing model ice sheets with the proper elastic modulus. With naturally formed prototype ice the ratio E/σ_f as the range

$$2000 < \frac{E}{\sigma_f} < 8000$$

Whereas in many models E/σ_f may be as low as 500, low values of E/σ_f yield ice which deflects an inordinate amount before breaking (Schwarz, 1977). Given the importance of this characteristic ratio to ice modeling, a few comments regarding the measurement of σ_f and E are appropriate. The flexural strength of ice both in the laboratory and field is usually determined by cutting a cantilever beam in the ice and testing it to failure. σ_f is then given by

$$\sigma_f = \frac{6P_L L}{bh^2} \tag{10.6.4}$$

where $\quad\quad\quad\quad P_f =$ applied load
$\quad\quad\quad\quad\quad\quad\quad L =$ length of beam
$\quad\quad\quad\quad\quad\quad\quad b =$ beam width
$\quad\quad\quad\quad\quad\quad\quad h =$ beam thickness

Although the elastic modulus is difficult to measure in the field environment, it can be conveniently measured in the laboratory by a combination of load and deflection measurements. For an ice sheet which is continuous and with boundaries sufficiently distant from the point of loading so that the ice sheet can be considered infinite, the deflection Δ for a load P_L is given by

$$\Delta = \frac{P_L}{8S_w l^2} \tag{10.6.5}$$

where S_w = specific gravity of water and l = a measure of the length of the infinite ice sheet over which the stresses are distributed. l may be estimated from the theory of infinite slabs supported by an elastic foundation or

$$l^4 = \frac{Eh^3}{12S_w \left(1 - n_p^2\right)} \tag{10.6.6}$$

where n_p^2 = Poisson's ratio.

An alternative to using real ice generated by refrigeration in a model is to use artificial ice which is composed of various chemicals bound in a plaster or wax matrix. Mod-Ice used by Arctec of Columbia, Maryland, employs five separate chemicals in a mixture, which is proprietary, to achieve appropriate controllable values of bending and crushing strength, elastic modulus, density and roughness (Kotras *et al*, 1977). In Table 10.4 the relevant properties of freshwater ice, seawater ice, and Mod-Ice are summarized. In models using Mod-Ice, the model scales are usually limited by the minimum value of the bending strength - approximately 2.5 lb/in^2 (13.8kN/m^2); however, this does not appear to be a significant limitation (Sharp, 1981). In addition to the proprietary formulations for ice, such as the one noted above, there are also formulations described and discussed in the open literature (see, for example, Timco, 1979).

If the model is designed to study ice drift, ice control structures, or ice jams, the ice is assumed to be perfectly rigid and thermal effects can be neglected; then the model study can usually be conducted using polyethylene, polypropylene, wood, or paraffin to simulate the ice blocks. In such a case, fluid motion is the primary concern; however, the study should be considered qualitative rather than quantitative. The difficulty in obtaining field data for the calibration of these models is a major problem.

Table 10.4.   Properties of freshwater ice, sea ice, and Mod-Ice (Kotras *et al.*, 1977).

Property	Freshwater ice	Seawater ice	Mod-Ice scale = 1/60
Ice thickness ft m	0. -2.0 0. - 0.6	0.1 - 11 0 - 3.5	0.3 - 20 0.1 - 6.0
Ice flexural strength lb/ft² kPa	10,000 - 21,000 500 - 1000	5,200 - 16,000 250 - 750	8,400 - 21,000 400 - 1,000
Crushing strength/flexural strength	2.0 - 5.0	2.0 - 5.0	0.7 - 7.0
Elastic modulus/flexural strength	1,000 - 5,000	1,000 - 8,000	500 - 11,000
Darcy Weisbach friction factor	0.3 - 0.7	----	0.3 - 0.8
Coefficient of sliding friction	0.1 - 0.4	0.1 - 0.4	0.1 - 0.55

10.7.   BIBLIOGRAPHY

Anonymous, "Hydraulic Models," *The American Society of Civil Engineers Manual of Practice*, No. 25, American Society of Civil Engineers, New York, 1942.

Blench, T., "Blench on Distorted Models," *Transactions of the American Society of Civil Engineers*, vol. 121, 1956, pp. 458-459.

Bogardi, J., "Hydraulic Similarity of River Models with Movable Bed," *Acta Tech Acad. Sci. Hung.*, vol. 14, 1959, pp. 417-445.

Chow, V.T., *Open Channel Hydraulics*, McGraw-Hill Book Company, New York, 1959.

Einstein, H.A., and Barbarossa, N.L., "River Channel Roughness," *Transactions of the American Society of Civil Engineers*, vol. 177, 1956a, pp. 440-457.

Einstein, H.A., and Chien, N., "Einstein-Chien on Distorted Models," *Transactions of the American Society of Civil Engineers*, Vol. 121, 1956b, pp. 459-462.

Fischer, H.B., "The Mechanics of Dispersion in Natural Streams," Proceedings of the *American Society of Civil Engineers, Journal of the Hydraulics Division*, vol. 93, no. HY6, November 1967, pp. 187-216.

Fischer, H.B. and Holley, E.R., "Analysis of the Use of Distorted Hydraulic Models for Dispersion Studies," *Water Resources Research*, vol. 7, no. 1, February 1971, pp. 46-51.

Henderson, F.M., *Open Channel Flow*, Macmillan, New York, 1966.

Kotras, T., Lewis, J. and Etzel, R., "Hydraulic Modeling of Ice-Covered Waters," *Proceedings of the 4th International Conference on POAC*, vol. 1, Memorial University of Newfoundland, St. John's, Newfoundland, Canada, 1977, pp. 453-463.

Moretti, P.M. and McLaughlin, D.K., "Hydraulic Modeling in Stratified Lakes," *Proceedings of the American Society of Civil Engineers, Journal of the Hydraulics Division*, vol. 103, no. HY4, April 1977, pp. 368-380.

Neill, C.R., "Dynamic Ice Forces on Piers and Piles: An Assessment of Design Guidelines in the Light of Recent Research," *Canadian Journal of Civil Engineering*, vol. 3, 1976, pp. 305-341.

Raudkivi, A.J., *Loose Boundary Hydraulics*, 2nd ed., Pergamon Press, New York, 1976.

Roberts, B.R., Findikakis, A.N. and Street, R.L., "A Program of Research on Turbulent Mixing in Distorted Hydraulic Models," The American Society of Civil Engineers, *Proceedings of the Specialty Conference on Conservation and Utilization of Water and Energy Resources*, 1979, pp. 57-64.

Roberts, B.R. and Street, R.L., "Impact of Vertical Scale Distortion on Turbulent Mixing in Physical Models," The American Society of Civil Engineers, *Proceedings of the Specialty Conference on Computer and Physical Modeling in Hydraulic Engineering*, 1980, pp. 261-271.

Rouse, H., *Engineering Hydraulics*, Wiley, New York, 1958.

Rouse, H. and Ince, S., *History of Hydraulics*, Dover Publications, New York, 1963.

Schwarz, J., "New Developments in Modeling Ice Problems," *Proceedings of the 4th International Conference on POAC*, vol. 1, Memorial University of Newfoundland, St. John's, Newfoundland, Canada, 1977, pp. 45-61.

Sharp, J.J., *Hydraulic Modeling*, Butterworth, London, 1981.

Streeter, V.L. and Wylie, E.B., *Fluid Mechanics*, 6th ed., McGraw-Hill Book Company, New York, 1975.

Timco, G.W., "The Mechanical and Morphological Properties of Doped Ice," *Proceedings of the 5th International Conference on POAC*, Norwegian Institute of Technology, Trondheim, Norway, 1979, pp. 719-739.

Walski, T.M., "Properties of Steady Viscosity-Stratified Flow to a Line Sink," Technical Report EL-79-6, U.S. Army Corps of Engineers, Waterways Experiment Station, Vicksburg, Miss., July 1979.

Zwamborn, J.A., "Reproducibility in Hydraulic Models of Prototype River Morphology," *La Houille Blanche*, no. 3, 1966, pp. 291-298.

Zwamborn, J.A., "Solution of River Problems with Moveable-Bed Hydraulic Models," MEG 597, Council for Scientific and Industrial Research, Pretoria, South Africa, 1967.

Zwamborn, J.A., "Hydraulic Models," MEG 795, Council for Scientific and Industrial Research, Stellenbosch, South Africa, 1969.

10.8.   PROBLEMS

1.   The Reynolds number may be defined as the ratio of
        a.   Viscous forces to inertial forces
        b.   Viscous forces to gravity forces
        c.   Inertial forces to viscous forces
        d.   Gravity forces to inertial forces
        e.   None of the above

2.   Select the situations in which inertial forces are not important
        a.   Flow over a spillway crest
        b.   Flow through an open channel transition
        c.   Flow through a long capillary tube
        d.   Flow through a half-open valve

3.   A model of a spillway is to be constructed. If $n_p = 0.013$ and the minimum roughness achievable in the model is $n_M = 0.009$, discuss the implications from a laboratory viewpoint.

4.   A fixed bed model is to be constructed of a channel reach that is rectangular in section and 100 m (328 ft) wide. The scale ratios to be used are $L_R = 1/100$ and $Y_R = 1/10$. Manning's n for the prototype is 0.025 and the slope is 0.001 m/m. Calculate the discharges of the model when the prototype depth of flow is 1 m (3.28 ft) and 2 m (6.56 ft). What are the required values of n in the model for these depths of flow?

5.   A fixed bed model is to be constructed of a channel reach that is triangular in section. In the prototype, $z = 2$ and $n = 0.030$. If $L_R = 0.005$ and $Y_R = 0.05$, find the required values of n in the model when the prototype depths of flow are 5 ft (1.5 m) and 10 ft (3.0 m).

6.   A fixed bed model is to be constructed of a channel reach that is approximately parabolic in section. When the depth of flow in the prototype channel is 4 ft (1.2 m), the top width is 100 ft (30.5 m). If $L_R = 1/150$ and $Y_R = 1/20$, estimate the required values of n in the model when the prototype depth of flow is 3.5 ft (1.1 m) if $n_p = 0.035$.

7.   For the data provided in Prob. 6, estimate the value of n required in the model when the depth of flow in the prototype is 6 ft (1.83 m).

8.   A fixed bed model of a wide channel reach is to be constructed. Space in the laboratory is such that $L_R = 1/200$ and the maximum flow to the laboratory is 0.04 m³/s (1.4 ft³/s). In the prototype, n_P can be no less than 0.04, and the laboratory materials available indicate that in the model n_M can be no less than 0.014. If the maximum discharge in the prototype is 2000 m³/s (70,000 ft³/s), calculate the range of Y_R and the corresponding values of n and maximum discharge in the model.

4.   A moveable bed model of a rectangular channel reach 60 m (197 ft) wide is to be constructed. The specific gravity of the material composing the prototype bed is 2.65 and that composing the model is 1.2. Use Eqs. (10.3.10) through (10.3.13) to determine suitable values of d_R, L_R, and Y_R.

APPENDIX 1
Table for determining *F(u,N)* for positive slopes (Chow, 1955).

					N					
u	2.2	2.4	2.6	2.8	3.0	3.2	3.4	3.6	3.8	4.0
0.00	0.000	0.000	0.000	0.000	0.000	0.000	0.000	0.000	0.000	0.000
0.02	0.020	0.020	0.020	0.020	0.020	0.020	0.020	0.020	0.020	0.020
0.04	0.040	0.040	0.040	0.040	0.040	0.040	0.040	0.040	0.040	0.040
0.06	0.060	0.060	0.060	0.060	0.060	0.060	0.060	0.060	0.060	0.060
0.08	0.080	0.080	0.080	0.080	0.080	0.080	0.080	0.080	0.080	0.080
0.10	0.100	0.100	0.100	0.100	0.100	0.100	0.100	0.100	0.100	0.100
0.12	0.120	0.120	0.120	0.120	0.120	0.120	0.120	0.120	0.120	0.120
0.14	0.140	0.140	0.140	0.140	0.140	0.140	0.140	0.140	0.140	0.140
0.16	0.161	0.161	0.160	0.160	0.160	0.160	0.160	0.160	0.160	0.160
0.18	0.181	0.181	0.181	0.180	0.180	0.180	0.180	0.180	0.180	0.180
0.20	0.202	0.201	0.201	0.201	0.200	0.200	0.200	0.200	0.200	0.200
0.22	0.223	0.222	0.221	0.221	0.221	0.220	0.220	0.220	0.220	0.220
0.24	0.244	0.243	0.242	0.241	0.241	0.241	0.240	0.240	0.240	0.240
0.26	0.265	0.263	0.262	0.262	0.261	0.261	0.261	0.260	0.260	0.260
0.28	0.286	0.284	0.283	0.282	0.282	0.281	0.281	0.281	0.280	0.280
0.30	0.307	0.305	0.304	0.303	0.302	0.302	0.301	0.301	0.301	0.300
0.32	0.329	0.326	0.325	0.324	0.323	0.322	0.322	0.321	0.321	0.321
0.34	0.351	0.348	0.346	0.344	0.343	0.342	0.342	0.342	0.341	0.341
0.36	0.372	0.369	0.367	0.366	0.364	0.363	0.363	0.362	0.362	0.361
0.38	0.395	0.392	0.389	0.387	0.385	0.384	0.383	0.383	0.382	0.382
0.40	0.418	0.414	0.411	0.408	0.407	0.405	0.404	0.403	0.403	0.402
0.42	0.442	0.437	0.433	0.430	0.428	0.426	0.425	0.424	0.423	0.423
0.44	0.465	0.460	0.456	0.452	0.450	0.448	0.446	0.445	0.444	0.443
0.46	0.480	0.483	0.479	0.475	0.472	0.470	0.468	0.466	0.465	0.464
0.48	0.514	0.507	0.502	0.497	0.494	0.492	0.489	0.488	0.486	0.485

u	2.2	2.4	2.6	2.8	N 3.0	3.2	3.4	3.6	3.8	4.0
0.50	0.539	0.531	0.525	0.521	0.517	0.514	0.511	0.509	0.508	0.506
0.52	0.565	0.557	0.550	0.544	0.540	0.536	0.534	0.531	0.529	0.528
0.54	0.592	0.582	0.574	0.568	0.563	0.559	0.556	0.554	0.551	0.550
0.56	0.619	0.608	0.599	0.593	0.587	0.583	0.579	0.576	0.574	0.572
0.58	0.648	0.635	0.626	0.618	0.612	0.607	0.603	0.599	0.596	0.594
0.60	0.676	0.663	0.653	0.644	0.637	0.631	0.627	0.623	0.620	0.617
0.61	0.691	0.678	0.667	0.657	0.650	0.644	0.639	0.635	0.631	0.628
0.62	0.706	0.692	0.680	0.671	0.663	0.657	0.651	0.647	0.643	0.640
0.63	0.722	0.707	0.694	0.684	0.676	0.669	0.664	0.659	0.655	0.652
0.64	0.738	0.722	0.709	0.698	0.690	0.683	0.677	0.672	0.667	0.664
0.65	0.754	0.737	0.724	0.712	0.703	0.696	0.689	0.684	0.680	0.676
0.66	0.771	0.753	0.738	0.727	0.717	0.709	0.703	0.697	0.692	0.688
0.67	0.787	0.769	0.754	0.742	0.731	0.723	0.716	0.710	0.705	0.701
0.68	0.804	0.785	0.769	0.757	0.746	0.737	0.729	0.723	0.718	0.713
0.69	0.822	0.804	0.785	0.772	0.761	0.751	0.743	0.737	0.731	0.726
0.70	0.840	0.819	0.802	0.787	0.776	0.766	0.757	0.750	0.744	0.739
0.71	0.858	0.836	0.819	0.804	0.791	0.781	0.772	0.764	0.758	0.752
0.72	0.878	0.855	0.836	0.820	0.807	0.796	0.786	0.779	0.772	0.766
0.73	0.898	0.874	0.854	0.837	0.823	0.811	0.802	0.793	0.786	0.780
0.74	0.918	0.892	0.868	0.854	0.840	0.827	0.817	0.808	0.800	0.794
0.75	0.940	0.913	0.890	0.872	0.857	0.844	0.833	0.823	0.815	0.808
0.76	0.961	0.933	0.909	0.890	0.874	0.861	0.849	0.839	0.830	0.823
0.77	0.985	0.954	0.930	0.909	0.892	0.878	0.866	0.855	0.846	0.838
0.78	1.007	0.976	0.950	0.929	0.911	0.896	0.883	0.872	0.862	0.854
0.79	1.031	0.998	0.971	0.940	0.930	0.914	0.901	0.889	0.879	0.870
0.80	1.056	1.022	0.994	0.970	0.950	0.934	0.919	0.907	0.896	0.887
0.81	1.083	1.046	1.017	0.992	0.971	0.954	0.938	0.925	0.914	0.904
0.82	1.110	1.072	1.041	1.015	0.993	0.974	0.958	0.945	0.932	0.922
0.83	1.139	1.099	1.067	1.039	1.016	0.996	0.979	0.965	0.952	0.940

					N					
u	2.2	2.4	2.6	2.8	3.0	3.2	3.4	3.6	3.8	4.0
0.84	1.171	1.129	1.094	1.064	1.040	1.019	1.001	0.985	0.972	0.960
0.85	1.201	1.157	1.121	1.091	1.065	1.043	1.024	1.007	0.993	0.980
0.86	1.238	1.192	1.153	1.119	1.092	1.068	1.048	1.031	1.015	1.002
0.87	1.272	1.223	1.182	1.149	1.120	1.095	1.074	1.055	1.039	1.025
0.88	1.314	1.262	1.228	1.181	1.151	1.124	1.101	1.081	1.064	1.049
0.89	1.357	1.302	1.255	1.216	1.183	1.155	1.131	1.110	1.091	1.075
0.90	1.401	1.343	1.294	1.253	1.218	1.189	1.163	1.140	1.120	1.103
0.91	1.452	1.389	1.338	1.294	1.257	1.225	1.197	1.173	1.152	1.133
0.92	1.505	1.438	1.351	1.340	1.300	1.266	1.236	1.210	1.187	1.166
0.93	1.564	1.493	1.435	1.391	1.348	1.311	1.279	1.251	1.226	1.204
0.94	1.645	1.568	1.504	1.449	1.403	1.363	1.328	1.297	1.270	1.246
0.950	1.737	1.652	1.582	1.518	1.467	1.423	1.385	1.352	1.322	1.296
0.960	1.833	1.741	1.665	1.601	1.545	1.497	1.454	1.417	1.385	1.355
0.970	1.969	1.866	1.780	1.707	1.644	1.590	1.543	1.501	1.464	1.431
0.975	2.055	1.945	1.853	1.773	1.707	1.649	1.598	1.554	1.514	1.479
0.980	2.164	2.045	1.946	1.855	1.783	1.720	1.666	1.617	1.575	1.536
0.985	2.294	2.165	2.056	1.959	1.880	1.812	1.752	1.699	1.652	1.610
0.990	2.477	2.333	2.212	2.106	2.017	1.940	1.873	1.814	1.761	1.714
0.995	2.792	2.621	2.478	2.355	2.250	2.159	2.070	2.008	1.945	1.889
0.999	3.523	3.292	3.097	2.931	2.788	2.663	2.554	2.457	2.370	2.293
1.000	4	4	4	4	4	4	4	4	4	4
1.001	3.317	2.931	2.640	2.399	2.184	2.008	1.856	1.725	1.610	1.508
1.005	2.587	2.266	2.022	1.818	1.679	1.506	1.384	1.279	1.188	1.107
1.010	2.273	1.977	1.757	1.572	1.419	1.291	1.182	1.089	1.007	0.936
1.015	2.090	1.807	1.602	1.428	1.286	1.166	1.065	0.978	0.902	0.836
1.020	1.961	1.711	1.493	1.327	1.191	1.078	0.982	0.900	0.828	0.766
1.03	1.779	1.531	1.340	1.186	1.060	0.955	0.866	0.790	0.725	0.668
1.04	1.651	1.410	1.232	1.086	0.967	0.868	0.785	0.714	0.653	0.600

u	\multicolumn{10}{c}{N}
	2.2	2.4	2.6	2.8	3.0	3.2	3.4	3.6	3.8	4.0
1.05	1.552	1.334	1.150	1.010	0.896	0.802	0.723	0.656	0.598	0.548
1.06	1.472	1.250	1.082	0.948	0.838	0.748	0.672	0.608	0.553	0.506
1.07	1.404	1.195	1.026	0.896	0.790	0.703	0.630	0.569	0.516	0.471
1.08	1.346	1.139	0.978	0.851	0.749	0.665	0.595	0.535	0.485	0.441
1.09	1.295	1.089	0.935	0.812	0.713	0.631	0.563	0.506	0.457	0.415
1.10	1.250	1.050	0.897	0.777	0.681	0.601	0.536	0.430	0.433	0.392
1.11	1.209	1.014	0.864	0.746	0.652	0.575	0.511	0.457	0.411	0.372
1.12	1.172	0.981	0.833	0.718	0.626	0.551	0.488	0.436	0.392	0.354
1.13	1.138	0.950	0.805	0.692	0.602	0.529	0.468	0.417	0.374	0.337
1.14	1.107	0.921	0.780	0.669	0.581	0.509	0.450	0.400	0.358	0.322
1.15	1.078	0.892	0.756	0.647	0.561	0.490	0.432	0.384	0.343	0.308
1.16	1.052	0.870	0.734	0.627	0.542	0.473	0.417	0.369	0.329	0.295
1.17	1.027	0.850	0.713	0.608	0.525	0.458	0.402	0.356	0.317	0.283
1.18	1.003	0.825	0.694	0.591	0.509	0.443	0.388	0.343	0.305	0.272
1.19	0.981	0.810	0.676	0.574	0.494	0.429	0.375	0.331	0.294	0.262
1.20	0.960	0.787	0.659	0.559	0.480	0.416	0.363	0.320	0.283	0.252
1.22	0.922	0.755	0.628	0.531	0.454	0.392	0.341	0.299	0.264	0.235
1.24	0.887	0.725	0.600	0.505	0.431	0.371	0.322	0.281	0.248	0.219
1.26	0.855	0.692	0.574	0.482	0.410	0.351	0.304	0.265	0.233	0.205
1.28	0.827	0.666	0.551	0.461	0.391	0.334	0.288	0.250	0.219	0.193
1.30	0.800	0.644	0.530	0.442	0.373	0.318	0.274	0.237	0.207	0.181
1.32	0.775	0.625	0.510	0.424	0.357	0.304	0.260	0.225	0.196	0.171
1.34	0.752	0.605	0.492	0.408	0.342	0.290	0.248	0.214	0.185	0.162
1.36	0.731	0.588	0.475	0.393	0.329	0.278	0.237	0.204	0.176	0.153
1.38	0.711	0.567	0.459	0.378	0.316	0.266	0.226	0.194	0.167	0.145
1.40	0.692	0.548	0.444	0.365	0.304	0.256	0.217	0.185	0.159	0.138
1.42	0.674	0.533	0.431	0.353	0.293	0.246	0.208	0.177	0.152	0.131
1.44	0.658	0.517	0.417	0.341	0.282	0.236	0.199	0.169	0.145	0.125

					N					
u	2.2	2.4	2.6	2.8	3.0	3.2	3.4	3.6	3.8	4.0
1.46	0.642	0.505	0.405	0.330	0.273	0.227	0.191	0.162	0.139	0.119
1.48	0.627	0.493	0.394	0.320	0.263	0.219	0.184	0.156	0.133	0.113
1.50	0.613	0.480	0.383	0.310	0.255	0.211	0.177	0.149	0.127	0.108
1.55	0.580	0.451	0.358	0.288	0.235	0.194	0.161	0.135	0.114	0.097
1.60	0.551	0.425	0.335	0.269	0.218	0.179	0.148	0.123	0.103	0.087
1.65	0.525	0.402	0.316	0.251	0.203	0.165	0.136	0.113	0.094	0.079
1.70	0.501	0.381	0.298	0.236	0.189	0.153	0.125	0.103	0.086	0.072
1.75	0.480	0.362	0.282	0.222	0.177	0.143	0.116	0.095	0.079	0.065
1.80	0.460	0.349	0.267	0.209	0.166	0.133	0.108	0.088	0.072	0.060
1.85	0.442	0.332	0.254	0.198	0.156	0.125	0.100	0.082	0.067	0.055
1.90	0.425	0.315	0.242	0.188	0.147	0.117	0.094	0.076	0.062	0.050
1.95	0.409	0.304	0.231	0.178	0.139	0.110	0.088	0.070	0.057	0.046
2.00	0.395	0.292	0.221	0.169	0.132	0.104	0.082	0.066	0.053	0.043
2.10	0.369	0.273	0.202	0.154	0.119	0.092	0.073	0.058	0.046	0.037
2.20	0.346	0.253	0.186	0.141	0.107	0.083	0.065	0.051	0.040	0.032
2.30	0.326	0.235	0.173	0.129	0.098	0.075	0.058	0.045	0.035	0.028
2.40	0.308	0.220	0.160	0.119	0.089	0.068	0.052	0.040	0.031	0.024
2.50	0.292	0.207	0.150	0.110	0.082	0.062	0.047	0.036	0.028	0.022
2.60	0.277	0.197	0.140	0.102	0.076	0.057	0.043	0.033	0.025	0.019
2.70	0.264	0.188	0.131	0.095	0.070	0.052	0.039	0.029	0.022	0.017
2.80	0.252	0.176	0.124	0.089	0.065	0.048	0.036	0.027	0.020	0.015
2.90	0.241	0.166	0.117	0.083	0.060	0.044	0.033	0.024	0.018	0.014
3.00	0.230	0.159	0.110	0.078	0.056	0.041	0.030	0.022	0.017	0.012
3.50	0.190	0.126	0.085	0.059	0.041	0.029	0.021	0.015	0.011	0.008
4.00	0.161	0.104	0.069	0.046	0.031	0.022	0.015	0.010	0.007	0.005
4.50	0.139	0.087	0.057	0.037	0.025	0.017	0.011	0.008	0.005	0.004
5.00	0.122	0.076	0.048	0.031	0.020	0.013	0.009	0.006	0.004	0.003
6.00	0.098	0.060	0.036	0.022	0.014	0.009	0.006	0.004	0.002	0.002
7.00	0.081	0.048	0.028	0.017	0.010	0.006	0.004	0.002	0.002	0.001

u	2.2	2.4	2.6	2.8	N 3.0	3.2	3.4	3.6	3.8	4.0
8.00	0.069	0.040	0.022	0.013	0.008	0.005	0.003	0.002	0.001	0.001
9.00	0.060	0.034	0.019	0.011	0.006	0.004	0.002	0.001	0.001	0.000
10.0	0.053	0.028	0.016	0.009	0.005	0.003	0.002	0.001	0.001	0.000
20.0	0.023	0.018	0.011	0.006	0.002	0.001	0.001	0.000	0.000	0.000

Table for determining $F(u,N)$ for positive slopes: *continued*

u	4.2	4.6	5.0	5.4	N 5.8	6.2	6.6	7.0	7.4	7.8
0.00	0.000	0.000	0.000	0.000	0.000	0.000	0.000	0.000	0.000	0.000
0.02	0.020	0.020	0.020	0.020	0.020	0.020	0.020	0.020	0.020	0.020
0.04	0.040	0.040	0.040	0.040	0.040	0.040	0.040	0.040	0.040	0.040
0.06	0.060	0.060	0.060	0.060	0.060	0.060	0.060	0.060	0.060	0.060
0.08	0.080	0.080	0.080	0.080	0.080	0.080	0.080	0.080	0.080	0.080
0.10	0.100	0.100	0.100	0.100	0.100	0.100	0.100	0.100	0.100	0.100
0.12	0.120	0.120	0.120	0.120	0.120	0.120	0.120	0.120	0.120	0.120
0.14	0.140	0.140	0.140	0.140	0.140	0.140	0.140	0.140	0.140	0.140
0.16	0.160	0.160	0.160	0.160	0.160	0.160	0.160	0.160	0.160	0.160
0.18	0.180	0.180	0.180	0.180	0.180	0.180	0.180	0.180	0.180	0.180
0.20	0.200	0.200	0.200	0.200	0.200	0.200	0.200	0.200	0.200	0.200
0.22	0.220	0.220	0.220	0.220	0.220	0.220	0.220	0.220	0.220	0.220
0.24	0.240	0.240	0.240	0.240	0.240	0.240	0.240	0.240	0.240	0.240
0.26	0.260	0.260	0.260	0.260	0.260	0.260	0.260	0.260	0.260	0.260

u	N									
	4.2	4.6	5.0	5.4	5.8	6.2	6.6	7.0	7.4	7.8
0.28	0.280	0.280	0.280	0.280	0.280	0.280	0.280	0.280	0.280	0.280
0.30	0.300	0.300	0.300	0.300	0.300	0.300	0.300	0.300	0.300	0.300
0.32	0.321	0.320	0.320	0.320	0.320	0.320	0.320	0.320	0.320	0.320
0.34	0.341	0.340	0.340	0.340	0.340	0.340	0.340	0.340	0.340	0.340
0.36	0.361	0.361	0.360	0.360	0.360	0.360	0.360	0.360	0.360	0.360
0.38	0.381	0.381	0.381	0.380	0.380	0.380	0.380	0.380	0.380	0.380
0.40	0.402	0.401	0.401	0.400	0.400	0.400	0.400	0.400	0.400	0.400
0.42	0.422	0.421	0.421	0.421	0.420	0.420	0.420	0.420	0.420	0.420
0.44	0.443	0.442	0.441	0.441	0.441	0.441	0.440	0.440	0.440	0.440
0.46	0.463	0.462	0.462	0.461	0.461	0.461	0.460	0.460	0.460	0.460
0.48	0.484	0.483	0.482	0.481	0.481	0.481	0.480	0.480	0.480	0.480
0.50	0.505	0.504	0.503	0.502	0.501	0.501	0.501	0.500	0.500	0.500
0.52	0.527	0.525	0.523	0.522	0.522	0.521	0.521	0.521	0.520	0.520
0.54	0.548	0.546	0.544	0.543	0.542	0.542	0.541	0.541	0.541	0.541
0.56	0.570	0.567	0.565	0.564	0.563	0.562	0.562	0.561	0.561	0.561
0.58	0.592	0.589	0.587	0.585	0.583	0.583	0.582	0.582	0.581	0.581
0.60	0.614	0.611	0.608	0.606	0.605	0.604	0.603	0.602	0.602	0.601
0.61	0.626	0.622	0.619	0.617	0.615	0.614	0.613	0.612	0.612	0.611
0.62	0.637	0.633	0.630	0.628	0.626	0.625	0.624	0.623	0.622	0.622
0.63	0.649	0.644	0.641	0.638	0.636	0.635	0.634	0.633	0.632	0.632
0.64	0.661	0.656	0.652	0.649	0.647	0.646	0.645	0.644	0.643	0.642

u	\multicolumn{10}{c|}{N}									
	4.2	4.6	5.0	5.4	5.8	6.2	6.6	7.0	7.4	7.8
0.65	0.673	0.667	0.663	0.660	0.658	0.656	0.655	0.654	0.653	0.653
0.66	0.685	0.679	0.675	0.672	0.669	0.667	0.666	0.665	0.664	0.663
0.67	0.697	0.691	0.686	0.683	0.680	0.678	0.676	0.675	0.674	0.673
0.68	0.709	0.703	0.698	0.694	0.691	0.689	0.687	0.686	0.685	0.684
0.69	0.722	0.715	0.710	0.706	0.703	0.700	0.698	0.696	0.695	0.694
0.70	0.735	0.727	0.722	0.717	0.714	0.712	0.710	0.708	0.706	0.705
0.71	0.748	0.740	0.734	0.729	0.726	0.723	0.721	0.719	0.717	0.716
0.72	0.761	0.752	0.746	0.741	0.737	0.734	0.732	0.730	0.728	0.727
0.73	0.774	0.765	0.759	0.753	0.749	0.746	0.743	0.741	0.739	0.737
0.74	0.788	0.779	0.771	0.766	0.761	0.757	0.754	0.752	0.750	0.748
0.75	0.802	0.792	0.784	0.778	0.773	0.769	0.766	0.763	0.761	0.759
0.76	0.817	0.806	0.798	0.791	0.786	0.782	0.778	0.775	0.773	0.771
0.77	0.831	0.820	0.811	0.804	0.798	0.794	0.790	0.787	0.784	0.782
0.78	0.847	0.834	0.825	0.817	0.811	0.806	0.802	0.799	0.796	0.794
0.79	0.862	0.849	0.839	0.831	0.824	0.819	0.815	0.811	0.808	0.805
0.80	0.878	0.865	0.854	0.845	0.838	0.832	0.828	0.823	0.820	0.818
0.81	0.895	0.881	0.869	0.860	0.852	0.846	0.841	0.836	0.833	0.830
0.82	0.913	0.897	0.885	0.875	0.866	0.860	0.854	0.850	0.846	0.842
0.83	0.931	0.914	0.901	0.890	0.881	0.874	0.868	0.863	0.859	0.855
0.84	0.949	0.932	0.918	0.906	0.897	0.889	0.882	0.877	0.872	0.868

u	4.2	4.6	5.0	5.4	5.8	6.2	6.6	7.0	7.4	7.8
					N					
0.85	0.969	0.950	0.935	0.923	0.912	0.905	0.898	0.891	0.887	0.882
0.86	0.990	0.970	0.954	0.940	0.930	0.921	0.913	0.906	0.901	0.896
0.87	1.012	0.990	0.973	0.959	0.947	0.937	0.929	0.922	0.916	0.911
0.88	1.035	1.012	0.994	0.978	0.966	0.955	0.946	0.938	0.932	0.927
0.89	1.060	1.035	1.015	0.999	0.986	0.974	0.964	0.956	0.949	0.943
0.90	1.087	1.060	1.039	1.021	1.007	0.994	0.984	0.974	0.967	0.960
0.91	1.116	1.088	1.064	1.045	1.029	1.016	1.003	0.995	0.986	0.979
0.92	1.148	1.117	1.092	1.072	1.054	1.039	1.027	1.016	1.006	0.999
0.93	1.184	1.151	1.123	1.101	1.081	1.065	1.050	1.040	1.029	1.021
0.94	1.225	1.188	1.158	1.134	1.113	1.095	1.080	1.066	1.054	1.044
0.950	1.272	1.232	1.199	1.172	1.148	1.128	1.111	1.097	1.084	1.073
0.960	1.329	1.285	1.248	1.217	1.188	1.167	1.149	1.133	1.119	1.106
0.970	1.402	1.351	1.310	1.275	1.246	1.319	1.197	1.179	1.162	1.148
0.975	1.447	1.393	1.348	1.311	1.280	1.250	1.227	1.207	1.190	1.173
0.980	1.502	1.443	1.395	1.354	1.339	1.288	1.262	1.241	1.221	1.204
0.985	1.573	1.508	1.454	1.409	1.372	1.337	1.309	1.284	1.263	1.243
0.990	1.671	1.598	1.537	1.487	1.444	1.404	1.373	1.344	1.319	1.297
0.995	1.838	1.751	1.678	1.617	1.565	1.519	1.479	1.451	1.416	1.388
0.999	2.223	2.102	2.002	1.917	1.845	1.780	1.725	1.678	1.635	1.596
1.000	4	4	4	4	4	4	4	4	4	4
1.001	1.417	1.264	1.138	1.033	0.951	0.870	0.803	0.746	0.697	0.651

u	N									
	4.2	**4.6**	**5.0**	**5.4**	**5.8**	**6.2**	**6.6**	**7.0**	**7.4**	**7.8**
1.005	1.036	0.915	0.817	0.737	0.669	0.612	0.553	0.526	0.481	0.447
1.010	0.873	0.766	0.681	0.610	0.551	0.502	0.459	0.422	0.389	0.360
1.015	0.778	0.680	0.602	0.537	0.483	0.440	0.399	0.366	0.336	0.310
1.020	0.711	0.620	0.546	0.486	0.436	0.394	0.358	0.327	0.300	0.276
1.03	0.618	0.535	0.469	0.415	0.370	0.333	0.300	0.272	0.249	0.228
1.04	0.554	0.477	0.415	0.365	0.324	0.290	0.262	0.236	0.214	0.195
1.05	0.504	0.432	0.374	0.328	0.289	0.259	0.231	0.208	0.189	0.174
1.06	0.464	0.396	0.342	0.298	0.262	0.233	0.209	0.187	0.170	0.154
1.07	0.431	0.366	0.315	0.273	0.239	0.212	0.191	0.168	0.151	0.136
1.08	0.403	0.341	0.292	0.252	0.220	0.194	0.172	0.153	0.137	0.123
1.09	0.379	0.319	0.272	0.234	0.204	0.179	0.158	0.140	0.125	0.112
1.10	0.357	0.299	0.254	0.218	0.189	0.165	0.146	0.129	0.114	0.102
1.11	0.338	0.282	0.239	0.204	0.176	0.154	0.135	0.119	0.105	0.094
1.12	0.321	0.267	0.225	0.192	0.165	0.143	0.125	0.110	0.097	0.086
1.13	0.305	0.253	0.212	0.181	0.155	0.135	0.117	0.102	0.090	0.080
1.14	0.291	0.240	0.201	0.170	0.146	0.126	0.109	0.095	0.084	0.074
1.15	0.278	0.229	0.191	0.161	0.137	0.118	0.102	0.089	0.078	0.068
1.16	0.266	0.218	0.181	0.153	0.130	0.111	0.096	0.084	0.072	0.064
1.17	0.255	0.208	0.173	0.145	0.123	0.105	0.090	0.078	0.068	0.060
1.18	0.244	0.199	0.165	0.138	0.116	0.099	0.085	0.073	0.063	0.055
1.19	0.235	0.191	0.157	0.131	0.110	0.094	0.080	0.068	0.059	0.051

u	4.2	4.6	5.0	5.4	5.8	6.2	6.6	7.0	7.4	7.8
					N					
1.20	0.226	0.183	0.150	0.2151	0.105	0.088	0.078	0.064	0.056	0.048
1.22	0.209	0.168	0.138	0.114	0.095	0.080	0.068	0.057	0.049	0.042
1.24	0.195	0.156	0.127	0.104	0.086	0.072	0.060	0.051	0.044	0.038
1.26	0.182	0.145	0.117	0.095	0.079	0.065	0.055	0.046	0.039	0.033
1.28	0.170	0.135	0.108	0.088	0.072	0.060	0.050	0.041	0.035	0.030
1.30	0.160	0.126	0.100	0.081	0.066	0.054	0.045	0.037	0.031	0.026
1.32	0.150	0.118	0.093	0.075	0.061	0.050	0.041	0.034	0.028	0.024
1.34	0.142	0.110	0.087	0.069	0.056	0.045	0.037	0.030	0.025	0.021
1.36	0.134	0.103	0.081	0.064	0.052	0.042	0.034	0.028	0.023	0.019
1.38	0.127	0.097	0.076	0.060	0.048	0.038	0.032	0.026	0.021	0.017
1.40	0.120	0.092	0.071	0.056	0.044	0.036	0.028	0.023	0.019	0.016
1.42	0.114	0.087	0.067	0.052	0.041	0.033	0.026	0.021	0.017	0.014
1.44	0.108	0.082	0.063	0.049	0.038	0.030	0.024	0.019	0.016	0.013
1.46	0.103	0.077	0.059	0.046	0.036	0.028	0.022	0.018	0.014	0.012
1.48	0.098	0.073	0.056	0.043	0.033	0.026	0.021	0.017	0.013	0.010
1.50	0.093	0.069	0.053	0.040	0.031	0.024	0.020	0.015	0.012	0.009
1.55	0.083	0.061	0.046	0.035	0.026	0.020	0.016	0.012	0.010	0.008
1.60	0.074	0.054	0.040	0.030	0.023	0.017	0.013	0.010	0.007	0.006
1.65	0.067	0.048	0.035	0.026	0.019	0.014	0.011	0.008	0.006	0.005
1.70	0.060	0.043	0.031	0.023	0.016	0.012	0.009	0.007	0.005	0.004
1.75	0.054	0.038	0.027	0.020	0.014	0.010	0.008	0.006	0.004	0.003

u	N									
	4.2	**4.6**	**5.0**	**5.4**	**5.8**	**6.2**	**6.6**	**7.0**	**7.4**	**7.8**
1.80	0.049	0.034	0.024	0.017	0.012	0.009	0.007	0.005	0.004	0.003
1.85	0.045	0.031	0.022	0.015	0.011	0.008	0.006	0.004	0.003	0.002
1.90	0.041	0.028	0.020	0.014	0.010	0.007	0.005	0.004	0.003	0.002
1.95	0.038	0.026	0.018	0.012	0.008	0.006	0.004	0.003	0.002	0.002
2.00	0.035	0.023	0.016	0.011	0.007	0.005	0.004	0.003	0.002	0.001
2.10	0.030	0.019	0.013	0.009	0.006	0.004	0.003	0.002	0.001	0.001
2.20	0.025	0.016	0.011	0.007	0.005	0.004	0.002	0.001	0.001	0.001
2.30	0.022	0.014	0.009	0.006	0.004	0.003	0.002	0.001	0.001	0.001
2.40	0.019	0.012	0.008	0.005	0.003	0.002	0.001	0.001	0.001	0.001
2.50	0.017	0.010	0.006	0.004	0.003	0.002	0.001	0.001	0.001	0.001
2.60	0.015	0.009	0.005	0.003	0.002	0.001	0.001	0.001	0.000	0.000
2.70	0.013	0.008	0.005	0.003	0.002	0.001	0.001	0.000	0.000	0.000
2.80	0.012	0.007	0.004	0.002	0.001	0.001	0.001	0.000	0.000	0.000
2.90	0.010	0.006	0.004	0.002	0.001	0.001	0.000	0.000	0.000	0.000
3.00	0.009	0.005	0.003	0.002	0.001	0.001	0.000	0.000	0.000	0.000
3.50	0.006	0.003	0.002	0.001	0.001	0.000	0.000	0.000	0.000	0.000
4.00	0.004	0.002	0.001	0.000	0.000	0.000	0.000	0.000	0.000	0.000
4.50	0.003	0.001	0.001	0.000	0.000	0.000	0.000	0.000	0.000	0.000
5.00	0.002	0.001	0.000	0.000	0.000	0.000	0.000	0.000	0.000	0.000
6.00	0.001	0.000	0.000	0.000	0.000	0.000	0.000	0.000	0.000	0.000
7.00	0.001	0.000	0.000	0.000	0.000	0.000	0.000	0.000	0.000	0.000

u	N									
	4.2	4.6	5.0	5.4	5.8	6.2	6.6	7.0	7.4	7.8
8.00	0.000	0.000	0.000	0.000	0.000	0.000	0.000	0.000	0.000	0.000
9.00	0.000	0.000	0.000	0.000	0.000	0.000	0.000	0.000	0.000	0.000
10.0	0.000	0.000	0.000	0.000	0.000	0.000	0.000	0.000	0.000	0.000
20.0	0.000	0.000	0.000	0.000	0.000	0.000	0.000	0.000	0.000	0.000

Note: This value is likely in error although it is the value given in Chow (1955 and 1959). From a historical perspective, the following statement in Chow (1955) is of interest "The preparation of this table (Appendix I) was undertaken and performed for the first time during 1914 to 1915 by the Research Board of the ten Russian Reclamation Service under the direction of Boris A. Bakhmeteff, then Professor of General and Advanced Hydraulics at Polytechnic Institute Emperor Peter the Great, St. Petersburg, Russia. It was told that the work involved a long and tedious procedure (Bakhmeteff, 1932). In the turmoil of the Russian Revolution in 1917, the table so computed became unavailable, so the task of computing was done over again by Professor Kholodovsky and partly by Dr. Pestrecov. The recomputed table was more precise and complete covering a range of N from 2.8 to 5.4. This table was published in 1932 (Bakhmeteff, 1932) when the author, Bakhmeteff, was Professor of Civil Engineering at the Columbia University." Finally, this author, while a graduate student at the University of California, was awarded the Boris A. Bakhmeteff Research Fellowship in Fluid Mechanics by Columbia University for the 1973-1974 academic year.

Table for determining $F(u,N)$ for positive slopes: *continued*

	N				
u	8.2	8.6	9.0	9.4	9.8
0.00	0.000	0.000	0.000	0.000	0.000
0.02	0.020	0.020	0.020	0.020	0.020
0.04	0.040	0.040	0.040	0.040	0.040
0.06	0.060	0.060	0.060	0.060	0.060
0.08	0.080	0.080	0.080	0.080	0.080
0.10	0.100	0.100	0.100	0.100	0.100
0.12	0.120	0.120	0.120	0.120	0.120
0.14	0.140	0.140	0.140	0.140	0.140
0.16	0.160	0.160	0.160	0.160	0.160
0.18	0.180	0.180	0.180	0.180	0.180
0.20	0.200	0.200	0.200	0.200	0.200
0.22	0.220	0.220	0.220	0.220	0.220
0.24	0.240	0.240	0.240	0.240	0.240
0.26	0.260	0.160	0.260	0.260	0.260
0.28	0.280	0.280	0.280	0.280	0.280
0.30	0.300	0.300	0.300	0.300	0.300
0.32	0.320	0.320	0.320	0.320	0.320
0.34	0.340	0.340	0.340	0.340	0.340
0.36	0.360	0.360	0.360	0.360	0.360
0.38	0.380	0.380	0.380	0.380	0.380
0.40	0.400	0.400	0.400	0.400	0.400
0.42	0.420	0.420	0.420	0.420	0.420
0.44	0.440	0.440	0.440	0.440	0.440
0.46	0.460	0.460	0.460	0.460	0.460
0.48	0.480	0.480	0.480	0.480	0.480
0.50	0.500	0.500	0.500	0.500	0.500
0.52	0.520	0.520	0.520	0.520	0.520
0.54	0.540	0.540	0.540	0.540	0.540
0.56	0.561	0.560	0.560	0.560	0.560
0.58	0.581	0.581	0.580	0.580	0.580

u	N 8.2	8.6	9.0	9.4	9.8
0.60	0.601	0.601	0.601	0.600	0.600
0.61	0.611	0.611	0.611	0.611	0.610
0.62	0.621	0.621	0.621	0.621	0.621
0.63	0.632	0.631	0.631	0.631	0.631
0.64	0.642	0.641	0.641	0.641	0.641
0.65	0.652	0.652	0.651	0.651	0.651
0.66	0.662	0.662	0.662	0.661	0.661
0.67	0.673	0.672	0.672	0.672	0.671
0.68	0.683	0.683	0.682	0.682	0.681
0.69	0.694	0.693	0.692	0.692	0.692
0.70	0.704	0.704	0.703	0.702	0.702
0.71	0.715	0.714	0.713	0.713	0.712
0.72	0.726	0.725	0.724	0.723	0.723
0.73	0.736	0.735	0.734	0.734	0.733
0.74	0.747	0.746	0.745	0.744	0.744
0.75	0.758	0.757	0.756	0.755	0.754
0.76	0.769	0.768	0.767	0.766	0.765
0.77	0.780	0.779	0.778	0.777	0.776
0.78	0.792	0.790	0.789	0.788	0.787
0.79	0.804	0.802	0.800	0.799	0.798
0.80	0.815	0.813	0.811	0.810	0.809
0.81	0.827	0.825	0.823	0.822	0.820
0.82	0.839	0.837	0.835	0.833	0.831
0.83	0.852	0.849	0.847	0.845	0.844
0.84	0.865	0.862	0.860	0.858	0.856
0.85	0.878	0.875	0.873	0.870	0.868
0.86	0.892	0.889	0.886	0.883	0.881
0.87	0.907	0.903	0.900	0.897	0.894
0.88	0.921	0.918	0.914	0.911	0.908
0.89	0.937	0.933	0.929	0.925	0.922

u	N 8.2	8.6	9.0	9.4	9.8
0.90	0.954	0.949	0.944	0.940	0.937
0.91	0.972	0.967	0.961	0.957	0.953
0.92	0.991	0.986	0.980	0.975	0.970
0.93	1.012	1.006	0.999	0.994	0.989
0.94	1.036	1.029	1.022	1.016	1.010
0.950	1.062	1.055	1.047	1.040	1.033
0.960	1.097	1.085	1.074	1.063	1.053
0.970	1.136	1.124	1.112	1.100	1.087
0.975	1.157	1.147	1.134	1.122	1.108
0.980	1.187	1.175	1.160	1.150	1.132
0.985	1.224	1.210	1.196	1.183	1.165
0.990	1.275	1.260	1.243	1.228	1.208
0.995	1.363	1.342	1.320	1.302	1.280
0.999	1.560	1.530	1.500	1.476	1.447
1.000	4	4	4	4	4
1.001	0.614	0.577	0.546	0.519	0.494
1.005	0.420	0.391	0.368	0.350	0.331
1.010	0.337	0.313	0.294	0.278	0.262
1.015	0.289	0.269	0.255	0.237	0.223
1.020	0.257	0.237	0.221	0.209	0.196
1.03	0.212	0.195	0.181	0.170	0.159
1.04	0.173	0.165	0.152	0.143	0.134
1.05	0.158	0.143	0.132	0.124	0.115
1.06	0.140	0.127	0.116	0.106	0.098
1.07	0.123	0.112	0.102	0.094	0.086
1.08	0.111	0.101	0.092	0.084	0.077
1.09	0.101	0.091	0.082	0.075	0.069
1.10	0.092	0.083	0.074	0.067	0.062
1.11	0.084	0.075	0.067	0.060	0.055
1.12	0.077	0.069	0.062	0.055	0.050
1.13	0.071	0.063	0.056	0.050	0.045

u	N				
	8.2	8.6	9.0	9.4	9.8
1.14	0.065	0.058	0.052	0.046	0.041
1.15	0.061	0.054	0.048	0.043	0.038
1.16	0.056	0.050	0.045	0.040	0.035
1.17	0.052	0.046	0.041	0.036	0.032
1.18	0.048	0.042	0.037	0.033	0.029
1.19	0.045	0.039	0.034	0.030	0.027
1.20	0.043	0.037	0.032	0.028	0.025
1.22	0.037	0.032	0.028	0.024	0.021
1.24	0.032	0.028	0.024	0.021	0.018
1.26	0.028	0.024	0.021	0.018	0.016
1.28	0.025	0.021	0.018	0.016	0.014
1.30	0.022	0.019	0.016	0.014	0.012
1.32	0.020	0.017	0.014	0.012	0.010
1.34	0.018	0.015	0.012	0.010	0.009
1.36	0.016	0.013	0.011	0.009	0.008
1.38	0.014	0.012	0.010	0.008	0.007
1.40	0.013	0.011	0.009	0.007	0.006
1.42	0.011	0.009	0.008	0.006	0.005
1.44	0.010	0.008	0.007	0.006	0.005
1.46	0.009	0.008	0.006	0.005	0.004
1.48	0.009	0.007	0.005	0.004	0.004
1.50	0.008	0.006	0.005	0.004	0.003
1.55	0.006	0.005	0.004	0.003	0.003
1.60	0.005	0.004	0.003	0.002	0.002
1.65	0.004	0.003	0.002	0.002	0.001
1.70	0.003	0.002	0.002	0.001	0.001
1.75	0.002	0.002	0.002	0.001	0.001
1.80	0.002	0.001	0.001	0.001	0.001
1.85	0.002	0.001	0.001	0.001	0.001
1.90	0.001	0.001	0.001	0.001	0.000
1.95	0.001	0.001	0.001	0.000	0.000

u	N				
	8.2	8.6	9.0	9.4	9.8
2.00	0.001	0.001	0.000	0.000	0.000
2.10	0.001	0.000	0.000	0.000	0.000
2.20	0.000	0.000	0.000	0.000	0.000
2.3	0.000	0.000	0.000	0.000	0.000
2.4	0.000	0.000	0.000	0.000	0.000
2.5	0.000	0.000	0.000	0.000	0.000
2.6	0.000	0.000	0.000	0.000	0.000
2.7	0.000	0.000	0.000	0.000	0.000
2.8	0.000	0.000	0.000	0.000	0.000
2.9	0.000	0.000	0.000	0.000	0.000
3.0	0.000	0.000	0.000	0.000	0.000
3.5	0.000	0.000	0.000	0.000	0.000
4.0	0.000	0.000	0.000	0.000	0.000
4.5	0.000	0.000	0.000	0.000	0.000
5.0	0.000	0.000	0.000	0.000	0.000
6.0	0.000	0.000	0.000	0.000	0.000
7.0	0.000	0.000	0.000	0.000	0.000
8.0	0.000	0.000	0.000	0.000	0.000
9.0	0.000	0.000	0.000	0.000	0.000
10.0	0.000	0.000	0.000	0.000	0.000
20.0	0.000	0.000	0.000	0.000	0.000

APPENDIX 2
Table for determining $F(u,N)$ for negative slopes (Chow, 1957).

u	N									
	2.0	**2.2**	**2.4**	**2.6**	**2.8**	**3.0**	**3.2**	**3.4**	**3.6**	**3.8**
0.00	0.000	0.000	0.000	0.000	0.000	0.000	0.000	0.000	0.000	0.000
0.02	0.020	0.020	0.020	0.020	0.020	0.020	0.020	0.020	0.020	0.020
0.04	0.040	0.040	0.040	0.040	0.040	0.040	0.040	0.040	0.040	0.040
0.06	0.060	0.060	0.060	0.060	0.060	0.060	0.060	0.060	0.060	0.060
0.08	0.080	0.080	0.080	0.080	0.080	0.080	0.080	0.080	0.080	0.080
0.10	0.099	0.100	0.100	0.100	0.100	0.100	0.100	0.100	0.100	0.100
0.12	0.119	0.119	0.120	0.120	0.120	0.120	0.120	0.120	0.120	0.120
0.14	0.139	0.139	0.140	0.140	0.140	0.140	0.140	0.140	0.140	0.140
0.16	0.158	0.159	0.159	0.160	0.160	0.160	0.160	0.160	0.160	0.160
0.18	0.178	0.179	0.179	0.180	0.180	0.180	0.180	0.180	0.180	0.180
0.20	0.197	0.198	0.199	0.200	0.200	0.200	0.200	0.200	0.200	0.200
0.22	0.216	0.217	0.218	0.219	0.219	0.220	0.220	0.220	0.220	0.220
0.24	0.234	0.236	0.237	0.238	0.239	0.240	0.240	0.240	0.240	0.240
0.26	0.253	0.255	0.256	0.257	0.258	0.259	0.259	0.260	0.260	0.260
0.28	0.272	0.274	0.275	0.276	0.277	0.278	0.278	0.279	0.280	0.280
0.30	0.291	0.293	0.294	0.295	0.296	0.297	0.298	0.298	0.299	0.299
0.32	0.308	0.311	0.313	0.314	0.316	0.317	0.318	0.318	0.319	0.319
0.34	0.326	0.329	0.331	0.333	0.335	0.337	0.338	0.338	0.339	0.339
0.36	0.344	0.347	0.350	0.352	0.354	0.356	0.357	0.357	0.358	0.358
0.38	0.362	0.355	0.368	0.371	0.373	0.374	0.375	0.376	0.377	0.377
0.40	0.380	0.387	0.390	0.392	0.393	0.394	0.395	0.396	0.396	0.42
0.42	0.397	0.401	0.405	0.407	0.409	0.411	0.412	0.413	0.414	0.414
0.44	0.414	0.419	0.423	0.426	0.429	0.430	0.432	0.433	0.434	0.435
0.46	0.431	0.437	0.440	0.444	0.447	0.449	0.451	0.452	0.453	0.454
0.48	0.447	0.453	0.458	0.461	0.464	0.467	0.469	0.471	0.472	0.473

u	N									
	2.0	2.2	2.4	2.6	2.8	3.0	3.2	3.4	3.6	3.8
0.50	0.463	0.470	0.475	0.479	0.482	0.485	0.487	0.489	0.491	0.492
0.52	0.479	0.485	0.491	0.494	0.499	0.502	0.505	0.507	0.509	0.511
0.54	0.494	0.501	0.507	0.512	0.516	0.520	0.522	0.525	0.527	0.529
0.56	0.509	0.517	0.523	0.528	0.533	0.537	0.540	0.543	0.545	0.547
0.58	0.524	0.533	0.539	0.545	0.550	0.554	0.558	0.561	0.563	0.567
0.60	0.540	0.548	0.555	0.561	0.566	0.571	0.575	0.578	0.581	0.583
0.61	0.547	0.556	0.563	0.569	0.575	0.579	0.583	0.587	0.589	0.592
0.62	0.554	0.563	0.571	0.578	0.583	0.578	0.591	0.595	0.598	0.600
0.63	0.562	0.571	0.579	0.585	0.590	0.595	0.599	0.603	0.607	0.609
0.64	0.569	0.579	0.586	0.592	0.598	0.602	0.607	0.611	0.615	0.618
0.65	0.576	0.585	0.592	0.599	0.606	0.610	0.615	0.619	0.623	0.626
0.66	0.583	0.593	0.600	0.607	0.613	0.618	0.622	0.626	0.630	0.634
0.67	0.590	0.599	0.607	0.614	0.621	0.626	0.631	0.635	0.639	0.643
0.68	0.597	0.607	0.615	0.622	0.628	0.634	0.639	0.643	0.647	0.651
0.69	0.603	0.613	0.621	0.629	0.635	0.641	0.646	0.651	0.655	0.659
0.70	0.610	0.620	0.629	0.637	0.644	0.649	0.654	0.659	0.663	0.667
0.71	0.617	0.627	0.636	0.644	0.651	0.657	0.661	0.666	0.671	0.674
0.72	0.624	0.634	0.643	0.651	0.658	0.664	0.669	0.674	0.679	0.682
0.73	0.630	0.641	0.650	0.659	0.665	0.672	0.677	0.682	0.687	0.691
0.74	0.637	0.648	0.657	0.665	0.672	0.679	0.684	0.689	0.694	0.698
0.75	0.643	0.655	0.664	0.671	0.679	0.686	0.691	0.696	0.701	0.705
0.76	0.649	0.661	0.670	0.679	0.687	0.693	0.699	0.704	0.709	0.713
0.77	0.656	0.667	0.677	0.685	0.693	0.700	0.705	0.711	0.715	0.719
0.78	0.662	0.673	0.683	0.692	0.700	0.707	0.713	0.718	0.723	0.727
0.79	0.668	0.680	0.689	0.698	0.705	0.713	0.719	0.724	0.729	0.733
0.80	0.674	0.685	0.695	0.703	0.712	0.720	0.726	0.732	0.737	0.741
0.81	0.680	0.691	0.701	0.710	0.719	0.727	0.733	0.739	0.744	0.749

	N									
u	2.0	2.2	2.4	2.6	2.8	3.0	3.2	3.4	3.6	3.8
0.82	0.686	0.698	0.707	0.717	0.725	0.733	0.740	0.745	0.751	0.755
0.83	0.692	0.703	0.713	0.722	0.731	0.740	0.746	0.752	0.757	0.762
0.84	0.698	0.709	0.719	0.729	0.737	0.746	0.752	0.758	0.764	0.769
0.85	0.704	0.715	0.725	0.735	0.744	0.752	0.759	0.765	0.770	0.775
0.86	0.710	0.721	0.731	0.741	0.750	0.758	0.765	0.771	0.777	0.782
0.87	0.715	0.727	0.738	0.747	0.756	0.764	0.771	0.777	0.783	0.788
0.88	0.721	0.733	0.743	0.753	0.762	0.770	0.777	0.783	0.789	0.794
0.89	0.727	0.739	0.749	0.758	0.767	0.776	0.783	0.789	0.795	0.800
0.90	0.732	0.744	0.754	0.764	0.773	0.781	0.789	0.795	0.801	0.807
0.91	0.738	0.750	0.760	0.770	0.779	0.787	0.795	0.801	0.807	0.812
0.92	0.743	0.754	0.766	0.776	0.785	0.793	0.800	0.807	0.813	0.818
0.93	0.749	0.761	0.772	0.782	0.791	0.799	0.807	0.812	0.818	0.823
0.94	0.764	0.767	0.777	0.787	0.795	0.804	0.813	0.813	0.824	0.829
0.950	0.759	0.772	0.783	0.793	0.801	0.809	0.819	0.823	0.829	0.835
0.960	0.764	0.777	0.788	0.798	0.807	0.815	0.824	0.829	0.835	0.841
0.970	0.770	0.782	0.793	0.803	0.812	0.820	0.826	0.834	0.840	0.846
0.975	0.772	0.785	0.796	0.805	0.814	0.822	0.828	0.836	0.843	0.848
0.980	0.775	0.787	0.798	0.808	0.818	0.825	0.830	0.839	0.845	0.851
0.985	0.777	0.790	0.801	0.811	0.820	0.827	0.833	0.841	0.847	0.853
0.990	0.780	0.793	0.804	0.814	0.822	0.830	0.837	0.844	0.850	0.856
0.995	0.782	0.795	0.806	0.816	0.824	0.832	0.840	0.847	0.753	0.859
1.000	0.785	0.797	0.808	0.818	0.826	0.834	0.842	0.849	0.856	0.862
1.005	0.788	0.799	0.810	0.820	0.829	0.837	0.845	0.852	0.858	0.864
1.010	0.790	0.801	0.812	0.822	0.831	0.840	0.847	0.855	0.861	0.867
1.015	0.793	0.804	0.815	0.824	0.833	0.843	0.850	0.858	0.864	0.870
1.020	0.795	0.807	0.818	0.828	0.837	0.845	0.853	0.860	0.866	0.872
1.030	0.800	0.811	0.822	0.832	0.841	0.850	0.857	0.864	0.871	0.877
1.040	0.805	0.816	0.829	0.837	0.846	0.855	0.862	0.870	0.877	0.883

u	N									
	2.0	2.2	2.4	2.6	2.8	3.0	3.2	3.4	3.6	3.8
1.05	0.810	0.821	0.831	0.841	0.851	0.859	0.867	0.874	0.881	0.887
1.06	0.815	0.826	0.837	0.846	0.855	0.864	0.871	0.879	0.885	0.891
1.07	0.819	0.831	0.841	0.851	0.860	0.869	0.876	0.883	0.889	0.896
1.08	0.824	0.836	0.846	0.856	0.865	0.873	0.880	0.887	0.893	0.900
1.09	0.828	0.840	0.851	0.860	0.870	0.877	0.885	0.892	0.898	0.9-4
1.10	0.833	0.845	0.855	0.865	0.874	0.881	0.890	0.897	0.903	0.908
1.11	0.837	0.849	0.860	0.870	0.878	0.886	0.894	0.900	0.907	0.912
1.12	0.842	0.854	0.864	0.873	0.882	0.891	0.897	0.904	0.910	0.916
1.13	0.846	0.858	0.868	0.878	0.886	0.895	0.902	0.908	0.914	0.919
1.14	0.851	0.861	0.872	0.881	0.890	0.899	0.905	0.912	0.918	0.923
1.15	0.855	0.866	0.876	0.886	0.895	0.903	0.910	0.916	0.922	0.928
1.16	0.859	0.870	0.880	0.890	0.899	0.907	0.914	0.920	0.926	0.931
1.17	0.864	0.874	0.884	0.893	0.902	0.911	0.917	0.923	0.930	0.934
1.18	0.868	0.878	0.888	0.897	0.906	0.915	0.921	0.927	0.933	0.939
1.19	0.872	0.882	0.892	0.901	0.910	0.918	0.925	0.931	0.937	0.942
1.20	0.876	0.886	0.896	0.904	0.913	0.921	0.928	0.934	0.940	0.945
1.22	0.880	0.891	0.900	0.909	0.917	0.929	0.932	0.938	0.944	0.949
1.24	0.888	0.898	0.908	0.917	0.925	0.935	0.940	0.945	0.950	0.955
1.26	0.900	0.910	0.919	0.927	0.935	0.942	0.948	0.954	0.960	0.964
1.28	0.908	0.917	0.926	0.934	0.945	0.948	0.954	0.960	0.965	0.970
1.30	0.915	0.925	0.933	0.941	0.948	0.955	0.961	0.966	0.981	0.975
1.32	0.922	0.931	0.940	0.948	0.955	0.961	0.967	0.972	0.976	0.980
1.34	0.930	0.939	0.948	0.955	0.962	0.967	0.973	0.978	0.982	0.986
1.36	0.937	0.946	0.954	0.961	0.968	0.973	0.979	0.983	0.987	0.991
1.38	0.944	0.952	0.960	0.967	0.974	0.979	0.985	0.989	0.993	0.996
1.40	0.951	0.959	0.966	0.973	0.979	0.984	0.989	0.993	0.997	1.000
1.42	0.957	0.965	0.972	0.979	0.984	0.989	0.995	0.998	1.001	1.004

u	\multicolumn{10}{c}{N}
	2.0	2.2	2.4	2.6	2.8	3.0	3.2	3.4	3.6	3.8
1.44	0.964	0.972	0.979	0.984	0.990	0.995	1.000	1.003	1.006	1.009
1.46	0.970	0.977	0.983	0.989	0.995	1.000	1.004	1.007	1.010	1.012
1.48	0.977	0.983	0.989	0.994	0.999	1.005	1.008	1.011	1.014	1.016
1.50	0.983	0.990	0.996	1.001	1.005	1.009	1.012	1.015	1.017	1.019
1.55	0.997	1.002	1.007	1.012	1.016	1.020	1.022	1.024	1.026	1.028
1.60	1.012	1.017	1.020	1.024	1.027	1.030	1.032	1.034	1.035	1.035
1.65	1.026	1.029	1.032	1.035	1.037	1.039	1.041	1.041	1.042	1.042
1.70	1.039	1.042	.1044	1.045	1.047	1.048	1.049	1.049	1.049	1.048
1.75	1.052	1.053	1.054	1.055	1.056	1.057	1.056	1.056	1.055	1.053
1.80	1.064	1.064	1.064	1.064	1.065	1.065	1.064	1.062	1.060	1.058
1.85	1.075	1.074	1.074	1.073	1.072	1.071	1.069	1.067	1.066	1.063
1.90	1.086	1.075	1.084	1.082	1.081	1.079	1.077	1.074	1.071	1.066
1.95	1.097	1.095	1.092	1.090	1.087	1.085	1.081	1.079	1.075	1.071
2.00	1.107	1.103	1.100	1.096	1.093	1.090	1.085	1.082	1.078	1.075
2.10	1.126	1.120	1.115	1.110	1.104	1.100	1.094	1.089	1.085	1.080
2.20	1.144	1.136	1.129	1.122	1.115	1.109	1.102	1.096	1.090	1.085
2.30	1.161	1.150	1.141	1.133	1.124	1.117	1.110	1.103	1.097	1.090
2.40	1.176	1.163	1.152	1.142	1.133	1.124	1.116	1.109	1.101	1.094
2.5	1.190	1.175	1.162	1.150	1.140	1.131	1.121	1.113	1.105	1.098
2.6	1.204	1.187	1.172	1.159	1.147	1.137	1.126	1.117	1.106	1.000
2.7	1.216	1.196	1.180	1.166	1.153	1.142	1.130	1.120	1.110	1.102
2.8	1.228	1.208	1.189	1.173	1.158	1.146	1.132	1.122	1.112	1.103
2.9	1.239	1.216	1.196	1.178	1.162	1.150	1.137	1.125	1.115	1.106
3.0	1.249	1.224	1.203	1.184	1.168	1.154	1.140	1.128	1.117	1.107
3.5	1.292	1.260	1.232	1.206	1.185	1.167	1.151	1.138	1.125	1.113
4.0	1.326	1.286	1.251	1.223	1.198	1.176	1.158	1.142	1.129	1.117
4.5	1.352	1.308	1.270	1.235	1.205	1.183	1.162	1.146	1.131	1.119
5.0	1.374	1.325	1.283	1.245	1.212	1.188	1.166	1.149	1.134	1.121

u	N									
	2.0	**2.2**	**2.4**	**2.6**	**2.8**	**3.0**	**3.2**	**3.4**	**3.6**	**3.8**
6.0	1.406	1.342	1.292	1.252	1.221	1.195	1.171	1.152	1.136	1.122
7.0	1.430	1.360	1.303	1.260	1.225	1.199	1.174	1.153	1.136	1.122
8.0	1.447	1.373	1.313	1.266	1.229	1.201	1.175	1.154	1.137	1.122
9.0	1.461	1.384	1.319	1.269	1.231	1.203	1.176	1.156	1.137	1.122
10.0	1.471	1.394	1.324	1.272	1.233	1.203	1.176	1.156	1.137	1.122

Table for determining $F(u,N)$ for negative slopes (Chow, 1957): *continued.*

u	N				
	4.0	**4.2**	**4.5**	**5.0**	**5.5**
0.00	0.000	0.000	0.000	0.000	0.000
0.02	0.020	0.020	0.020	0.020	0.020
0.04	0.040	0.040	0.040	0.040	0.040
0.06	0.060	0.060	0.060	0.060	0.060
0.08	0.080	0.080	0.080	0.080	0.080
0.10	0.100	0.100	0.100	0.100	0.100
0.12	0.120	0.120	0.120	0.120	0.120
0.14	0.140	0.140	0.140	0.140	0.140
0.16	0.160	0.160	0.160	0.160	0.160
0.18	0.180	0.180	0.180	0.180	0.180
0.20	0.200	0.200	0.200	0.200	0.200
0.22	0.220	0.220	0.220	0.220	0.220
0.24	0.240	0.240	0.240	0.240	0.240
0.26	0.260	0.260	0.260	0.260	0.260
0.28	0.280	0.280	0.280	0.280	0.280
0.30	0.300	0.300	0.300	0.300	0.300
0.32	0.320	0.320	0.320	0.320	0.320

u	N				
	4.0	4.2	4.5	5.0	5.5
0.34	0.339	0.340	0.340	0.340	0.340
0.36	0.359	0.360	0.360	0.360	0.360
0.38	0.378	0.379	0.380	0.380	0.380
0.40	0.397	0.398	0.398	0.400	0.400
0.42	0.417	0.418	0.418	0.419	0.420
0.44	0.436	0.437	0.437	0.439	0.440
0.46	0.455	0.456	0.457	0.458	0.459
0.48	0.474	0.475	0.476	0.478	0.479
0.50	0.493	0.494	0.495	0.497	0.498
0.52	0.512	0.513	0.515	0.517	0.518
0.54	0.531	0.532	0.533	0.536	0.537
0.56	0.549	0.550	0.552	0.555	0.558
0.58	0.567	0.569	0.570	0.574	0.576
0.60	0.585	0.587	0.589	0.593	0.595
0.61	0.594	0.596	0.598	0.602	0.604
0.62	0.603	0.605	0.607	0.611	0.613
0.63	0.612	0.615	0.616	0.620	0.622
0.64	0.620	0.623	0.625	0.629	0.631
0.65	0.629	0.632	0.634	0.638	0.640
0.66	0.637	0.640	0.643	0.647	0.650
0.67	0.646	0.649	0.652	0.656	0.659
0.68	0.654	0.657	0.660	0.665	0.668
0.69	0.662	0.665	0.668	0.674	0.677
0.70	0.670	0.673	0.677	0.682	0.686
0.71	0.678	0.681	0.685	0.690	0.694
0.72	0.686	0.689	0.694	0.699	0.703
0.73	0.694	0.698	0.702	0.707	0.712
0.74	0.702	0.705	0.710	0.716	0.720

	N				
u	4.0	4.2	4.5	5.0	5.5
0.75	0.709	0.712	0.717	0.724	0.728
0.76	0.717	0.720	0.725	0.731	0.736
0.77	0.724	0.727	0.733	0.739	0.744
0.78	0.731	0.735	0.740	0.747	0.752
0.79	0.738	0.742	0.748	0.754	0.760
0.80	0.746	0.750	0.755	0.762	0.768
0.81	0.753	0.757	0.762	0.770	0.776
0.82	0.760	0.764	0.769	0.777	0.783
0.83	0.766	0.771	0.776	0.784	0.790
0.84	0.773	0.778	0.783	0.791	0.798
0.85	0.780	0.784	0.790	0.798	0.805
0.86	0.786	0.791	0.797	0.804	0.812
0.87	0.793	0.797	0.803	0.811	0.819
0.88	0.799	0.803	0.810	0.818	0.826
0.89	0.805	0.810	0.816	0.825	0.832
0.90	0.811	0.816	0.822	0.831	0.839
0.91	0.817	0.821	0.828	0.837	0.845
0.92	0.823	0.282	0.834	0.844	0.851
0.93	0.829	0.833	0.840	0.850	0.857
0.94	0.835	0.840	0.846	0.856	0.864
0.950	0.840	0.845	0.852	0.861	0.869
0.960	0.846	0.861	0.857	0.867	0.875
0.970	0.851	0.866	0.863	0.972	0.881
0.975	0.854	0.859	0.866	0.875	0.883
0.980	0.857	0.861	0.868	0.878	0.886
0.985	0.859	0.863	0.870	0.880	0.889
0.990	0.861	0.867	0.873	0.883	0.891

u	N				
	4.0	4.2	4.5	5.0	5.5
0.995	0.864	0.869	0.876	0.885	0.894
1.000	0.867	0.873	0.879	0.887	0.897
1.005	0.870	0.874	0.881	0.890	0.899
1.010	0.873	0.878	0.884	0.893	0.902
1.015	0.875	0.880	0.886	0.896	0.904
1.020	0.877	0.883	0.889	0.898	0.907
1.030	0.882	0.887	0.893	0.902	0.911
1.040	0.888	0.893	0.898	0.907	0.916
1.05	0.892	0.897	0.903	0.911	0.920
1.06	0.896	0.901	0.907	0.915	0.924
1.07	0.901	0.906	0.911	0.919	0.928
1.08	0.905	0.910	0.916	0.923	0.932
1.09	0.909	0.914	0.920	0.927	0.936
1.10	0.913	0.918	0.923	0.931	0.940
1.11	0.917	0.921	0.927	0.935	0.944
1.12	0.921	0.926	0.931	0.939	0.948
1.13	0.925	0.929	0.935	0.943	0.951
1.14	0.928	0.933	0.938	0.947	0.954
1.15	0.932	0.936	0.942	0.950	0.957
1.16	0.936	0.941	0.945	0.953	0.960
1.17	0.939	0.944	0.948	0.957	0.963
1.18	0.943	0.947	0.951	0.960	0.965
1.19	0.947	0.950	0.954	0.963	0.968
1.20	0.950	0.953	0.958	0.966	0.970
1.22	0.956	0.957	0.964	0.972	0.976
1.24	0.962	0.962	0.970	0.977	0.981
1.26	0.968	0.971	0.975	0.982	0.986
1.28	0.974	0.977	0.981	0.987	0.990

u	N				
	4.0	4.2	4.5	5.0	5.5
1.30	0.979	0.978	0.985	0.991	0.994
1.32	0.985	0.986	0.990	0.995	0.997
1.34	0.990	0.992	0.995	0.999	1.001
1.36	0.994	0.996	0.999	1.002	1.005
1.38	0.998	1.000	1.003	1.006	1.008
1.40	1.001	1.004	1.006	1.009	1.011
1.42	1.005	1.008	1.010	1.012	1.014
1.44	1.009	1.013	1.014	1.016	1.016
1.46	1.014	1.016	1.017	1.018	1.018
1.48	1.016	1.019	1.020	1.020	1.020
1.50	1.020	1.021	1.022	1.022	1.022
1.55	1.029	1.029	1.029	1.028	1.028
1.60	1.035	1.035	1.034	1.032	1.030
1.65	1.041	1.040	1.039	1.036	1.034
1.70	1.047	1.046	1.043	1.039	1.037
1.75	1.052	1.051	1.047	1.042	1.039
1.80	1.057	1.055	1.051	1.045	1.041
1.85	1.061	1.059	1.054	1.047	1.043
1.90	1.065	1.060	1.057	1.049	1.045
1.95	1.068	1.064	1.059	1.051	1.046
2.0	1.071	1.068	1.062	1.053	1.047
2.1	1.076	1.071	1.065	1.056	1.049
2.2	1.080	1.073	1.068	1.058	1.050
2.3	1.084	1.079	1.071	1.060	1.051
2.4	1.087	1.081	1.073	1.061	1.052
2.5	1.090	1.083	1.075	1.062	1.053
2.6	1.092	1.085	1.076	1.063	1.054

u	N				
	4.0	4.2	4.5	5.0	5.5
2.7	1.094	1.087	1.077	1.063	1.054
2.8	1.096	1.088	1.078	1.064	1.054
2.9	1.098	1.089	1.079	1.065	1.055
3.0	1.099	1.090	1.080	1.065	1.055
3.5	1.103	1.093	1.082	1.066	1.055
4.0	1.106	1.097	1.084	1.067	1.056
4.5	1.108	1.098	1.085	1.067	1.056
5.0	1.110	1.099	1.085	1.068	1.056
6.0	1.111	1.100	1.085	1.068	1.056
7.0	1.111	1.100	1.086	1.068	1.056
8.0	1.111	1.100	1.086	1.068	1.056
9.0	1.111	1.100	1.086	1.068	1.056

SUBJECT INDEX

A

A profiles, 265-267
Absolute viscosity (see Viscosity),
Accessibility, 67-75
Adverse slope, 265-267, 286-288
Alluvial channels, 147, 392-402
Angle of repose, 362-364, 375, 384
Average velocity, 15

B

Backwater (see Gradually varied
    flow)
Bakhmeteff varied flow function,
    (See also Gradually varied
    flow)
Bed formation, 147, 566, 569, 575
Bends, effect of on:
    permissible tractive force, 365-368
    permissible velocity, 358, 360, 368,
      413
    resistance, 152, 157
    transverse mixing, 445-446
    water surface, 343-344
Bernoulli equation (see Equations,
    Bernoulli)
Best hydraulic section, 339-340
Blasius equation, 149
Blasius solution, 31
Boundary:
    hydraulically rough, 32-33, 36, 38
    hydraulically smooth, 32-33, 36, 38
Boundary condition, 263-264, 434,
    441, 520, 537-538, 540
Boundary layer,
    development of, 30
    laminar, 31, 32
    separation, 34
    turbulent, 31, 32, 35
Boussinesq assumption, 24
Boussinesq coefficient (see
    viscosity, eddy)
Bresse function, 278

Buckingham Λ theorem, 552

C

C profiles, 265-267
Calibration, 586-588
Canal (see channel)
Channel:
    best hydraulic, 340
    circular:
      best hydraulic section, 340
      critical depth, 64
      geometric elements, 13
      gradually varied flow in, 280-
        286
      hydraulic jump in, 99-103
      uniform flow in, 225-226
    elliptical, 13, 64
    exponential, 64
    parabolic:
      best hydraulic section, 340
      critical depth, 64
      geometric elements, 13
      hydraulic jump in, 103, 104
    rectangular:
      best hydraulic section, 340
      critical depth, 63
      geometric elements, 13
      hydraulic exponents, 276
      hydraulic jump in, 97-98, 103
      uniform flow in, 224-229
    trapezoidal:
      best hydraulic section, 340
      critical depth, 63
      geometric elements, 13
      hydraulic exponents, 276, 279-
        281
      hydraulic jump in, 103-104
      uniform flow in, 222-223, 225,
        229-230
    triangular:
      best hydraulic section, 340
      critical depth, 63
      geometric elements, 13
      hydraulic exponents, 276, 280

hydraulic jump in, 103, 104
uniform flow in, 222-225, 229-230

Channel alignment, 152, 157

Channel section geometric elements, 13

Channels:
alignment, 150, 155-156, 343, 366-367, 374
alluvial, 392-402
artificial, definition of, 9
canal, definition of, 9
chute, definition of, 10
compound section, 28, 86, 251-252
design of:
erodible, 339, 357-411
grass-lined, 339, 411-424
lined, 339, 345-357
nonerodible, 345-357
ephemeral, 404-409
erodible, 357-424
flume, definition of, 9
ice-covered, 243-251, 452
lined, 339, 342
natural, 9
prismatic and nonprismatic, definition of, 9

Characteristics, method of, 516-520

Chezy coefficient, definition of, 33

Chezy equation, definition of, 33

Chezy's C (see Chezy coefficient)

Choke
at bridge piers, 89-90, 140
definition of, 74

Circular channels (see Channel, circular)

Coefficient:
Boussinesq (see Viscosity, eddy)
of drag, 133
of energy (see Energy, coefficient of)
of friction (see Friction, coefficient of)
of momentum, 306
of pressure distribution, 56, 57, 306
of roughness (see Chezy coefficient; Manning's n)
of turbulent mixing and transfer, 445-476

Cohesive material:
permissible tractive force, 366
permissible velocity for, 364-365

Complete mixing, 448

Complementary error function, 442

Composite roughness, 231-239, 246-251

Compound section (see Channels, compound section)

Concave flow, 56

Conservation:
of energy (see Energy, conservation of)
of mass (see Continuity equation)
of momentum (see Momentum, conservation of)

Continuity equation, 16, 26, 493, 499, 504, 513

Contraction, 83-86, 89-90, 140, 155-156

Control:
definition of, 72-73

Convergence, 268, 277-278, 293-294, 332, 536-537

Convex flow, 56

Conveyance, 86, 221, 239, 241, 274, 300

Courant stability condition, 520-521, 524

Critical depth (see Depth, critical)

Critical discharge, definition of, 59-60

Critical flow, 63

Critical slope, definition of, 226-230

Critical velocity, definition of, 29-60

Culvert, 10, 224

Curve (see Bends)

Curvilinear flow, 55-56

D

Darcy-Weisbach friction factor,
Dead zone, 458-460
Densimetric Froude number (see Number, Froude, densimetric)

Density stratification, 1, 5
Depth:
  critical, 59, 60, 61, 62, 99, 102
    curves for estimation of, 65
    definition of, 59
    equations for estimation, 63
  of flow, definition of, 10
  hydraulic, definition of, 6, 13, 58
  normal:
    curves for estimation, 225
    definition of, 143
    determination of, 221-225
  sequent, 57, 95, 97, 103, 127-131
    circular section, 99-103
    parabolic section, 103-105
    rectangular section, 97-98
    triangular section, 103-105
    trapezoidal section, 103-105
  uniform, definition of, 143
Diffusion:
  advective, 436
  Fickian, 432, 434, 436
  molecular, 434
  numerical, 477-480
  turbulent, 431
Diffusion coefficient, 432-438
Dimensional analysis, 552
Dirac delta function, 433
Direct integration method (see
  Gradually varied flow)
Direct step method (see Gradually
  varied flow)
Discharge:
  critical, definition of, 59-60
  in ice-covered channels, 244-251
  normal, definition of, 221-222
Dispersion, 431
  coefficient of, 440, 561
  initial period, 440
  numerical, 477-480
  physical model, 560-563
Divided flow, 326-333
Drowned hydraulic jump, 107-108,
  114-115, 326
Dye, rhodamine, 469-473
Dynamic equation:
  gradually varied flow, 261-262
  spatially varied flow, 304-318

  unsteady flow, 493-494
Dynamic similarity, 39
Dynamic viscosity, 555

E
Economic design of channels, 350-
  357
Eddy loss, 262, 289-290
Eddy viscosity (see Viscosity,
  eddy)
Elliptic integral, 387-389
Energy:
  coefficient of, 59, 86, 241, 314
    definition of, 27, 8
    nonprismatic channels, 86, 294
  conservation of, 26, 53
  dissipation of, 73, 107-110
  minimum, 57, 96
  in nonprismatic channels,
  specific, 315
    in channels of compound
      section, 86
Equations:
  Bernoulli, 26, 53, 70
  Blasius, 146
  boundary layers, 30
  Chezy, definition of, 33, 142-144
  Colebrook, 149
  of continuity (see Continuity
    equation)
  energy (see Energy)
  Euler, 55
  Manning, definition of, 41, 144-145
  of momentum (see Momentum,
    conservation of)
  Navier-Stokes, 28, 30
  Prandtl-von Karman, 36
  Saint Venant, 493-494, 527-537
Equivalent n value, 231-239
Erodible channels, 357-424
Error function, 441, 442
Euler Equation (see Equations,
  Euler)
Expansion, 151, 155-156, 262, 289-
  290

F
f-R relationship:

rough channels, 149-150
smooth channels, 148-149
Fick's law, 432, 434, 436
Finite differences, 477-480, 514-
516, 520-521, 523, 529
First order second moment
(FOSM), 41-44
Flood plain, roughness of, 150,
163, 210-217, 251
Flow:
area, 11, 13
concave, 56
convex, 56
critical, 6
gradually varied (see Gradually
varied flow)
laminar, 4,5, 15, 16, 28, 38, 40, 550
passing islands, 326-333
profiles:
adverse slope, 267, 286
analysis of, 261-286
classification of, 263-267
gradually closing crown, 280,
283-285
in horizontal channels, 286
in prismatic channels, 268-288
in nonprismatic channels, 289-
303
of spatially varied flow, 2, 303-
317
rapidly varied, 2, 3
secondary, 39
spatially varied, 2, 3, 303-317
steady, definition of, 2, 3
stratified, 1, 125-129
subcritical, 6, 8, 59, 125
supercritical, 6, 8, 59, 125
transitional, 5
turbulent, 5, 15, 16, 40, 55, 550
uniform:
computation of, 221-253
criterion for, 2, 3, 141
establishment of, 143
unsteady (see Gradually varied
unsteady flow)
Fluorescence, 469
Force, specific (see also Specific
momentum)
hydrodynamic, 132-134

impact, 132-134
Frazil ice, 245
Freeboard, 342-343, 350
Friction:
coefficient of, 145, 147, 235-237,
214, 362, 417-424
force of, 22
slope, 147, 295, 301-302
velocity, definition of,
Friction factor (see Friction,
coefficient of)
Froude law model, 40, 553
Froude number (see Number,
Froude)

G
Gaussian distribution, 433, 459
Geometric elements:
of circular section, 13
of parabolic section, 13
of rectangular section, 13
of trapezoidal secion, 13
of triangular section, 13
Geometric similarity, 39, 549
Gradient Richardson number (see
Number, Richardson)
Gradually varied flow, 2
steady:
basic assumptions of, 261-262
characteristics of, 262-267
classification of, 262-267
definition of, 2
discharge computation, 319-326
equation, 261
flow past islands, 326-333
profile behind a dam, 317-318
solutions:
Bakhmeteff-Chow, 276, 286
direct integration, 271
HEC-2, 299
HEC-RAS, 299-303
nonprismatic channels, 288
prismatic channels, 268
step method, 269, 279, 290
unsteady:
characteristics, method of, 516-
520
continuity, 493, 499, 513

definition of, 2
HEC-RAS, 299, 529-537
momentum, 494, 497, 513
numerical solution of problems:
    explicit finite differences, 473-
    480, 514-516
Grain roughness, 147-150
Grassed channels, 410-424
    design procedure, 414
    n - $\overline{u}R$ curves, 410, 418-424
    retardance classification, 410-
    411, 418-424
    selection of grass, 412

H

H profiles, 265-267
Headloss:
    in contractions and expansions, 290
    in hydraulic jump, 108-111, 120-
    126, 131-133
HEC-RAS, 299-303, 529-538
Horizontal channels, flow profiles
    in,
Horizontal slope (see Slope,
    horizontal)
Hydraulic conductivity (effective),
    406-408
Hydraulic depth, 11, 58
Hydraulic exponent M, 274-275
Hydraulic exponent N, 275-276
Hydraulic jump, 95, 96
    applications of, 96
    in circular channel, 99, 113
    drowned, 106-107
    energy loss, 107-110, 119, 121, 123,
    130-131
    internal, 125-129
    length of, 110-114, 116, 118-119,
    121, 123
    in parabolic channel, 109,112
    in rectangular channel, 107-108,
    120, 122, 124
    sequent depth, 97-106
    in sloping channels, 114-125
    submerged, 107-108
    in trapezoidal channel, 109, 113
    in triangular channel, 109, 112

Hydraulic radius, definition of, 11,
    13
Hydrologic routing, 499-513
Hydrostatic pressure distribution,
    55

I

Ice:
    characteristics of, 243-244
    formation of, 244-245
    frazil, 245
    models of, 588-592
Ice-covered channels, 243-252, 452
Image source, 446-450
Implicit numerical scheme, 480
Initial condition, 441, 520, 537-
    538, 540
Initial period, 440-441, 452, 473-
    476
Interface, 9, 127-131
Internal Froude number (see
    Number, Froude, densimetric)
Internal hydraulic jump, 127-131
Islands in rivers, 326-333

K

Karman's turbulence coefficient, 35
Kinematic similarity, 549
Kinematic viscosity (see Viscosity,
    kinematic)
Kinetic energy correction factor,
    27, 55

L

Laminar boundary layer, 31
Laminar flow, criterion for, 4-5
Laminar sublayer, 31
Lateral inflow, 304, 406
Lateral mixing (see Mixing,
    transverse)
Limit slope, 226
Lined channels, design of, 345-357
Link node model, 326-333
Longitudinal dispersion (see
    Dispersion)

Longitudinal profiles:
 computation of, 261-333
 general behavior, 262-267
Looped rating curves, 498, 511-515
Loses, energy:
 at transitions, 290
 eddy, 290

M

M profile, 265-267
M value (hydraulic exponent M)
Manning equation, definition of, 41
Manning's n, 44, 147-217
 in closed conduits, 159-160
 composite roughness, 231-239, 246-
  249, 569-574
 determination of, 147-218
 factors affecting, 150-153
 on flood plains, 152, 162-163, 211-
  218
 for ice-covered channels, 247-248
 photographs of channels for, 163-
  204
 relation to roughness height, 147-
  150
 table for, 159-163
Maximum permissible velocity,
  357-360, 364-368
Method of characteristics, 516-520
Mild slope, 264
Minimum permissible velocity,
Mixing,
 length, 25, 35
 transverse:
  in rivers, 445-446
 vertical:
  in rivers, 444-446
  in stratified environments, 480-
   484
Models, physical, 39-41, 549-592
 calibration, 586-588
 distorted, 553-554, 557, 564, 568
 fixed-bed, 557-565, 586
 Froude law, 40, 553, 556
 ice, 588-592
 materials for, 582-586
 methods of construction, 5, 584-586
 moveable-bed, 553, 565-581, 586
 regime method, 580-581

 Reynolds law, 40, 555-556
 scale effects, 556
 undistorted, 557
 verification, 586-588
 Weber law, 41, 555-556
Momentum:
 conservation of, 21,26, 494, 497,
  513
 correction coefficient, 27, 93, 306
 specific, definition of, 93, 95, 97
 transport coefficient:
  homogeneous, 25-26
  stratified, 25-26

N

N value (hydraulic exponent N)
n-uRe curves, 410
Navier-Stokes equations, 30
Negative slope, 265-267, 286-288
Noncohesive material:
 angle of repose, 363-364
 tractive force, 360, 365-384
Nonerodible channels, design of,
  357-402
Nonsilting and noneroding velocity,
  357-402
Normal depth (see Depth, normal)
Normal distribution (see, Gaussian
  distribution)
Normal slope, 226-230
Number:
 Froude, 6, 30, 39-40, 59, 73, 554
  densimetric, 8, 9, 41
 Reynolds, 4, 15, 30, 31, 39- 41, 145-
  148, 235-237, 418, 552, 555
 Richardson, 5, 8
 Weber, 41, 555
Numerical dispersion, 477-480

O

Over-bank flow, 164, 294-299, 303

P

Parabolic channels (see Channel,
  parabolic)
Permissible maximum velocities,
  358-360

Permissible tractive force, 357-402
Permissible velocity:
  corrections for depth and sinuousity, 367, 368
  in grassed channels, 413
  maximum, 358-360
  method of, 358-360
  minimum, 340-341
Piers, bridges, 83, 90
Prandtl-von Karman universal velocity distribution law, 36, 38
Pressure in curvilinear flow, 57
Pressure coefficient, 57

R

Radius of curvature, 55-57, 343-344
Rectangular channel (see Channel, rectangular)
Regime theory, 393-402
Retardance, coefficient of, 411-412, 423-424
Reynolds equations, 23
Reynolds law models, 555
Reynolds number (see Numbers, Reynolds)
Reynolds stresses, 38
Rhodamine dye, 469-473
Richardson number, 5, 8, 481
Roughness:
  composite, 230-239, 247
  due to ice, 247
  height, relative, 32, 34, 147-148, 205
  scales for, in model studies, 39-41
Rugosity, 392

S

S profiles, 265-267
Saint Venant equations (see, Equations, Saint Venant)
Scale in model studies:
  distortion, 553
  effect, 556

time, hydraulic, 40, 555-556, 559, 561
Scaling equations, 28
Scour, 149-150
Secondary flow, 38
Section factor:
  for critical flow, 62
  for uniform flow, definition of, 13, 62, 221
Sedimentation, 150
Sediment transport, 397-402
Seepage losses:
  methods of calculation, 402-409
  methods of measurement, 402-404
Series expansion, 98, 277-278
Shape factor, 222
Shear, interfacial, 128
Shear forces, 53
Shear stress, 35, 423
Shear velocity, definition of, 36
Side slopes of channel, 342, 346, 361, 369
Silt factor, 392
Similarity (see Similitude)
Similitude:
  dynamic, 39, 549
  geometric, 39, 549
  hydroelastic, 589-591
  kinematic, 549
Sinuousness, 358, 365, 446
Slope:
  adverse, 265-267, 287
  critical, 226, 265, 287
  friction, 141, 209, 239, 261, 292, 300-302, 328
  horizontal, 265-267
  limiting, 226
  mild, 264, 265-267, 320
  negative, 265-267, 287
  normal, 226
  subcritical, 226-230
  supercritical, 226-230
  in uniform flow, 143
  of water surface, 143
Slope-area methods, 239-243
Snow ice, 244-251
Spatially varied flow (see Flow, spatially varied)

Specific energy, 55
Specific force (see Specific Momentum)
Specific momentum, 93-96
Stable hydraulic section, 384-402
Stage, 10, 150, 163
Standard step method (see Gradually varied flow)
State of flow:
    laminar, definition of, 5
    subcritical, definition of, 8
    supercritical, definition of, 8
    transitional, definition of, 5
    turbulent, definition of, 5
Steady flow (see Gradually varied flow, steady)
Step method (see Gradually varied flow)
Stratified flow (see Flow, stratified)
Stream tube, 14
Streamline, 14, 53-55
Strickler's equation, 207-208, 569
Subcritical slope, 226-230
Sublayer, laminar, 31, 32
Supercritical slope, 226-230
Superelevation, 343-344
Superposition, 434-435
Surface roughness:
    hydraulically rough, 33, 146-147, 205
    hydraulically smooth, 33, 146-147, 236-237
Surface tension, 1, 41, 555-556

T
Threshold of movement, 377-384
Top width, definition of, 11, 13
Tractive force:
    critical, 367
    distribution of, 359-362
    method of design, 357-402
    permissible, 365-367
Tractive force ratio, 363
Transition loss, 290,
Transverse mixing (see Mixing, transverse)

Trapezoidal channel (see Channel, trapezoidal)
Triangular channel (see Channel, triangular)
Turbulent boundary layer, 31
Turbulent diffusion, 431
Turbulent flow (see Flow, turbulent)

U
Uncertainty, 41-44
Uniform flow (see Flow, uniform)
Unit tractive force (see Tractive force)
Unlined channels, design of, 357-424
Unsteady flow, 2, 493-548
    diffusion model,496-497, 525-527
    full dynamic (St. Venant), 527-540
    kinematic wave model, 497-498, 521-525
    level pool routing, 499-503
    Muskingum, 503-513

V
Varied flow, 2, 261-335
Vegetation, 149, 152-153, 210-217, 340-341, 348
Velocity:
    average, definition of, 2, 4
    critical, definition of, 61
    fall, 382
    friction, definition of, 146
    minimum permissible, 340-341, 348
    maximum permissible, 358-360, 366, 368, 413
    measurement, 205
    nonerodible, 357-360
    nonsilting, 340
    permissible (see Permissible velocity)
    shear, definition of, 33
Velocity distribution, turbulent, 35-36, 205
Velocity distribution coefficients (see Energy, coefficient of)

Velocity distribution law, 35, 205
Verification, 538-540
Vertical mixing (see Mixing, vertical)
Viscosity, 40
   dynamic, 555
   eddy, 25
   kinematic, 4, 36, 40
Von Karman's turbulence coefficient, 35

W
Waves,
   celerity, 6
   elementary, 6, 8
   kinematic, 521-525
Weber law models, 41
Weber number, 41
Wetted perimeter, 11, 13

AUTHOR INDEX

A

Abt, S. R., 213
Ackers, P., 150
Akan, A. O., 374-376, 418, 423
Albertson, M. L., 393-395
Aldridge, B.N., 213, 214
Amein, M., 527
Anderson, E. R., 481, 483, 486
Arcement, G. J., 151, 152, 212, 214
Argyropoulos, P. A., 105, 113, 115, 126
Ashton, G. D., 249, 250, 251

B

Babcock, H. A., 77, 80
Baltzer, R. A., 539, 540
Banks, R. B., 231
Barbarossa, N. L., 567
Barnes, H. H., 163 - 204
Bakhmeteff, B. A., 111, 111, 319
Barkau, R. L., 300, 528, 529
Benjamin, J. R., 41, 42, 434
Benson, M. A., 152, 240
Blench, T., 393, 578
Bogardi, J., 572
Bonham, C. D., 213
Bosmajian, G., 214
Bouwer, H., 402
Bradley, J. N., 111, 112, 114, 115, 118, 119, 120
Brater, E. F., 71, 329
Brooks, N. H., 234, 431, 442, 445, 446 458, 459
Brownlie, W. R., 381
Buchanan, T. L., 409

C

Calkins, D. J., 250, 251
Carlson, E. J., 209
CCRFCD, 506
Chang, H. H., 393, 397, 398
Chaurasia, S. R., 131
Chen, Y. H., 418, 422
Chien, N., 564, 570, 578, 579

Chiu, C. L., 39
Chu, H. H., 280,284
Chu, H.L., 527
Chow, V. T., 34, 36, 37, 86, 108, 110, 126, 152, 151, 158, 159, 207, 247, 277, 288, 307, 315, 316, 317, 319, 342, 359, 364, 366, 367, 567, 595
Cornell, C. A., 41, 42, 434
Cotton, G. K., 418, 422
Cowan, J. J., 212
Cox, R. G., 232
Coyle, J. J., 211, 410-413, 416
Crow, F. R., 214
Curtis, S., 409

D

Dalrymple, T., 152, 240
Davis, C., 403
Dawdy, D. R., 257
duBoys, P., 359

E

Edwards, I., 215
Edwards, R. W., 454
Einstein, H. A., 38, 231, 564, 567, 570, 578, 579
Elder, J. W., 444, 452, 454
Engmann, E. O., 452
Etzel, R., 589, 590

F

Fang, C.S., 527
Fathi-Moghadam, M., 215, 216, 218
Feldman, A. D., 299, 303
FEMA, 134, 136
FHWA, 375
Findikakis, A. N., 563
Fischenich, J. C., 213
Fischer, H. B., 431, 442, 445, 446, 453, 454, 458, 459, 460, 461, 463, 465, 473, 561, 562
Flippin-Dudley, S. J., 214
Florey, Q. L., 384
Forsythe, G. E., 478

Fortier, S., 341, 358, 360, 364, 368
Franz, D. D., 11, 12
Fread, D. L., 527, 528, 538, 540
Frederick, B. J., 454
French, R. H., 207, 257, 409, 431, 482,
   483, 484, 538

G

Garcia, M., 380
Garde, R. J., 145, 208, 208
Gardner, G. B., 127
Garrett, J.M., 213, 214
Gibbs, J. W., 454
Glancy, P. A., 240
Glover, R. E., 384, 454
Godfrey, R. G., 454
Goldberg, D. E., 539, 540
Govinda Rao, N. S., 107, 114
Guha, C. R., 127, 129

H

Hager, W.H., 418, 423
Harbaugh, T. E., 521
Harden, T. O., 476
Harleman, D. R. F., 480
Harmsen, L., 240
Hayakawa, N., 127, 130
Henderson, F. M., 86, 208, 295, 307,
   319, 567, 572, 573
Hinds, J., 310
Hinze, J. O., 35
Hokett, S., 409
Holley, E. R., 435, 436, 439, 561, 562
Holmes, P., 105, 113, 114
Holzman, B., 481
Horton, R. E., 231
Houk, I. E., 344, 346, 368, 369
Hsing, P. S., 105, 113
Hsiung, D. E., 39

I

Imberger, J., 431, 442, 445, 446, 458,
   459
Ince, S., 547

J

Jarrett, R. D., 210, 258
Julien, P. Y., 158

K

Keefer, T., 454
Kennedy, J. F., 249
Kennedy, R. G., 392
Kennison, K. R., 126
Kent, R. E., 481
Keulegan, G. H., 36, 38, 308
Kiefer, C. J., 280, 284
Kilpatrick, F. A., 469, 470, 471
Kindsvater, C. E., 113, 114, 115, 118
King, H. W., 71, 329
Knapp, J. W., 521
Koh, R.C.Y., 431, 442, 445, 446, 458,
   459
Kotras, T., 589, 590
Kouwen, N., 211, 215, 216, 218, 417
Krenkel, P. A., 431, 468, 538
Krishnappen, B. G., 445

L

Lacey, G., 392
Lai, C., 538
Lane, E. W., 209, 359, 362-366, 378
Lane, L. J., 404-407, 409
Larsen, P. A., 247
Lau, Y. L., 445
Laurenson, E. M., 493
Lee, T. S., 39
Levenspiel, O., 463
Levi, E., 144
Lewis, G. L., 521
Lewis, J., 522, 589, 590
Li, H., 38,
Li, R., 211
Li, R. M., 497, 498
Liggett, J. A., 39
Limerinos, J. T., 209
Lin, H. C., 39
List, E.J., 431, 442, 445, 446, 458, 459

M

Martens, L. A., 469, 470, 471
Martin, J. L., 431
Matzke, A. E., 111
Mays, L. W., 41
McCarthy, G. T., 503
McCutcheon, S. C., 207, 431, 483, 484

McLaughlin, D. K., 563
McQuivey, R. S., 454
Mehrota, S. C., 367
Meyer-Peter, E., 209
Miao, L. S., 39
Michel, B., 245
Mizumura, K., 39
Montgomery, R. B., 480, 481, 482
Moretti, P. M., 563
Muller, R., 209
Munk, W. H., 481, 483, 486
Murphey, J. B., 406
Myers, W. R. C., 252

N
Nagaratnam, S., 111, 112
Neill, C. R., 586, 587
Nelson, J. E., 483, 484
Nezhikhovskiy, R. A., 251
Novotny, V., 431, 468
Nowell, A. R. M., 127

O
Odd, N. V. M., 482, 483, 484
Osterkamp, T. E., 245
Owens, M., 454

P
Peterka, A. J., 111, 112, 114, 115, 117,
    118, 119
Petryk, S., 214
Phelps, H. O., 418
Ponce, V. M., 497, 498, 506, 527
Posey, C. J., 252
Press, M. J., 105, 113, 114
Pritchard, D. W., 481

R
Ramseier, R. O., 245
Ranga Raju, K. G., 145, 208, 209
Rajaratnam, N., 111, 111, 112, 115,
    117, 118, 126
Raudkivi, A. J., 208, 572
Raupach, M. R., 215
Ree, W. O., 214
Reed, J. R., 301, 302
Richardson, E. V., 159
Roberts, B. R., 563

Rodger, J. G., 482, 483, 484
Rossby, C. G., 480, 481, 482
Rouse, H., 111, 112, 547, 577
Rubey, W. W., 382

S
Safranez, K., 111, 112
Sandover, J. A., 105, 113, 114
Sayre, W. W., 454
Schaffranek, R. W., 539, 540
Schiebe, F. R., 127, 130
Schlichting, H., 25, 30, 31, 33, 34, 35
Schneider, V. R., 151, 152, 212, 214
Schuster, J. C., 454
Schwarz, J., 588
Scobey, F. C., 341, 358, 360, 364, 368
Senturk, F., 146, 147, 207
Sharp, J. J., 244, 580, 581, 585, 587,
    589
Shearman, J. O., 295, 299, 302
Shen, H. R., 476
Shen, H. T., 251
Shields, A., 377, 378
Shukry, A., 344
Siao, T. T., 111, 112
Silvester, R., 82, 103, 104, 105, 109,
    112
Simons, D. B., 146, 147, 159, 207, 394-
    395, 497, 498
SLA, 159
Slotta, L. S., 35
Smith, G. F., 540
Smith, J. D., 127
Smith, R. H., 528
Smith, W. K., 463
Sorenson, K. E., 403
Stefan, H., 127, 130
Stepanov, P. M., 114
Straub, L. G., 398
Straub, W. O., 61, 63, 99
Street, R. L., 477, 563
Streeter, V. L., 16, 57, 147, 339, 550
Subramanya, K., 209
Sumer, S. M., 486

T
Taylor, B. D., 380
Taylor, G. I., 439

Temple, D. M., 211
Thatcher, M., 480
Thom, A. S., 215
Thomas, I. E., 454
Timco, G. W., 589
Tracy, H. J., 39
Trout, T. J., 349-351, 353, 355
Tsang, G., 246, 248, 249, 250, 251
Tung, Y., 41
Turner, J. S., 127

U
Unny, T. E., 417
Urquhart, W. J., 152, 153
USACE, 295, 299-303, 499, 529, 536
Uzuner, M. S., 246, 247

V
Vanoni, V. A., 35, 234, 380
Viessman, W., 521

W
Wallace, D. E., 406

Walski, T. M., 550
Wasow, W. R., 477
Watson, C. C., 213
Weinmann, D. E., 495
Wenzel, H. G., 306
White, C. M., 378
Willison, R. J., 345
Wilson, J. F., 469, 470, 471
Wolfkill, A. J., 301, 302
Woodward, S. M., 344
Wylie, E. B., 16, 57, 147, 328, 331,
   332, 339, 550

Y
Yang, C. T., 381, 383
Yapa, P. N. D. D., 251
Yen, B. C., 306
Yih, C. S., 127, 129
Yotsukura, N., 454

Z
Zegzhda, A. P., 148
Zwamborn, J. A., 564, 574, 575, 576,
   577

CPSIA information can be obtained
at www.ICGtesting.com
Printed in the USA
BVHW050931310720
585064BV00003B/5